Managing
Hotel Front Office
Operations

Managing Hotel Front Office Operations

Rajeev R Mishra

Assistant Professor
Department of Hotel Management
Rawal Institute of Management
Faridabad, Haryana

CBS Publishers & Distributors Pvt Ltd

New Delhi • Bengaluru • Chennai • Kochi • Kolkata • Mumbai
Hyderabad • Nagpur • Patna • Pune • Vijayawada

Managing
Hotel Front Office
Operations

ISBN: 978-93-85915-76-5

Copyright © Author and Publisher

First Edition: 2016

Published by Satish Kumar Jain and produced by Varun Jain for

CBS Publishers & Distributors Pvt Ltd
4819/XI Prahlad Street, 24 Ansari Road, Daryaganj, New Delhi 110 002, India.
Ph: 23289259, 23266861, 23266867   Website: www.cbspd.com
Fax: 011-23243014   e-mail: delhi@cbspd.com; cbspubs@airtelmail.in.

Corporate Office: 204 FIE, Industrial Area, Patparganj, Delhi 110 092
Ph: 4934 4934   Fax: 4934 4935   e-mail: publishing@cbspd.com; publicity@cbspd.com

Branches

- **Bengaluru:** Seema House 2975, 17th Cross, K.R. Road,
  Banasankari 2nd Stage, Bengaluru 560 070, Karnataka
  Ph: +91-80-26771678/79   Fax: +91-80-26771680   e-mail: bangalore@cbspd.com
- **Chennai:** 7, Subbaraya Street, Shenoy Nagar, Chennai 600 030, Tamil Nadu
  Ph: +91-44-26680620, 26681266   Fax: +91-44-42032115   e-mail: chennai@cbspd.com
- **Kochi:** Ashana House, No. 39/1904, AM Thomas Road, Valanjambalam,
  Ernakulam 682 016, Kochi, Kerala
  Ph: +91-484-4059061-62-64-65   Fax: +91-484-4059065   e-mail: kochi@cbspd.com
- **Kolkata:** 6/B, Ground Floor, Rameswar Shaw Road, Kolkata-700 014, West Bengal
  Ph: +91-33-22891126, 22891127, 22891128   e-mail: kolkata@cbspd.com
- **Mumbai:** 83-C, Dr E Moses Road, Worli, Mumbai-400018, Maharashtra
  Ph: +91-22-24902340/41   Fax: +91-22-24902342   e-mail: mumbai@cbspd.com

Representatives

- **Hyderabad**   0-9885175004
- **Nagpur**   0-9021734563
- **Patna**   0-9334159340
- **Pune**   0-9623451994
- **Vijayawada**   0-9000660880

Printed at: Rashtriya Printer, Dilshad Garden, Delhi-110092

to

_____

my source of inspiration and strength
my parents
Smt Shashikala Mishra and Shri Sreekant Mishra
for going through untiring efforts to ease my life
and
my wife and my son
for their unstinted support and help

Foreword

Foreword

I congratulate Mr Rajeev R Mishra on the inaugural edition of his book *Managing Hotel Front Office Operations*. Mr Rajeev has long and varied experience of teaching, particularly, front office, to the students of hotel management at various levels. I appreciate the efforts made by him in including chapters on domestic airlines and major hotel chains in India. Also, a section of glossary of terms used in the front office would help the reader in better conceptualization of the subject.

This book is replete with the systems and procedures required for front office operations and the author has done a good job by penning his professional experience in black and white. The book has been written in a lucid style. Each chapter is peppered with homilies and examples that enhance the presentation of the book.

Since the book ably covers all aspects of front office, I am quite confident that it is comprehensive and will stand the students in good stead.

With best wishes to the author for success in his future endeavors.

Satvir Singh

Principal
Institute of Hotel Management
Catering Technology and
Applied Nutrition
Faridabad

Preface

India holds a special place in the international world of hospitality. Culturally the country might very well be the most diverse place in the world. It is a vivid kaleidoscope of landscapes, magnificent historical sites and royal cities, misty mountain retreats, colorful people, rich cultures, and festivities. Luxurious and destitute, hot and cold, chaotic and tranquil, ancient and modern—India's extremes rarely fail to leave a lasting impression.

The hospitality industry is defined as 'hosts offering services to guests', which includes reception, entertainment, and other services for travelers and tourists. Hospitality is a long running tradition in India. From the majestic Himalayas and the stark deserts of Rajasthan, over beautiful beaches and lush tropical forests, to idyllic villages and bustling cities, India offers unique opportunities for every individual preference. From Kashmir to Kanyakumari, from Gujarat to Assam, there are different cultures, languages, life styles, and cuisines. This variety is increasingly reflected by the many forms of accommodation available in India, ranging from the simplicity of local guesthouses and government bungalows to the opulent luxury of royal palaces and five star deluxe hotel suites.

The Indian tourism and hospitality industry has emerged as one of the key drivers of growth among the services sector in India. Tourism in India has significant potential considering the rich cultural and historical heritage, variety in ecology, terrains and places of natural beauty spread across the country. According to the World Travel and Tourism Council (WTTC) forecasts, travel and tourism has the potential to contribute 46 million jobs to the Indian economy by 2025. The industry should rise by 6.5% per annum over the next 10 years to 4,337.8 billion in 2025 or 6.9% of the total. Expected to be the second-largest employer in the world, the hospitality sector will employ close to 5 million people in India by 2019, according to a report by the WTTC.

Tourism has shown its presence in 'Make in India' initiative because of huge possibility to bring socioeconomic reforms that include infrastructure development, attract investment, creation of jobs and entrepreneurial possibilities. One top agenda of government is to increase India's share in world tourist arrivals from the present 0.68% to 1% by 2020 and further increase it to 2% by 2025.

Taking this growth into consideration, the world's leading tourism and hospitality companies are looking for skilled professionals to become tomorrow's leaders. Tourism sector provides a wide range of career opportunities both at home and overseas in many and varied industry sectors such as tour operators, airlines, tourist attractions and hotel chains. On the other hand, hospitality industry is in continuous need of managerial staff in hotel management, restaurant management, public house management, nightclub management and even includes working on cruise liners and in theme parks. Both the industries are truly international with career opportunities at home and abroad.

Furthermore, the events industry is flourishing and the management of events has become increasingly important within the hospitality, tourism, leisure and sports sectors. Event managers can be found working on a wide range of events from weddings to conferences, meetings to product launches, at festivals and major sporting occasions.

Some high profile events include the Olympic Games, the Commonwealth Games, ICC Cricket World Cup, Grand Prix, FIFA World Cup, etc.

The front office department is the nerve centre, the hub and the heart of the hotel. Within a hotel no department is as vital and as visible as the front office. Front office personnel have more contact with guests than staff in other departments. The front desk is usually the focal point of activity for the front office and is prominently located in the hotel's lobby. A hotel's front office is where guests are greeted when they arrive, where they are registered and assigned rooms, provided information, their luggage handled, their accounts settled at departure, and their problems, complaints, and suggestions are looked after. The front desk is the link between the guest and the hotel and represents the hotel to the guest and is a liaison between the hotel management and the coordination of all the guest services.

Rajeev R Mishra

Acknowledgments

Atter a long stint in the hospitality industry and later teaching at the Rawal Institute of Management, the idea of writing a book on front office germinated in my mind. It has taken approximately two years since its inception to complete the work, and it gives me immense pleasure to have achieved this milestone, my contribution to the enrichment of the hotel industry.

I am indebted to all those who very kindly extended their support to me in the preparation of this book. No endeavor achieves success without the advice and cooperation of others. I would like to acknowledge the people and organizations who have either directly or indirectly contributed towards the conceptualization and compilation of this book. I would like to acknowledge all my students for their queries that helped me in the realization of the fact that there is a need for a quality textbook on front office.

I am thankful to Mr Tushar Abrol (Manager—Learning and Development, Vivanta by Taj, Gurugram) for providing free access to the hotel to understand the various procedures and practices used in the front office operations

I deeply acknowledge the patience, support and understanding of my wife Sudha, and son Akshat, who had taken all pains and strains in managing our household during the entire period of writing this book. Without her encouragement and inspiration, I would not have this achievement. I record my gratitude to my parents and sisters for their love, encouraging pat and unstinted support which they provided while I was working on this book. I would also like to acknowledge the assistance of Dr Pawan Kumar who was instrumental in providing feedback and critical analysis.

I thankfully acknowledge all the authors whose research papers, books, and articles have been referred to. I also extend my heartfelt thanks to the editorial team at CBS Publishers & Distributors Pvt Ltd for its coordination and support from the beginning and for timely suggestions and encouragement, which have made this textbook more logical in its approach and presentation.

Finally, I would like to thank all my well-wishers, friends, colleagues, and academicians, who have made helpful suggestions for the improvement of this book.

The author will be grateful to the teachers and the readers for pointing out errors and giving constructive suggestions which will be incorporated in the next edition of this book. Your suggestions and feedbacks to improve the book are always welcomed at: *rajeevmishra1976@rediffmail.com.*

Rajeev R Mishra

rajeevmishra1976@rediffmail.com

Contents

Part IV: Front Office Management

Part I

The Travel and Tourism Industry

Introduction to Travel and Tourism

Learning Objectives

After reading this chapter, you will be able to understand the following:

- Overview of tourism
- Evolution of tourism
- Reasons for travel
- Significance of tourism
- Constituents of the travel and tourism industry
- Categorization of tourism—international tourism (inbound and outbound) and domestic tourism
- Terms and terminology related to travel and tourism

OVERVIEW OF TOURISM

Tourism may bring forth pictures of cool hill stations, snow clad mountains, warm sunny beaches or long scenic drives. Enjoyment, pleasure, excitement, packing of bags, carrying documents and credit cards, shopping, and spending money are some of the things which may come to your mind when you think of tourism.

Tourism is a structured break from routine life. It involves a separation from everyday life and offers an entry into another moral and mental state, where expressive and cultural needs become more important. Hence it may be identified with recreation or renewal of life. It can be considered as a modern ritual in which people 'get away from it all'; particularly the usual workplace.

William F. Theobald (1994) suggested that etymologically, the word *tour* is derived from the Latin, *tornare* and the Greek, *tornos*, meaning *a lathe or circle; the movement around a central point or axis*. This meaning changed in modern English to represent 'one's turn'. The suffix -ism is defined as 'an action or process; typical behavior or quality', whereas the suffix -ist denotes one that performs a given action. When the word tour and the suffixes -ism and -ist are combined, they suggest the action of movement around a circle. One can argue that a circle represents a starting point, which ultimately returns back to its beginning. Therefore, like a circle, a tour represents a journey that is a round trip, i.e. the act of leaving and then returning to the original starting point, and therefore, one who takes such a journey can be called a Tourist.

Tourism is an activity and is very generic in nature and as such defies a common and a standard definition. There is no single universally accepted, clear cut definition of tourism. Many people and many organizations have defined tourism in many different ways over the years. Some of the common yet important definitions are given below:

- Tourism may be defined as the movement of people from their normal place of residence to another place (with the intention to return) for a minimum period of twenty-four hours to a maximum of one year for the sole purpose of leisure and pleasure.
- Tourism as a product can be defined as an amalgam of three components—attractions of the destination, the facilities of destination and the accessibility of it.
- Tourism is the temporary short-term movement of people to destinations outside of the place of where they normally live and work, and includes the activities they indulge in at the destination as well as all facilities and services specially created to meet their needs. Tourism does not only mean traveling to a particular destination but also includes all activities undertaken during the stay. It includes day visits and excursions.
- Tourism is a collection of activities, services and industries that delivers a travel experience, including transportation, accommodations, eating and drinking establishments, retail shops, entertainment businesses, activity facilities and other hospitality services provided for individuals or groups traveling away from home.
- The study of man away from his usual habitat; of the industry which responds to his needs, and of the impacts that both he and the industry have on the host's socio-cultural, economic, and physical environments.
- An activity of persons traveling to and staying in places outside their usual environment for not more than one consecutive year for leisure, business and other purposes not related to the exercise of an activity remunerated from within the place visited.
- Tourism is an activity in which money earned by a person in his normal domicile is spent at the place visited by him.

EVOLUTION OF TOURISM

For the early man, the term 'travel' was not associated with the words pleasure or leisure as it is today. The word travel has originated from the word 'travail' meaning painful or laborious.

To establish any specific period or era for the origin of tourism is a very difficult task. As tourism is involved with the movement so it can be said that tourism activities started with the development of mankind. In ancient and pre-historical period, people used to move in search of food, and for shelter from climatic conditions such as rainy season, winters and summer season and also for protection from other people.

The history of tourism can be divided into 6 different stages as discussed below.

Roman Empire Period

During the Roman Empire period (from about 27 BC to AD 476), travel developed for military, trade and political reasons, as well as for communication of messages from the central government to its distant territories. Travel was also necessary for the artisans and architects 'imported' to design and constr ct the great palaces and tombs. In ancient Greece, people traveled to Olympic Games. Both the participants and spectators required accommodations and food services. Wealthy Romans, in ancient times, traveled to seaside resorts in Greece and Egypt for sightseeing purpose.

Middle Age Period

During the Middle Age (from about AD 500 to 1400), there was a growth of travel for religious reasons. It had become an organized phenomenon for pilgrims to visit their 'holy land', such as Muslims to Mecca, and Christians to Jerusalem and Rome for wish fulfillment or to purify the body.

16th Century

In the 16th century, the growth in England's trade and commerce led to the rise of a new type of tourists—those traveled to broaden their own experience and knowledge.

17th Century

In the 17th century, the sons and daughters of the British aristocracy traveled throughout Europe (such as Italy, Germany and France) for periods of time, usually 2 or 3 years, to improve their knowledge. This was known as the Grand Tour, which became a necessary part of the training of future administrators and political leaders.

Travel for treatment at natural springs or Spas was gaining popularity in the mid-seventeenth century when doctors advocated the healing powers of mineral water. They soon became important meeting places for the gentry or elite.

Industrial Revolution Period

The Industrial Revolution (from about AD 1750 to 1850) in Europe created the base for mass tourism. This period turned most people away from basic agriculture into the town/factory and urban way of life. As a result, there was a rapid growth of the wealth and education level of the middle class, as well as an increase of leisure time and a demand for holiday tourism activities. At that time, travel for health became important when the rich and fashionable Europeans began to visit the spa towns (such as Bath in England and Baden—Baden in Germany) and seaside resorts in England (such as Scarborough, Margate and Brighton).

19th to 20th Centuries

In the 19th and 20th centuries, the social and technological changes have had an immense impact on tourism. Great advances in science and technology made possible the invention of rapid, safe and relatively cheap forms of transport: The railways were invented in the 19th century and the passenger aircraft in the 20th century. World War II (AD 1939–1945) was also the impetus for dramatic improve-ments in communication and air transportation, which made travel much easier today than in earlier times.

1980s

The 1980s were called the boom years. Business and leisure travel expanded very rapidly. The baby-boomers were coming of age and had the money to spend. These travelers were looking for a variety of travel products from exciting vacation options such as adventure travel, eco-tourism and luxurious travel.

There was not only a significant expansion in the travel market but also in tourist destinations. The fall of the Berlin Wall in Germany in 1989 signified the doom of communism in Europe. Countries such as Russia and the Czech Republic became new tourist destinations both for vacation and business travelers.

1990s

The Aviation Industry was facing high operational costs, including wage, oil prices, handling fee of Central Reservation System (CRS), landing charge of the aircrafts and advertising fee, etc. During this decade, CRS also marched towards more sophisticated technology. It became possible for agents to book a huge inventory of travel products, such as hotels, car rentals, cruises, rail passes, and theatre tickets from the CRS.

The introduction of 'ticketless traveling' (electronic ticket) brings benefits to the airlines by cutting the amount of paperwork and cost of tickets. At the same time, passengers do not have to worry about carrying or losing tickets. Although, electronic ticketing does not bypass the travel agents as intermediaries, it makes it easier for the airline to deal directly with consumers.

The advance in technology also allows the airlines and other travel suppliers to sell directly to travelers through the Internet and interactive kiosks at airports. The kiosks at the airport usually sell hotel accommodation, transfer tickets such as bus tickets between airport and downtown areas and coach tickets from one city to another.

Travelers can now log on to the internet easily, reach for travel information, book a simple ticket or hotel room through their personal computer at home. There are thousands of new destinations, tour products and discounted airfares for travelers to choose from.

In its early stage, travel was a luxury available to the privileged class as transport costs were very high. Today, tourism is no longer the privilege of the rich and famous exclusively, but it is an activity to be enjoyed by people from all strata of society. It is ingrained into the daily lives of many people across the globe.

REASONS FOR TRAVEL

Before discussing the travel motivations of tourists, one question that should be answered first is: *Why do people travel?*

Tourism has witnessed considerable changes in the twenty-first century from its previous motivations of travel, which were mainly visiting places of religious interest or travel for trade purposes.

Travel motivators can be defined as those factors that create a desire in people to travel. Motivators are the internal psychological influences affecting individual choices.

It is important to study the factors which promoted tourism during all periods. Figure 1.1 presents an overview of the travel motivators.

Curiosity and culture: People are curious and eager to learn about other countries, their people, and their culture. Tourists visit places of historical interest, fairs, festivals, museums, dances, folk art, etc. to know more about the lifestyle of people from different countries.

Interpersonal reasons: This includes people's desire to visit their friends or relatives (VFR), families, ancestral homelands, and also for meeting new people and seek new experiences.

Relaxation and refreshment of body and mind: Due to the increased industrialization and hectic modern lifestyle, there is a need for rest and relaxation, to de-stress the body and mind. This desire for relaxation varies from individual to individual. People undertake travel simply to escape from their mundane day-to-day routine.

Health: Since the medieval days, people have been visiting spas and bathing in hot sulfur springs for specialized medical treatment. Several spas and health resorts have developed over time in most of the countries which attract visitors because of their curative aspects.

Status and prestige motivators: These are identified with one's personal esteem and status symbol.

Professional or business reason: People need to travel for business-related reasons and this involves both domestic as well as international travel. People travel to expand their business, attend meetings, conferences, and exhibitions.

Spiritual purpose: A large number of people are motivated to travel because of spiritual motives, i.e. visiting holy places, shrines, etc. The number of people who seek solace in such places is increasing dramatically.

Posterity: It is mainly in search of roots, i.e. to find decendency.

Pleasure: Satisfying an individual's need for pleasure is the most predominant of all individual travel motivations. A person's need for pleasure is very deep-rooted and travel can satisfy this desire.

Sports: The participants and spectators have been traveling for World Cups, Commonwealth Games, Olympics, Asian Games, Grand Prix races, Formula One, etc.

Leisure time and disposable income: The concept of work for 5 days a week gave more leisure time to people.

Education: Education has always been a great motivator for the travelers. There have

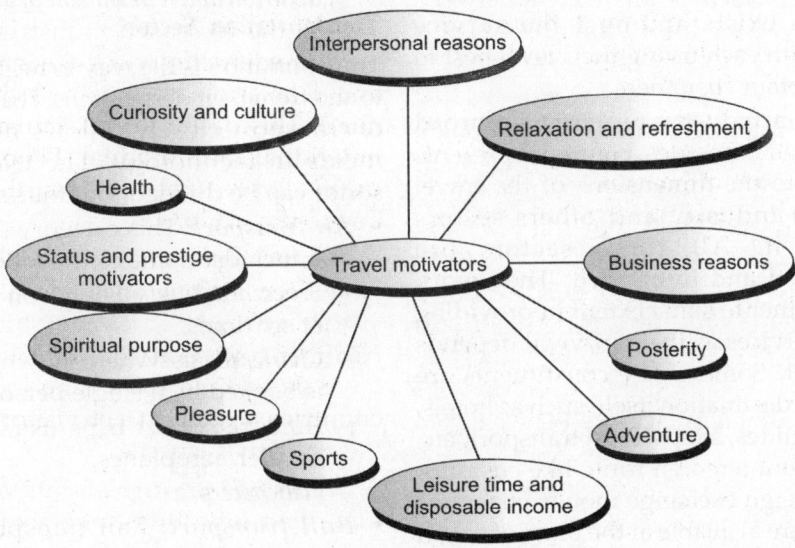

Fig. 1.1: Travel motivators

been great universities all across the globe that attracts the students from far and near.

Adventure: The thrill always lies in new and unexplored like voyages undertaken by Marco Polo, Christopher Columbus, Vasco da Gama, etc.

SIGNIFICANCE OF TOURISM

Tourism is a global phenomenon. Tourism is an activity essential to the life of nations because of its direct effects on the social, cultural, educational, and economic sectors of national societies and on their international relations.

Tourism is one of the world's largest and fastest growing industries and a major source of foreign exchange and employment generation for many countries. It is regarded as one of the most remarkable economic and social phenomenon of the past century. The travel and tourism sector has developed into an industry with an annual economic report (direct, indirect and induced) of around US$ 6.5 trillion worldwide. United Nations World Tourism Organization (UNWTO) forecasts 1.6 billion international tourist arrivals worldwide by 2020.

Tourism has also an important role to play in the bringing of prosperity to those under-developed parts of the country, which for various reasons, are relatively unsuited to industrial development or agriculture.

CONSTITUENTS OF THE TRAVEL AND TOURISM INDUSTRY

The tourism industry is a vast industry made up of businesses and organizations that provide goods and services to meet the distinctive needs of tourists. These businesses and organizations are related to virtually all areas of the economy making tourism a very huge industry.

The tourism industry is characterized by constant change and development and is a highly dynamic industry offering innovative products, new destinations, and technologically advanced transportation every year. The latest in this range is the world's largest cruise liner Freedom of the Seas which can accommodate over 5500 passengers. Stiff competition exists amongst the service providers, with each trying their level best to attract and retain customers.

The tourism industry represents a broad range of related industries. Figure 1.2 presents an insight into the dimensions of the travel and tourism industry and others sectors related to it. All these sectors are interconnected and integrated. They work with one another to some extent in providing goods and services as their survival depends on each other. Some of the constituents are located at the destination itself, such as hotels, attractions, guides, shops, local transport, etc. some are encountered en route like customs, transport, foreign exchange money changers; while others are available at the place of origin of the journey, such as consulate for visa and travel agents.

Accommodation Sector

Accommodation is one of the most important constituents of tourism industry because the type of accommodation available at a place will, of course, decide what types of tourists go to that place. For this reason, accommodation should actually come before any other type of development at a destination.

These may range from 5 star luxury hotels to circuit house or guest house.

Transportation Sector

Transportation is the way to help travelers get to and from their destinations. The development of transport closely follows new developments in technology. The transportation sector can be divided into four main groups:

- *Air transport*: This category of transportation includes some of the following:
  - *Scheduled flights*: Flights on large airlines at set times.
  - *Charter flights*: Where the whole aeroplane is booked by a single person or group.
  - *Air taxis*: Short trips usually taken on smaller aeroplanes.
  - *Helicopters*
- *Rail transport*: Rail transport is not as popular today as it was centuries ago. It is seldom used as a way to travel long distances.
- *Sea transport*: Sea transport is also not a very popular mode of transport. The reason for this is that it is expensive and takes a long time for people to get to their destinations. It is probably the most expensive mode of transport. Nowadays, it is growing in popularity as a way of combining several tourism experiences in one. For example, a trip on the cruise liner Symphony combines accommodation, entertainment and

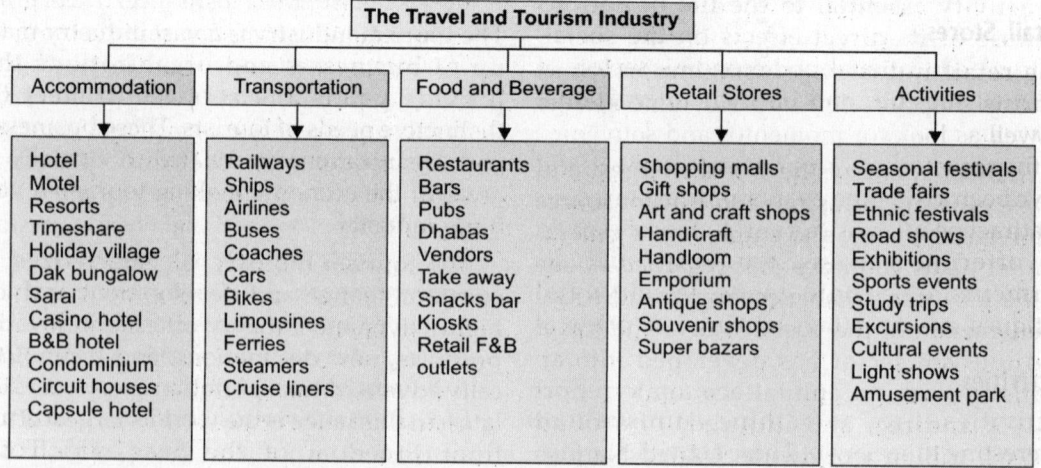

The Travel and Tourism Industry				
Accommodation	Transportation	Food and Beverage	Retail Stores	Activities
Hotel Motel Resorts Timeshare Holiday village Dak bungalow Sarai Casino hotel B&B hotel Condominium Circuit houses Capsule hotel	Railways Ships Airlines Buses Coaches Cars Bikes Limousines Ferries Steamers Cruise liners	Restaurants Bars Pubs Dhabas Vendors Take away Snacks bar Kiosks Retail F&B outlets	Shopping malls Gift shops Art and craft shops Handicraft Handloom Emporium Antique shops Souvenir shops Super bazars	Seasonal festivals Trade fairs Ethnic festivals Road shows Exhibitions Sports events Study trips Excursions Cultural events Light shows Amusement park

Fig. 1.2: Dimensions of travel and tourism industry

transport all in one vessel. Some types of sea transport include:

- *Cruise liners*: Usually involves travel for several days or even weeks.
- *Day trips on cruise boats*
- *Ferries*

• *Road transport*: This mode of transport is by far the most commonly used. It is a convenient means of transport that gets you where you need to go with the least amount of difficulties and delays. It is fairly cheap and quick. Road transport includes the following:

- *Private cars*
- *Rented cars*
- *Coaches/tour buses*
- *Motorbikes*
- *Taxis*

Food and Beverage Sector

The food service industry engages itself in the provision of food and beverages, mainly to the people who are away from their homes for different reasons. Such people need accommodation with food and beverages if they are away for more than a day and only food and beverages if they are away for a short duration of time. The basic needs of customers for food and beverages are met by the food service industry, which has been associated with lodging ever since people started traveling.

Retail Stores

The retail industry is very important as tourists shop for their day-to-day necessities as well as look for momentos and souvenirs. In the recent years, several cities in the world have been promoted as shopping destinations to attract people with a penchant for shopping by offering various products, such as garments, electronic goods, jewelry, and antiques, among others, at very low prices.

Activities

Activities are something unusual or interesting to see or do. These activities appeal so much that many people travel so that they can be a part of these activities.

Travel Organizers Sector

This sector makes it easier for tourists to plan their trips. Some of the service providers found here includes:

• *Travel agent*: A travel agent is one who makes arrangements of tickets for travel by air, rail, ship, etc. It may also arrange accommodation, tours, entertainment and other tourism related services like passport and visa.

A travel agency is a retailing business that sells travel-related products and services, particularly package tours, to customers on behalf of suppliers such as airlines, car rentals, cruise liners, hotels, railways or sightseeing and tour operators. But, unlike other retail businesses, they do not keep a stock in hand. They do not buy a holiday package or a ticket from a supplier unless a customer requests it. Most travel agencies operate on a commission-basis, implying that the supplier offers a fixed percentage of the sale to the agencies as commission for booking clients. The agencies may offer a discount on a holiday package or ticket to the customers by shrinking their commission. Some travel agencies undertake other commercial operations, such as the sale of in-house insurance, travel guide books, time tables, car rentals, and the services of an on-site bureau de change (dealing in the most popular currencies).

Travel agencies play a very important role as they plan out the itinerary of their clients and make the necessary arrangements for their travel, stay, and sightseeing, besides facilitating their passport, visa, etc.

Some of the famous travel agencies in India are Abercrombie and Kent India Pvt Ltd, ACME Tours and Travels Pvt Ltd, Adventure Tours, Balmer Lawrie and Co Ltd, Concord Travels and Tours, Cox and Kings Ltd, Thomas Cook, Jetair Tours

Limited, Kuoni Travel (India) Pvt Ltd, Le Passage To India Tours and Travels, Mercury Travels Limited, Orient Express Pvt Ltd, Swagatam Tours Private Limited, Travelite, etc.

Functions of a travel agency

The travel agency performs a number of functions such as:

- *Providing travel-related information:* Information concerning destination such as climate, location, culture, sightseeing, mode of transport, clothing, visa and other such formalities, currency, customs formalities, necessary geographical information about destination, local foreign exchange rules, health regulation, etc.
- *Insurance*: Insurance cover for death, disablement and loss or damage of baggage, etc.
- *Accommodation*: Negotiates with hotels for reasonable room and meal package plans.
- *Ticketing (Domestic and International)*: Negotiates with airlines and railways for cheap rates of transportation by the most economical route. Negotiates with surface transporters to provide taxi/coach for transport from airport to hotel, sightseeing tours, etc.
- *Ancillary services*: Travel agents help in providing ancillary services such as passport, visa, foreign exchange, etc.
- *Packaging and selling of tour*: Planning and selling of tour package to fit the requirements of traveler keeping in mind his pleasure, convenience and security.
- *Plans tour itineraries*: The tour professionals after receiving the client's preference of destinations to be visited, his approximate date of travel and duration, the mode of transport, hotels, etc. has to plan the itinerary to suit the client's needs.
- **Tour operator:** A tour operator assembles the various elements of a tour. It typically combines tour and travel components to create a holiday. The most common example of a tour operator's product would be a seat on a charter airline plus a transfer from the airport to a hotel, and the services of a local representative, all for one price.

The reason for existence of tour operators was the difficulty of making arrangements in far-flung places, with problems of language, currency, and communication. Although the Internet has made self-packaging of holidays easier now, tour operators still have their competence in arranging tours for those who do not have the time to do so. Also, tour operators still exercise contracting power with suppliers and influence over other entities (tourism boards and other government authorities) in order to create packages and special departures for destinations that are otherwise difficult and expensive to visit.

- *Inbound tour operators*: An inbound tour operator is one who makes arrangements for transport, accommodation, sightseeing, entertainment and other tourism related services for foreign tourists
- *Tourist transport operators*: A tourist transport operator is one who provides tourist transport like cars, coaches, boats, etc. to tourists for transfers, sightseeing and journeys to tourist places, etc.
- *Adventure tour operators*: An adventure tour operator is one who is engaged in activities related to adventure tourism in India, namely water sports, aero sports, mountaineering, trekking and safaris of various kinds, etc. In addition to that he may also make arrangements for transport, accommodation, etc.
- *Domestic tour operators*: A domestic tour operator is one who makes arrangements for transport, accommodation, sightseeing, entertainment and other tourism related services for domestic tourists.

Table 1.1: Total number of approved service providers of travel trade as on 31st December 2014

Travel agents	280
Inbound tour operators	486
Tourist transport operators	127
Adventure tour operators	34
Domestic tour operators	90
Total	**1017**

CATEGORIZATION OF TOURISM

Tourism can be categorized as international tourism and domestic tourism.

- *International tourism*: It involves people traveling from one country to another country, crossing national borders or through immigration. In order to travel to a foreign country, one needs a valid passport, visa, health documents, foreign exchange, etc. International tourists may be classified into the following categories:
  - *Inbound tourism*: This refers to incoming tourists or tourists entering a country. For example, Japanese citizens traveling to India would be considered as inbound tourists for India and outbound tourists for Japan.
  - *Outbound tourism*: This refers to outgoing tourists or tourists leaving their country of origin to travel to another country.
- *Domestic tourism*: The tourism activity of people within their own country is known as domestic tourism. It involves residents of a country traveling within the borders

of that country. A person from Delhi going for a holiday to Mumbai is a domestic tourist. Traveling within the same country is easier because it does not require formal travel documents and tedious formalities like compulsory health checks and foreign exchange. In domestic tourism, a traveler generally does not face much language problem or currency exchange issues. However, domestic tourism has become an important money-spinner and job generator for the hospitality industry (Fig. 1.3).

TERMS AND TERMINOLOGY

Tourists (overnight visitor)

The UNWTO defines tourists as 'people who travel to and stay in places outside their usual environment for more than 24 hours but not more than one consecutive year for leisure, business and other purposes not related to the exercise of an activity remunerated from within the place visited'.

A tourist is a temporarily leisured person who voluntarily visits a place for the purpose of experiencing a change.

It must be kept in mind that tourists are short-term, temporary visitors and should not be confused with people who migrate to a country to settle permanently and become residents.

Classification of Tourists

Tourists are classified, according to their needs and their reasons for traveling, into four broad categories:

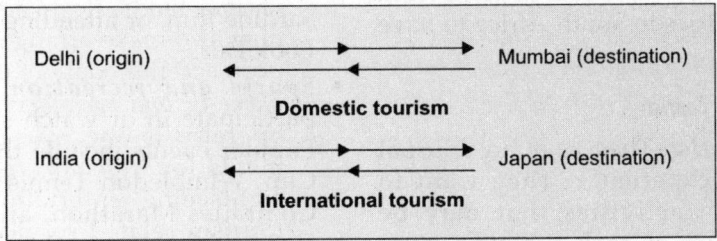

Fig. 1.3: Domestic and international tourism

Business and Professional Tourist

- **Business tourist:** Travel is related to business and the world of work. Business tourism involves meetings, conferences and conventions, exhibitions. These are all part of the business world. All major cities have conference centers that cater for the needs of business tourists. An example of a business tourist would be a salesman who travels to another city to attend a trade show and promote the products he sells.

- **Educational tourist:** They travel to attend a place of learning in another town, city, or country, in order to study for or improve a qualification. They may also be people that attend workshops to learn new skills or improve existing ones. A clinic nurse who travels to another province to attend a workshop about infectious diseases is an example of an educational tourist.

- **Incentive tourist:** They are people who are rewarded in the form of a company paid holiday for their hard work, or for achieving goals set by their company. This incentive to travel motivates employees to work harder, improves work relationships, and builds team spirit. A salesman who receives a holiday package for achieving the most sales in the company is an example of an incentive tourist.

- **Health or medical tourist:** They travel because they want to visit a holiday spa, needs medical special treatment that is only available away from home, undergo procedures that are cheaper in another country, or are recovering from an illness in a healthier climate. Many tourist come from oversees countries to South Africa to have plastic surgery.

Leisure and Holiday Tourist

- **Adventure tourist:** They want an unusual and exciting experience. They want to participate in activities that may be dangerous, such as rock climbing, river rafting, skydiving, shark cave diving and bungee jumping.

- **Cultural tourist:** They want to experience different cultures, such as San rock art, or culture-related festivals such as the National Art Festival in Grahamstown, or the International Jazz Festival in Cape Town. They would also want to experience the World Heritage Sites in the country.

- **Eco-tourists:** They travel to experience nature such as traveling to Bonita Gardens in Bloemfontein South Africa.

- **Leisure tourist:** They want to rest and relax and have a break from the usual routine. Examples of this type of tourism are a cruise on a cruise liner, a trip on a Blue train, attending a special music performance or relaxing on the beach.

- **Religious tourist:** They want to see and experience places of religious importance. There are many religious destinations in the world such as the Hajj in Mecca, Jerusalem in Israel, Varanasi in India, and the Vatican in Rome. During Easter the largest Christian pilgrimage to Zion City, Moria, Limpopo, takes place. More than a million pilgrims travel to Moria to every Easter.

- **Shopping tourist:** They travel to shopping malls, shopping centers, factory shops, crafts market, festivals, and touring shopping routes such as the Midlands in KwaZulu-Natal. Their main purpose is to buy items.

- **Special interest tourist (SIT):** They have particular interest such as bird watching, food and wine, flowers, fishing during the Sardine Run, or attending the Cape Town Book Fair.

- **Sports and recreation tourist:** They participate in or watch sporting events. Popular events include the Soccer World Cup, Wimbledon Tennis Championship, Comrades Marathon, and Fisher River Canoe Marathon. Surfing, mountain climbing, cricket, swimming, golf and tennis are popular sports.

Tourists traveling to visit friends and relatives (VFR)

Tourist visiting friends and relatives (VFR) want to stay in contact with friends and relatives and travel away from home to visit them. These tourists may travel to attend a wedding, funeral, or birthday celebration of friends or relatives.

Youth tourists, including backpackers and gap year travelers

- *Backpacking or youth tourist*: They generally have little luggage, are on a budget, want to experience adventure and excitement, tend to travel independently, enjoy meeting other traveler, and have flexible travel schedules. A group of young tourists on a weekend walking tour in the mountains, or a student touring around the country by bus are examples of this group of tourist.
- *Gap year travelers*: They do not study further or enter job opportunity after school; instead they take break called a gap year. They travel, work and earn money, learn new skills or do volunteer work in another country. During this time they gain skills and life experience before starting tertiary education. These young people are also known as 'gappers'.

Excursionists (same-day visitor or day tripper)

An excursionist is a day visitor who stays for less than 24 hours at a place. Excursionists do not stay overnight in the place visited. For example, if a group of students from Pune go to the nearby hill station Lonavala early in the morning and return late in the evening, they are called excursionists.

Passport

A Passport is an official document issued by the government of a country to one of its citizen, authenticating his identity for the purpose of international travel and right to re-enter his native country.

A passport is a document issued by a government to allow its citizens to travel abroad, and requests other governments to facilitate their passage and provide protection on a reciprocal basis. Without a valid passport a person is not permitted to move in the territory of a foreign country.

A passport does not of itself entitle the passport holder entry into another country, nor to consular protection while abroad or any other privileges, in the absence of any special agreement which cover the situation.

Normally a passport is valid for a period of 10 years. However, it can be renewed for another 5 years on first expiry.

The public authorities competent to issue passports and other identity documents vary from one country to another. The external affairs ministry issues passports to the citizens of India after verifying the details of the applicants from various quarters.

All passports generally bear the following information of the passport holder, though the format may vary from country to country:

- Family name/surname
- Given name
- Nationality
- Date of birth
- Place of birth
- Gender
- Date of issue
- Place of issue
- Holder's signature
- Validity for certain countries as well as restrictions for travel to others
- Holder's photograph
- Any visible distinguishing mark
- Name of father/legal guardian
- Name of mother

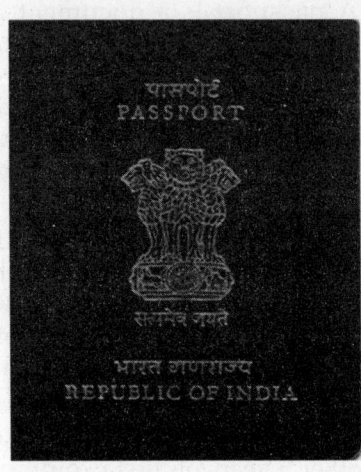
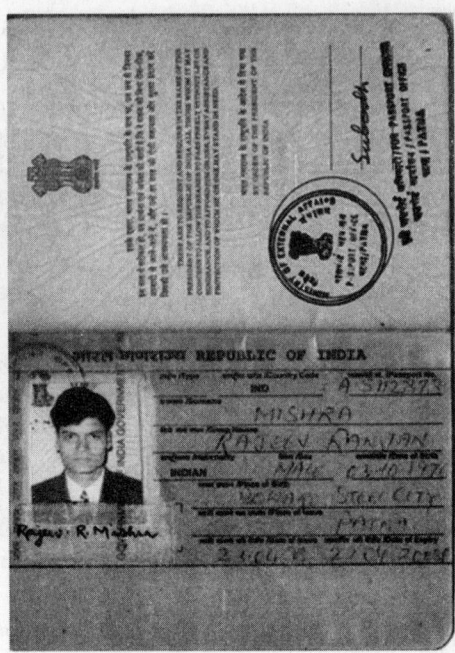

Fig. 1.4: Specimen copy of a Passport

- Name of spouse
- Address
- Validity period (expiry date)
- Children
- Endorsement
- Emigration status

In India, the following types of passport are issued under the provisions laid in the Passport Act, 1967.

Ordinary passport

It is dark blue in color and is issued to any ordinary citizen of India.

Official passport

It is white/grey in color and is issued to government officials or persons on government missions.

Diplomatic passport

It is red in color and is issued to persons with a diplomatic or consular status as per international laws and customs.

Visa

Foreign nationals visiting India are required to possess a valid passport and a valid Indian visa. Visa is an endorsement on the passport, allowing the holder to enter the territory of the issuing country. It is a document or, more frequently, a stamp in a passport, authorizing the bearer to visit a country for specific purposes and for a specific length of time. However, the issuance of a visa may not be treated as a guarantee to enter into the foreign territory. The bearer may be subjected to inspection at the port of entry and may be asked to produce the documents presented at the time of the procurement of the visa. A visa does not generally give a person any right beyond the right to enter a country and remain there.

The Consular Passport and Visa (CPV) Division of the Ministry of External Affairs issues visas to foreign nationals through various Indian missions abroad. Depending upon the nature of visit, the following types of visa may be issued.

Immigrant/Permanent Visa

It authorizes the holder of the visa to settle permanently in the county issuing the visa. This type of visa is rarely issued by countries and there are some countries that never issue such type of visa.

Temporary/Non–immigrant Visa

This type of visa is issued for a specific duration only. The person holding such a visa will have to return to the home country after the expiry of the term of the visa. Temporary visas are of the following types:

- *Tourist visa*: Issued for a limited period for leisure travel only (up to 6 months), no business activities allowed.
- *Student visa*: Issued to students who have got admission in an institution located in the issuing country. It is issued for the duration of the course of study.
- *Business visa*: Issued for business related activities, but precludes permanent employment.
- *Work visa*: Issued for approved employment in the host country, with longer validity than a business visa.
- *Transit visa*: Issued for passing through the countries that fall en route the final destination; valid for 15 days or less.

Itinerary

Itinerary is a plan for a journey, listing different places in the order in which they are to be visited. It is a tour program which details the date, time and other services such as sightseeing, etc. for the tour from the beginning to the end of the tour. It also includes the details such as flight number/train number and their departure time, etc. It is usually made by the travel agent in consultation with the tour leader.

The itinerary is a tour program in sequential order which is designed daywise to identify the origin points, destinations en route points, hotel, meals, mode of transport, sightseeing, car/coach and other relevant details related to the tour.

Guide

A guide is an authorized and licensed local person of the destination who is hired to explain the details of the destination such as historical places, culturally important places, etc. A guide should therefore be knowledgeable about history, geography, sociocultural practices, etc. related to his area of concern, so as to inform the tourists accordingly. He should have knowledge of different languages.

Tour Escort

The tour escort has to accompany the tourist right from commencement till the end of the tour. He has to perform the role of a tour leader or a tour manager. The escort may accompany the tourists to historical sites, rural areas, pilgrimage places, shopping, museums, etc. The escort has also to look after the facilitation of the tour such as the check-in formalities, customs clearances, etc. The escorts have to take care of the tourists, throughout the tour and at the destination.

The tour escort has to plan the tour properly with a schedule of events and live the tour day by day. He has to be prepared with alternative arrangements; in case of any unforeseen circumstances due to weather, transport strike, accident, etc. Some frequently encountered problems during a tour which may arise are loss of money, loss of passport, sickness of any tour member, missing tour members, etc.The tour escort should be well prepared in advance with all the travel arrangements such as checklists, tour itinerary, etc. The tour escort should be able to handle the tour members in an effective way and advise them accordingly so that time schedules are maintained and there are no missed trains or flights.

REVIEW QUESTIONS

1. Define tourism and enumerate its importance in overall growth of a nation.
2. Write in brief about the growth and history of tourism.
3. Discuss in detail the various reasons why people travel for tourism purpose.
4. What are the various documents required to be completed for undertaking overseas travel?
5. What is a passport? Write any four details mentioned in a passport.
6. What is the function of passport? Explain the types of passport.
7. Write short notes on:
   - Excursionist
   - Tourist
   - Visa
   - Passport
   - Tourism
   - VFR
   - Guide
   - Itinerary
   - Tour Escort
   - Significance of tourism
8. Differentiate between:
   - International tourism and Domestic tourism
   - Inbound tourism and Outbound tourism
   - Travel agent and Tour operator
   - Tour escort and Guide
   - Passport and Visa
9. Explain in short the primary and secondary constituents of tourism.
10. Explain the different modes of transport.
11. What is the importance of travel agent to hotel industry?
12. Define tour operator. Discuss different types of tour operators.
13. Explain the functions performed by a travel agent.
14. Discuss classification of tourists.
15. Discuss different types of visa in detail.

Five 'A's of Tourism

FIVE 'A's OF TOURISM

Tourist destination is a geographical unit which the tourist visits and where he stays. The success of a tourist destination depends upon the interrelationship of the following five 'A' factors:

- Attractions
- Accessibility
- Accommodation
- Amenities
- Activities

Developing a suitable combination of these factors is at the heart of tourism planning.

Attractions

Attractions are key elements that need to be considered in assessing the tourism potential of an area. Attraction means anything that creates a desire in any person to travel in a specific tourist destination. A tourist attraction is a place of interest that tourists visit, typically for its inherent or exhibited cultural value, historical significance, natural or built beauty, or amusement opportunities.

Some examples include historical places, monuments, zoos, museums and art galleries, botanical gardens, buildings and structures (e.g. castles, libraries, former prisons, sky-scrapers, bridges), national parks and forests, theme parks and carnivals, ethnic enclave communities, historic trains and cultural events.

Attractions are classified basically into four categories which are as follows:

- Natural attractions such as pristine beaches, waterfalls, scenic views, climate, heavy rainfall, snow clad mountains, flora and fauna, islands, etc.
- Human-made attractions such as theme parks, amusement parks, casinos, sports complex, zoo, etc.
- Cultural attractions in the form of fairs, festivals, celebrations, theatre and museums, monuments which depict the history and culture of a country.

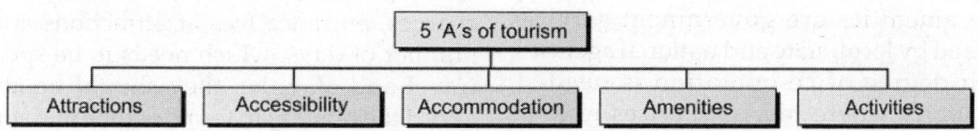

Fig. 2.1: Five 'A's of tourism

- Social attractions where one can meet and interact with the locals at a destination as well as meet friends and relatives.

Accessibility

Accessibility means reach ability to the place of destination through various means of transportation. Transportation should be regular, comfortable, economical and safe. Various means of transportation are airlines, railways, road and water. Globally, air transport dominates the movement of international visitors.

Sometimes modes of transport can be attractions in their own right. Ferries, steam trains and scenic roads with viewing areas both provide access and can be attractions in themselves.

Accommodation

Although day-trippers spend money on consumable items, overnight visitors have a variety of needs, including accommodation, dining and are also more likely to spend on attractions and activities. A variety in style and quality of accommodation in a region provides an important means to increase the economic impact of visitors. There are various types of accommodation ranging from star hotels to budget hotels.

Amenities

Amenities are the services that are required to meet the needs of tourists during their visit to a destination. They include public toilet, signage, retail shopping, restaurant and cafes, visitor information centers, telecommunications, emergency services, drinking water, local transport, automatic teller machines (ATMs), proper garbage and sewage disposal systems, medical facilities, etc. Because many of the amenities are government services delivered by local, state and national agencies, a high degree of co-operation is needed, particularly where tourist services may be seen to be competing with the needs of local residents.

Activities

Activities provide interesting or entertaining diversions for people once they are in the area. Activities add variety and can make a visit more enjoyable, but it is important to understand that they cannot replace attractions. For example, elephant bath, jungle walking, etc. are some of the common activities.

Apart from the classic five 'A's of tourism, we suggest two more 'A's which are extremely vital to the success of any destination.

Awareness

Having the best attractions, access, accommodation and amenities in the world is totally useless if the awareness factor is missing. Awareness in this sense has three meanings.

Firstly, the local population must have a positive attitude towards tourism. If the local community sees 'tourists as terrorists', then this will have a negative impact.

Secondly, those in the front line of tourism, that is, those who directly interface with tourists must have strong, positive attitudes towards tourists. This includes the shops, post offices, road houses and the many other businesses that come in contact with tourists, not just the hotels and restaurants. In all a local community must be made aware of the value of tourism. The third plank in the awareness platform is market awareness. The destination or more importantly, the destination's image must be a strong, positive one and firmly implanted in the tourist's mind.

Affordability

Tourists should be able to afford the trip in terms of transport costs, accommodation charges, entrance fees at attractions and the number of days, which needs to be spent for travel and stay, i.e. they should be able to afford the holiday in terms of time and money.

Tour operators prepare package tours keeping affordability in mind. These group tours work out cheaper than individuals booking their own tickets and making itineraries for themselves.

A successful destination would have a good balance between above-mentioned 'A's and ensure that there is something to see and do for people of different ages and backgrounds so that a large number of tourists visit the place.

REVIEW QUESTIONS

1. What are the 5 'A's of tourism?
2. Write short notes on:
   - Attractions
   - Accessibility
   - Accommodation
   - Amenities
   - Activities
   - Awareness
   - Affordability

Tourism Products

Learning Objectives

After reading this chapter, you will be able to understand the following:
- Tourism products
- Characteristics of tourism products

INTRODUCTION

Anything that is offered in a market for use or consumption by the consumers as per the market requirement is called a product. According to Philip Kotler, 'a product is anything that can be offered to a market for attention, acquisition, use or consumption that might satisfy a want or need'.

Products which fulfil or satisfy the customers' leisure, pleasure, or business needs at places other than their own places of residence are known as tourism products. The product in tourism industry is the complete experience of the tourist from the time tourist leaves his home till the time he returns.

In the tourism industry, the basic raw material used in the formulation of a tourism product is the country's natural beauty, its climate, history, culture, and the people. The other essential elements are the existing facilities or the infrastructure, which are necessary for the stay to be comfortable and it includes water supply, electricity, roads, transport, communication, services and other ancillary services.

Thus, we can understand that the tourism product is the sum total of a country's tourist attractions, transport systems, hospitality,

entertainment, and infrastructure which is offered to the tourist, and if well designed and developed, will result in consumer satisfaction.

A tourism product can be either a tangible item, for example, a comfortable seat in an aircraft or the food served in a restaurant or an intangible item, for example, the quality of services provided by a cruise liner or scenic beauty at a hill resort. In general, in almost all the cases, the tourism product is a combination of both tangible and intangible items. This combination of different components results in giving the tourist the total travel experience and satisfaction.

CHARACTERISTICS OF TOURISM PRODUCTS

Tourism products are intangible, are produced and consumed at the same time, can be differentiated from each other, are non-storable, are mutually complementary, and the ownership rights in them are non-transferable, etc. Each characteristic will influence consumer behavior (Fig. 3.1).

Intangibility

Commodities are tangible products which have physical dimensions and attributes

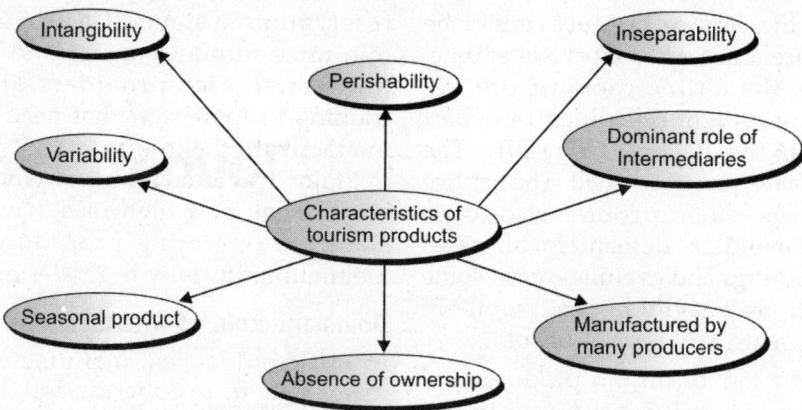

Fig. 3.1: Characteristics of tourism products

which can be seen, touched or, tasted while service products are intangible and cannot be seen, touched, or measured, but can only be experienced. The tourism product can be anything like a package tour, an airline ticket, or a stay in a hotel. The purchase of a package tour to the Far East is nothing but purchasing an experience or buying a dream—as we cannot show the product we have purchased to our friends. After enjoying such services, tourists can only retain these experiences in their memory, and have no way of acquiring physical ownership. The tangible items from a tour are in the form of souvenirs.

The timely performance, efficiency of service, getting the baggage quickly is the intangible items of tourism product. These aspects are very important for the business travelers who travel frequently. Similarly, noisy air-conditioners, other guests talking loudly in the restaurant, etc. are all part of the meal experience which is real but intangible and even though the tangible aspect, i.e. the meal served is good, the overall experience is poor. The product has both ta..gible and intangible elements.

Inseparability

Consumer products, for example, a motorcycle manufactured at New Delhi can be brought to a Mumbai outlet and sold there. In case of tourism industry the products are mostly services which cannot be separated from the person or the company that provides it.

This can best be explained by a guide or escort who provides the services to the tourist. The guide has particular skills which are used along with the infrastructure such as any monument or place of attraction. The guide has to go physically with the group to explain the monument. Here the service is inseparable and the product exists only when the consumption takes place.

The production and consumption of the product occurs simultaneously and cannot be separated. The tourist has to go to the site of production to utilize the product. For example, the courtesy shown by an airhostess while serving a meal on board can only be experienced in the aircraft by the passenger and not before or afterwards as the production and consumption cannot be separated.

Perishability

Any tangible good or product can be manufactured and stored for a certain period of time and sold or used at a later date. For example, pens can be manufactured, stored in the warehouse for a few months, and can be sold when there is a demand.

However, the service product cannot be stored in a warehouse and sold at a later date. For example, the airline cannot store 100 unsold seats of a flight scheduled to depart on 12 May 2016 to sell it on 13 May 2016. The unsold seats have no value at all. The service industry faces such problems due to fluctuating demands, as demand for air travel is more in mornings and evenings, or in some sectors which are heavily booked, while in other sectors, seats may remain unsold.

To avoid the loss of unsold products and overcome the losses incurred due to the perishable nature of the product, the airlines offer last minute sales or standby rates at drastically reduced rates. These rates indicate that although they are not getting profit; they are minimizing losses and at least earning some revenue.

It is due to perishability that the hotel industry also offers heavy discounts along with transport operators especially during off season.

Variability / Heterogeneity

In tourism industry, services are rendered by humans to humans. These services have a high level of variability, when producer and consumer interact. The human element makes standardization of the product a difficult task. The services rendered vary from person to person and from time to time.

The guide's or escort's behavior may not be consistent every single day. Family problems, ill health, or stress may affect his interest in the job, concentration in his work and ultimately his performance. Good tidings, minimum personal problems, and an interested audience help in boosting one's performance. The waiter in a restaurant will not be uniformly efficient on all days of the week for similar reasons.

To avoid variation in services and to maintain the standards in delivery of the products, the hotel industry, tour operators, and airlines have introduced computerized

reservation systems (CRS). Such systems minimize human contact and errors. The tourism service providers also provide training to their staff that needs to directly interact with the tourist.

Major travel companies, who have their branches all over the world, have developed standard operating procedures (SOP) to maintain uniformity in rendering services.

Dominant Role of Intermediaries

In most industries, manufacturers play a major role in product design, distribution, promotion, and th pricing. On the other hand, in tourism, sales intermediaries such as tour operators, travel agents, reservation services and hotel brokers play a very dominant role in tourism marketing. They are the ones who decide to a large extent which services will be sold and to whom. The types of services to be offered as well as the pricing policies and promotion strategies to be adopted by tourist enterprises are, therefore, determined not only by the needs and preferences of the customers but also by the decisions of the travel sales intermediaries.

Absence of Ownership

If you buy a computer, the ownership of the computer is transferred to you but when you hire a car, you buy the right to be transported to a predetermined destination at a pre-determined price. The tourist or consumer cannot own the car or the driver of the car. The same is the case with the hotel industry. The hotel rooms can be used by a tourist during the hotel stay and the tourist acquires the right to certain benefits that the seller or hotel offers, but the ownership of the rooms remains with the hotel.

This can be further explained by a tourist purchasing a ticket of Deccan Odyssey, the exotic journey showcasing the rich cultural heritage of Maharashtra, or an aircraft which brings the tourist to the destination. The ticket allows them to use the services on board which they offer, but the customers do not own the product.

The tangible product can be bought and ownership can be transferred to the buyer, whereas the tourism product, being a service product, services can be bought only for consumption. The ownership remains with the person or organization which is providing the service. Similarly, Kathakali, the famous dance from Kerala can be enjoyed by viewing it, but the dancer cannot be owned.

Manufactured by Many Producers

In case of tangible products, one manufacturer produces a total product. In tourism industry, the tourist product cannot be provided by a single enterprise. Each of the components of the tourist product is highly specialized and when combined together makes the final product.

The hotel industry produces guest nights or hotel rooms, and airlines fly passengers as their products. The travel agent's products are the bookings done on that day and in case of a museum or an archaeological site, the product is measured on the basis of the number of visitors who visited the site on that day.

But from the tourist's point of view, the product he purchased was a single product— a Kerala package tour which covers the complete experience of his visit to a destination. The Kerala package tour is a total product for the tourist.

In other words, to a tourist, tourist product is not an airline seat or a hotel room or ticket to museum or guide services but it is a combination of all the above stated services manufactured by different producers which makes a complete product.

This is peculiar to tourism product and hence requires greater co-ordination in marketing the product.

Highly Unstable Demand/Seasonal Product

The demand for tourist products depends on many factors such as season, economy of the destination, political factors, social factors, etc. Except for the seasonal factor all other factors can be made favorable. Season is a factor which affects the tourism industry greatly.

Seasonality means the time period when the tourist destination is frequently visited by tourists which is for a limited period of the year. Almost all tourist areas have a short season which is called peak season which often may be as short as three months.

This seasonal usage of the product creates unemployment and also has an impact on transportation and hospitality services as well as most other services.

Along with unemployment, investment is greatly affected by seasonality.

Political unrest and economic instability caused by currency fluctuation and inflation have an impact on tourism demand but this may be temporary.

To tackle the problem of seasonality and unstable demand, the suppliers of tourism services have different pricing strategies. For example, the hotel tariff will be higher in peak season, in mid season it will be moderate and in lean or off season the rates of the product will be very low, i.e. off season discounts are offered to combat the problem of seasonality. Sometimes the product is also offered in combination with other products like a package tour.

REVIEW QUESTIONS

1. What do you understand by a tourism product?
2. Discuss the characteristics of tourism products.

Impacts of Tourism

Learning Objectives

After reading this chapter, you will be able to understand the following:
- Socioeconomic impacts of tourism
- Environmental impacts of tourism
- Cultural impacts of tourism

SOCIOECONOMIC IMPACTS OF TOURISM

Tourism industry has several positive and negative impacts on the economy and society. These impacts are highlighted below.

Positive Impacts

- *Generation of national income*: Tourism has proved to be successful in generating national income. Being a multi-segment industry, the hotel and restaurants, transportation services, tourist resorts, amusement parks, entertainment centers, sales outlets of curios, handicrafts, jewelry, etc. provide services to both tourists and non-tourists. The modest contribution of tourism industry to the NNP (Net National Product) is a staunch testimony to this proposition that it contributes a lot to the process of national income generation. To be more specific in the Indian context, we find domestic tourism contributing a major share to the NNP.

- *Employment generation*: Tourism has emerged as an instrument of income and employment generation, poverty alleviation and sustainable human development.

Tourism employs more than 230 million people in the world. It creates a large number of jobs among direct service providers (such as hotels, restaurants, travel agencies, tour operators, guide and tour escorts, etc.) and among indirect service providers (such as, suppliers to hotels and restaurants, supplementary accommodation, etc.). It contributes 6.23% to the national GDP (Gross Domestic Product) and 8.78% of the total employment in India. Almost 20 million people are now working in the India's tourism industry.

- *Infrastructure development*: Tourism spurs infrastructure development. In order to become an important commercial or pleasure destination, any location would require all the necessary infrastructure, like good connectivity via rail, road, and air transport, adequate accommodation, high-end restaurants, a well-developed telecommunication network, and healthcare facilities, among others.

- *Source of foreign exchange earnings*: The people who travel to other countries spend a large amount of money on accommodation, transportation, sight-seeing, shopping, etc. Thus, an inbound tourist is an important source of foreign exchange for any country. This has favorable impact on the balance of payments of the country. The tourism industry in India generated about US$100 billion in 2008 and that is expected to increase to US$275.5 billion by 2018 at a 9.4% annual growth rate.

- *Preservation of national heritage and environment*: Tourism helps to preserve several places which are of historical importance by declaring them as heritage sites. For instance, the Taj Mahal, the Qutab Minar, Ajanta and Ellora temples, etc. would have been decayed and destroyed had the Tourism Department not made efforts to preserve them. Likewise, tourism also helps in conserving the natural habitats of many endangered species.

- *Promoting peace and stability*: The tourism industry can also help promote peace and stability in developing country like India by providing jobs, generating income, diversifying the economy, protecting the environment, and promoting cross-cultural awareness. However, key challenges like adoption of regulatory frameworks, mechanisms to reduce crime and corruption, etc. must be addressed if peace-enhancing benefits from this industry are to be realized.

- *Transformation of regional economy*: The development of Khajuraho in MP, Kovalam in Kerala and Gulmarg in Jammu & Kashmir give positive evidences for the contribution of tourism to the development of backward areas in India. If we turn our eyes on their past, it is apparent that till a few years back, all of them were relatively unknown villages, with dismal economic activities, inhabited by traditional rural folk or were sleeping in far-off places. Of late, these places are internationally known. A good number of skilled and semi-skilled or even un-skilled local people are found

employed in tourism or other ancillary industries furthering socioeconomic justice with the help of equitable distribution of economic benefits.

- *Development of art, handicrafts and monuments*: A society endowed with art and culture along with lively customs and cheerful public life is found helpful in promoting world tourism. Tourism industry provides sufficient motivation for the promotion and preservation of art, craft and culture which is found a matter of appreciation to be more specific by the external community.

Negative Impacts

- *Undesirable social and cultural change*: Tourism sometimes led to the destruction of the social fabric of a community. The more tourists coming into a place, the more the perceived risk of that place losing its identity. A good example is Goa. From the late 60's to the early 80's when the Hippy culture was at its height, Goa was a heaven for such hippies. Here they came in thousands and changed the whole culture of the state leading to a rise in the use of drugs, gambling, prostitution and human trafficking. This had a ripple effect on the country.

- *Increased tension and hostility*: Tourism can increase tension, hostility, and suspicion between the tourists and the local communities when there is no respect and understanding for each other's culture and way of life. This may further lead to violence and other crimes committed against the tourists. The recent crime committed against Russian tourist in Goa is a case in point.

- *Creating a sense of antipathy*: Tourism brought a little benefit to the local community. In most *all-inclusive package tours* more than 80% of travelers' fees go to the airlines, hotels and other international companies, not to local businessmen and workers. Moreover, large hotel chain restaurants often import food to satisfy foreign visitors and rarely employ local staff for senior management positions, preventing local farmers and workers from reaping the benefit of their presence. This has often created a sense of antipathy towards the tourists and the government.

- *Adverse effects on environment and ecology*: One of the most important adverse effects of tourism on the environment is increased pressure on the carrying capacity of the ecosystem in each tourist locality. Increased transport and construction activities led to large scale deforestation and destabilisation of natural landforms, while increased tourist flow led to increase in solid waste dumping as well as depletion of water and fuel resources. Flow of tourists to ecologically sensitive areas resulted in destruction of rare and endangered species due to trampling, killing, disturbance of breeding habitats. Noise pollution from vehicles and public address systems, water pollution, vehicular emissions, untreated sewage, etc. also have direct effects on bio-diversity, ambient environment and general profile of tourist spots.

ENVIRONMENTAL IMPACTS OF TOURISM

The tourism industry in India can have several positive and negative impacts on the environment which are discussed below.

Positive Impacts

- *Direct financial contributions*: Tourism can contribute directly to the conservation of sensitive areas and habitat. Revenue from park, entrance fees and similar sources can be allocated specifically to pay for the protection and management of environmentally sensitive areas. Special fees for park operations or conservation activities can be collected from tourists or tour operators.

- *Contributions to government revenues*: The Indian government through the tourism department also collect money in more far-reaching and indirect ways that are not linked to specific parks or conservation areas. User fees, income taxes, taxes on sales or rental of recreation equipment, and license fees for activities such as rafting and fishing can provide governments with the funds needed to manage natural resources. Such funds can be used for overall conservation programs and activities, such as park ranger salaries and park maintenance.

- *Improved environmental management and planning*: Sound environmental management of tourism facilities and especially hotels can increase the benefits to natural environment. By planning early for tourism development, damaging and expensive mistakes can be prevented, avoiding the gradual deterioration of environmental assets significant to tourism. The development of tourism has moved the Indian government towards this direction leading to improved environmental management.

- *Raising environmental awareness*: Tourism has the potential to increase public appreciation of the environment and to spread awareness of environmental problems when it brings people into closer contact with nature and the environment. This confrontation heightens awareness of the value of nature among the community and lead to environmentally conscious behavior and activities to preserve the environment.

- *Protection and preservation of environment*: Tourism can significantly contribute to environmental protection, conservation and restoration of biological diversity and sustainable use of natural resources. Because of their attractiveness, pristine sites and natural areas are identified as valuable and the need to keep the attraction alive can lead to creation of national parks and wildlife parks.

In India, new laws and regulations have been enacted to preserve the forest and to protect native species. The coral reefs around the coastal areas and the marine life that depend on them for survival are also protected.

Negative Impacts

- *Depletion of natural resources*: Tourism development can put pressure on natural resources when it increases consumption in areas where resources are already scarce.
  - *Water resources*: Water, especially fresh water, is one of the most critical natural resources. The tourism industry generally overuses water resources for hotels, swimming pools, golf courses and personal use of water by tourists. This can result in water shortages and degradation of water supplies, as well as generating a greater volume of waste water. In dryer regions like Rajasthan, the issue of water scarcity is of particular concern.
  - *Local resources*: Tourism can create great pressure on local resources like energy, food, and other raw materials that may already be in short supply. Greater extraction and transport of these resources exacerbates the physical impacts associated with their exploitation. Because of the seasonal character of the industry, many destinations have ten times more inhabitants in the high season as in the low season. A high demand is placed upon these resources to meet the high expectations tourists often have (proper heating, hot water, etc.).
  - *Land degradation*: Important land resources include minerals, fossil fuels, fertile soil, forests, wetland and wildlife. Increased construction of tourism and recreational facilities has increased the pressure on these resources and on scenic landscapes. Direct impact on natural resources, both renewable and non-

renewable, in the provision of tourist facilities is caused by the use of land for accommodation and other infrastructure provision, and the use of building materials.

Forests often suffer negative impacts of tourism in the form of deforestation caused by fuel wood collection and land clearing, e.g. the trekking in the Himalayan region, Sikkim and Assam.

- *Pollution*: Tourism can cause the same forms of pollution as any other industry—air emissions, noise, solid waste and littering, releases of sewage, oil and chemicals, even architectural/visual pollution.
  - *Air and noise pollution*: Transport by air, road, and rail is continuously increasing in response to the rising number of tourist activities in India. Transport emissions and emissions from energy production and use are linked to acid rain, global warming and photochemical pollution. Air pollution from tourist transportation has impacts on the global level, especially from carbon dioxide (CO_2) emissions related to transportation energy use. And it can contribute to severe local air pollution. Some of these impacts are quite specific to tourist activities where the sites are in remote areas like Ajanta and Ellora temples. For example, tour buses often leave their motors running for hours while the tourists go out for an excursion because they want to return to a comfortably air-conditioned bus.

  Noise pollution from airplanes, cars, and buses, as well as recreational vehicles is an ever-growing problem of modern life. In addition to causing annoyance, stress, and even hearing loss for humans, it causes distress to wildlife, especially in sensitive areas.
  - *Solid waste and littering*: In areas with high concentrations of tourist activities and

appealing natural attractions, waste disposal is a serious problem and improper disposal can be a major despoiler of the natural environment—rivers, scenic areas, and roadsides.

In mountain areas of the Himalayas and Darjeeling, trekking tourists generate a great deal of waste. Tourists on expedition leave behind their garbage, oxygen cylinders and even camping equipment. Such practices degrade the environment particularly in remote areas because there have a few garbage collection or disposal facilities.

– *Sewage*: Construction of hotels, recreation and other facilities often leads to increased sewage pollution. Wastewater has polluted seas and lakes surrounding tourist attractions, damaging the flora and fauna. Sewage runoff causes serious damage to coral reefs because it stimulates the growth of algae, which cover the filter-feeding corals, hindering their ability to survive. Changes in salinity and siltation can have wide-ranging impacts on coastal environments. And sewage pollution can threaten the health of humans and animals. Examples of such pollution can be seen in the coastal states of Goa, Kerala, Maharashtra, Tamil Nadu, etc.

• *Destruction and alteration of ecosystem*: An ecosystem is a geographic area including all the living organisms (people, plants, animals, and micro-organisms), their physical surroundings (such as soil, water, and air), and the natural cycles that sustain them. Attractive landscape sites, such as sandy beaches in Goa, Maharashtra, Kerala, Tamil Nadu; lakes, riversides, and mountain tops and slopes, are often transitional zones, characterized by species-rich ecosystems. The threats to and pressures on these ecosystems are often severe because such places are very

attractive to both tourists and developers. Examples may be cited from Krushedei Island near Rameshwaram. What was once called paradise for marine biologists has been abandoned due to massive destruction of coral and other marine life. Another area of concern which emerged at Jaisalmer is regarding the deterioration of the desert ecology due to increased tourist activities in the desert.

Moreover, habitat can be degraded by tourism leisure activities. For example, wildlife viewing can bring about stress for the animals and alter their natural behavior when tourists come too close. Safaris and wildlife watching activities have a degrading effect on habitat as they often are accompanied by the noise and commotion created by tourists.

CULTURAL IMPACTS OF TOURISM

Positive Impacts

Revival and strengthening of cultural heritage is found to be an outstanding benefit of tourism industry. Cultural purity is a virtue of the society. Wherever we find cultural exchange, the give and take of plus points cannot be ruled out. This offers opportunities to the general masses to accelerate their learning cycles. We find a positive change in their action and behavior. On the other hand, the catalytic role of tourism in boosting growth of art, protecting and maintaining monuments and heritage contribute a lot to the process of cultural transformation. A society endowed with art and culture, lively customs and cheerful public life naturally attract tourists. Some of the monuments are supposed to be permanent structure of cultural heritages. The development of tourism industry helps in protecting these monuments and cultural heritages. The protection of architectural wonders and landmarks of the glorious ancient past help in promoting cultural tourism.

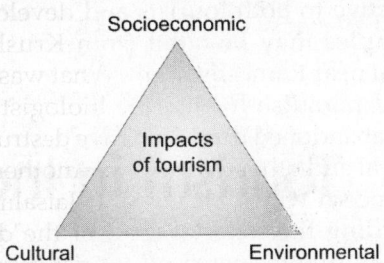

Fig. 4.1: Impacts of tourism

Negative Impacts

The development of tourism industry instrumentalizes the process of cultural transformation but the aggressive development of active tourism introduces an element of cultural damage. It is very natural since the old forms are transformed with the influx of a large number of tourists. The Western culture influences youths of Eastern which may likely to result in confusion in their minds resulting into turmoil or conflict with the traditional segment of the society.

Local residents especially young people get attracted by the tourists' clothing, eating habits, spending patterns and their lifestyles. Eventually, they are adopting tourist behaviors, and this is called the demonstration effect. The consequences associated with demonstration effect are often local residents' feelings of frustration, antagonism and resentment.

REVIEW QUESTIONS

1. Discuss the socioeconomic, environmental and cultural benefits of tourism.
2. Discuss the negative impacts of tourism.

Tourism in India

Learning Objectives

After reading this chapter, you will be able to understand the following:

- Development of tourism in India—historical perspectives
- Initiatives by the Government
- Growth drivers for tourism
- Notable trends in the tourism industry in India
- Future prospects
- Ministry of Tourism—an overview
- Indian tourism at a glance

DEVELOPMENT OF TOURISM IN INDIA— HISTORICAL PERSPECTIVES

Tourism development in India has passed through many phases (Fig. 5.1). The first conscious and organized efforts to promote tourism in India were made in 1945 when a committee was set up by the Government under the Chairmanship of Sir John Sargent, the then Educational Adviser to the Government of India. The main objective of the committee was to encourage and develop tourist traffic both internal and external by all possible means. Thereafter, the development of tourism was taken up in a planned manner in 1956 coinciding with the Second Five Year Plan. The approach has evolved from isolated planning of single unit facilities in the Second and Third Five Year Plans. The Sixth Plan marked the beginning of a new era when tourism began to be considered a major instrument for social integration and economic development.

But it was only after the 80's that tourism activity gained momentum. The Government took several significant steps. A *National Policy on Tourism* was announced in 1982. Later in 1988, the *National Committee on Tourism* formulated a comprehensive plan for achieving a sustainable growth in tourism. In 1992, a *National Action Plan* was prepared and in 1996 the *National Strategy for Promotion of Tourism* was drafted. In 1997, the *New Tourism Policy* recognized the roles of Central and State governments, public sector undertakings and the private sector in the development of tourism. The need for involvement of Panchayati Raj institutions, local bodies, non-governmental organizations and the local youth in the creation of tourism facilities had also been recognized. The other major development that took place was the setting up of the *India Tourism Development Corporation* (ITDC) in 1966 to promote India as a tourist destination and the *Tourism Finance Corporation* in 1989 to finance tourism projects.

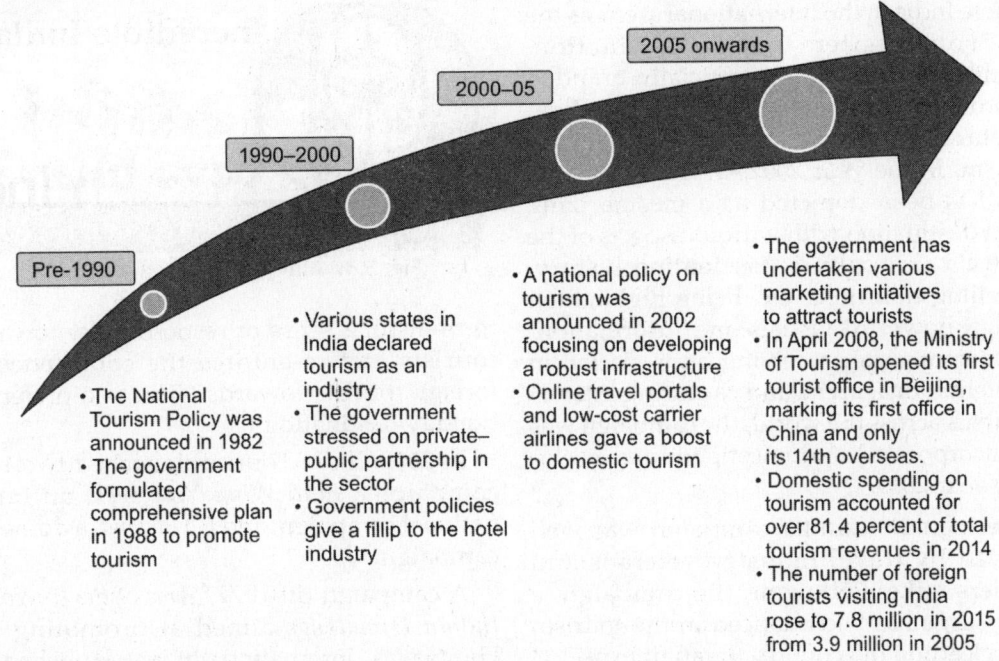

Pre-1990
- The National Tourism Policy was announced in 1982
- The government formulated comprehensive plan in 1988 to promote tourism

1990–2000
- Various states in India declared tourism as an industry
- The government stressed on private–public partnership in the sector
- Government policies give a fillip to the hotel industry

2000–05
- A national policy on tourism was announced in 2002 focusing on developing a robust infrastructure
- Online travel portals and low-cost carrier airlines gave a boost to domestic tourism

2005 onwards
- The government has undertaken various marketing initiatives to attract tourists
- In April 2008, the Ministry of Tourism opened its first tourist office in Beijing, marking its first office in China and only its 14th overseas.
- Domestic spending on tourism accounted for over 81.4 percent of total tourism revenues in 2014
- The number of foreign tourists visiting India rose to 7.8 million in 2015 from 3.9 million in 2005

Fig. 5.1: Evolution of the Indian tourism and hospitality sector (*Source*: WTTC, Ministry of Tourism, TechSci Research)

INITIATIVES BY THE GOVERNMENT

India can always boast of its rich cultural heritage. Travel and tourism in India is an integral part of its tradition and culture. The beauty of India's cultural heritage and the richness of nature's endowments make India tourists' paradise. Vibrant culture, awe-inspiring tourist spots, nature's bounty, gourmet delights and cordial people—there are thousands of reasons that have made India a traveler's delight. India is a place known for its hospitality around the world. The hospitality spreads from Kashmir in the North to Kanyakumari in the South and from Manipur in the East to the Gujarat in the West.

Pandit Jawaharlal Nehru often remarked, *'Welcome a tourist and send back a friend'*. That was the essence of India's approach to tourism in the post-independence era. Tourism was seen as an important instrument for national integration and international understanding.

The Indian government has realised the country's potential in the tourism industry and has taken several steps to make India a global tourism hub. Some of the major initiatives taken by the Government of India to give a boost to the tourism and hospitality sector of India are as follows:

Incorporated on October 1, 1966, *India Tourism Development Corporation (ITDC)* has been playing a key role in the develoment of tourism infrastructure in the country. Apart from developing the largest hotel chain in India, ITDC offered tourism related facilities like transport, duty free shopping, entertainment, production of tourist publicity literature, consultancy, etc.

The first-ever *Indian Tourism Day* was celebrated on January 25, 1998. In 2002, Government of India launched an international marketing campaign named *Incredible India* to give a boost to India's tourism sector and promote India in the International arena as the most sought-after tourist destination. Incredible India title was officially branded and promoted by Amitabh Kant, the then Joint Secretary under the Union Ministry of Tourism, in the year 2002. In the campaign, India has been depicted as a mesmerizing tourist destination with various aspects of the country's rich culture, fascinating history, enthralling traditions, etc. being highlighted through powerful visuals and information-rich content. After receiving appreciation in the international arena and capturing interest of tourists across the world, the campaign was also incorporated in domestic tourism sector in the year 2009.

The Incredible India campaign was well received by travel industry veterans and travelers alike. Following the campaign, a major surge has been noticed in the tourism sector, leading the country to tap unexpected growth with regard to international tourist spending.

Fig. 5.2: Incredible India campaign

In line with its *Incredible India* campaign, the Ministry of Tourism also launched a campaign titled *Atithi Devo Bhava* in the year 2008 targeting the local population to educate them regarding good behavior and etiquettes while dealing with foreign tourists. Indian actor Aamir Khan was commissioned to endorse the campaign. 'Atithi Devo Bhava' aimed at creating awareness about the effects of tourism and sensitizing the local population about preservation of India's heritage, culture, cleanliness and hospitality. It also attempted

Atithi Devo Bhava means "Guest is God"

Fig. 5.3: Atithi Devo Bhava campaign

to re-instill a sense of responsibility towards tourists and re-enforce the confidence of foreign tourists towards India as a preferred holiday destination.

In 2012, India Tourism launched two new campaigns: *Find What You Seek*, an international campaign; and *Go Beyond*, a domestic campaign.

A campaign titled *777 days of the Incredible Indian Himalayas* aimed at promoting the Himalayas, internationally was launched on 27th September 2013. The objective of the campaign was to attract more international tourists to India during the lean summer season and to remind the world that 73% of the Himalayan region is in India.

The Ministry of Tourism, Government of India, the Departments of Tourism of Governments of Bihar and Uttar Pradesh and International Finance Corporation (World Bank Group) had entered into an agreement in October 2013 to cooperate in upgrading the quality of services and goods provided for tourist along the 'Buddhist Circuit' in India.

For promotion of tourism two new schemes have been announced in the Budget 2014–15. With a view to beautify and improve amenities and infrastructure at pilgrimage centers of all faiths, a National Mission on Pilgrimage Rejuvenation and Spiritual Augmentation Drive (*PRASAD*) has been announced and an amount of ₹100 crore has been provided for this initiative. During the Union Budget 2015–16, an investment of US$ 16.4 million was allocated.

India's rich cultural, historical, religious and natural heritage provides a huge potential for development of tourism and job creation. In due recognition of this potential, a scheme on *Swadesh Darshan* has been initiated. US$ 98.3 million has been allocated for Swadesh Darshan under the Union Budget 2015–16.

To promote River Cruise, it is decided on 21st June 2014, that the Ministry of Shipping and Ministry of Tourism would jointly identify the routes for carrying out cruise tours on waterways and also take measures to develop necessary infrastructure.

With the objective of visa facilitation, the Government of India launched the *Tourist Visa on Arrival (TVoA)* enabled with *Electronic Travel Authorisation (ETA)* scheme on 27 November 2014 for national of 43 countries to travel to India for tourism for a short stay of 30 days whose objective of visiting India is recreation, sightseeing, casual visit to meet friends or relatives, short duration medical treatment or casual business visit. In the year 2015, this scheme was extended to 150 countries.

It has been the endeavour of Ministry of Tourism to put in place a system of training and professional education, with necessary infrastructure support, capable of generating manpower sufficient to meet the needs of the tourism and hospitality industry both quantitatively and qualitatively. As of now, there are 54 Institutes of Hotel Management (*IHMs*), comprising 21 Central IHMs and 19 State IHMs, 1 Public Sector Undertaking IHM and 13 Private IHM affiliated with NCHMCT (National Council for Hotel Management and Catering Technology). Nine Food Craft Institutes located in different parts of the country offer Diploma programs in specific operational areas.

In order to ensure safety and security of tourists especially to the women tourists, to curb the activities of touts and to prevent cheating of tourists, to provide one point source of information accessible 24 × 7, to provide an authentic information to tourists and also to guide them during emergencies, once they are in India, the Ministry of Tourism has set up the *Incredible India Helpline* on a pilot basis.

The Government of India has set aside ₹ 500 crore (US$ 79.17 million) for the first phase of the National Heritage City Development and Augmentation Yojana (*HRIDAY*). The 12 cities in the first phase are Varanasi, Amritsar, Ajmer, Mathura, Gaya, Kanchipuram, Vellankani, Badami, Amaravati, Warangal, Puri and Dwarka.

Under *Project Mausam* the Government of India has proposed to establish cross cultural linkages and to revive historic maritime cultural and economic ties with 39 Indian Ocean countries.

Formulation of *National Tourism Policy 2015* that would encourage the citizens of India to explore their own country as well as position the country as a 'Must See' destination for global travelers.

Ministry of Tourism has identified 29 mega projects that would help India to develop in a holistic manner. These mega projects include a judicious mix of heritage, spiritual, cultural and eco-tourism sites that would provide tourists a holistic glimpse of India.

Some of the recent initiatives taken by the Government to accelerate tourism include grant of export house status to the tourism sector and incentives for promoting private investment in the form of Income Tax exemptions, interest subsidy and reduced import duty. The hotel and tourism-related industry has been declared a high priority industry for foreign investment which entails automatic approval of direct investment up to 51 percent of foreign equity and allowing 100 percent non-resident Indian investment and simplifying rules regarding the grant of approval to travel agents, tour operators and tourist transport operators.

GROWTH DRIVERS FOR TOURISM

Reasons for the rapid pace of growth of domestic tourism in India are directly linked to the following factors:

- Greater disposable incomes with women entering the workforce, which is spent on recreation and leisure.
- Employees in many organizations are entitled to a minimum number of days paid leave per annum. Leave travel allowance (LTA) facility is given to the employees which covers self and family.
- Availability of low-cost airlines. Growth in low-cost airlines is expected to lower tourism costs and increase domestic spending on tourism.
- Rapidly increasing purchasing power of the middle class and evolving lifestyle.
- Better road connectivity.
- Greater awareness about travel and tourism, through the Internet, as well as through articles and advertisements published in leading magazines and dailies and on television.
- Increase in discretionary time—shorter work weeks and longer vacations.
- Greater credit availability through credit cards—'travel now, pay later' stimulates travel.

Reasons for increase in international tourist traffic in India:

- Value for money/economical holiday destination.
- Business-cum-pleasure destination
- Development of Kerala and Rajasthan as the most popular tourist destinations in India with the distinctive brand image
- Opening of the sectors of the economy to private sector/foreign investment
- Reform in the aviation sector such as Open Skies Policy has led to better connectivity with many countries with India
- Success of 'Incredible India' campaign and other tourism promotion measures
- An unquenchable thirst or desire to travel which has always existed in humankind.

NOTABLE TRENDS IN THE TOURISM INDUSTRY IN INDIA

Online Travel Operators

Over 70 percent of air tickets are now being booked online in the country. A number of online travel and tour operators, which provide better prices and options to consumers, has emerged in India.

Fig. 5.4: Tourism drivers

Wellness Tourism

The widespread practice of ayurveda, yoga, siddha and naturopathy that is complemented by the nation's spiritual philosophy makes India a famous wellness destination.

Cruises

India attracted 163,000 cruise visitors in 2013. Government of India has estimated that India would emerge with a market size of 1.2 million cruise visitors by 2030–31.

Adventure

Adventure tourism is one of the most popular segments of tourism industry. Owing to India's enormous geo-physical diversity, it has progressed well over the years. Part of India's tourism policy, almost every state has definite program to identify and promote adventure tourism.

Camping Sites

Promotion of camping sites has been encouraged with adequate acknowledgement of its adverse effects on environment. Besides providing unique rewarding experiences, responsible conduct of camping can be a major source for both additional economic opportunities in remote areas as well as an instrument of conservation.

Spiritual Tourism

India has been known as the seat of spiritualism and India's cosmopolitan nature is best reflected in its pilgrim centers. India has been recognised as a destination for spiritual tourism for domestic and international tourists.

FUTURE PROSPECTS

Tourism in India has grown in leaps and bounds over the years, with each region of India contributing something to its splendour and exuberance. The travel and tourism industry has emerged as one of the fastest growing sectors contributing significantly to the Indian economic growth and develop-

ment. India has significant potential to become a preferred tourist destination globally. Its rich and diverse cultural heritage, abundant natural resources and biodiversity provides numerous tourist attractions. Despite an impressive growth in the Foreign Tourist Arrivals (FTAs) witnessed in the recent years, it is felt that India still has a vast untapped potential in tourism.

India is ranked 11th among 184 countries in terms of travel and tourism's total contribution to GDP in 2015. In India, the sector's direct contribution to GDP is expected to grow 7.2 percent per annum from 2015–25 to US\$ 88.6 billion in 2025.

The medical tourism market in India is projected to hit US\$ 3.9 billion mark in the year 2015 having grown at a compounded annual growth rate (CAGR) of 30 percent over the last three years, according to a joint report by FICCI and KPMG. Also, inflow of medical tourists is expected to cross 320 million by 2015 compared with 85 million in 2012.

Electronic tourist authorizations, known as E-Tourist Visa, launched by the Government of India is likely to see a spurt growth of 7.5 percent in the tourism sector in 2015.

According to the World Travel and Tourism Council (WTTC) report, India's travel and tourism industry contributed ₹7.64 trillion and 36.7 million jobs to the Indian economy in 2014. By the end of 2015, the travel and tourism sector will contribute ₹8.22 trillion or 7% of India's gross domestic product (GDP) and 37.4 million jobs—almost 9% of total employment, the report said.

According to WTTC forecasts, travel and tourism has the potential to contribute 46 million jobs to the Indian economy by 2025. The industry should rise by 6.5% per annum over the next 10 years to ₹4,337.8 billion in 2025 or 6.9% of the total.

The UNWTO predicts that India will account for 50 million outbound tourists by 2020; the 'Kuoni Travel Report India 2007' predicts that total outbound spending will cross the US$ 28 billion mark in 2020.

Meanwhile, the recent measures taken by the Government, such as FDI in aviation and the new visa laws, are further going to provide impetus to the Indian tourism industry. It is estimated that improved visa facilitation could bring in up to 6 million more international visitors to India, which could in turn create 1.8 million more travel jobs in the country over three years.

MINISTRY OF TOURISM—AN OVERVIEW

The Ministry of Tourism is the nodal agency for the formulation of national policies and programs and for the co-ordination of activities of various Central Government Agencies, State Governments/UTs and the private sector for the development and promotion of tourism in the country. This Ministry is headed by the Union Minister of State for Tourism (Independent Charge).

The administrative head of the Ministry is the Secretary (Tourism). The Secretary also acts as the Director General (DG) Tourism. The office of the Director General of Tourism provides executive directions for the implementation of various policies and programs. Director General of Tourism has a field formation of 20 offices within the country and 14 offices abroad and one sub-ordinate office/project, i.e. Indian Institute of Skiing and Mountaineering (IISM)/Gulmarg Winter Sports Project. The overseas offices are primarily responsible for promotion of Indian tourism in the market abroad. The domestic field offices in India are responsible for providing information service to tourists and to monitor the progress of field projects by the State Governments in their respective jurisdictions.

The Ministry of Tourism has under its charge a public sector undertaking, the India Tourism Development Corporation (ITDC) and the following autonomous institutions:

- Indian Institute of Tourism and Travel Management (IITTM) and National Institute of Water Sports (NIWS)
- National Council for Hotel Management and Catering Technology (NCHMCT) and the Institutes of Hotel Management.

Role and Functions of the Ministry of Tourism

The ministry has the following main functions:
- *All policy matters, including*:
  - Development policies
  - Incentives
  - External assistance
  - Manpower development
  - Promotion and marketing
  - Investment facilitation
- Planning
- Co-ordination with other ministries, departments, state/UT governments
- *Regulation*:
  - Standards
  - Guidelines
- *Infrastructure and product development*:
  - Central assistance
  - Distribution of tourism products
- Research, analysis, monitoring and evaluation
- *International co-operation and external assistance*:
  - International bodies
  - Bilateral agreements
  - External assistance
  - Foreign technical collaboration
- Legislation and parliamentary work
- Establishment matters
- Vigilance matters
- Budget co-ordination and related matters
- Plan-coordination and monitoring
- Implementation of official language policy
- Plan coordination
- Overseas marketing (OM) work
- Integrated finance matters
- Welfare, grievances and protocol

INDIAN TOURISM AT A GLANCE

Table 5.1: The contribution of tourism to total gross domestic product (GDP) and employment of the country

Year	Contribution of tourism in GDP of the country (%)			Contribution of tourism in employment of the country (%)		
	Direct	Indirect	Total	Direct	Indirect	Total
2009-10	3.68	3.09	6.77	4.37	5.8	10.17
2010-11	3.67	3.09	6.76	4.63	6.15	10.78
2011-12	3.67	3.09	6.76	4.94	6.55	11.49
2012-13	3.74	3.014	6.88	5.31	7.05	12.36

Table 5.2: Foreign tourist arrivals (FTAs) in India according to mode of travel

Year	Arrivals	% Distribution by mode of travel		
		Air	Sea	Land
1996	2287860	98.5	0.1	1.4
1997	2374094	98.5	0	1.5
1998	2358629	98.5	0	1.5
1999	2481928	98.4	0	1.6
2000	2649378	98.5	0	1.5
2001	2537282	87.1	0.9	12
2002	2384364	81.9	0.6	17.5
2003	2726214	83.1	0.5	16.4
2004	3457477	85.6	0.5	13.9
2005	3918610	6.5	0.4	13.1
2006	4447167	87.1	0.6	12.3
2007	5081504	88.4	0.6	11
2008	5282603	89.1	0.7	10.2
2009	5167699	89.8	1	9.2
2010	5775692	91.8	0.7	7.5

Table 5.3: FTAs in India according to purpose of visit

Year	FTAs (Numbers)	Business and Professional	Leisure, holiday and recreation	Visiting friends and relatives	Medical treatment	Education	Others
2009	5167699	15.1	57.5	17.6	2.2	0	7.6
2010	5775692	18.6	24	27.5	2.7	0	27.2
2011	6309222	22.5	26	24.9	2.2	0	24.3
2012	6577745	22.5	27.1	27.2	2.6	0	20.6
2013	6967601	20.9	30.3	25.9	3.4	1.9	17.6

Table 5.4: Foreign tourist arrivals (FTAs) in India

Year	FTAs from tourism in India (in million)	Percentage (%) change over the previous year
1997	2.37	3.8
1998	2.36	−0.7
1999	2.48	5.2
2000	2.65	6.7
2001	2.54	−4.2
2002	2.38	−6
2003	2.73	14.3
2004	3.46	26.8
2005	3.92	13.3
2006	4.45	13.5
2007	5.08	14.3
2008	5.28	4
2009	5.17	−2.2
2010	5.78	11.8
2011	6.31	9.2
2012	6.58	4.3
2013	6.97	5.9
2014	7.7	10.6

Source: (i) Bureau of Immigration, Govt. of India, (ii) Ministry of Tourism, Govt. of India

Table 5.5: Foreign exchange earnings (FEE), in crore, from tourism in India

Year	FEE from tourism in India	Percentage (%) change over the previous year
1997	10511	4.6
1998	12150	15.6
1999	12951	6.6
2000	15626	20.7
2001	15083	−3.5
2002	15064	−0.1
2003	20729	37.6
2004	27944	34.8
2005	33123	18.5
2006	39025	17.8
2007	44360	13.7
2008	51294	15.6
2009	53700	4.7
2010	64889	20.8
2011	77591	19.6
2012	94487	21.8
2013	107671	14
2014	120083	11.5

Source: (i) Bureau of Immigration, Govt. of India
(ii) Ministry of Tourism, Govt. of India

Table 5.6: Foreign exchange earnings (FEE), in US$ million, from tourism in India

Year	FEE from tourism in India	Percentage (%) change over the previous year
1997	2889	2
1998	2948	2
1999	3009	2.1
2000	3460	15
2001	3198	−7.6
2002	3103	−3
2003	4463	43.8
2004	6170	38.2
2005	7493	21.4
2006	8634	15.2
2007	10729	24.3
2008	11832	10.3
2009	11136	−5.9
2010	14193	27.5
2011	16564	16.7
2012	17737	7.1
2013	18445	4

Source: (i) Bureau of Immigration, Govt. of India
(ii) Ministry of Tourism, Govt. of India

Table 5.7: Top 15 source countries for foreign tourist arrivals (FTAs) in India during 2012 and 2013

Rank in 2012	Country	FTAs in India in 2012	% Share in 2012	Rank in in 2013	Country	FTAs in India in 2013	% Share in 2013
1	USA	1039947	15.81	1	USA	1085309	15.58
2	UK	788170	11.98	2	UK	809444	11.62
3	Bangladesh	487397	7.41	3	Bangladesh	524923	7.53
4	Sri Lanka	296983	4.51	4	Sri Lanka	262345	3.77
5	Canada	256021	3.89	5	Russian Fed.	259120	3.72
6	Germany	254783	3.87	6	Canada	255222	3.66
7	France	240674	3.66	7	Germany	252003	3.62
8	Japan	220015	3.34	8	France	248379	3.56
9	Australia	202105	3.07	9	Malaysia	242649	3.48
10	Malaysia	195853	2.98	10	Japan	220283	3.16
11	Russian Fed.	177526	2.70	11	Australia	218967	3.14
12	China (Main)	168952	2.57	12	China(Main)	174712	2.51
13	Singapore	131452	2.00	13	Singapore	143025	2.05
14	Nepal	125375	1.91	14	Thailand	117136	1.68
15	Rep. of Korea	109469	1.66	15	Nepal	113790	1.63
Total top 15 countries		4694722	71.37	**Total top 15 countries**		4927307	70.72
Other countries		1883023	28.63	**Other countries**		2040294	29.28
Grand total		6577745	100.00	**Grand total**		6967601	100.00

Source: Bureau of Immigration, Govt. of India

Table 5.8: Domestic tourist visits (DTVs) and foreign tourist visits (FTVs) to states/UTs during 2012 and 2013

S.No.	State/ UT	2012		2013		Growth rate 13/12		Rank 2013	
		Domestic	Foreign	Domestic	Foreign	Domestic	Foreign	Domestic	Foreign
1.	Andaman and Nicobar Islands	238699	17538	243703	14742	2.10	–15.94	30	25
2.	Andhra Pradesh	207217952	292822	152102150	223518	–26.60	–23.67	3	14
3.	Arunachal Pradesh	132243	5135	125461	10846	–5.13	111.22	32	27
4.	Assam	4511407	17543	4684527	17638	3.84	0.54	21	24
5.	Bihar	21447099	1096933	21588306	765835	0.66	–30.18	11	8
6.	Chandigarh	924589	34130	936922	40124	1.33	17.56	24	22
7.	Chhattisgarh	15036530	4172	22801031	3886	51.64	–6.86	10	30

Contd.

Table 5.8: Domestic tourist visits (DTVs) and foreign tourist visits (FTVs) to states/UTs during 2012 and 2013 (Contd.)

S.No.	State/ UT	2012		2013		Growth rate 13/12		Rank 2013	
		Domestic	Foreign	Domestic	Foreign	Domestic	Foreign	Domestic	Foreign
8.	Dadar and Nagar Haveli	469213	1234	481618	1582	2.64	28.20	28	33
9.	Daman and Diu	803963	4607	819947	4814	1.99	4.49	25	29
10.	Delhi	18495139	2345980	20215187	2301395	9.30	−1.90	14	3
11.	Goa	2337499	450530	2629151	492322	12.48	9.28	22	10
12.	Gujarat	24379023	174150	27412517	198773	12.44	14.14	8	16
13.	Haryana	6799242	233002	7128027	228200	4.84	−2.06	20	13
14.	Himachal Pradesh	15646048	500284	14715586	414249	−5.95	−17.20	16	11
15.	Jammu and Kashmir	12427122	78802	13642402	60845	9.78	−22.79	17	19
16.	Jharkhand	20421016	31909	20511160	45995	0.44	44.14	13	20
17.	Karnataka	94052729	595359	98010140	636378	4.21	6.89	4	9
18.	Kerala	10076854	793696	10857811	858143	7.75	8.12	18	7
19.	Lakshadweep	4417	580	4784	371	8.31	−36.03	35	35
20.	Madhya Pradesh	53197209	275930	63110709	280333	18.64	1.60	6	12
21.	Maharashtra	74816051	2651889	82700556	4156343	10.54	56.73	5	1
22.	Manipur	134541	749	140673	1908	4.56	154.74	31	32
23.	Meghalaya	680254	5313	691269	6773	1.62	27.48	26	28
24.	Mizoram	64249	744	63377	800	−1.36	7.53	33	34
25.	Nagaland	35915	2489	35638	3304	−0.77	32.74	34	31
26.	Odisha	9052871	64719	9800135	66675	8.25	3.02	19	18
27.	Puducherry	981714	52931	1000277	42624	1.89	−19.47	23	21
28.	Punjab	19056143	143805	21340888	204074	11.99	41.91	12	15
29.	Rajasthan	28611831	1451370	30298150	1437162	5.89	−0.98	7	5
30.	Sikkim	558538	26489	576749	31698	3.26	19.66	27	23
31.	Tamil Nadu	184136840	3561740	244232487	3990490	32.64	12.04	1	2
32.	Tripura	361786	7840	359586	11853	−0.61	51.19	29	26
33.	Uttar Pradesh	168381276	1994495	226531091	2054420	34.53	3.00	2	4
34.	Uttarakhand	26827329	124555	19941128	97683	−25.67	−21.57	15	17
35.	West Bengal	22730205	1219610	25547300	1245230	12.39	2.10	9	6
	Total	**1045047536**	**18263074**	**1145280443**	**19951026**	**9.59**	**9.24**		

Source: State/UT tourism departments

Table 5.9: Number of domestic tourist visits (DTVs) to all states/UTs in India

Year	No. of domestic tourist visits to states/UTs (in million)	Percentage (%) change over the previous year
1997	159.28	14.1
1998	168.2	5.2
1999	190.67	13.4
2000	220.11	15.4
2001	236.47	7.4
2002	269.6	14
2003	309.04	14.6
2004	366.27	18.5
2005	392.01	7
2006	462.32	17.9
2007	526.56	13.9
2008	563.03	6.9
2009	668.8	18.8
2010	747.7	11.8
2011	864.53	15.6
2012	1045.05	20.9
2013	1145.28	9.6
2014	1282	12

Source: State/UT tourism departments

Table 5.10: Number of foreign tourist visits (FTVs) to all states/UTs in India

Year	No. of foreign tourist visits to states/UTs (in million)	Percentage (%) change over the previous year
1997	5.5	9.3
1998	5.54	0.7
1999	5.83	5.3
2000	5.89	1.1
2001	5.44	–7.8
2002	5.16	–5.1
2003	6.71	30.1
2004	8.36	24.6
2005	9.95	19
2006	11.74	18
2007	13.26	12.9
2008	14.38	8.5
2009	14.37	0.1
2010	17.91	24.6
2011	19.5	8.9
2012	18.26	–6.3
2013	19.95	9.2

Source: State/UT Tourism departments

Table 5.11: Number of Indian national departures (INDs) from India

Year	No. of Indian national departures (in million)	Percentage (%) change over the previous year
1997	3.73	7.6
1998	3.81	2.3
1999	4.11	8
2000	4.42	7.3
2001	4.56	3.4
2002	4.94	8.2
2003	5.35	8.3
2004	6.21	16.1
2005	7.18	15.6
2006	8.34	16.1
2007	9.78	17.3
2008	10.87	11.1
2009	11.07	1.8
2010	12.99	17.4
2011	13.99	7.7
2012	14.92	6.7
2013	16.63	11.4

Source: Bureau of Immigration, Govt. of India

Table 5.12: Nationality-wise visa on arrivals (VoAs) in India

S. No.	Source country	Visa on arrivals		
		2011	2012	2013
1.	Cambodia	149	157	120
2.	Finland	1335	914	1030
3.	Indonesia	2063	2426	2758
4.	Japan	2344	4604	6448
5.	Laos	14	10	19
6.	Luxemburg	74	110	145
7.	Myanmar	71	109	148
8.	New Zealand	2762	3150	3968
9.	Philippines	1956	2444	2967
10.	Singapore	1848	1974	2486
11.	Vietnam	145	186	205
	Total	**12761**	**16084**	**20294**

Source: Bureau of Immigration

Table 5.13: Share of top 10 countries of the world and India in international tourism receipts in 2013

Rank	Country	International tourist receipts (in US$ billion)	Percentage share
1	USA	139.6	12.04
2	Spain	60.4	5.21
3	France	56.1	4.84
4	China	51.7	4.46
5	Macao (China)	51.6	4.45
6	Italy	43.9	3.79
7	Thailand	42.1	3.63
8	Germany	41.2	3.55
9	UK	40.6	3.5
10	Hong Kong	38.9	3.36
Total of top 10 countries		566.1	48.83
India		1.4	1.59
Others		574.5	49.58
Total		**1159**	**100**

Source: UNWTO barometer April 2014 and Ministry of Tourism (MOT)

Table 5.14: Share of top 10 countries of the world and India in international tourist arrivals in 2013

Rank	Country	International tourist arrivals (in million)	Percentage share
1	France	NA	NA
2	USA	69.8	6.74
3	Spain	60.7	5.86
4	China	55.7	5.38
5	Italy	47.7	4.61
6	Turkey	37.8	3.65
7	Germany	31.5	3.04
8	United Kingdom	31.2	3.01
9	Russian Federation	28.4	2.74
10	Thailand	26.5	2.56
Total of top 10 countries		389.3	37.59
India		6.97	0.67
Others		638.7	61.74
Total		**1035**	**100**

Source: UNWTO barometer April 2014 and Bureau of Immigration (BOI)

REVIEW QUESTIONS

1. Discuss the initiatives taken by Government of India to boost tourism.
2. Discuss the factors that led to the increase in domestic tourism in India.
3. Discuss the reasons for increase in international tourism in India.
4. Discuss the role and functions of Ministry of Tourism.
5. Write short notes on:
   - ITDC
   - Incredible India
   - Atithi Devo Bhava

Types of Tourism

Learning Objectives

After reading this chapter, you will be able to understand the following:
- Types of tourism
- Alternative forms of tourism

TYPES OF TOURISM

Adventure Tourism

The urge for adventure is there in every human being. Perhaps due to this innate nature of man that adventure tourism is one of the most popular niche segments of tourism industry. It is both a leisure pastime and serious hobby and involves exploration of remote areas and exotic places to indulge in learning and experiencing through activities with offering of higher risk and thrills. Owing to India's enormous geo-physical diversity, it has taken a big shape over the years.

Adventure tourism has various forms: Land adventure, water adventure and aerial adventure. The various kinds of adventure tourism in India are: Rock climbing, skiing, camel safari, paragliding, mountain climbing, rafting in white water, trekking.

For *trekking*, tourists prefer to go to places like Ladakh, Sikkim, and Himalaya.

Manali, Shimla, Nainital, and Mussoorie are popular for the *skiing* facilities they offer. White water *rafting* is also catching on in India and tourists flock to places such as Uttranchal, Assam, and Arunachal Pradesh for this adrenalin-packed activity. Due to the presence of climbing rocks in large numbers throughout the country, rock climbing as a kind of adventure tourism in India is taking off in a big way. The various places in India where tourists can go for *rock climbing* are Badami, Kanheri Caves, Manori Rocks, and Kabbal. The most famous destinations in India for *camel safaris* are Bikaner, Jodhpur, and Jaisalmer. Tourists can go to Garhwal, Himachal Pradesh, and Jammu and Kashmir for indulging in *mountain climbing*.

Educational Tourism

In educational tourism, the main focus of the tour visiting another country is to learn about the culture, such as in student exchange programs and study tours, or to work and apply skills learned inside the classroom in a different environment. This type of tourism has developed since time immemorial, because of the growing popularity of teaching and learning of knowledge and the enhancing of technical competency outside of the classroom environment.

Medical Tourism

Medical tourism (also known as *health tourism*) is a term used to describe the rapidly-growing practice of traveling across international borders to obtain healthcare. It is an age-old concept that has gained popularity in the recent times. It is a type of tourism in which people travel from one country to another for various medical treatment and surgery. India is adequately equipped with state-of-the-art hospital infrastructure and facilities to treat many critical illnesses. By now, India has been able to establish as a favorite destination for many complex surgeries like cosmetic surgery, joint replacement surgery, cardiac surgery, dental surgery and like at very low cost in comparison to the developed countries. There are several medical institutes in the country that cater to foreign patients and impart top-quality healthcare at a fraction of what it would have cost in developed nations such as USA and UK. The city of Chennai attracts around 45% of medical tourists from foreign countries. This indeed gives great momentum for the hospitals in particular and tourism in general. Many specialized hospitals and tour operators have already come up to promote the medical tourism. 'Medical Visa' has been introduced, which can be given for specific purpose to foreign tourists coming to India for medical treatment. India's earnings from medical tourism could exceed US$ 200 billion by 2015.

Wellness Tourism

Wellness tourism may be described as traveling for the purpose of revitalizing one's health and spiritual well-being especially through alternative healing practices. It is about being proactive in discovering new ways to promote a healthier, less stressful lifestyle or finding balance in one's life. The core of wellness tourism in India is the ancient medical system of Ayurveda combined with the system of Yoga. Indeed, many states in the country have already taken great strides to promote wellness as a tourist product. Wellness tourism includes massages, body treatments, facial treatments, exercise facilities and programs, weight loss programs, nutrition programs, pre- and post-operative Spa treatments and mind/body programs.

Eco-tourism

'Eco-tourism' (also known as *ecological tourism*) is responsible travel to fragile, pristine, and usually protected areas that strives to be low impact and (often) small scale.

The International Eco-tourism Society (TIES) defines eco-tourism as 'a form of tourism that entails responsible travel to natural areas and which conserves the environment and sustains the well-being of local people.'

'Eco-tourism is a form of tourism that fosters learning experiences and appreciation of the natural environment, or some component thereof, within its associated cultural context. It is managed in accordance with industry best practice to attain environmentally and socioculturally sustainable outcomes as well as financial viability'.

Eco-tourism pertains to a conscious and responsible effort to preserve the diversity of a naturally endowed region and sustaining its beauty and local culture. Indians have been known since ages to worship and conserve nature.

Responsible eco-tourism includes programs that minimize the negative aspects of conventional tourism on the environment and enhance the cultural integrity of local people. Therefore, in addition to evaluating environmental and cultural factors, an integral part of eco-tourism is the promotion of recycling, energy efficiency, water conservation, and creation of economic opportunities for local communities. Eco-friendly measures adopted by the hotels in India are sewage treatment plant (STP), rain water harvesting plant, waste management system, pollution control, etc.

Advantages of eco-tourism

- It is low impact tourism, where people make a conscious effort to appreciate the environment, conserve the natural resources, and re-invest sufficient amount of revenues in protecting natural habitat.
- Eco-tourism is good for the local community, as it provides employment, services and stimulates the economy. It helps in energy conservation and protects the plants and animals from the effects of traditional tourism.
- Eco-tourism offers new opportunities for small-scale investments and increases the national responsibility in protecting biological resources. Especially in developing countries, it is a means of socioeconomic and environmental upliftment.
- Eco-tourism provides recreational and educational travel without disturbing the harmony of the natural environment. It helps in reaping the benefits by the participation of people in the conservation of the flora and fauna.
- Visiting a national park or protected area will contribute towards the park maintenance in the form of fees for the staff taking care of it.
- One can admire the craftwork made by local artisans and buying these from them will help their economy as well as conserve the local heritage.
- One can go bird watching or walk through the forests exploring the different natural wonders, and stopping by villages to enjoy their cuisine, and lifestyle.
- Eco-tourism can bring you closer to nature, open up to new ideas, take you to places less traveled, and give you a wonderful new experience without harming the environment.
- Eco-tourism is a rapidly growing business and can provide the best of all the worlds. With the business of eco-tourism growing, it will provide employment to the local people and will ensure that fewer people leave for the cities. The local population will gain new skills, so they will not be totally dependent on the limited natural resources.
- Eco-tourism is becoming very popular nowadays especially in adventure trips such as mountain climbing, white water rafting, bird watching, etc. In such travels, the visitors are expected to clean up the mess before leaving and hence natural habitat is kept undisturbed.

Religious Tourism

Religious tourism (also known as *Pilgrimage tourism*) is a form of tourism, where people travel individually or in groups for pilgrimage. The world's largest form of mass religious tourism takes place at the annual Hajj pilgrimage in Mecca, Saudi Arabia. The most famous holy cities in world are Jerusalem and Mecca. The various places for tourists to visit in India for pilgrimage are Vaishno Devi, Golden temple, Char Dham, Mathura-Vrindavan, Varanasi and so on.

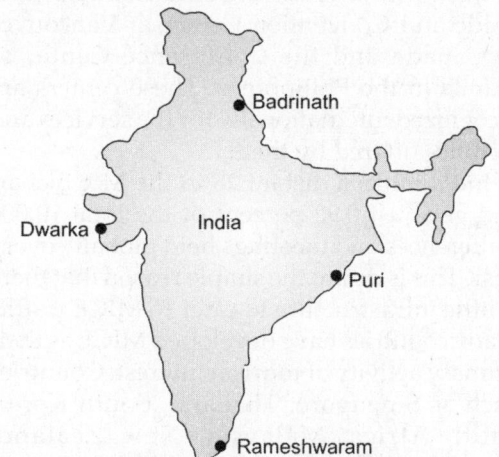

Fig. 6.1: Four Dhams of India

Sports Tourism

Sports tourism refers to travel which involves either viewing or participating in a sporting event. Normally these kinds of events are the motivators that attract visitors to visit the events like Olympic Games, FIFA World Cup.

Sports tourism can be adventurous also. *Golf tourism* in India is gaining popularity. India has several golf courses of international standards.

Business Tourism

Business tourism can be defined as travel for the purpose of business. Business tourists visit a particular destination for various reasons pertaining to his work such as attending a business meeting, conferences or convention associated with the business, meeting clients.

Business tourism involves meetings, conferences and conventions where information is exchanged, incentive travel is offered to motivate or reward staff, exhibitions are organized to promote corporations, lavish events are organized to launch new products, etc. That is why business tourism is popularly called *MICE tourism (meetings, incentives, conferences, and exhibitions).*

Many hotels, resorts, and countries have developed facilities that are uniquely devoted to this form of tourism. Some examples are: Trade and Convention Center at Vancouver in Canada and the Conference Center at Manila in the Philippines. These centers are recognized internationally for the services and facilities offered by them.

India ranks a distant 28 in the MICE chart and gets just 0.92 percent of the total 10,000 conferences and meetings held globally every year. This is due to the simple reason that there is little infrastructure to cater to MICE traffic. Many countries have developed MICE as their primary activity of tourism interest. Countries such as Singapore, Thailand, South Korea, South Africa, Malaysia, New Zealand, Australia, Spain, and even Nepal, etc. are trying to promote themselves as MICE destinations. Similarly, places like Dubai are projecting themselves as major exhibition and event centers.

The characteristics of business tourism are:

- Business tourists frequently travel to destinations not usually seen as tourist destinations. Cities such as London, Frankfurt, New York, Tokyo and Hong Kong are important destinations for the business travelers.
- Business travel is relatively price-inelastic; business people cannot be encouraged to travel more frequently by the offer of lower prices, nor will an increase in price discourage them from traveling.
- Business travel is not greatly affected by seasonal factors such as variation in climate or holidays.
- Business tourists take relatively short but frequent trips to major business destinations.
- Business tourists may require different services, such as communication facilities or secretarial service.
- Business travelers expect, and generally receive, a higher standard of service. Much business travel is first-class or business-class, and thus receives higher levels of service from the suppliers, including the travel agencies.

Leisure Tourism

Tourists, who seek break from the stress of day-to-day life, devote their holiday to rest, relaxation and refresh themselves. These tourists prefer to stay in some quiet and relaxed destination preferably at hill stations, beaches, waterfalls, zoological parks, etc.

Highway Tourism

Highways are like veins in the symbolic body of a country. Even in the most ancient times, the kings have well envisaged the importance of highways and constructed numerous inns and wells to facilitate the travelers. In modern times too, initiatives have been taken to develop tourism infrastructure along the highways so that it caters the travelers and provides income and employment opportunities along the hinterland. Haryana has been pioneering this concept in successful fashion.

Rail Tourism

Indian Railway Catering and Tourism Corporation (IRCTC), a public sector enterprise under Indian railway, promotes rail tourism in India. From luxury trains to steam locomotive trains, hill charters and Char Dham trains; it offers the tourists with attractive options and at the reasonable costs. Maharaja Express, Mahaparinirvan Express (Buddhist circuits), Bharat Darshan and Bharat Tirth are some of its famous train journeys. It also has provisions for charters and exclusive tour packages. Besides IRCTC initiatives, state-level corporations and private operators also operate tourist trains. Indeed, the world famous Palace on Wheels and Deccan Odyssey are part of successful rail tourism initiatives in the world.

Heliport Tourism

In India, there are plenty of exotic places but reaching them out those is an uphill task. Viewing this, the state and union territories with the help of central financial assistance, identifies such places where helipads could be constructed to harness the tourism potential.

Visiting Friends and Relatives (VFR)

Some tourists travel abroad because they want to visit their friends and relatives. It is sometimes said that those visiting their friends or relatives (VFR) are not really tourists at all in the conceptual sense. They do not usually buy accommodation or much food or drink or other services at the destination; but they do consume food and beverages from the supermarkets used by their hosts, and household consumption of other services (e.g. electricity, water) is increased during their stay.

Some of the VFR tourists' destinations may not be tourist attractions, but they are the population centers where friends or relatives live.

ALTERNATIVE FORMS OF TOURISM

Alternative tourists are different from the regular tourists. Alternative tourism aims at seeking a transition from impersonal, traditional tourism to establishing cordial

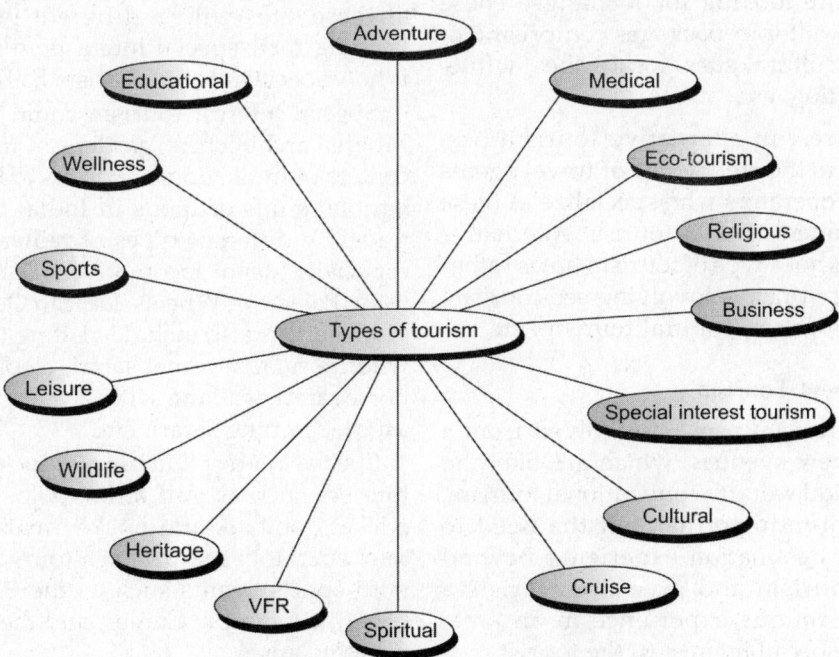

Fig. 6.2: Different types of tourism

rapport between visitors and the local community. These tourists normally avoid the services that are used by tourists such as accommodation, transport, and other services. They prefer to use or share the services of the local people. Their main motive is to experience and get an insight into their way of living.

Alternative tourism is nowadays regarded as a key to sustainable development:

- While mass tourism can have a negative impact on a destination, alternative tourism promotes a balanced growth form, more in line with local environmental and socio-cultural concerns.

- Tourism development activities such as building of infrastructure, etc. are not only costly but also affect the environment and are minimal in this form of tourism.

- Many of the western travelers have expressed their dissatisfaction towards the sun-based holidays. In fact many of the tourists are looking for a change. These tourists want tour packages comprising of wildlife, cultural sites, local tribes, white-water rafting, etc.

- This interest in alternative tourism has, thus, led to the emergence of travel agents and tour operators who specialize in these different interests of the tourists. Alternative tourism is not a type of tourism but is rather a guiding principle involving eco-tourism, heritage tourism, cultural tourism, etc.

Special Interest Tourism

Special interest tourism has evolved from a search for new avenues, which are likely to attract and add value to conventional tourism. The tourism industry has felt the need to expand the destination experience beyond pleasure tourism and give the tourists a completely unique experience in an area, which is of special interest to the tourist.

The special interest tourist looks for the unusual and not for the routine itineraries, which include attractions, which cater to the needs of the mass tourist. Special interest tourism can be defined as people traveling to a particular destination with the purpose of fulfilling a particular interest, which can be pursued only at that destination.

Special interest tourism offers many alternative forms. It is developed keeping tourist preferences in mind. It covers diverse market segments and equally diverse tourism products ranging from historical, culinary, archaeological, and other interests such as golf, fishing, etc.

In India, from the special interest point of view, culture is the most popular of a range of special interests that a tourist may pursue. A large number of tourists visit cultural sites in India such as the Red Fort, the Taj Mahal, temples, and palaces to understand their historical value. Although a majority of mass tourists in India travel to these historical sites, a large number of tourists nowadays are looking at these sites with a different perspective, keeping their special interests in mind. This is how special interest tourism has developed.

Special interest tourism could be visiting Mughal architecture, gardens of India, textile centers of India, gourmet tours to learn about various kinds of foods in India, rail tours—travel by different types of railway systems, especially steam locomotives, Darjeeling toy train, Palace on Wheels, Deccan Odyssey, etc. The interests also include visiting tribal areas, wildlife safaris, camel safari, elephant safari, horse safari, walking safari, cycling safari, jeep safari, camping safari, etc.

Special interest tourism also includes sports tourism such as golf tours, polo, car racing, cricket, football, and hockey matches. Some tour operators sell package tours to coincide with sports events such as the FIFA World Cup, the Olympic Games, and the Common-wealth Games.

Rural Tourism

Tourists nowadays are shifting their interest of travel to new destinations to explore and experience the destination and have first-hand knowledge of the local people, their cuisine and actual way of living.

Of late, rural tourism has gained importance in India. The Indian government is marketing rural tourism through its 'Incredible India' promotional campaign.

India is a country of villages and showcasing the rich rural life, art, culture and heritage in villages in responsible manner would be mutually beneficial since tourism and conservation complement each other. Rural tourism schemes in India envisage encouraging and promoting the villages that have inherent strengths in art and craft, handloom, and textiles. The main purpose is to benefit the local community economically and socially, diversifying the local development opportunities as well as enable interaction between tourists and local population for a mutually enriching experience. The promotion of rural tourism is also aimed to arrest the migration from rural to urban areas.

Ethnic Tourism

Ethnic tourism involves travel for the purpose of observing the cultural expressions of lifestyles and customs of the indigenous and exotic people. This type of tourism focuses directly on the local people. It involves direct intimate contact with the authentic culture of the indigenous people. The tourists visit the local homes, observe, and participate in their traditional rituals, ceremonies, dances, festivals, etc.

This type of tourism is also referred to as a combination of culture and nature tourism. In ethnic tourism, the tourist is mainly interested in having direct contact with the local people.

The tourist's main aim is to gain first-hand experience of the way of life and cultural artifacts of the local people, whereas in cultural tourism the contact with the local people is done indirectly, i.e. these tourists will view the culture but not experience it. Some examples of ethnic tourism are the Pushkar fair, Bikaner fair, and Nagaur fair of Rajasthan.

Senior Citizen Tourism

This is a fairly new emerging trend in tourism meant for the senior citizens or old people. Many tour operators nowadays specialize in package tours specially designed for the elderly. They provide them with a tour escort or tour leader who keeps them occupied with entertainment and activities suitable for their age and at the same time making sure they do not get tired. Extreme care has to be taken while planning their tour. Due to their age factor and other health problems they should have a relaxed and enjoyable tour.

This type of tourism is common in the West, because of the nuclear family concept. It is now gaining popularity in India as well. World over, the number of senior citizens is on the rise because of a longer lifespan attributed to developments in medicine and technology.

Wildlife Tourism

Wildlife is a term used to refer to both the floral and faunal components of a natural environment. Wildlife tourism has gained popularity in the last decade. Many young enthusiasts and nature lovers as well as adventure seekers are exploring this new area of tourism. Realizing the economic benefits of tourism, the governments of many countries are promoting wildlife tourism in a big way. Wildlife tourism is also considered an important element in wildlife protection. On one hand, the tourists can help in the conservation of wildlife while on the other conversely their presence can affect the wildlife. Thus, each area has to be assessed according to the number of people it can sustain. Disruptive human presence in the parks can have a negative impact on the number of wildlife and, thus, cause a drop in the number of tourists also. Due to the increased poaching and hunting activities, many wildlife species are on the verge of extinction, like the Asiatic Lion at Gir National Park in Gujarat.

Many tour operators specialize only in wildlife tourism. Luxury safaris, wilderness backpacking, zoos, aquaria, and safari parks all form part of the increasingly successful wildlife tourism industry. Examples of well-known wildlife sanctuaries and national parks are Corbett National Park in Uttarakhand, Bandhavgarh and Kanha National Park in Madhya Pradesh, and Kaziranga National Park in Assam, etc.

Cultural Tourism

People are curious to know about foreign lands and their cultures. Culture is most important factors which attracts tourists to a destination. Cultural tourism gives insight to:

- Way of life of the people of distant land
- Dress, food, language, dance, music, architecture
- Customs and traditions
- Fairs and festivals
- Religions
- Culinary delights

India is known for its rich cultural heritage and an element of mysticism, which is why tourists come to India to experience it for themselves. The various fairs and festivals that tourists can visit in India are the Pushkar fair, Taj Mahotsav, and Surajkund mela. Cultural tourism is the predominant factor behind India's meteoric rise in the tourism segment in recent years, because from time immemorial, India has been considered the land of ancient history, heritage, and culture. India's glorious past and cultural diversity make a potent blend which attracts millions of tourists each year to its tourist attractions.

Heritage Tourism

Heritage tourism has registered an immense growth in the last few years, ever since additional initiatives were taken by the Government of India to boost India's image as a destination for heritage tourism.

India has always been famous for its rich heritage and ancient culture. So the onset of heritage tourism in India was long anticipated. The Government of India and the Ministry of Tourism encourage heritage tourism in India by offering several benefits to the Indian states that are particularly famous for attracting tourists.

India's rich heritage is amply reflected in the various temples, palaces, monuments, and forts that can be found everywhere in the country. This has led to the increase in India's heritage tourism.

The most popular heritage tourism destinations in India are:

- Taj Mahal in Agra
- Mandawa castle in Rajasthan
- Mahabalipuram in Tamil Nadu
- Madurai in Tamil Nadu
- Lucknow in Uttar Pradesh
- Red Fort, Jama Masjid, Humayun's Tomb, and Tughlaqabad Fort in Delhi

Fig. 6.3: Heritage destinations

REVIEW QUESTIONS

1. Explain the various types of tourism.
2. What do you understand by sustainable tourism?
3. Mention the important features of medical tourism.
4. Differentiate between:
   - Adventure tourism and Sports tourism
   - Religious tourism and Cultural tourism
   - Business tourism and Leisure tourism
5. Write short notes on:
   - Eco-tourism
   - VFR
   - Special interest tourism
   - Heritage tourism
   - Wildlife tourism
   - Senior citizen tourism

Emerging Trends in Tourism

Learning Objectives

After reading this chapter, you will be able to understand the following:
- Emerging trends of tourism in the world
- Emerging types of tourism

EMERGING TRENDS OF TOURISM IN THE WORLD

Customers are always looking at services that give them a unique experience along with comfort and value for money. With the increasing competition, hotels are redefining the concept of hospitality and are going all out to make the customer's stay an exciting, special and memorable one. They are pampering travelers with the latest and the best in services and products. Some emerging trends in tourism are as follows:

- *Increasing choices of destinations*: For several decades, Western Europe has been a popular destination for international tourists. However, as tourists have got used to visiting Western Europe, they become curious about the less explored parts of the world such as Eastern Europe, the Asia-Pacific area and the less developed parts of the world including Africa. In general, there appears to be a slow shift of tourist arrivals from the economically advanced countries to the less developed ones.

  Nowadays, the experienced traveler looks for rare, authentic vacations in remote and less well-known places, as against luxurious five star vacations. The trend has shifted towards ethnic and rural tourism.

  The high-end luxury traveler looks for spa facilities, luxury cruises, wildlife safaris, premium holiday packages, wellness holidays, spiritual getaways, etc.

- *Mercurial responses to changing economic environment*: The potential for tourism growth is enormous throughout the world. As the production of goods and services increases, people have more disposable income and more leisure time. At the same time, a better-educated population would like to travel for different purposes such as rest, relaxation, recreation, health and wellness. Although there may be economic setbacks that will discourage tourism development, tourism has always found new ways to flourish. For example, many people would rather change their travel destinations or spending patterns than give up their vacation. Also, there are tour packages to suit every taste and income level.

- *Governments encouraging tourism development*: As many countries recognize the potential contribution of tourism to their economy, there is increasing competition in the development and promotion of tourism among countries.

- *Sustainable forms of tourism*: In future, tourism development will no longer be determined solely by economic consideration. It is suggested that tourism development should not abuse the natural environment. As environmental issues are becoming a worldwide concern, there will be new forms of tourism such as 'eco-tourism', 'agri-tourism' and 'green tourism'.

- *'Special-interest tourism' changes forms of tourism*: Due to cultural and social changes, there have been significant changes in the pattern of international tourism. 'Special-interest tourism' (such as weight-losing and mind-broadening) has been developed to cater for the wide range of interests of tourists.

- *Increasing ability to travel of young people and the elderly*: It is suggested that in the next decade, the number of tourists of the following two age groups will increase faster than that of the others: Senior citizens and young people. Due to changes in socio-economic conditions such as better retirement benefits, more senior citizens can afford to travel after retirement. Moreover, better education and new travel opportunities enable young people to travel more.

- *Information technology contributing to tourism development*: Information technology will become all-powerful in influencing destination choice and distribution. Travel suppliers and promoters are using information technology to identify and communicate with travelers through promotion and information supply, and to assist the travelers in their choice of destinations. Travelers that are familiar with surfing on the internet for information and reservations could make their travel arrangements by themselves. As a result, the traditional distribution channels of delivery through intermediaries are being affected.

- *Service of intermediaries professional and personalized*: The role of travel agents is now changing from that of intermediaries to that of a provider of personal service and professional expertise. Travel agencies and other travel professionals are merging with each other with the objective that 'bigger is better'.

- *Theme-based tourism product diversification*: Theme-based tourism product is being developed with a combination of the three Es—entertainment, excitement and education.

- *Terrorist attack enhanced concern of travel safety*: Air traffic control systems play a major role in overall air travel safety such as collision avoidance, precision landing aids and ground obstacle avoidance. Air security issues such as security screening at airports, permanent reinforcement of cockpit doors, public safety are also being major concerns especially after the September 11 terrorist attack in New York and Washington.

- *SMERFs*: The dominant segment of MICE travelers is facing competition from a fast emerging segment to and around Asia, who travel for social, military, education, religious, and fraternity reasons (SMERFs). The SMERFs are the resilient groups, who are budget conscious and do not mind gathering during non-peak times if expenses can be saved. The SMERFs collectively form a huge market and have vast untapped potential for the developing or recovering Asian travel markets. SMERFs travel for a purpose and not just to see places. They are willing to travel abroad despite the economic cycle, travel off-season and off the beaten track to save on transport and accommodation. Social travel includes people participating in sports teams, talent and dance organizations, or as volunteer workers for events, etc. Asia's military needs civilian transport for its estimated 32 million soldiers on the move and their proceeding on leave itself is a highly significant market. The education travel market specially studying or visiting Singapore as part of a study tour has tremendous potential. The Singapore Tourism Board is targeting 1,50,000 international students particularly from Asia by 2016. Indian students are discovering India's cultural heritage and school groups are emerging as a valuable market because of repeat tours. Asia is recognized as the birthplace of Hinduism and Buddhism and religious tourism is on the rise.

- *Tourists with special needs and the differently-abled tourist*: Differently-abled guests are a growing segment of the traveling public and often need wheelchairs, walkers, canes or crutches. As per the guidelines of the Ministry of Tourism for classification/re-classification of hotels, all star hotels shall provide at least one room for the differently-abled guest. Awareness for the need of accessibility for all including the disabled and making the hotel disabled-friendly can be done by providing:

  – Minimum door width of one meter to allow wheelchair.

  – Suitable low height furniture, low peep hole, cupboard and sliding doors with low cloths hangers, etc.

  – Room should have audible and visible blinking light alarm system.

  – Bathroom should be wheelchair accessible with sliding door, suitable fixtures like low wash basin, low height WC, grab bars, etc.

  – Ramp with anti-slip floors at the entrance of the hotel to allow wheelchair access.

  – Free accessibility in all public areas and to at least one restaurant in 5 star and 5 star deluxe hotel.

  – In public restrooms (unisex), wheelchair should be accessible with low height urinal (24" maximum) with grab bars.

  – Braille on elevator panels, restrooms, and directional signs

  – Disabled-friendly transport such as buses and trains for a person on wheelchair

  – Separate counters for disabled travelers

  – Designated parking facilities

  – Soft skills training on how to transfer guests with mobility problems, how to speak to the disabled depending on the nature of the disability, for example, one should not shout while talking to guests with hearing impairment

Differently-abled or disabled tourists with special needs should have accessibility to most destinations.

For example, Vivekananda Rock Memorial in Kanyakumari has been re-designed to suit the needs of the disabled during its recent renovation. Basilica of Bo Jesus at Goa has wheelchair facility available at the church. Arland Sagar of Shri Gajanan Maharaja Sansthan at Shegaon in Maharashtra has been designed keeping the needy, old, infirm, and disabled in mind.

- *Online travel agents (OTA)*: Also known as virtual travel agents (VTAs) or E-travel agents, they are the new breed of travel agents who are taking over a major share of reservations online and offline from the traditional travel agents. Over 70 percent of air tickets are now being booked online in India. A number of online travel and tour operators, which provide better prices and options to consumers, have emerged in India. Some of them are: MakeMyTrip.com, Yatra.com, Cleartrip.com,Travelocity.com, ezeegol.com, Travelguru.com, and last-minute.com. These web portals allow consumers to access information and make online bookings.

EMERGING TYPES OF TOURISM

Many new forms and types of tourism are being developed to meet the needs of the growing tourist market. All forms of innovation in tourism are being promoted keeping specific needs of different tourism in mind such as culinary tourism, tea tourism, film tourism, highway tourism, etc. Some of the emerging forms are discussed below:

Cruise Tourism

Cruising to exotic locales in different parts of the world is no longer a niche activity limited to the upper echelons of society. The cruising culture has spread from Europe and the Americas to India. Cruise holidays on luxury floating resorts are gaining popularity as cruise operators are offering affordable packages for all budgets ranging from one day to a number of days for all age groups.

India with its vast and beautiful coastline, virgin forests and undisturbed idyllic islands, rich historical and cultural heritage, can be a fabulous tourist destination for cruise tourists.

The Cruise Shipping Policy of the Ministry of Shipping was approved by the Government of India on 26th June 2008. The objectives of the policy are to make India as an attractive cruise tourism destination with the state-of-the-art infrastructural and other facilities at various ports in the country to attract the right segment of the foreign tourists to cruise shipping in India and to popularize cruise shipping with Indian tourists.

India attracted 163,000 cruise visitors in 2013. Government of India has estimated that India would emerge with a market size of 1.2 million cruise visitors by 2030-31.

The government has identified 8 tourist circuits along National Waterways (NW-1 and NW-2):

National Waterway-1 (river Ganga)
- Allahabad circuit
- Varanasi circuit
- Patna circuit
- Bhagalpur circuit
- Kolkata circuit

National Waterway-2 (river Brahmaputra)
- Guwahati circuit
- Tezpur circuit
- Neamati circuit

Spiritual Tourism

Spiritual tourism is one of the most popular forms of tourism today. It involves travel to places for spiritual benefit. Many people today follow the path of their gurus, and find solace in their preaching and discourses. Tour operators are developing special packages for the spiritual tourists.

Floating spiritual villages in the tranquil backwaters of Kerala, offer corporate executives an unforgettable experience and choice of spiritual and health therapies and discourses from spiritual gurus, all inside houseboats and floating cottages.

Space Tourism

One of the most advanced technological developments to be witnessed by humankind and one of the costliest types of tourism is a trip to space which at present costs approximately US$20 million. This is a new upcoming form of tourism where ordinary people will buy tickets to travel to space and back. Space tourism as defined by the space tourism society covers:

- Travel to the earth's orbit and sub-orbit
- Travel to planets beyond the earth's orbit, for example, to mars
- Earth-based simulated experiences at ASH center and entertainment based experiences
- Cyberspace tourism experiences.

Some of the companies promoting space tourism are Virgin Galactic, Space Adventures, Star Chaser, Blue Origin, Bigelow Aerospace, etc. Space tourism will give the space tourist the unique and thrilling experience of viewing the earth from outer space. More affordable sub-orbital flights are currently priced at US$20 million, flying at an altitude of 100 to 160 kilometers in space and letting the tourists experience weightlessness for a few minutes, be amongst the stars, and view the curved earth below.

Space Adventures Ltd is working on circumlunar missions to the moon at a passenger price of US$ 100,000,000. Space stations are being set-up and old space stations are being converted to space hotels. Bigelow Aerospace has launched the first inflatable model called Genesis I in July 2006 and Genesis II in June 2007.

Wine Tourism

Wine tourism (also called *enotourism*, *oenotourism* or *vinitourism*) involves tasting, consumption or purchase of wine, visits to wineries and vineyards, organized wine tours, wine festivals and other special wine-related events. Although relatively new, it is gaining popularity by competing with other beverages and tourism themes.

Many wine regions having perfected the art of superior wine making are extensively marketing wine tourism for monetary gains. Although the old world producers as Spain, Hungary, Portugal, Italy, France and Germany have been the forerunners, the New World wine regions of Australia, Argentina, USA and South Africa are equally poised.

Although India is experimenting with wines, the Government's proactive steps are aiming to catch up with the global leads. Wine making in India is age-old with references in the Vedas to 'soma'—an intoxicating beverage. European travelers to the Mughal court praised wines from the royal vineyards of Hyderabad, Kashmir and Surat.

Wine sector has shown significant growth in the last five years in India and the related tourism activity has also increased, particularly in Maharashtra followed by Karnataka, Delhi and Goa.

Some of the best vineyards in India are:
- Sula Vineyards, Nasik, Maharashtra
- Chateau Indage Estate Vineyards, Narayangaon, Maharashtra
- Chateau d'Ori, Dindori, Madhya Pradesh
- Grover Vineyards, Nandi Hills, Karnataka
- Zampa Wines, Nasik, Maharashtra
- Fratelli Wines, Akluj, Maharashtra

Culinary Tourism

Culinary tourism or *food tourism* is the exploration of food as the purpose of tourism. It is now considered a vital component of the tourism experience. Dining out is common among tourists and 'food is believed to rank alongside climate, accommodation, and scenery' in importance to tourists.

Culinary or food tourism is the pursuit of unique and memorable eating and drinking experiences, both near and far. Culinary tourism is not limited to gourmet food.

In recent years, food tourism has grown considerably and has become one of the most dynamic and creative segments of tourism. Both destinations and tourism companies are aware of the importance of gastronomy in order to diversify tourism and stimulate local, regional and national economic development. Furthermore, food tourism includes in its discourse ethical and sustainable values based on the territory, the landscape, the sea, local culture, local products, authenticity, which is something it has in common with current trends of cultural consumption.

Gastronomic tourism applies to tourists and visitors who plan their trips partially or totally in order to taste the cuisine of the place or to carry out activities related to gastronomy.

Underwater Tourism

The deep seas have always been an attraction for tourists whether it is fun and frolic on the beach, viewing the coral reefs on the seabed or marine creatures in ocean parks. The latest addition to deep sea attractions are restaurants and hotels submerged in the sea. The Hydropolis in Quingdao, China is a hotel under the Yellow Sea, which offers rates comparable to five star hotels.

Another Hydropolis hotel under construction is the 220 suite; luxurious underwater hotel in Dubai situated 66 feet beneath the Persian Gulf. In keeping with the theme, the hotel will be shaped like a jelly fish with bubble-shaped suites and will be connected to land via a submerged transparent train tunnel.

The Jules Undersea Lodge at Key Largo, Florida which was once an authentic underwater research station now offers overnight packages to tourists from 13:00 hours to 10:00 hours at US$475 which include diving gear, a gourmet dinner, and breakfast. The access to the hotel is via scuba diving 21 feet below the surface of the sea.

Underwater tourism also includes visiting wreckages of famous ocean liners which lie deep down on the ocean bed and traveling in submarines to study the marvels of the water kingdom.

Perpetual Tourism

This is a term used to describe people who are perpetually on the move and stay in one country for a set period of time only so that they can avoid the legal obligations which arise out of permanent residency, for example, paying of income tax which is mandatory if a person stays for 122 days in one country. They adopt this lifestyle to be free from the laws that govern citizens of a country.

Perpetual travelers (PTs) are people who live in such a way that they are not considered legal residents of any of the countries in which they spend time. By lacking a legal permanent residence status, they seek to avoid the legal obligations which may accompany residency, such as income and asset taxes, jury duty, and military service. For example, while PTs may hold citizenship in one or more countries that impose taxes based solely on residency, their legal residence will most likely be in a tax haven. PTs may spend the majority of their time in other countries, never staying long enough to be considered as residents.

Virtual Tourism

This allows a tourist to visit a destination sitting comfortably in an armchair, confined to the home. Virtual reality helps them to explore different regions of the world, visit sites without having to book tickets, apply for visas or spend money, i.e. without having to physically travel.

It is a boon to those who do not get an opportunity to travel or do not have the time and money to travel.

It also helps tourists wishing to travel, to decide on a destination for a holiday and shows them what they can expect to see. Technology is used to take tourists on a panoramic tour to give an all-round unbroken view of the destination, which makes them virtually feel they are at the place. People are made to sit in the centre of a dome-shaped room which has screens all around. The Internet, multimedia packages, and the television along with travel literature are used for the virtual experience.

Dark Tourism

Dark tourism or *black tourism* or *grief tourism* is a pilgrimage to places where people gave up their life for the nation or where famous personalities breathed in their last. The Cellular Jail at Port Blair, the death site of St. Peter in Rome, the crash site where Princess Diana lost her life, the Nazi Holocaust, World War II sites, etc. are places people visit to pay homage to the departed souls or are curious to see.

Disaster Tourism

Tourism professionals are cashing in on destruction caused by various disasters to satisfy the curiosity of people to witness the extent of damage caused by various natural and human-made disasters by organizing guided bus tours to such sites. Prominent among such sites are Ground Zero in New York, where the Twin Towers of World Trade Center were razed to the ground following the

9/11 terrorist attack, the sites of destruction which bore the brunt of the 2004 Tsunami and Hurricane Katrina, etc.

Extreme Tourism

Extreme tourism or *shock tourism* involves traveling to risky places or participating in dangerous events. This is a niche tourism product for the physically fit, daredevil tourist who is aware of the risk, yet is inclined to experience what majority tourists would not dare to imagine and includes ice diving in the white sea, trekking through dense jungles, etc.

Experiential Tourism

Experiential travel or immersion travel is one of the major market trends in the modern tourism industry. It is an approach to traveling which focuses on experiencing a country, city or particular place by connecting to its history, people, food and culture.

REVIEW QUESTIONS

1. Discuss in detail the emerging trends of tourism in the world.
2. Write in detail about the emerging types of tourism.
3. Write short notes on:
   - OTA
   - SMERFs
   - Differently-abled tourist
4. Differentiate between:
   - Cruise tourism and Underwater tourism
   - Dark tourism and Disaster tourism
   - Food tourism and Wine tourism
   - Extreme tourism and Experiential tourism

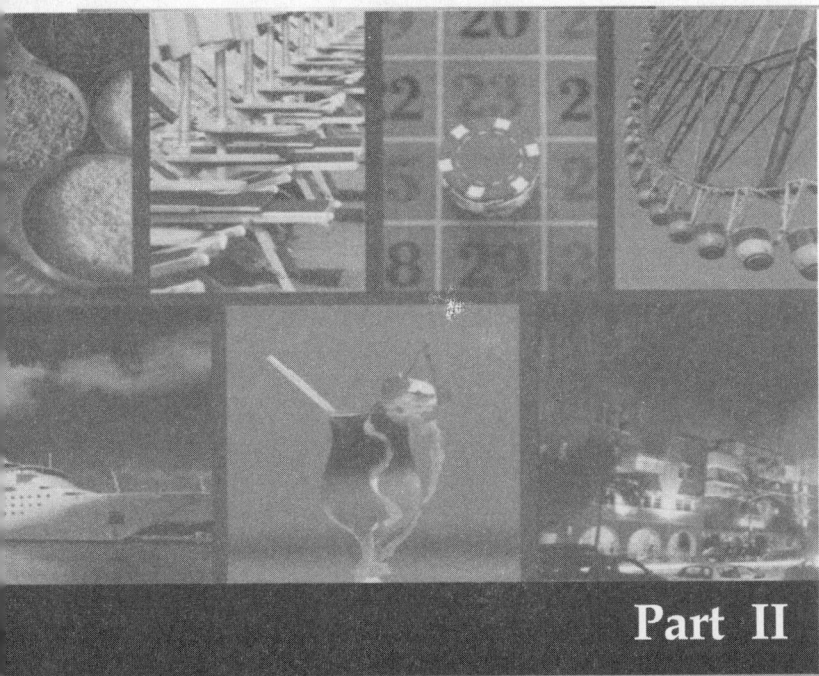

Part II

The Hospitality and Lodging Industry

Introduction to the Hospitality Industry

Learning Objectives

After reading this chapter, you will be able to understand the following:

- Concept of hospitality
- Hospitality industry—scope and constituents
- Evolution and growth of the hotel industry in the world
- Some important events in the hotel industry
- Evolution and growth of the hotel industry in India
- Hotel industry—an overview
- Hotel as a service provider in the modern day

CONCEPT OF HOSPITALITY

Hospitality refers to the relationship between a guest and a host, and it also refers to the act or practice of being hospitable. It includes cordial reception and entertainment of guests, visitors, or strangers. Hospitality is also known as the act of generously providing care and kindness to whoever is in need.

Some of the common yet important definitions are given below:

Hospitality is 'providing your guests with the same amount of attention and service as you would expect if you were in their shoes.'

Hospitality is—to be invited and made to feel genuinely welcomed and relaxed; to be treated with a disposition of cordiality, reception and support; to be professionally guided by the host to meet and exceed the guest's expectations.

'Hospitality is treating others with warmth and generosity, authentically'.

Hospitality in the commercial context refers to the activity of hotels, restaurants, catering, inn, resorts or clubs who make a vocation of treating tourists.

Hospitality can be seen in a genuinely smiling face whilst providing attentive and courteous services, facilities, and amenities to a traveler, by meeting, greeting and providing him with efficient and caring services, thereby creating an environment of 'a-home-away-from-home', and making his visit a memorable and pleasant experience.

Hospitality is a long running tradition in India. From the majestic Himalayas and the stark deserts of Rajasthan, over beautiful beaches and lush tropical forests, to idyllic villages and bustling cities, India offers unique opportunities for every individual preference. Every day in the southern part of the country, ladies make rangoli of rice, flowers and flour, etc. It is believed that by doing so they are inviting guests and giving a message that the house is open to welcome guests. Greeting guests with folded hands, touching their feet in respect, aarti, puja, garlanding, applying tilak and ringing of bells, are numerous other ways of providing warm reception and welcome and hence hospitality—in various parts of our country. Beating of drums, dhols, blowing of trumpets, exotic dances and shows such as puppetry, etc. are all different ways of expressing warm hospitality in our country.

Fig. 8.1: Hospitality

HOSPITALITY INDUSTRY—SCOPE AND CONSTITUENTS

Many people's definition of hospitality industry extends only to restaurants and hotels. In reality, it goes far beyond this and includes any organization that provides food, shelter and other services to people away from home. When viewed in this light, the hospitality industry can be quite large and far reaching.

Many academics, industrialists and policy-makers have attempted to define the hospitality industry—yet there is still no one commonly accepted definition.

The hospitality industry is a broad category of fields within the service industry that includes lodging, food and beverages sector, event planning, theme parks, transportation, cruise line, and additional fields within the tourism industry.

Hospitality industry is the range of for-profit and not-for-profit organizations that provide lodging and/or accommodations including food services for people when they are away from their homes.

The hospitality industry is a multi-billion dollar industry that depends on the availability of leisure time and disposable income. One of the most defining aspects of this industry is that it focuses on customer satisfaction. While this is true of nearly every business, this industry relies entirely on customers' being happy.

Hospitality management involves the planning, organizing, directing and controlling of human and material resources within lodging, restaurants, travel and tourism.

Tourism and hospitality are among the major revenue earning enterprises in the world and also happen to be among the highest priority industries and employers.

The global hotel industry generates approximately between US$400–500 billion in revenue each year, one-third of that revenue is attributable to the United States. The multi-faceted hospitality industry is one of the largest industries in the world and is the single largest employer, supporting 1 in every 11 jobs worldwide, according to the World Travel and Tourism Council. Hospitality industry offers about 8.7% of worldwide employment, with growth projection of 337 million jobs by 2023.

Expected to be the second-largest employer in the world, the hospitality sector will employ close to 5 million people in India by 2019, according to a report by the World Travel and Tourism Council (WTTC).

The number of hotel beds in the country is expected to increase to 501,000 by 2016 from 269,000 in 2011. The number of hotel rooms in the country is expected to increase to 220,000 by 2016 from 121,000 in 2011.

The sector expects more than 50,000 hotel rooms to be added in India by 2020, according to a survey by Cushman and Wakefield. This will lead to a rise of over 65% in total hotel inventory in India.

According to WTTC statistics, the market size of the tourism and hospitality industry in India is anticipated to touch over US$418 billion by 2022. The sector has been growing at a cumulative annual rate of 14% every year, adding significant amount of foreign exchange to the economy.

The industry is very diverse and global, but is very vulnerable to the fluctuations of national economies and various happenings across the world, especially terrorist attacks that have at times dealt severe blows to the business. Figure 8.2 presents an insight into the segments of the hospitality industry.

EVOLUTION AND GROWTH OF THE HOTEL INDUSTRY IN THE WORLD

The origin and development of the hospitality industry is a direct outcome of travel and tourism. There are many reasons for which a person may travel: Business, pleasure, leisure, recreation, further studies, cultural activities, medical treatment, pilgrimage, religion, adventure, curiosity and posterity, fairs and festivals, sports, visiting friends and relatives (VFR) and so on. When a person travels for a few days, he may carry his clothes with him, but it is not possible for him to carry his food and home. Thus, two of his three basic needs —food and shelter—are not taken care of while he is traveling. This is where the hospitality industry steps in.

With the *establishment of money and the invention of wheel* sometime in the 6th century BC, came the first real impetus for people to

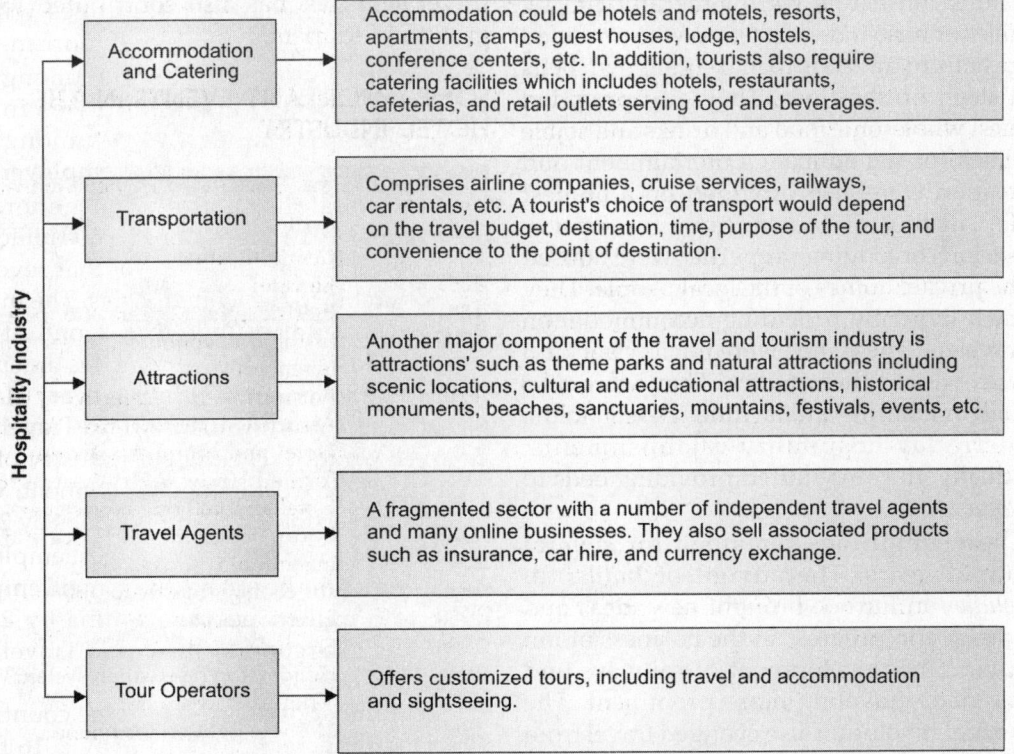

Fig. 8.2: Segments of hospitality industry

trade and travel, since prior to that it was difficult due to lack of a standardized medium of exchange. However, there was a limit to the distance they could cover in a day. At nightfall they avoided travel due to the fear of wild animals and bandits, and also because of animal fatigue.

Thus, for the night halt, they looked for a place that could provide them with water, fuel to cook food, and above all, security from wild animals and bandits.

The primitive lodging houses originated essentially to cater to these needs of the travelers. Throughout the world they were known by different names such as *Dharamshala and Sarai* in India, *Ryokans* in Japan, *Paradors* in Spain, *Pousadas* in Portugal, *Coffee Houses* in America, *Taverns and Inns* in Europe, *Cabarets* and *Hotelleries* in France, *Mansionis* and *Hospitia* in Switzerland, *Phatnal* in Greece, *Relay Houses* in China.

The earliest inns were generally run by families or husband–wife teams who provided large halls to travelers to make their own beds and sleep on the floor. They also provided modest wholesome food and drinks and stable facilities for the animals. Entertainment and recreation were also provided on a modest scale. All this, of course, came for a price. These inns or lodging properties were housed in the private homes of the local people. They made a living by providing accommodation to travelers. These inns were not as clean and tidy as we see them today. They were also devoid of the frills and facilities as seen in the modern-day hospitality establishments. Gradually, the inns started providing beds to travelers.

These conditions remained for several hundred years. The advent of *Industrial Revolution* in Europe brought new ideas and processes and progress in the business of inn keeping. The development of railways and ships made traveling more prominent. The industrial revolution also changed travel from social to business travel. There was an urge for quick and clean service because the inns were basically self-service institutions.

The present-day tourist, who has higher levels of disposable income, international exposure, and refined tastes, wants specialized versions of products and services, such as quieter resorts, family-oriented holidays, or commercial hotels. This has led to a demand for better quality products and services, mainly regarding accommodation and traveling, thus feeding the growth of the hospitality industry as a whole.

The developments in technology and transport infrastructure, such as jumbo jets, low-cost airlines, and more accessible airports have made tourism affordable and convenient. There have also been changes in lifestyle—for example, now retiree-age people sustain tourism round the year. The sales of tourism products on the Internet, besides the aggressive marketing of tour operators and travel agencies, has also contributed to the growth of tourism.

SOME IMPORTANT EVENTS IN THE HOTEL INDUSTRY

Table 8.1: Great Firsts in the US hotel industry

Year	Event
1846	Central heating
1859	Elevator
1881	Electric lights (2 years after patent)
1907	In-room telephones (31 years after invention)
1910	Formation of American Hotel Association (later AHMA—American Hotel and Motel Association) was formed, (now AHLA—American Hotel and Lodging Association)
1927	Radio in rooms (21 years after invention)
1940	Air cooling (mostly in public areas)
1950	Electric elevator
1958	Free television
1964	Holiday Inn reservation system with centralized computer
1965	Message light on telephone

Contd.

Table 8.1: Great firsts in the US hotel industry (*contd.*)

Year	Event
1965	Initial front office systems followed by room status
1970	Color TV (invented in 1954)
1970 (Early)	ECR (electronic cash register)
1970 (Mid)	POS (point of sales) system and key less locks
1973	Free in room movies (Sheraton)
1983	In room personal computers

Table 8.2: International landmarks in the growth of hotel industry

Year	Event
1650	Pascal opened a café in Paris and Coffee House in London
1794	The first commercial building for the hotel purposes called the 'The City Hotel' (73 rooms) was constructed in New York, USA
1829	'Tremont House' (170 rooms), the first class hotel, was constructed in Boston, USA
1875	The most expensive hotel of its day called 'The Palace' was built in San Francisco
1908	Ellsworth Statler started the first chain of hotels, called 'The Buffalo Statler'
1915	The international chain of hotels were started
1930	The Economic Depression in USA and Europe led to the decline of the hotel industry
1945	There was an increase in the number of hotels after World War II
1950	Upsurge in hotels. New concepts developed such as motels, boatels, floatels, rotels, and loatels. Kemmons Wilson formed 'Holiday Inn' and the first Holiday Inn was made in 1952.
1960	Individual hotels merged themselves with hotel chains like Sheraton, Hyatt, Holiday Inns, Ramada Inns, etc. Later part of the 60s offered Budget Hotels and Motels.
1970	The hotel industry took once again the frenzy of hotel construction.

Contd.

Hilton, Sheraton, and Western Corp, opened convention oriented hotels. Airport Hotel locations also appeared and the Marriotts started locating their properties outside the central city area. Aided by financers money became available and so also the franchisees.

Current Scenario	Since then a lot of changes in the technological and management approach have taken place and today's hotel industry is progressing day by day. International business is rapidly developing and with the development of air travel, a lot of business executives are traveling out. Modern hotels are rendering facilities according to the needs and wants of the tourists and the business class. Some international hotel chains of repute that rendering a world class services are:

Accor, American International Hotel and Travel Lodge, CEDOK, Club Meridian, Friendship Inn, Golden Tulip, Hilton, Holiday Inn, Howard Johnson, Hyatt, Imperial, Inter Continental, Marriott, Meridian, Motel 6, Park Royal, Quality Inns, Ramada Inn, Red Carpet Inn, Red Roof Inns, Sheraton, Sofitel, SRS Hotels, Super 8 Motels, Topeka Inns, Trust House Fort, United Inn, Utell International, Western International Hotel, Wolfe International, etc.

EVOLUTION AND GROWTH OF THE HOTEL INDUSTRY IN INDIA

The origin and growth of hotel industry in India can be broadly categorized in the following three periods.

Ancient and Medieval Era (from Indus Valley Civilization to AD 1600)

The beginning of the hospitality sector in India stand rooted in the Hindu philosophy of *atithi devo bhava*, implying that an unannounced guest is to be accorded the status of God. While it is not clear when hospitality emerged as a commercial activity in ancient India, there is evidence of accommodation facilities for travelers and guests, though not as organized as we see them today. The lodging houses during those times were known as *Dharamshalas, Musafir Khanas and Sarais*, provided by rich people such as Rajas, Kings, Zamindars, etc. Usually free accommodation and food for travelers was given. During medieval period it was mandatory for the state authorities to provide food and shelter to the wayside traveler. Ancient texts and literature, as also Hindu mythology, have many references to travel and the provision of accommodation facilities for traveling pilgrims and traders by the authorities of those days. Records of many foreign visitors and philosophers who came to India speak highly of the hospitality facilities. There are several localities in Delhi, such as Katwaria Sarai, Lado Sarai, Sarai Kale Khan, that have retained their names till date, although the medieval constructions are hard to find.

Colonial Era (1601–1947)

The organized existence of the hotel industry in India started taking shape during the colonial period, with the advent of Europeans in the seventeenth century. The early hotels were mostly operated by people of foreign origin to cater to the needs of the European colonizers and later officials of the Raj. Among the first such properties were taverns like Portugese Georges, Paddy Goose's, and Racquent Court, which opened in Mumbai between 1837 and 1840. However, within a period of about 10 years, most of the taverns disappeared and more respectable hotels like Hope Hall Family Hotel began to make an appearance. Other famous properties included the Victoria Hotel, more famous as British Hotel, by Pallanjee Pestonjee in 1840; Esplanade Hotel in 1871; Watson's, which was exclusively for Europeans; Auckland Hotel (1841) in Kolkata, which went on to become the Great Eastern Hotel in post-independent India, and so on.

In December 1903, Jamsetji Nusserwanji Tata, inaugurated the Taj Mahal Palace and Tower Hotel, overlooking the Gateway of India in Mumbai, following a racial discrimination incident wherein he was refused entry into the Watson's Hotel for being an Indian. In 1923, Shapurji Sorabji built the Grand Hotel in Mumbai.

The two World Wars brought a fresh lot of hotels to Mumbai, an important port city of the times. The Ritz, The Ambassador, West End, and Airlines, which opened during these years, are fondly referred to as 'war babies' by industry historians.

Modern Era (1947 onwards)

Post-independence, there were big leaps in the hotel trade in the country. The Oberoi Group of Hotels (founded by Raj Bahadur Mohan Singh Oberoi) and the Taj Group of Hotels took over several British properties, maintained high standards of service and quality, and expanded their business overseas. The later decades saw corporate like the ITC (Indian Tobacco Company) also join the hotel industry with properties under ITC WelcomGroup. The year 1955 saw the emergence of Federation of Hotels and Restaurants in India (FHRAI). The federation serves as an interface between the hospitality industry, political leadership, government, international associations, and other stakeholders in the trade.

Famous as Clarks Group of Hotels, UP Hotels and Restaurants Ltd. was established in 13th February, 1961. It started with its flagship hotel Clarks Shiraz Agra.

Realizing the importance and potential of the tourism and hospitality industry, the government constituted India Tourism Development Corporation (ITDC) in 1966, which opened many large and small hotels across the country. The most popular face of ITDC is the Ashoka group of hotels that provides a wide range of hospitality-related services.

Apeejay Surendra Group (Park Hotel) started with its first hotel Park Calcutta in

1967. Over the last few decades, various well-known international hotel chains have come to India. These include Hyatt Hotels and Resorts, InterContinental Hotels and Resorts, Marriott International, Hilton Hotels, Best Western International, Shangri-La Hotels and Resorts, Four Seasons Hotels and Resorts, Carlson Hotels worldwide, Sheraton, Ramada, Sofitel, Meridien, Marriott, etc.

HOTEL INDUSTRY—AN OVERVIEW

Hotels are defined in numerous ways from early times to today. Some of the important definitions of hotels are as follows.

Hotel is defined as a place where a bonafide traveler can receive food and shelter, provided he is in a position to pay for it and is in a fit condition to be received. Hence, a hotel must provide food and beverage and lodging to travelers on payment and has, in turn, the right to refuse if the traveler is drunk, disorderly, unkempt, or is not in a position to pay for the services.

Hotel is defined as an establishment whose primary business is to provide lodging facilities to a genuine traveler and which furnishes one or more of the following services in return of payment from the guest provided that he is in a fit condition to be received:

- Food and beverage service
- Room service
- Uniformed service
- Housekeeping service
- Use of furniture and fixture

Hence, hotel can also be called home multiplied by commercial activities. It is also called 'A house of Taxes and Endless Luxury'.

Hotel refers to a house of entertainment of travelers.

Reader's Digest Dictionary

Hotel is a place where all who conduct themselves properly, and who being able to pay and ready to pay for their entertainment, are received, if there be accommodation for them, and who without any stipulated engagement as to the duration of their stay or as to the rate of compensation, are while there, supplied at a reasonable cost with their meals, lodging and other services and attention as are necessarily incident to the use as a temporary home.

Common Law

A hotel is an establishment held out by the proprietor as offering food, drink and if so required, sleeping accommodation, without special contract to any traveler presenting himself who appears able and willing to pay a reasonable sum for the services and facilities provided and who is in a fit state to be received.

Hotel Proprietors Act, 1956

Building that provides lodging, meals, and other services to the traveling public on a commercial basis.

Encyclopedia Britannica

Hotel or inn is defined as a place where a bonafide traveler can receive food and shelter, provided he is in a position to pay for it and is in a fit condition to be received.

British law

HOTEL AS A SERVICE PROVIDER IN THE MODERN DAY

Gone are the days when people looked upon a hotel for a bed and food. Now it provides almost everything that a guest needs. Hotels have become service providers for the guests in all possible ways. Competition has set into providing these services in order to woo the guests. Every big hotel chain/group has been spending crores of rupees in order to stay in this race.

These services are either wholly owned and run by the hotel, or owned by hotel and run on franchise by experts in the field, or owned and run by outsiders but attached to the hotels. These services complement each other and also help in improving the occupancy rate of the hotels.

Facilities provided in a room

- Round the clock room service (in-room dining)
- Hot and cold running water
- Satellite TV
- Channel music
- Direct dial telephone
- Mini bar—soft drinks, mineral water, beer, chocolate, dry fruits (may be chargeable)
- Tea coffee maker (TCM)
- Safe deposit box/in-room safe
- Hair dryer
- Electronic ving card lock
- Iron with ironing board
- Wi-Fi internet access

Facilities provided in a hotel

- 24-hr coffee shop
- Bars and cocktail lounges
- Restaurant—multi-cuisine/specialty/fine dining
- Discotheque/night club
- Banquet and conference facility
- Beauty parlor/barber shop
- Shopping arcade/retail shop
- Health club, gym, fitness center—indoor heated pool, jacuzzi, sauna, steam bath, spa treatments and massages, etc.
- Swimming pool
- Games/sports—table tennis, cricket, badminton, golf, billiard, snooker, board games such as chess and carom, cycling, mountain-climbing, shooting, jogging track, etc.
- Confectionary shop/pastry shop
- Florist
- ODC
- Book shop

Services offered in 5-star hotel

- Business center—internet access, scanner, facsimile, courier service, personal computers, laptops, mobile phones, portable printers, LCD projector, video conferencing, white boards and flip charts, DVD players, photocopying, typing and printing service, etc.
- Secretarial service
- Interpreter
- Doctor on call
- In-house laundry
- Valet service
- Forex/money changer
- Banking
- Safe deposit
- Baby sitting/child care
- Parking space
- Travel desk—railway/air booking, car rental service, airport transfer
- Centrally air-conditioned
- Gift shop/Souvenir shop
- Chemist shop
- Limousine service
- Laundry and dry cleaning
- Smoking and non-smoking rooms
- Barbeque
- Private chef and butler service
- Chauffer service
- Private helicopter and jet services

Banquet: Hotels provide a wide range of banquet menus, weddings, parties, busine з gathering, all of which help in improving food sales and also work in attracting new customers.

Convention centre: Meetings, seminars, conventions and other social gatherings are arranged which in turn attract group bookings, good occupancy and food service.

Restaurant: Restaurants serving different specialty cuisine like Chinese, Korean, Italian, Continental, Mexican, Indian (South Indian, Mughlai, etc.) are set up by hotels with the interiors suiting those places are run to cater to the different tastes of domestic as well as international tourists. Catering services are also undertaken at off-campus locations.

Secretarial service: This is an essential service for corporate clients. The CEOs, MDs and Chairpersons of different companies need this arrangement for expediting their notes, letters and agreements, communications, etc.

Corporate Service: This works as an extension to the corporate offices of the clients. All the services needed for handling corporate affairs are provided to make the corporate guest feel at—office away from his office. Internet service is the latest addition to the corporate service list. This is available even in small and medium-sized hotels. The guest can connect his laptop to this service and perform his regular work/tasks.

Money changers: Where there is high proportion of foreign tourists/visitors; there will be the need of money changing service. Here the hotel exchanges the foreign currency for local currency. However, it cannot sell foreign currency. This service must function as per the foreign exchange rules and guidelines of the Reserve Bank of India from time to time.

Travel desk: Provides car rental services, air and train ticketing and looks after the other travel needs of the guest. In some hotels Concierge looks after this service too.

Butler service: A personalized service provided to the guest and his visitors exclusively. He provides the food and beverage and also wine service, looks after the visitors to the guest and helps in keeping the room tidy from time to time and also arranges the wardrobe for the guest.

Valet service: A personalized service, but limited to help at car parking and laundry facility, etc. in hotels.

Health club and sports and games: This is provided not only in resort or leisure hotels but also in downtown and commercial hotels. The present corporate guest even though a busy person wants some time out for himself for health and pleasure reasons. Health club, spa, gym, massage parlor, swimming pool, billiards, bowling alley, tennis court, mini golf course are some of the popular facilities. In many hotels, health spas and ayurvedic treatments are provided. Once you participate in the whole body treatment programs, you can eliminate all the disease-causing toxic elements in the body and induce health and vitality to your body and soul. And when you come back, your mind and body are rejuvenated and they are ready to take on challenges that your lifestyle gives to them.

REVIEW QUESTIONS

1. What do you understand by the hospitality industry? Discuss its constituents.
2. Define the term 'Hotel'. Discuss the various types of facilities and services provided by a five star hotel to guests for repeat business.
3. Outline the evolution and growth of the hotel industry, giving important developments, dates and people.
4. Briefly explain the growth of the hotel industry in India.
5. Discuss historical developments and technological advances in hospitality industry.
6. Define a hotel as per British law.
7. Trace the growth of hotels in the following countries:
   - London
   - France
   - America
8. Trace the origins of the developments of Inns in/ after the 6th century BC.

Hotel Guest and Guest Relation

Learning Objectives

After reading this chapter, you will be able to understand the following:

- Guest classification
- Guest expectation
- Insight into guest satisfaction and dissatisfaction
- Customer relations

WHO IS A GUEST?

A hotel customer or client is called a guest because the hotel offers homely and professional service to him, and establishes an intimate relationship with him during his stay. Hotel guests may be defined as—people who have, who are, or who will be availing the services of hotel, for a particular period in order to satisfy their demand for accommodation, food and beverage or entertainment, for which they are willing and able to pay.

A hotel guest is the one who buys rooms, meeting space, food and beverage services and avails services and facilities from the hotel.

He is an important person to the hotel and is not just a cold statistics. He is someone who, in his own mind, is always right, and is the reason for business. He is a valued customer who is willing to pay fair price for a quality product and wants to be neither overcharged nor undeserved.

Guest relation is an integral part of the hotel industry. It is a means for the management to reach out to the guest and convey him the feeling of warmth and welcome.

Guest Classification

A guest may be classified on the following basis:

On the basis of presence in the hotel
- Expected guest
- In-house guest
- Checked-out guest

On the basis of recognition
- Regular guest
- VIP
- Special attention (SPATT)
- Distinguished guest (DG)
- Handle with care (HWC)
- Commercially important person (CIP)
- Politically important person (PIP)
- New guests

On the basis of revenue
- Paying guests
- Complimentary guests

On the basis of country of origin
- *Domestic traveler*: He is a resident of a country and travels within it.

- *International traveler*: He is a resident of a foreign country who visits another country.

On the basis of purpose of travel

- Business guest
- Leisure guest

GUEST EXPECTATION

A guest coming to a hotel expects a unique and attractive property with friendly, courteous and competent service. He expects clean and well-appointed rooms and public areas as per the standard, safe quality food with personal touch and safety to him and his belongings. He also expects facilities and amenities in his room. He appreciates hassle free and speedy check-in and check-out. Further, some added frills such as efficient room service, newspaper, shoe shine, laundry, valet service, car parking, etc. are also expected by guest.

The personality of the person coming in contact with the guest is very important. He should have friendly and positive attitude, i.e. he should be tolerant, willing to accept the responsibility and should be people-oriented.

GUEST SATISFACTION

A hotel guest's significance can be seen from the fact that all activities and innovations in a hotel revolve around one term—guest satisfaction. Basically guest needs can be classified as follows:

- *Physiological needs*: Related to protection against heat, cold, rain, etc. in the form of room and related to his appetite/thirst and special food requirement.
- *Economic needs*: Value for money, i.e. is he getting returns worth the value of money spent by him? How much can he afford?
- *Social needs*: Attending functions and gatherings, to meet others, going out with friends.
- *Psychological needs*: Enhancement of self-pride, the need for variety.

GUEST DISSATISFACTION

This can be usually because of two reasons classified as follows:

- **Under the control of the hotel**: Such as unhelpful staff, cramped conditions and lack of choice of product, i.e. type of accommodation and other service available for offer to the guest.
- **Beyond the control of the front office of the hotel**: Factors such as location, the weather, other customers and transport problems.

Guest satisfaction shall ensure repeat business, increased clientele and maximized revenue for the establishment.

CUSTOMER RELATIONS

Better known as *guest relation* in hotel industry, it undoubtedly cannot be anything but providing hospitality to the guest of the hotel. It can be achieved by using procedures which will improve guest sales factor. It is important in this connection that we understand the basic principle that service is a two-sided coin and we must also understand their marketing concept which involves identification and satisfaction of guest needs.

Service is a Two-sided Coin

There are two sides to service: Operational side and people side.

Most of the hospitality operations do understand the importance of providing basic levels of physical services such as good food (food to be served hot, etc.), clean rooms and standard facilities, etc.

But it is the 'people side' where the major shortfall tends to be. For a serious understanding of marketing concept and hence satisfaction to guest, it may be said that the service provider (staff) is some kind of an amateur psychologist. Human needs can be arranged according to priorities. Needs like hunger and thirst come first and must be taken care of first and other needs follow this. Coldness, rudeness or being ignored leaves some powerful human needs—the needs to be accepted, to belong to, feel secure, to have status, to have recognition, and to be wanted—unsatisfied.

Delivery of Psychological Service

It requires understanding of the following:

- *Recognition of customer's needs*: It is vital to pay attention to customers and to demonstrate that you are paying attention. Aim at individualizing the customer, for example, address the guest by his name.
- *Smile*: Smile is contagious like anger. No smile goes waste. No matter how bad you feel, you must keep smiling. Staff must be friendly. Hire right staff with right attitude for your customers.
- *Think positive*: Make sure your staff has positive attitude. They take their lead from you. Treat your staff in a friendly and fair

way. Do not discuss your problems with your guests; it does not help in building a rapport of your establishment.

- *Genuineness*: Genuineness cannot be preached and taught. It comes from within like positive attitude. Wherever possible address the guest by name. Genuineness means accepting the guest for what he is without judgment or censor.
- *Customer's expectation*: It is a very important because normally customer dissatisfaction is not related to the actual level of service provided but to the fact that the customer did not get what he expected and perceived.

REVIEW QUESTION

1. Discuss the term 'guest satisfaction', 'guest expectations' and 'customer relations'.

Classifications of Hotels

Learning Objectives

After reading this chapter, you will be able to understand the following:

- Accommodation industry
- The need and the criteria for the classification of hotels
- Classification of hotels—on the basis of star ratings, size, location, type of clientele, duration of stay, level of service, and ownership, management and affiliation
- Alternative accommodation
- Intermediate accommodation

ACCOMMODATION INDUSTRY

Accommodation industry or lodging industry can be defined as a dimension of the hospitality industry which strives on the vision and mission of providing lodging services to the bonafide travelers. Figure 10.1 presents an overview of the accommodation industry.

Fig. 10.1: An overview of the accommodation industry

Many countries have recognized the vital importance of accommodation industry in relation to tourism and their governments have taken initiatives by providing attractive incentives and concessions in the form of long-term loans, liberal import licenses and tax relief, cash grants for construction and renovation of buildings, and similar other concessions to the accommodation industry.

THE NEED AND THE CRITERIA FOR THE CLASSIFICATION OF HOTELS

Since the hotel industry comprises numerous hotels providing different services and facilities, it lacks a single identity. In order to identify themselves in the vast hotel market, hotels need to place themselves into a specific group so as to develop its share of the particular market based on its characteristics and assets and gain more and more of its recognition. Though hotels provide with the basic concept of hospitality, food and accommodation, placing them into particular groups is not easy because of the diversity of services, facilities and amenities provided or not provided. Even though most hotels do not fit into any specific well-defined category, several general classification criteria do exist based on:

- Star ratings
- Size
- Location
- Type of clientele (target market)
- Length of stay/duration of stay/visitors stopover

- Level of services
- Ownership, management and affiliation

In a nutshell, hotel classification serves the following purposes:

- Lends uniformity in services and sets general standards of a hotel.
- Provides an idea regarding the range and type of hotels available within a geographical location.
- Acts as a measure of control over hotels with respect to the quality of services offered in each category.

CLASSIFICATION OF HOTELS

Table 10.1: Classification of hotels

Star ratings (by HRACC)	Size	Location	Clientele (target markets)	Length of stay	Level of services	Ownership, management and affiliation	Alternative accommodation
One Star	Small	Downtown	Commercial	Transient	Upmarket	Independent	Sarai
Two Star	Medium	Suburban	Transient	Semi-residential	Mid-market	Chain	Dharamshala
Three Star	Large	Airport	Suite	residential	Budget	Management	Dak bungalow
Four Star	Very	Resort	Residential	Residential		contract	Circuit house
Five Star	large	Motel	Bed and			Franchise	Lodge
Five Star Deluxe		Boatel	breakfast			Referral	Youth hostel
Heritage		Floatel	Timeshare			group	Yatri niwas
Heritage Classic		Rotel	Condominium				Forest lodge
Heritage Grand			Casino				Dormitory
			Conference				Railway retiring room
							Guest house

On the Basis of Star Ratings

This system is one of the most commonly understood, accepted and recognized system in India. With the aim of providing standardized, world class services to the tourists, the Government of India, Department of Tourism classifies functioning hotels under the star system, into five categories from *1-star to 5-star deluxe and heritage hotels*. For this purpose a permanent committee, the Hotel and Restaurant Approval and Classification Committee (HRACC) has been set up which inspects and assesses the new and functioning hotels and classifies them based on services and facilities offered. The committee also inspects once in 5 years, the existing hotels

which have been classified to confirm that the hotels are sticking to the prescribed standards. Based on the recommendations of the HRACC, deserving hotels are awarded the appropriate star category and are placed on the approved list of the department.

Approved hotels become eligible to various fiscal reliefs and benefits. The department intercedes on behalf of such hotels whenever necessary to ensure that their needs get priority consideration from various concerned authorities. These hotels also get worldwide publicity through tourist literature published by the Department of Tourism and distributed by the Government of India Tourist Offices in India and abroad. Approved hotels become eligible for obtaining foreign exchange for their import of essential equipment and provisions and for their overseas advertising, publicity and promotion under the Hotel Incentive Quota Scheme.

Eligibility for Star Classification

To be eligible to apply for classification, a functioning hotel must fulfill the following minimum basic requirements:

- The hotel must have at least 10 lettable bedrooms.
- Carpet areas in respect of rooms and areas of bathrooms should by and large adhere to the following limits as given in Table 10.2.

Table 10.2: Eligibility for star classification	
Category of hotel	Area standards for bedroom/bathroom
5-Star/5-Star Deluxe Hotels	
Single room	180 sq ft
Double room	200 sq ft
Bathrooms	45 sq ft
4-Star and 3-Star Hotels	
Single room	120 sq ft
Double room	140 sq ft (extra area may be provided if twin beds are to be provided)
Bathrooms	36 sq.ft
2-Star and 1-Star Hotels	
Single Room	100 sq ft
Double Room	120 sq ft (all rooms should have proper ventilation and ceiling fans)
Bathrooms	30 sq ft or subject to local by laws

Failure to satisfy these conditions will disqualify a hotel for consideration.

Procedure of Classification of Hotels

For classification, hotels need to apply to the Regional Director of the concerned Government of India Tourist Office at Delhi, Mumbai, Kolkata, or Chennai for 1-, 2-, and 3-star category classifications, and to the Additional Director General of Tourism, Govt. of India/Chairperson (HRACC) or a representative nominated by him in the case of 4-, 5-, 5-Star deluxe and heritage category classifications. The application fee (by means of a demand draft drawn in favor of the Pay and Accounts Officer, Department of Tourism, New Delhi) which varies from 1-star to 5-star deluxe is also sent along with the application. The hotel is supposed to fill up a questionnaire containing details of facilities, features, amenities, and services and their standards in the form of questions which are divided into 3 categories—desirable, necessary and essential—and have marks allotted to them for a hotel to be considered for classification. All items marked under essential category must be there in the hotels; two items marked under necessary may be deferred while the items marked under desirable may or may not be there.

- *Essential*: Features those are vital to hotel operations. The absolute basic on which there can be no compromise.
- *Necessary*: Features those are essential but may be deferred.
- *Desirable*: Features those are not necessary but desirable for enhanced guest services.

On a pre-notified date the HRACC team members visit the hotel and inspect the hotel, and finally on the basis of the report of the committee and the marks scored by the hotel, the hotel is either approved or rejected for the applied star category.

Composition of HRACC

The members of HRACC comprises:

- Officials from Central and State Tourism Ministry
- Members of FHRAI (Federation of Hotel and Restaurant Association of India)
- Members of HAI (Hotel Association of India)
- Members of IATO (Indian Association of Tour Operators)

- Members of TAAI (Travel Agents Association of India)
- Principal of the nearest IHM (Institute of Hotel Management Catering Technology and Applied Nutrition)
- In case of Heritage category, a representative of IHHA (Indian Heritage Hotel Association)

Classification Criteria of Star Hotels

The general features, facilities and services expected of hotels in the different star categories are broadly described below.

Five Star Deluxe Category

A hotel which applies for 5-star deluxe category has basically the same number of features as a five star hotel but is superior in quality of service, amenities and facilities, etc. to a five star hotel's requirement.

Five Star Category

General features: The façade, architectural features and general construction of the hotel building should have the distinctive qualities of a luxury hotel of this category. The locality, including the immediate approach and environs should be of highest standards and should be suitable for a luxury hotel of this category. There should be adequate parking space for cars. The hotel should have at least 25 lettable bedrooms, all with attached bathrooms with long baths or the most modern shower chambers. All public rooms and private rooms should be fully air-conditioned and should be well equipped with superior quality carpets, curtains, luxurious furniture of high standards, fittings, etc. in good taste. It would be advisable to employ the services of professionally qualified and experienced interior designers of repute for this purpose. There should be an adequate number of efficient lifts in the building of more than two storey (including the ground floor) with 24 hours service. There should be a well-appointed lobby and ladies' and gentlemen's cloakrooms equipped with fittings and furniture of the highest standard, adequate parking space and swimming pool.

Facilities: There should be a reception, cash and information counter attended by highly qualified, trained and experienced personnel and conference facilities in the form of one each or more of the conference rooms, banquet halls and private dining rooms. There should be a shopping arcade and bookstall, beauty parlor, barber shop, recognized travel agency, money change and safe deposit facilities, left luggage room, florist and a shop for toilet requisites and medicines on the premises. There should be a telephone in each room and house phone for use of guests and visitors and provision for a radio or relayed music and a TV set in each room. There should be a well-equipped, well-furnished and well-maintained dining room/restaurant on the premises and wherever permissible by law, there should be an elegant, well-equipped bar/permit room. The kitchen, pantry and cold storage should be professionally designed to ensure efficiency of operation and should be well-equipped. There should be dancing facility and orchestra in dining hall.

Services: The hotel should offer both International and Indian Cuisine and the food and beverage service should be of the highest standard. There should be professionally qualified, highly trained, experienced, efficient and courteous staff in smart, clean uniforms, and the staff coming in contact with guests should understand English. The supervisory and senior staff should possess good knowledge of English and staff knowing at least one international language should be rotated on duty at all times. There should be 24-hour service for reception, information and telephones. There should be provision for reliable laundry and dry cleaning services. Housekeeping, at the hotel, should be of the highest possible standard and there should be plentiful supply of linen, blankets, towels, etc. which should be of the highest quality available. Each bedroom should be provided with a good vacuum jug/thermos flask with ice cold, boiled drinking water except where centrally chilled purified drinking water is provided. Glassware, cutlery, silver, tableware and all necessary accessories should be of best quality and standard.

Four Star Category

General features: The façade, architectural features and general construction of the building should be distinctive and the locality including the immediate approach and the environs should be suitable for a hotel of this category. There should be adequate parking facilities for cars. The hotel should have at least 25 lettable bedrooms, all with attached bathrooms. At least 50% of the bathrooms must have long baths or the most modern shower chambers, with 24-hour service of hot and cold running water. All public rooms and private rooms should be fully air-conditioned and should be well furnished with carpets, curtains, furniture, fittings, etc. in good taste. It would be advisable to employ the services of professionally qualified and experienced interior designer of repute for this purpose. There should be an adequate number of efficient lifts in building of more than 2 storey (including the ground floor). There should be a well-appointed lobby and ladies' and gentlemen's cloakrooms equipped with fittings of standard befitting a hotel of this category.

Facilities: There should be a reception, cash and information counter attended by trained and experienced personnel. There should be a bookstall, recognized travel agency, money changing and safe deposit facilities and a left luggage room on the premises. There should be a telephone in each room and house phone for use of guests and visitors and provision for a radio or relayed music in each room. There should be a well-equipped, well-furnished and well-maintained dining room/restaurant on the premises and wherever permissible by law, there should be an elegant,

well-equipped bar/permit room. The kitchen, pantry, cold storage should be professionally designed to ensure efficiency of operation and should be well equipped.

Services: The hotel should offer both International and Indian cuisine and food and beverage service should be of the highest standards. There should be professionally qualified, highly trained, experienced, efficient and courteous staff in smart, clean uniform and the staff coming into contact with the guests should possess good knowledge of English. It will be desirable for some of the staff to possess knowledge of foreign language and staff knowing at least one international language should be rotated on duty at all times. There should be 24 hours service for reception, information and telephones. There should be provision of reliable laundry and dry cleaning services. Housekeeping at the hotel should be of the highest possible standard and there should be plentiful supply of linen, blankets, towels, etc. that should be of the highest quality available. Similarly, the cutlery and glassware should be of the highest quality available. Each bedroom should be provided with a vacuum jug/flask with ice cold, boiled drinking water. There should be a special restaurant/dining room where facilities for dancing, orchestra are provided.

Three Star Category

General features: The architectural features and general construction of the building should be of a very good standard and the locality, including the immediate approach and environs should be suitable for a very good hotel and there should be adequate parking facilities for cars. The hotel should have at least 20 lettable bedrooms, all with attached bathrooms with bathtubs and/or showers and should be modern in design and equipped with fittings of a good standard, with hot and cold running water. At least 50% of the rooms should be air-conditioned and the furniture and furnishings such as carpets, curtains, etc.

should be of a very good standard and design. There should be adequate number of lifts in the building with more than two storey (including the ground floor). There should be a well-appointed lounge and separate ladies' and gentlemen's cloakrooms equipped with fittings of a good standard.

Facilities: There should be a reception and information counter attended by qualified and experienced staff, and a bookstall, recognized travel agency, money changing and safe deposit facilities on the premises. There should be a telephone in each room (except in seasonal hotels where there should be a call bell in each room and a telephone on each floor for the use of hotel guests) and a house phone for the use of guests and visitors to the hotel. There should be a well-equipped and well-maintained air-conditioned dining room/restaurant and wherever permissible by law, there should be a bar/permit room. The kitchen, pantry and cold storage should be clean and organized for orderliness and efficiency.

Services: The hotel should offer good quality cuisine both Indian as well as continental, and the food and beverage service should be of good standard. There should be qualified, trained, experienced, efficient and courteous staff in smart and clean uniforms and the supervisory staff coming in contact with the guests should understand English. The senior staff should possess good knowledge of English. There should be provision for laundry and dry cleaning service. Housekeeping at the hotel should be of a very good standard and there should be adequate supply of linen, blankets, towels, etc. of good quality. Similarly, cutlery, crockery, glassware should be of good quality. Each bedroom should be provided with vacuum jug/thermos flask with cold, boiled drinking water. The hotel should provide orchestra and ballroom facilities and should attempt to present specially choreographed Indian dance.

Two Star Category

General features: The building should be well constructed and the locality and environs including the approach should be suitable for a good hotel. The hotel should have at least 10 lettable bedrooms, all with attached bathrooms with showers and should be with modern sanitation and running cold water with adequate supply of hot water, soap and toilet papers. At least 25% of the rooms should be air-conditioned (where necessary there should be heating arrangements in all the rooms) and all rooms must be properly ventilated, clean, and comfortable with all the necessary items of furniture. There should be a well-furnished lounge.

Facilities: There should be a reception counter with a telephone. There should be a telephone or call bell in each room and there should be a telephone on each floor unless each room has a separate telephone. There should be a well-maintained and well-equipped dining room/restaurant serving good, clean, wholesome food and a clean, hygienic and well-equipped kitchen and pantry.

Services: There should be experienced, courteous and efficient staff in smart and clean uniforms. The supervisory staff coming in contact with guests should understand English. There should be provision for laundry and dry cleaning services. Housekeeping at the hotel should be of a good standard and good quality linen, blanket, towels, etc. should be provided. Similarly, crockery, cutlery and glassware should be of a good quality.

One Star Category

General features: The general construction of building should be good and locality and environs including immediate approach should be suitable. The hotel should have at least 10 lettable bedrooms, all with attached bathrooms with showers and should have western style WC. All bathrooms should have modern sanitation and running cold water with adequate supply of hot water, soap and toilet paper. The rooms should be properly ventilated and should have clean and comfortable bed and furniture.

Facilities: There should be a reception counter with a telephone and a house phone for the use of guests and visitors. There should be a clean and moderately well equipped dining room/restaurant serving clean, wholesome food and there should be a clean, well-equipped kitchen and pantry.

Services: There should be experienced, courteous and efficient staff in smart and clean uniforms and the senior staff coming in contact with guests should possess working knowledge of English. Housekeeping at the hotel should be of a good standard and clean and good quality linen, blankets, towel, etc. should be supplied. Similarly, crockery, cutlery and glassware should be of good quality.

Heritage Hotels

Heritage hotels are properties set in palaces, castles, forts, havelis, hunting lodges, the mansions of erstwhile royal and aristocratic families. They have added a new dimension to cultural tourism. They have a very grandeur atmosphere, giving you the feeling of living in a bygone era of Kings and Queens. In India, many such palaces, castles, forts, havelis, hunting lodges have been converted into hotels to preserve their rich culture and heritage. The façade, architectural features and general construction should have the distinctive qualities and ambience in keeping with the traditional way of life of the area. Any extension, improvement, renovation, change in the existing structure should be in keeping with the traditional architectural styles and constructional technique harmonizing the new with the old.

After expansion/renovation, the newly built up area added should not exceed 50% of the total built up (plinth) area including the old and new structures. *Example—Taj Lake Palace (Udaipur), Taj Umaid Bhawan Palace (Jodhpur), Rambagh Palace (Jaipur).*

According to the Ministry of Tourism, the heritage hotels are further classified as follows:

Heritage: This category will cover hotels in residences/havelies/hunting lodges/castles/forts/palaces built between 1935 and 1950. The hotel should have a minimum of 5 rooms (10 beds).

Heritage classic: This category will cover hotels in residences/havelies/hunting lodges/castles/forts/palaces built between 1920 and 1935. The hotel should have a minimum of 15 rooms (30 beds).

Heritage grand: This category will cover hotels in residence/havelies/hunting lodges/castles/forts/palaces built prior to 1920. The hotel should have minimum of 15 rooms (30 beds).

Room and Bath Size

No room or bathroom size is prescribed for any of the categories. However, general ambience, comfort and imaginative re-adaptation would be considered while awarding sub-classification 'classic' or 'grand'.

Special Features

Heritage: General features and ambience should conform to the overall concept of heritage and architectural distinctiveness.

Heritage Classic: General features and ambience should conform to the overall concept of heritage and architectural distinctiveness. The hotel should provide at least one of the under mentioned sporting facilities.

Heritage Grand: General features and ambience should conform to the overall concept of heritage and architectural distinctiveness. However, all public and private areas including rooms should have superior appearance and decor. At least 50% of the rooms should be air-conditioned (except in hill stations where there should be heating arrangements). The hotel should also provide at least two of the under mentioned sporting facilities.

Sporting Facilities

Swimming pool, health club, lawn tennis, squash, riding, golf course, provided the ownership vests with the concerned hotel. Apart from these facilities, credit would also be given for supplementary sporting facilities such as golf, boating, sailing, fishing or other adventure sports such as ballooning, parasailing, wind-surfing, safari excursions, trekking, etc. and indoor games.

Cuisine

Heritage: The hotel should offer traditional cuisine of the area.

Heritage Classic: The hotel should offer traditional cuisine but should have 4 to 5 items which have close approximation to continental cuisine.

Heritage Grand: The hotel should offer traditional and continental cuisine.

Bar

In the case of Heritage Grand and Heritage Classic, bar is necessary and desirable in the case of Heritage.

On the Basis of Size

Classification on the basis of size refers to the number of guest rooms and should not be confused with the building height or the area of property or the sales/turnover.

On the Basis of Location

Downtown Hotel/City–Center Hotel

These are located in the heart of the city, within a short distance of the shopping areas, malls and multiplex, theatre, business center

Table 10.3: Classification of hotels on the basis of size

Sizes	American standards	Indian standards
Small	Under 150 rooms	Under 25 rooms
Medium	150–300 rooms	25–100 rooms
Large	300–600 rooms	100–300 rooms
Very large	More than 600 rooms	More than 300 rooms

etc. Room rates in these hotels are normally high due to their locational advantage. They are generally preferred by business clientele as they find it convenient to stay close to the place of their business activities. They have restaurants, coffee shop, bar, discotheque, health club, swimming pool, 24 hours room service, laundry and offer uniformed services and business center with secretarial, fax, internet and photocopying facilities along with conference rooms, cocktail lounges. *Example—Taj Palace Hotel (New Delhi), Oberoi Grand (Kolkata), Hotel Le Meridien (Pune).*

Suburban Hotels

The suburban hotels as the name suggests are hotels which are located on the outskirts of city. These hotels provide extensive accommodation and food and beverages facilities and have huge conference room and seminar halls. These are ideal for the guests who want peaceful and quieter surroundings and prefer to stay away from the hustle and bustle of a city. *Example—Jaypee Palace (Agra), Trident Hotel (Gurgaon).*

Airport Hotels

These hotels are located in the close vicinity of international and domestic airports of major cities. They cater primarily to the passengers with cancelled and delayed flights and airlines crew members. Generally the guests in these hotels stay for a very short duration. These hotels have well furnished guest rooms with restaurants and coffee shops and offer various

भारतपर्यटन, मुंबई
(पर्यटन मंत्रालय, भारत सरकार)

सत्यमेव जयते

INDIATOURISM, MUMBAI
(Ministry of Tourism, Govt. of India)

ISO 9001 : 2008 Certified
BN 2262 /1659 : 0410

No. ITM(245)(HRACC)(536)/2013   28.07.2014

The recommendation of the Regional Classification Committee of HRACC (WR) in respect of classification of hotels for star category has been examined by the Ministry of Tourism, Government of India, whose decision is hereby announced as under:

THREE STAR (***)		W.E.F.
La Paz Gardens	Classification	15.07.2014
(A Unit of La Paz Gardens Pvt. Ltd.)	5 years	to
Swatantra Path, Vasco Da Gama	(Rooms-65)	14.07.2019
Goa - 403 802		

2.   The above hotel has been awarded the said rating for a period of five years with effect from the date mentioned above, subject to the condition that the management of the hotel concerned should at all times comply with all the regulatory conditions for classification of hotels and other "Terms and Conditions" to be introduced by the Ministry of Tourism, Government of India from time to time.

3.   The hotel would require to apply for re-classification, within six months prior to the expiry of this classification as per the Terms and Conditions laid down by the Ministry of Tourism, Govt. of India.

(Vikas Rustagi)
Regional Director
Western & Central India

Copy to :

i.   The Asstt. Director General (H &R), Ministry of Tourism, Govt. of India, C-1 Hutments, Dalhousie Road, New Delhi-11011

ii.   The President, Hotel and Restaurant Association (Western Region),4,Candy House, Mandalik Road, Colaba, Mumbai-400 001

iii.   The President, Travel Agent Association of India (Western Region),2-D, Lawrence & Mayo House, 276 Dr. D.N.Road, Fort, Mumbai-400001

iv.   The Chairman (Maharashtra Chapter), Indian Association of Tour Operator , C/o Garha Tours & Travels Pvt. Ltd., 104 Atlantic Apartments, A-Wing, Lokhandwala Complex, Swami Samarth Nagar, Andheri (W),Mumbai-400053

(Vikas Rustagi)
Regional Director
Western & Central India

१२३, महर्षि कर्वे मार्ग, मुंबई-४०० ०२०. / 123, Maharshi Karve Road, Mumbai - 400020
ई-मेल/E-mail : touristoffice-mum@nic.in, indiatourism-mum@nic.in, regdir.indtour@gmail.com
फैक्स /Fax : 91-22-2201 4496 ● टेलिफोन /Telephone : 2203 3144, 2203 3145, 2207 4333, 2207 4334
वेबसाइट / Website : www. incredibleindia.org / www. tourism.gov.in
अतुल्य ! भारत

BY SPEED POST

Government of India
Ministry of Tourism
(HRACC)

C-1, Hutments
Dalhousie Road
New Delhi - 110011
Tele/Fax: 011-23012810

No. 23- HRACC (06)/2013   Dated 23.07.2013

To,

The General Manager
Jaypee Palace Hotel
Fatehabad Road
Agra - 282004

Sir,

The recommendation of the Hotel & Restaurants Approval and Classification Committee (HRACC) in respect of **Re-Classification** of the hotel has been examined by the Government, whose decision is hereby announced as under:

2.	**FIVE(*****) STAR DELUXE CATEGORY**		**PERIOD**
	Jaypee Palace Hotel	Re-Classification	15.09.2013
	Fatehabad Road	Five years	to
	Agra – 282004	Rooms – 341	14.09.2018

3. The rating has been awarded to the hotel as mentioned above for a period of five years, subject to the condition that the management of the hotel should at all times comply with all the regulatory conditions for classification of hotels and other terms and conditions introduced by this Ministry from time to time.

4. The hotel will apply for Re-classification six months before the expiry of this approval on the Terms and Conditions laid down in this Ministry's Circular No. 8-TH.I (03)/ 07 Vol./ IV dated June, 2012 as and when due.

5. The **Re-Classification** of "Jaypee Palace Hotel, Fatehabad Road, Agra - 282004" by the Ministry of Tourism is no substitute to other NOCs / permissions / clearances which shall be taken by the Hotel.

Yours faithfully,

Singh

(S. V. Singhh)
Assistant Director General (H&R)
& Member Secretary (HRACC)

(एस. वी. सिंह / S. V. SINGH)
सहायक महानिदेशक / Asstt. Director General
पर्यटन मंत्रालय / Ministry of Tourism
भारत सरकार / Government of India
नई दिल्ली / New Delhi

other facilities such as airport pick up and drop through the hotel owned cars and buses which is very important for passenger and crew members.

The occupancy rate is usually very high going up to more than 100%, since rooms can sometimes be sold more than once in a day. *Example—Airport Ashok (Kolkata), Hotel Centaur (New Delhi), JW Marriott Hotel (New Delhi).*

Resorts

These cater to the needs of holiday makers, honeymooners, pilgrims who want to relax and enjoy themselves in exotic locations such as hill station, near the sea beach, island and other areas abounding in natural beauty. They also cater to those people who by reasons of health desire a change of atmosphere. Therefore, resorts may further be classified as health resorts, hill resorts, beach resorts, summer resorts, winter resorts, etc. depending on their location.

The resorts have various entertainment and recreational facilities like beauty contest, fashion show, fancy dress competition as well as tennis courts, golf course, badminton, billiards, amusement parks, etc. Usually guests stay for a week or even more in a resort.

A resort comprises individual bungalow, cottages or small blocks of houses spaced up. Since resorts are usually spread out over a large area, commuting from one cottage to another takes more time and effort. Most resorts work to full capacity only during seasonal periods and hence undergo fluctuations in sales revenue from season to season. There are peak seasons and off seasons. In order to boost business during the off season, attractive discounts and packages are offered to the guests. *Example—Vivanta by Taj-Fort Aguada (Goa), The Oberoi Vanyavillas (Ranthambhore)*

Motels

The term motel is an American concept and has originated from the word 'Motor Hotel'. These hotels are located primarily on highways or at road junctions and mostly cater to the various passerby on the highways. These hotels are not very large and generally have a fewer number of rooms and provide modest lodging facility. They do provide parking space, garage, service station and washing facilities. The length of stay is usually overnight and they may be located near the petrol pump.

Boatels

A houseboat hotel is referred to as a boatel. The 'Shikaras' of Kashmir and the 'Kettuvalams' of Kerala are houseboats that offer small but luxurious accommodation.

Floatels / Floating Hotel

These are cruise liners taking passengers on a weeklong trip around famed locations connected with water. The ships are five star hotels with every conceivable luxury including several restaurants, suites, ballrooms, shopping arcades, etc. Guests are served by well-trained and talented personnel. Such hotels provide exclusive and exotic atmosphere. *Example— The Oberoi Motor Vessel Vrinda Cruise (Kerala), Sunderbans Luxury Cruise (Kolkata), M.V. Mahabaahu Cruise (Assam).*

Rotels

These are hotels on wheels. They are normally used by a small group of travelers to visit various places by road. Facilities include centrally air-conditioning, wall to wall carpeting, channel music, well stocked bar, restaurant serving different cuisine. The guests are fed traditional Indian food of royalty and served by liveried waiting staff. *Example—Palace on Wheel, Orient Express, Deccan Odyssey.*

On the Basis of Clientele (Target Markets)

Target markets are distinctly defined groups of people which the hotel aims to retain or attract as guests. These are distinctly defined groupings of potential buyers (market segments) at which the hotels aim or target their marketing efforts.

Commercial Hotel / Business Hotel

These hotels primarily cater to businessmen and commercial executives. They are situated in the heart of the city in busy commercial areas so as to get increased business.

Generally duration of stay is a few days only and weekend business is slack. Best possible facilities of high standards are provided in these hotels. *Example—Taj Mahal Palace (Mumbai), The Oberoi (New Delhi), ITC Maurya (New Delhi).*

Transient Hotels

They cater to the needs of people who are on the move and need a stopover en route their journey. Located in the close proximity of ports of entry, such as sea port, airport, and major railway stations, these hotels are normally patronized by transient travelers. They cater primarily to the passengers with cancelled and delayed flights and airlines crew members. Generally the guests in these hotels stay for a very short duration. These hotels have well furnished guest rooms with restaurants and coffee shops and offer various other facilities such as airport pick up and drop through the hotel owned cars and buses which is very important for passenger and crew members.

The occupancy rate is usually very high going up to more than 100%, since rooms can sometimes be sold more than once in a day. *Example—Hilton Hotel (Mumbai), Swissotel (Kolkata).*

Suite Hotels

Suite hotels are among the newest and fastest-growing segments of the lodging industry.

These hotels according to the name have either all suite rooms or majority of the rooms as suites. Some suites include a compact kitchenette complete with utensils and refrigerator, microwave oven and a sink for washing dishes. They have fewer guest services and limited public areas than other hotels. Suite hotels appeal to several different market segments. People who are relocating transform suites into temporary living quarters; frequent travelers enjoy the comforts of a 'home-away-from-home' and vacationing families discover the privacy and convenience of non-standard hotel accommodation designed for extended stays. Professionals such as accountants, lawyers and executives find suite hotels attractive as they get two rooms instead of one and can work and entertain in an area which is separate from the bedroom. The privacy of the guest is maintained. *Example—Om Niwas Suite Hotel (Jaipur), Fraser Suites (New Delhi).*

Residential Hotel / Apartment Hotel

These hotels can be described as an apartment house complete with hotel services. They provide accommodation for a longer duration. These hotels are generally patronized by people who are on a temporary official deputation to a city where they do not have their own residential accommodation. The tariff in these hotels is charged on monthly, half yearly or yearly basis. Advance room charges are usually collected while other charges are billed weekly. Almost all residential hotels operate a restaurant, offer telephone service and laundry service. *Example—Ramee Hotel Apartments (Dubai), The Emerald Hotel & Executive Apartments (Mumbai), The Treehouse Blue (Goa).*

Bed and Breakfast Hotels

These are usually converted residences having 20–30 rooms. The owner usually lives on the premises. Most B&B hotels offer only lodging and limited board, or, as the name implies, breakfast only. Main meals are not served. Since meeting rooms, laundry, recreational facilities are usually not offered, the tariff is generally lower than full-service hotels. They are suitable for budget travelers. *Example— Bentleys Hotel (Mumbai), Treebo Midaas Comfort (Mumbai).*

Timeshare Hotel

These hotels are also referred to as *vacation-interval* hotels. The concept of timeshare hotels was introduced in Europe in the year 1970. According to this concept, individual guest will purchase the ownership of a particular unit of a hotel for a particular time slot in a year (*typically one week, and almost always the same time every year*) and will occupy the unit during that period or rent the unit to other vacationers if they cannot avail the facility. It is a popular choice for persons who wish to secure a long-term commitment to a particular location. Purchaser has to make a one-time payment for the time slot and an annual fee to cover the maintenance costs and related expenses. Thus, in timeshare hotels, there are multiple owners for a single unit of a hotel.

Let us suppose that there is a hotel in Goa with twenty rooms. The various rooms of the hotel can be sold to different people for different periods of time for a specific number of years. The total number of one week slots may be calculated as under:

No. of one week slot owners

= No. of guest rooms × No. of weeks in a year

= 20 × 52

= 1040

Thus, the same property can be sold to 1040 individual owners for specific time slots during the year. These individuals are the owners of the room for that time duration.

Types of Timeshare

Fixed week ownership:

- The purchaser is entitled to use the resort's facilities at a set week and every year for the length of the agreement.
- The resort will have a calendar enumerating the weeks starting with the first calendar week of the year. An owner may own a deed to use a unit for a single specified week.

Floating:

- In the 1970s and 1980s, consumers began demanding more flexibility in how they could use their purchase, and the industry answered this concern with a 'float week' offering. Simply, a floating week offering

means that the consumer is entitled to resort access rights within a specified range of weeks within a calendar year or as specified within the contract.

- An example of this may be a floating summer week where the owner may request any week during the summer season. Some floating contracts exclude major holidays so they may be sold as fixed weeks.

Rotating:

- The units are sold on rotating basis commonly referred to as *flex weeks*.
- In an attempt to give all owners a chance for the best weeks, the weeks are rotated forward or backward through the calendar.
- One year the owner may have use of week 25, then week 26 the next year and then week 27 the year after that.
- This method does give each owner a fair opportunity for prime weeks but it is not flexible.

Vacation clubs:

- Major international hotel chains such as Hilton, Accor and Marriott have introduced their own Vacation Ownership Programs which are based on point systems.
- The consumer simply purchases enough points to satisfy his annual vacation needs.
- They are sold both as deeded and with right to use the club's services for a certain number of years.
- There are also vacation clubs that may own units in multiple resorts in different locations, offering services to a private customer base for exclusivity.

Points programs: Points programs annually give the owner an amount of points equal to the level of ownership. These points can then be used to make travel arrangements within the resort group. The number of points required to stay at the resort will vary based on a points chart. The points chart will allow for factors such as:

- The popularity of the resort
- The size of the accommodations

- The number of nights
- The popularity of the season
- The specific nights requested

Deeded contract: The purchaser obtains legal ownership of the villa for a weekly interval that grants him the right to use the property for the week specified in the deed. Under this deeded type of conveyance, the purchaser has the legal right to use his week, give the week to a family member, or sell the week to another prospective buyer.

Right to use contracts: The individual is given contractual rights to use the timeshare facilities for a specific number of years. Upon expiration of this specific period (e.g. twenty years), the purchaser's rights of usage terminate unless he purchases additional time.

Scope of Timeshare in India

In India, the concept arrived quite late and wholesomely welcomed since it meant buying future vacations at today's price.

There exists a tremendous potential for timesharing in the Indian market and only 0.069% of the market are members. The market for timeshare models in India is huge, and more importantly, domestic traffic is adopting a lifestyle that supports the timeshare model. Much of it still remains untapped and developers keep coming up with new, innovative and attractive deals.

The timeshare concept in India was introduced by Dalmia Resorts in 1985 after which Sterling Resorts were fast to catch up which came into existence in 1986.

During the years following incorporation, Sterling has built a network of 19 resorts in 12 holiday destinations in India and is having a membership base of over 100,000 Vacation Ownership members. In addition, the resorts in the Sterling network also offer vacationers in India, the option of staying as a one-time hotel guest.

In today's date one of the major players in timeshare industry in India is Club Mahindra which has over the years evolved a position for itself and currently the company has 38 resorts in India and 5 resorts abroad. The company was also able to sell 91,997 Club Mahindra Holiday Vacation Ownership memberships. The company was incorporated in September, 1996 in Chennai as Mahindra Holidays and Resorts India Private Limited. The status of company was changed to a public limited company by a special resolution of the members passed at a shareholders' meeting held on January 29, 1998.

According to India Report (2009) impact of branded hospitality players and reputed conglomerates is the need of the hour. Potential consumers, while agreeing to the benefits of the product, have often cited the lack of branded players as their reason for not purchasing timeshare, thus indicating a requirement for both credibility and glamour in the product.

Every business has its success and failure stories and timeshare is no exception. From the time this concept was introduced in India, a lot of companies have entered this segment out of which most have been successful and a few unsuccessful. Timeshare schemes are being run by both public and private sector companies in India. Statistics from the latest survey of the timeshare industry confirm that the consumer satisfaction index is reaching a high of 85%. There are 40 timeshare companies, 80 resorts, 200,000 memberships with 15,000 annual additions and 4000 units. The timeshare industry is growing manifold with big brands such as Resort Condominiums International (RCI), Ramada Hotels and Resorts, Hyatt Vacation Club, etc. entering the business.

Even though the customers are quite happy with the quality of timeshare resorts that they are offered; still they feel that the number of destinations (resorts) is quite less and they do not get much on their platter. However this problem can be overcome by having affiliation with an exchange company such as RCI, II, Dial an exchange, etc. Having an affiliation from an exchange company is also crucial because practically it is not feasible for any of the timeshare companies to offer a bunch of resorts to the customers. Moreover, having an affiliation also opens the international gateway for the customers.

There appears to be a potential in targeting the growing Indian middle class as the trend now is towards more fun-filled holidays rather than visiting friends and relatives (VFR). The company may improve the condition of its resorts and furnishing of its rooms as time-share owners are the persons who will not come to the resort only once but they are the people who are going to visit the resorts year after year and that too at least for a week's time in one go.

Criteria for the Classification of Timeshare Resorts (TSR) in India

Timeshare resorts (TSR) is increasingly becoming popular for leisure holidays and family holidays, etc. With the aim of providing standardized world class services to tourists, the Government of India, Ministry of Tourism has a voluntary scheme for classification of fully operational timeshare resorts in the following categories: 5 Star, 4 Star and 3 Star.

The Hotel and Restaurant Approval and Classification Committee (HRACC) inspects and assesses the TSR based on facilities and services offered. Classification is valid for 5 years from the date of issue of orders.

The fees payable for the project approval and subsequent extension, if required, are as follows. The demand draft may be payable to 'Pay and Accounts Officer, Department of Tourism, New Delhi'.

Star category	Amount in ₹
5 Star	15,000
4 Star	12,000
3 Star	8,000

The application fees payable for classification/re-classification are as follows:

Star category	Classification/Re-classification fees in ₹
5 Star	20,000
4 Star	15,000
3 Star	10,000

Guidelines for Approval of Timeshare Resorts

General	3*	4*/ 5*/H	Comments
24-hour lifts for buildings higher than ground plus two floors	N	N	Mandatory for new TSRs. Local laws may require a relaxation of this condition.
Parking	N	N	Adequate parking space should be provided.
Guest rooms Minimum no. of apartments available for year round (10). All rooms with outside window/ventilation	N	N	No. of apartment weeks available should not be less than eligible members to holiday.
Minimum floor area studio including verandah, sleeping, living, bathing, cooking and dining—sq. ft.	250	251–350	
Minimum floor area 1 bedroom including sleeping, living, bathing, cooking and dining —sq. ft.	450–650	550–650	Living, dining, bedroom and kitchen areas are separate with doors.
Minimum floor area 2 bedrooms including sleeping, living, bathing, cooking and dining —sq. ft.	650–850	750–850	Living, dining, bedroom and kitchen areas are separate with doors.
Minimum floor area 3 bedrooms including sleeping, living, bathing, cooking and dining —sq. ft.	1000	1250	Living, dining, bedroom and kitchen areas are separate with doors.

Dining area	N	N	Separate dining table and chairs to accommodate maximum bedding.
Air-conditioning	N	N	Applicable for resorts/hotels at locations less than 2000 ft. above sea level. Air-conditioning/heating depends on climatic conditions and architecture. Room temp. should be between 20 and 28°C. For 4*, 5* between 20 and 24°C. For 3 star minimum 50% of the apartments should be air-conditioned as applicable.
Iron with ironing board	—	—	Should be available on request.
15 Amp earthed power socket	N	N	
Television	N	N	
Internet connection	D	N	For 3 star and 4 star internet facility be made available in the business center.
Telephone in the room	N	N	
Wardrobe with minimum 12 clothes hangers per bedding	N	N	
Shelves or drawer space	N	N	
Bathrooms			
Number of dedicated (private) bathrooms—Studio	-**	**	
Number of dedicated (private) bathrooms—1 Bedroom	1	1	
Number of dedicated (private) bathrooms—2 Bedrooms	2	2	
Number of dedicated (private) bathrooms—3 Bedrooms	2	3	
Minimum size of bathroom in sq. ft.	36	40	
Western WC toilet to have a seat and lid, toilet paper	N	N	
Floors and walls to have non-porous surfaces	N	N	
Furniture	N	N	Twin sofa-cum-bed, chairs and other furniture as necessary.
Water saving taps/shower	N	N	
Kitchen			
Kitchenettes for studios	N	N	Defined area—two burner stove top, no open flame, microwave oven or OTG, fridge, utensils, crockery and cutlery, tea/coffee maker, sink, exhaust fan or central exhaust.
Kitchen for 1 bedroom and larger	N	N	Dedicated kitchen—2 burner stove, microwave oven, tea/coffee maker, fridge, sink, exhaust fan, utensils, cutlery, crockery.
Washing machines/dryers	D	D	Arrangements be made available for laundry/dry cleaning services.
Public areas			
A lounge or seating in the lobby area	N	N	
Reception	N	N	Manned minimum 16 hours, Call service 24 hrs.

Heating and cooling to be provided in enclosed public rooms	D	N	Temperatures to be between 20 and 28°C
Restaurant/dining room	N	N	Multi cuisine for all 3 meals.
Garbage room (wet and dry)	N	N	
Room for left luggage facilities	N	N	
Health fitness facilities		N	Necessary for 4 star and above, desirable for 3 star
Guest services			
Utility shop	N	N	
Acceptance of common credit cards	N	N	
A public telephone on premises, unit charges made known	N	N	
Messages for guests to be recorded and delivered	N	N	
Name, address and telephone number of doctors with front desk	N	N	
Assistance with luggage on request	N	N	
Stamps and mailing facilities	N	N	
Safekeeping facilities available	N	N	
Smoke/heat detectors	N	N	These can be battery operated as per prevailing building laws.
Fire and emergency procedure notices displayed in rooms behind door	N	N	
Fire exit sign on guest floors with emergency power	N	N	
Public liability insurance	D	D	
Swimming pool	D	N	This can be relaxed for hill destinations.
Indoor games activity room	N	N	
Outdoor games like tennis, badminton	D	N	To be relaxed for urban timeshare properties, hill resorts and others, where site conditions do not permit.

Condominium Hotels/Condos

Condominium hotels are similar to timeshare hotels, except that condominium hotels have a single owner instead of multiple owners. According to this concept, individual guests are encouraged to purchase the ownership of the individual unit of a hotel and share the cost common to the whole hotel such as taxes, maintenance, upkeep of building, swimming pool, park, tennis court, etc. The concept relieves the owner of maintenance and upkeep worries. Each owner can occupy or sell his unit independently. Management of the hotel looks after the unit in the absence of the owner and if required, let it to provide income to the owner. *Example—RCI (Resorts and Condominiums and Inns) Group of Singapore, Hotel* *Metro Palace (Mumbai), 9 Star Hospitalities (Mumbai).*

Casino Hotels

The focus in this type of hotels is on state-of-the-art gambling facilities and provision of casino which operates 24 hours a day throughout the year. These hotels frequently provide specialty restaurants and extravagant floor shows and may offer charter flights for guests. Casino hotels are very popular in the USA, particularly Las Vegas. Top artists give their live performances and entertain the guests. They primarily cater to leisure and vacation travelers and offer luxurious rooms and the other top class services and amenities. *Example —The Palazzo Resort Hotel Casino (Las*

Vegas), *Marina Bay Sands (Singapore), Park Hyatt Mendoza (Argentina)*.

Conference Centers/Convention Hotels

These are very large hotels especially constructed to cater specifically to the MICE customers (meetings, incentives, convention and exhibitions). They are located outside metropolitan areas and may provide extensive leisure facilities like golf course, swimming pool, fitness centers, spa, etc. Most hotels of this category offer overnight accommodation to the conference delegates to make the meeting a success. They usually operate on special tariff for groups such as an all inclusive tariff which includes room, meals, conference hall/state-of-the-art convention center, audio-visual equipments and other related services. These hotels provide various audio-visual equipments required for conference and conventions such as overhead projectors, LCD projector with screen, white board with markers, television, DVD player, computer and laptop, public address system, business centers with fax, photocopying, internet facilities, secretarial assistance, various seating arrangements, display screens, flipcharts and other important technical assistance required during the meetings. They have 1000 to 3000 rooms to be able to accommodate a large number of guests coming to attend meetings. *Example—Hotel Novotel Convention Center (Hyderabad), Renaissance Convention Center (Mumbai)*.

On the Basis of Length of Stay/Visitors Stopover

Transient Hotels/Brief Stay Hotels

They cater to the needs of people who are on the move and need a stopover en route their journey. Located in the close proximity of ports of entry, such as seaport, airport, and major railway stations, these hotels are normally patronized by transient travelers. They cater primarily to the passengers with cancelled and delayed flights and airlines crew

members. Generally the guests in these hotels stay for a very short duration. These hotels have well-furnished guest rooms with restaurants and coffee shops and offer various other facilities such as airport pick up and drop through the hotel owned cars and buses which is very important for passenger and crew members.

The occupancy rate is usually very high going up to more than 100%, since rooms can sometimes be sold more than once in a day. *Example—Hotel Hyatt Regency (Mumbai), Novotel Hotel and Residences (Kolkata)*.

Semi-residential Hotels

These hotels incorporate the features of both the transient and residential hotels which means that in addition to taking guests on transit basis they also provide accommodation to guests on long term basis like residential hotels.

Residential Hotels/Apartment Hotels/Extended Stay Hotels

These hotels can be described as an apartment house complete with hotel services. They provide accommodation for a longer duration. These hotels are generally patronized by people who are on a temporary official deputation to a city where they do not have their own residential accommodation. The tariff in these hotels is charged on monthly, half yearly or yearly basis. Advance room charges are usually collected while other charges are billed weekly. Almost all residential hotels operate a restaurant, offer telephone service and laundry service. *Example—La Sunila Clarks Inn Suites (Goa), Residency Sarovar Portico (Mumbai)*.

On the Basis of Level of Services

Upmarket/Luxury/World Class Services Hotels/Full-Service Hotel

These hotels offer world-class service—sometimes called luxury service and target top business executives, celebrities, high-ranking political figures, commercially important persons and wealthy clientele as their primary markets who are affluent people demanding luxury and are not price sensitive. They provide guest rooms with world class/exquisite décor and finish, upscale restaurants and private lounges and dining rooms, concierge service, valet service, uniformed services and opulent meeting halls. Guests may find larger sized bath towels, bath robes, soap bars, shampoo, shower caps, clock radios, and more expensive furnishings, decor, and artwork in the hotel's guest rooms. Top-end recreational facilities, such as golf course, tennis courts, designer swimming pools with trained life guards, and other sports facilities, shopping arcades, beauty salons, health spas with saunas and jacuzzi are a regular feature. One of the special features of these hotels is the 'Club Floor' or 'Tower' which are separate floors or buildings constructed for very important guests to provide them with more comfortable accommodation and are more personalized and above all provide security as these floors or buildings have restricted entry for the staff as well as visitors to ensure complete privacy to the guests. Staff requirement is higher as personalized services are given to each guest. *Example—The Oberoi Udaivilas (Udaipur), ITC Hotel Grand Maratha Sheraton and Towers (Mumbai).*

Mid-market/Mid-range Services Hotels/Middle Class Service Hotels

These hotels offer modest services without the frills and personalized attention of luxury hotels, and appeal to the largest segment of travelers which consist of businessmen, families and free individual travelers. The guest rooms of these hotels are equipped with the basic amenities and supplies required for a comfortable stay. *Example—Taj Residency (Lucknow), Trident Hotel (Jaipur).*

Budget/Economy Hotels/Limited Service Hotels

Budget hotels focus on meeting the most basic needs of guests by providing clean, comfortable, and inexpensive rooms. These hotels mainly cater to the budget minded travelers such as families with children, retired persons, tour groups as well as some traveling business people who require clean and inexpensive guest rooms with least amenities required for a modest stay.

On the Basis of Ownership, Management and Affiliation

Independent Hotels

They do not have any affiliation or contract with any other hotels. And also they do not have any tie-up with any other hotels with regard to policy, procedures or financial obligations. A typical example of an independent hotel is a family owned-and-operated hotel (sole proprietorship or partnership) that is not required to conform to any corporate policy or procedure. They are usually autonomous. The advantage in this type of hotel is that they need not maintain a particular image and they are not bound to maintain any set targets, but can independently adapt quickly to changing trends. An independent hotel, however, may not enjoy the broad exposure or purchasing power or management insight of a chain hotel. *Example—Bristol (Gurgaon), The Landmark Towers (Kanpur).*

Chain Hotels

The chain hotels are hotels which are affiliated with other hotels. Chains usually impose certain minimum standards, rules, policies and procedures on their affiliates. The chain hotels enjoy large economies of scale as:

- All the properties of the chain enjoy the benefit of advertisement and promotion as it is the brand which is being promoted rather than the individual hotel.

- They have the advantage of operating under a large professional organization providing reservation network access, management services, financial strength, expertise, specialized manpower, merchandises and promotional help.

Several different structures exist for chain hotels. Some chains own their hotels (chain owned and operated properties), but many do not (management contract or franchise).

Management Contract (chain operated properties which are independently owned)

Management contract, as the name suggests, is a contract between the owner of the hotel (who does not have fair knowledge about the management of hotel) and a hotel operator (hotel chain with reputed name and market image, such as, Hilton, Best Western, Choice Hospitality, etc.) by which the owner employs the operator as an agent to assume the full responsibility for operating and managing the hotel. In return, the hotel operator would get a management fee and sometimes a percentage of the gross revenue. Although a management contract gives total control of the hotel property to the hotel operator, but the legal and financial responsibility lie ultimately with the owner. A chain acting as hotel operator for a particular hotel has total control over the standards and quality. The owner gets the advantage of brand name and efficient staff while the hotel operator gets wider network and popularity.

Franchise (chain affiliated properties which are independently owned and operated)

In franchising process, a hotel owner (franchisee) enters into an agreement with the hotel chain (franchisor) to use the chain's brand name by paying an initial franchising fee which is a lump sum amount and then royalty fee which is a certain percentage of the total revenue. Though the franchisee is the legal owner of the hotel, it must conduct the business according to the pattern established by the franchisor; otherwise, it stands the risk of losing the franchise and all associated rights and privileges. The owner gains a lot from the franchising agreement as the hotel enjoys the brand name and reputation of the franchisor, receive assistance from the company regar-

ding the technical know-how to run the hotel and above all enjoys CRS, advertising and sales promotion. The best known franchising companies are Holiday Inn, Choice International, Quality Hotels and Inns, etc.

Referral groups

The concept of referral groups was developed by the independent hotels in order to compete with other hotels. Referral groups consist of independent hotels which have banded together for some common purpose. While each hotel in a referral group is not an exact replica of the others, there is sufficient consistency in quality of service to satisfy guest expectations. It is a mutual understanding between two or more independent hotels to help each other where required by referring their guests when sold out or proceeding to another city. Through this approach, an independent hotel may gain a much broader level of exposure. The benefits that each hotel in the group may enjoy are a more extensive reservation system and expanded advertising through pooled resources.

ALTERNATIVE ACCOMMODATION

Alternative accommodation or *supplementary accommodation* can be simply defined as 'all those types of accommodation that are available outside the formal or organized accommodation sector'. These establishments provide the basic services of accommodation and may not provide food or other services. It may be of the following types:

- *Youth hostels*: YMCA (Young Men Christian Association)/YWCA (Young Women Christian Association). They generally offer clean and inexpensive accommodation to the young people who are either out for exploring the country or for educational or sports purpose.

- *Circuit house*: They cater to the high government officials.

- *Dak bungalow*: These establishments are constructed primarily to cater to the needs of traveling officials on government duty.
- *Dormitory*: It is a big hall with a large number of individual beds, mainly for students and low class tourists.
- *Railway retiring room*: These are owned by the railways and are situated in the railway stations for passengers holding tickets.
- *Lodge*: These are situated in the places of tourist interest catering to budget travelers.
- *Paying guest accommodation/private house holds*: Many people owning houses provide accommodation and sometimes even food to the travelers in return for money.
- *Guesthouse*: These establishments are generally managed by small companies for their employees staying on official visit.
- *Forest lodge*: They generally cater to the tourists on a visit to forest or wildlife sanctuary.
- Sarai
- Dharamshala
- Camping ground/tourist camp/open air hostels/caravan camping sites
- Holiday camps
- Yatri niwas
- Boarding houses
- Tourist bungalows

INTERMEDIATE ACCOMMODATION

These include all the establishments of the semi-organized sector of the accommodation industry which do provide accommodation, food and beverage services and some other services but not as elaborate as the services that are provided by the organized hotel industry. It may be of the following types.

Eurotel

The Eurotels are very similar to the apartment houses and are very popular in the USA and Europe. An important characteristic of the Eurotels is that they have multiple owners or co-owners and the co-owners can use another apartment in another place and another building through an exchange system between them agreed well in advance. These establishments are designed in such a way that the occupants are provided with all the facilities of a pronounced hotel nature.

Apart Hotels

The name of the apart hotels has been derived from apartments which is also a sort of accommodation. This concept of intermediate accommodation was developed in Spain in 1970. These establishments are hotels because the accommodation consists of an apartment which the guest who are also the owners can sell when desired.

Villas, Chalet Bungalows

Villa is an Italian concept which is a large country residence with an estate around it or adjacent to it. Chalet is a Swiss concept which means the farmers' residences which are generally wooden cottages with overhanging roofs. The chalets are generally smaller in size as compared to the villas.

Cabins in Mountains and Alpine Clubs

These types of establishments are generally situated in the mountains and other hilly terrains. They cater to the requirements of the mountaineers and tourists on a visit to mountains.

Sanitaria

These are hotels built and equipped appropriately for persons requiring rest during medical and therapeutic treatment. The rooms are specifically equipped for therapy including sauna, Turkish bath and Jacuzzi. Meals are personalized diet regimes for each guest. These hotels have proper dieticians, doctors and medical arrangements. This is ideal for those recouping from surgeries and illnesses. It is also a place for attaining physical fitness.

Hospices

The hospices were the primitive type of hotels which were built primarily to cater to the needs of the religious travelers. The owners of these establishments offered free accommodation and food to the religious travelers and thus these establishments were totally built for socially noble cause. Followers of Islam provide Madrassas while Hindus have their Dharamshalas for the pilgrims of their faith.

REVIEW QUESTIONS

1. 'Hotel classification is important for maintaining hotel standards'. Keeping in mind the above statement, classify hotels on the basis of location, size, length of stay, clientele, facilities and services.

2. How can hotels be classified based on their ownership and affiliation?

3. What are the important criteria considered whilst categorizing hotels as per the star ratings system?

4. Explain 5-star classification of hotels in detail. What are the general features, facilities and services for a five star hotel?

5. What are independent hotels?

6. What is a unique advantage of an independent hotel? How might independent hotels be at a disadvantage?

7. What is a referral group? How does it function?

8. What do you understand by the term franchise?

9. What do you mean by chain hotels?

10. Explain the concept of timeshare and briefly describe the various types of timeshare.

11. Highlight the historical development of time-share and condominium business in India.

12. Discuss the growth of timeshare in India and highlight the contribution of the industry.

13. Discuss the advantages of timeshare.

14. Throw light on the government's role towards enhancing the timeshare business in India.

15. Write an essay on the Resort Condominium International.

16. What are the approval and norms to be met at project stage of a star hotel?

17. Discuss the need for classification of hotels.

18. Mention the details for classification and re-classification of a hotel.

19. Distinguish between the following:
    • One star category hotel and three star category hotels
    • Resort hotel and Commercial hotel
    • Management contracted hotel and Franchise hotel
    • Chain hotels and Independent hotels
    • Floatels and Boutique hotels
    • Timeshare and Condominium

20. Explain the following with suitable examples:
    • Heritage hotel
    • Timeshare resort
    • Boutique hotel
    • Franchise hotel
    • Resort
    • Motel
    • Airport hotel
    • Casino hotel
    • Spa
    • Supplementary accommodation

11 | Emerging Concepts in Hotel Industry

Learning Objectives

After reading this chapter, you will be able to understand the following:

- Emerging concepts in hotels in the world
- Global trends that will impact hospitality industry

EMERGING CONCEPTS IN HOTELS IN THE WORLD

Boutique Hotels

Boutique hotel is a small but very expensive and exclusive hotel. Boutique hotels differentiate themselves from larger chains or branded hotels by providing exceptional and personalized levels of accommodation, services and facilities. They are sometimes known as *design hotels* or *lifestyle hotels*. Professional but at the same time very personal and intimate services by staff are prominent features to give guest a richer experience; each restaurant of a boutique hotel has a different entertainment concept which includes lighting that changes with the mood of the guest.

For more personalized services and attention, restaurants are kept small. Boutique hotels are furnished in a themed and stylish manner and everything from decoration to food and service must be in tune with the theme. Of the total travel market, small percentages are discerning travelers, who place a high importance on privacy, luxury and service delivery. As this market is typically corporate travelers, the market segment is non-seasonal, high-yielding and

repeat, and therefore one which boutique hotel operators target as their primary source. *Example—The Park (Bengaluru), Le Meridien (Bengaluru).*

Spa Hotel

Spa was a name of the city in Belgium. The city had abundance of medicated hot water springs and mineral water springs. Spa—'salus per aquae' (Latin phrase)—actually means 'health through water'. Since ancient times, a spa helps in healing, rejuvenation and relaxation. In olden times, a spa symbolized luxury but nowadays, it has become almost a necessity.

A spa resort is a resort hotel providing therapeutic baths and massages along with other features of a luxury hotel. Every spa treatment includes relaxation, usage of water in any form, medical treatment, cosmetic treatment, dietary recommendation, stress management, healing certain ailments, detoxification, gymnasium, yoga, meditation, steam, sauna and massage, facial, body massage, body polish (scrub), body wraps (mask), hydrotherapy and wellness program. Spas have professional staff that often include dieticians, therapists, masseurs, exercise physiologists, and in some cases, physicians. A spa is a facility that operates under the full-time, on-site supervision of a licensed healthcare professional. *Example—Park Hyatt Goa Resort and Spa (Goa), Oberoi Amarvilas (Agra), Ananda Spa (Rishikesh).*

Different types of spa are:

- *Destination spa*: It offers a short-term residential facility. The primary purpose of guiding individual spa-goers is to develop healthy habits. The program includes spa services, physical fitness activities, wellness education, healthy cuisine and special interest programming. Guests reside and participate in the program at a destination spa instead of just visiting for a treatment or pure vacation.

- *Resort spa*: These are basically spas in resort settings. They allow you to combine a variety of recreational activities with your spa experience. Spa vacation resorts feature spa treatments and services while offering activities such as golf, tennis, horseback riding, skiing, and water sports. Evenings allow guests to enjoy resort pastimes such as dancing and live entertainment. Many resort spas have children's programs that are completely separate from their parents.

- *Urban spa*: It gives the road warrior that much relaxation to prepare them for another strenuous business day. One can expect to find services such as steam rooms, saunas, pools, exercise equipment. Fitness classes may be offered. Urban spas are found in metropolitan hotels.

- *Ayurveda spa*: It has more well-being oriented treatments with a lot of herbal products used in keeping with the ancient science of Ayurveda. Treatment for ailments like arthritis, paralysis, obesity, sinusitis, premature ageing and general health are offered. *Example—Kairali Ayurvedic Health Resorts (Kerala), The Coconut Grove (Goa) and Somatheeram Ayurveda Resort (Kerala).*

- *Medi spa*: Medical spa features traditional and complementary medical services supervised or administered by medical professionals. These spas are more cosmetic makeover oriented through invasive methods like surgeries. The spa's specialty may be diagnoses and testing, preventive care, cosmetic procedures, or some a combination. Modern technology is also used for various beauty and skin related enhancements.

- *Other types of spa*: These include chocolate spa, wine spa, tea spa, coffee spa, super luxury spa.

ECOTELS/Eco-friendly Hotels/Green Hotels

ECOTELS, also called 'green hotels', are earth-friendly or eco-sensitive hotels that feature innovative and imaginative programs for conserving natural resources, reducing waste, minimizing pollution, and maximizing sustainability.

These hotels have made important improvements to their architecture and overall façade in order to minimize its impact on the environment. Some ECOTELS consist of recycled or renovated buildings with upgrades to conserve energy and water, minimize waste, incorporate natural landscaping, or utilize recovered building materials. Other hotels support local environmental efforts and/or offer environmental education or excursions.

Organizations like the 'Green Hotels Association' bring together hotels interested in environmental issues. All ECOTEL-certified hotels are inspected annually by environmental experts from Hospitality Valuation Services (HVS) International, a global ISO-certified body. To get an ECOTEL certification, a hotel must adhere to at least two of the 'five globes':

- *Energy efficiency*: This requires the existence of a formalized framework to reduce the energy consumption of the hotel. For instance, the hotel should have been designed and constructed keeping in mind maximum energy conservation. Factors such as whether only the minimum of lighting required at a given time is being used and the extent of involvement of the guests as well as the employees also make a difference to the final score.

- *Water conservation*: The effective conservation of water in all departments of the hotel, across the levels, must be evaluated. The extent to which water is recycled and re-utilized is also considered an important factor. All employees are expected to be well versed in the water-conservation operations of the hotel.

- *Employee environmental education and community involvement*: How involved the employees are in the efforts of the hotel to contribute towards the environment is crucial to the fulfilment of its mission. The hotel should have training modules in place for employees at all levels to familiarize them with the eco-friendly initiatives of the hotel.

- *Solid waste management*: The hotel must effectively recycle and manage waste, wherever generated. Proper systems for the collection, recycling, and disposal of these wastes in all its departments are a must for ECOTEL certification. All employees of the hotel must undergo training in the basic techniques of solid waste management.

- *Environmental commitment*: The hotel must demonstrate the existence of a formalized commitment towards the preservation and enhancement of the natural environment. It must, through all its operations, activities, and written statements, communicate its commitment to the environment. For instance, the mission statement of the hotel must mention its environmental dedication. Additionally, every hotel should have a 'green team', headed by a member of the top management, ensuring that all departments are working in consonance with the hotel's mission of environmental responsibility (Fig. 11.1).

Fig. 11.1: 5 Globes of ECOTELS

The first ECOTEL certified hotel was Hotel New York Vista in the year 1994. Mr Vithal Kamat of Kamat Group of Hotels brought the concept of ECOTEL to India through his hotel known as Hotel Orchid. Hotel Orchid was the first ECOTEL certified hotel not only in Mumbai and India but also in Asia. Other ECOTELS in India are: *The Orchid (Mumbai), Uppal's Orchid (New Delhi), The Lotus Suites (Mumbai), The Raintree (Chennai), Hotel Rodas (Mumbai), Yogamagic Canvas (Goa).*

Matels

These are fully-automated hotels which require minimum human contact. A guest books online and gets an immediate confirmation. On arrival, he punches in his reservation number and name in a machine which initiates a dialogue with a virtual receptionist who registers the guest through a close-circuit TV and issues a key to the room through a slot machine. Outsourced cleaning crews come during the day to clean the room and make it fresh and impeccably clean. This concept was started in Japan and is likely to be the future of hotels in extremely populated locations.

Auberge/Gasthof/Herberge

These are the counterpart of Inn in various countries. They represent a smaller unit and may have bar and restaurant for travelers.

Pop-up Hotel

It is a hotel which is temporarily established at a location for a short time before being moved. Such hotels may be built from pre-fabricated modules which are joined together on site or from collapsible structures such as tents or they may be fully mobile, being built on a large vehicle.

Capsule Hotel

It is a type of hotel found in Japan with a large number of extremely small rooms (capsules) with an average size of 2 m × 1 m × 1.25 m which are made of modular plastics of fiberglass. These capsules are stacked side by side and two units top to bottom, with steps providing access to the second level rooms. Luggage is stored in a locker, usually some-where outside of the hotel. Washrooms are communal and most hotels include restau-rants and other entertainment facilities.

Fig. 11.2: Capsule hotel

Destination Hotel

It is a type of hotel whose location, facilities and amenities make the hotel itself a destination for tourists, rather than merely a convenient place to stay while traveling through or visiting the area for other reasons.

The destination hotels have extremely well designed and beautiful rooms, fine dining restaurants, recreational and entertainment facilities and overall has a beautiful garden with stunning landscaping.

Modular Hotels

It is not just hotel but a movable property and not a temporary structure. This approach develops hotels in much faster way compared to traditional brick and cement process. It involves assembling hotel room at factory level and transported to hotel site in containers. A hotelier can save up to 40% of project cost. Construction time is usually six weeks after the plinth work is done. Excess rooms can be dismantled in case of oversupply. Same can be transferred to other location to bridged demand–supply gap.

Tree Hotels

These are hotels which are built with living trees as structural elements.

Lotel

A Lotel is a hotel equipped with helipad for helicopter landings.

GLOBAL TRENDS THAT WILL IMPACT HOSPITALITY INDUSTRY

Hotels have long been an integral part of travel, whether they are booked for business trips, family vacations or anything in between. But with the rise of home-sharing and vacation rental businesses like Airbnb and HomeAway, the industry is being forced to improve upon the status quo and find new ways to appeal to travelers.

As a hospitality sector business, one would like to stay in-sync with the latest trends in the global hospitality industry. This is rather important since trends that are increasingly engaging attention soon can become a standard demand from your hotel guests.

Some distinct trends have manifested themselves, which shows that the industry is perfectly poised to change with the times and offer state-of-the-art services to customers.

A quick glimpse below will provide a look at some of the latest trends that are making their presence felt across the globe in the hospitality industry.

Millennials the New Power Segment

Exploration, interaction, and emotional experience are the hallmark of Millennials, the fastest growing customer segment in the hospitality industry, expected to represent 50% of all travelers by 2025. They currently represent 32% of all US travelers. With the rise of millennial consumers, businesses will need to be more transparent and tech savvy, with a strong focus on empathy and customer connection. Technology is essential for this demographic and they will expect technology to power check-in, paying their restaurant and bar bills and looking up places to eat, shop and play to name a few.

In addition to wanting technology, millennials have no problems speaking up. If what they are seeking is not handled to their liking, they will turn to Twitter, Facebook, Yelp or TripAdvisor to voice their complaints.

From glamping and gourmet to high-tech ambiences, the hospitality industry has to prepare for this new generation of clients. Hotel and travel will be positively affected by their spending power and in the long run, the profile of their businesses will also shape up their service portfolios.

Disruption and the Sharing Economy

Emerging new business models including peer-to-peer networks life Airbnb, Uber, and Lyft, multi-sided platforms such as Google and eBay, or free business models such as Skype and Flickr will change the business landscape. As peer-to-peer networks expand and grow, they will become more professional and pose stronger direct competition to traditional travel services. Further, the growing popularity of meta search engines from big players like Google and Microsoft and the rapid growth of firms like Kayak may alter the user experience, define the mobile experience, lead to consolidation and impact partnerships with Online Travel Agencies (OTAs) and hotels. As OTAs consolidate and expand their relationship with customers, the costs of distribution will become increasing critical.

Political Uncertainty and Terrorism

Around the world, citizens have responded to increased government involvement with distrust and have begun to challenge entrenched political parties. Punishing economic policies and austerity measures along with ethnic, cultural and religious tensions have resulted in the rise in civil unrest. A megatrend found in Europe and likely to spread is the rise in populist movements that seek to regain national identity. The ability to efficiently deliver social services will be an ongoing challenge for governments. Countries and states with ethnic and religious tensions along with poor governance, and weak economies will breed terrorism. Transnational and free-wheeling terrorism enabled by information technology will replace state-supported political terrorism. In spite of collective actions to prevent, protect, and respond to terrorism, the threat will remain high in Europe and the US.

Health and Wellness Trends

Hotel chains are re-branding around wellness, recognizing the concept and emphasizing on fitness. Health and wellness trends will continue to drive customer decisions. There is a need to balance health and wellness with tasty food options that are cost effective. Working out and eating well may not be on every globetrotter's agenda, but many business travelers and health nuts appreciate staying fit while on the go and being able to find a well-balanced meal—and hotels are taking note. For example, Fairmont's Lifestyle Cuisine program provides a wealth of healthy culinary options (and can even cater to specialized diets) to choose from in the chain's restaurants, room service and even for business meeting meals.

Some brands boast fitness-focused programs to help guests maintain their exercise regimen during their stays. A prime example is Westin, which touts the RunWESTIN program where running concierges lead guests on 3- to 5-mile routes through town (Westin even partnered with New Balance to lend shoes and workout gear to lodgers for $5), in addition to its WestinWORKOUT gyms stocked with weight-training equipment and cardio machines. Meanwhile, Kimpton hotels provide yoga mats in every room, along with workout routines accessible on the television, as well as complimentary bikes. Many hotels are using scheduled renovations to improve their fitness centres with additional machines and newer equipment. And still, some properties house sprawling gyms with not-just-your-average cardio and weight-lifting gear. The Houstonian Hotel, Club & Spa's private health club has more than 300 machines and offers daily exercise classes, while The Venetian Las Vegas is home to a 40-foot rock climbing wall and a cycling studio in addition to a Canyon Ranch Spa Club outpost. Trump Hotels' Wellness Program, Westin's Super Foods Rx program and IHG's wellness-focused EVEN Hotels brand provide further evidence of hotels embracing a renewed interest in health and wellness on the road.

Air purification, energizing lighting, a yoga space, in-room exercise equipment, and vitamin infused shower water are some of the innovative wellness options.

Technology Driven Self-Sufficient Travelers

Innovative technologies on a mobile platform will be expected as more individuals rely on digital concierge services. Mobile check-in and seamless connectivity across platforms and devices are no longer the future, they are the present. With geo-location software easily available, selling locally with a focus on content marketing is expected. Connectivity is a key as more individuals are relying on information delivered through social software from virtual networks. Technology is better and smarter, and more integrated user experiences are likely.

Hotels are appealing to tech-obsessed travelers by offering the latest and greatest amenities to make a stay more convenient for guests. A few examples include in-room touchscreen controls that operate everything from the television and lighting to the curtains and thermostat and complimentary iPads available for use in hotels like the Weekapaug Inn, XV Beacon and SLS South Beach, among others. Some properties (including some Hyatt and Starwood hotels) have even taken a cue from airports, offering check-in kiosks to streamline and simplify the process for travelers, while Starwood has gone even further by introducing SPG Keyless. This smartphone app allows Starwood Preferred Guest members to use their phone to unlock their room door. Just 10 of the brand's nearly 1,200 properties currently participate, though all W Hotels, Aloft Hotels and Element Hotels are expected to offer the program in 2016.

Booking more Profitable Business

Booking more profitable business is critical as more revenues result from strong increases in occupancy levels, average rates and revenue per available room (RevPAR). This may suggest more profits, but the growth in distribution costs as well as other operating costs such as healthcare and the minimum wage increases can stunt profit growth. While the revenues are coming first and foremost from RevPAR growth, there are additional ways to increase both revenue and net income.

One is by less reliance on the online travel agencies (OTAs). By directing guests to your hotel's website and telephones, the savings are abundant. The digital distribution costs are soaring and the number of players entering the market to compete with OTAs is rapidly rising (think Google, Facebook, Apple, TripAdvisor, Amazon and more). The key is to negotiate with your distribution team (yes, the OTAs can be an integral part of your team) and reduce your commissions. Then make certain that you have a strategy in place to earn the repeat business of every single guest…and get them to book direct next time. Think incentives!

Rankings will Rule

To ride and survive the new-age hospitality wave, businesses must respect and win the ranking war.

The Internet has changed our lives in more ways than we can imagine. Just as we spend hours researching options, we spend equal amount of quality time on weighing the pros and cons of each of those options.

Industry reports have found that 93% of hotel consumers are seriously influenced by reviews. With sites like Yelp and TripAdvisor, these are all organic reviews and user-generated online ratings that can make or break a business.

Some users will leave bad reviews, but more often than not they are inclined to leave realistic and positive feedback as well. This means hotels must beef up services to match (or beat) the competition to harness the power of these rankings.

Social Reputation will Matter

Online presence in this digital age is important. So along with user-generated content—which is slated to be the next big thing in social—there also needs to be a dynamic, interactive and attractive online profile that will beguile and impress effortlessly. A static page with a few links will no longer work. Instead businesses need to work with developers to create platforms where guests can post their own content and help build up your social reputation. A brand with a more flexible site and platform will bring on more visitors, more loyalty and more information sharing. Needless to say, the content needs to be rich and powerful to keep up with this social storm in 2016.

Customer Service should Create a Wow Factor

To reach the apex of this organic brand-building process, hotels need to re-evaluate their services and beef them to match the highest standards. If you expect your customers to give you genuine positive feedback, then the customer service—in every aspect of the hospitality business—has to create a definite wow factor.

While no-frills services are coming up to combat the economy and the competition, there has also been a trend for the distinctive. We can see these in the many innovative offers like online ordering, destination specials that partner with local wineries and eateries, digital in-room dining experiences for busy executives, having famous chefs as brand ambassadors and strategic tie-ups with various service providers to make their experience enjoyable.

Sustainability is the Keyword

Eco-friendly practices are becoming the norm, and most hotels must have an attractive 'green policy', as travelers expect hotels to have some type of environmental program in place, while a few are still willing to pay more for eco-features. Critical resources such as water and power are under increasing strain leading to price increases, volatility and even shortages. Global warming and energy use are affecting how we consume and live on a societal scale. Water scarcities and allocation pose challenges to governments in the Middle East, Sub-Saharan Africa, South Asia, and northern China. Renewable energy resource and innovative projects will shape the future of resource use, while regional tensions over water will be heightened in 2016. Falling oil prices, show how easy resource constrains can change, with a dampening effect on the power of countries such as Russia and Iran, while lowering prices for jet fuel, impacting growth in air travel, even as airlines acquire new fuel-efficient jets from Boeing and Airbus and replace old fleets.

The good news is that 62 % of users now expect hotels to be environmentally responsible and have solid environmental programs in place. These programs have to make deep-seated impact to lessen their carbon footprints and not just make cosmetic changes to impress the crowd. One must remember that the consumer is aware and informed these days, and only the best efforts will bring in results.

Global Outlook is Important

Global travel has opened up, and some of the best revenue figures can be anticipated from this arena. According to The US Department of Commerce, global travel to the US in 2014 resulted in approximately 84.6 million visitors, a 4 % growth from 2013.

The travel industry is among the largest and fastest-growing industries worldwide, forecasted to support 328 million jobs, or 10% of the workforce, by 2022 according to the World Travel and Tourism Council. Citizens of Finland, Sweden and the United Kingdom have the best passports for global travel (may enter 173 countries without a visa). In general, passport holders in North America and Europe have the most freedom of travel, while passport holders in Africa, the Middle East and South Asia have the least. Chinese tourists still encounter difficulty traveling abroad with only 50 countries and territories offering visa-free or visa-on-arrival access for this group of travelers.

Providing a Pleasant Stay for Pets, Too

Sure, plenty of hotels allow pets. But some properties have really upped the ante, welcoming pets as what they are—an extension of the family—and treating them to top-notch customer service. Loews Hotels' Loews Loves Pets, Kimpton Hotels' Very Important Pets and W Hotels' Pets Are Welcome (PAW) programs are a few of the brand-backed offerings for travelers and their furry companions. These particular programs provide animals with everything from welcome treats, bowls, and beds, to specialized room service menus and pet massages. Pet stays at Kimpton hotels are free; Loews and W hotels charge an extra fee—but the hotels' added amenities sweeten the deal.

Smaller hotels are even getting in on the action: New York City's Soho Grand welcomes pets free of charge, and with perks like toys, bedding and leashes, plus a park that features a dog run and small garden with fire hydrant water stations. Across the country, the Peninsula Beverly Hills pampers pets with special menus, walks, personalized towels and doggie beds. The AH&LA's 2015 survey found that more hotels reported allowing pets than ever before, so it is likely Americans will continue traveling with their pets in coming years.

REVIEW QUESTIONS

1. Define ECOTEL. What important factors make an ECOTEL different from other hotels?
2. What are boutique hotels?
3. Describe 5 globes of ecotels.
4. Explain different types of spa.
5. Discuss top five trends in hospitality industry.

Hotel Organization

THE NEED OF ORGANIZATION IN HOTELS

When we stay n a hotel as a guest and enjoy its services and 1 cilities, we seldom think how the hotel is able t provide such flawless and streamlined servi es. Every hotel, irrespective of its size, type, and mode of operation, is an organization that utilizes all its resources in a definite way to realize its business objectives. The word 'organization' can be defined as a group of people who form business together in order to achieve a particular objective. Organizations are social units (or human groupings) deliberately constructed and reconstructed to seek specific goals. Organization is an activity which provides mechanism or apparatus for purpose of integrated and cooperative action by two or more people with a view to implementing any plan.

The hotel guests receive a wide variety of services and facilities from the hotel. To carry out all the functions effectively and efficiently, the hotel should have a well-organized structure. Such a structure has the following advantages:

- It facilitates managerial action.
- It encourages and improves efficiency.
- It makes communication easier, faster, and more effective.
- It ensures the optimal use of resources.
- It stimulates creativity and adherence to conformity.
- It creates job satisfaction in employees, thus motivating them to excel.
- It leads to quality services, nurturing brand loyalty in guests, which would ensure the growth of business.

Characteristics of Organization

- *Division of labor*: Each man in an organization has got a definite purpose which he should fulfil in order that things may go as desired.
- *Presence of power control centers* which control the concerned efforts of organization and direct them towards its goals.

These power centers must also review continuously the organization's performance and re-pattern its structure, when necessary, to increase its efficiency.

- *Span of control*: The limit of number of subordinates—whose activities are inter-related—that a person can effectively supervise is called span of control.

VISION

Corporate vision is a short, succinct, and inspiring statement of what the organization intends to become and to achieve at some point in the future, often stated in competitive terms. It concretely describes how a company sees itself in the future, and therefore must be realistic and attainable. Vision refers to the category of intentions that are broad, all-inclusive, and forward-thinking. It is the image that a business must have of its goals before it sets out to reach them. It describes aspirations for the future, without specifying the means that will be used to achieve those desired ends.

MISSION

A business organization, like a hotel, is a deliberate and purposive creation, which strives for certain end results. It has a purpose for existence, which is known as its mission. Mission, which is an abstract idea, has an external orientation and relates the organization to the society in which it exists. According to Vern McGinnis (1981), a mission should perform the following seven functions:

- Define what the organization is.
- Define what the organization aspires to be.
- Be limited to exclude some ventures.
- Be broad enough to allow for creative growth.
- Distinguish the firm from all others.
- Serve as framework to evaluate current activities.
- Be stated clearly so that it is understood by all.

Mission Statement

A mission statement is the statement of purpose that identifies the scope of a hotel's operations in product and market terms, and distinguishes the hotel from other competing hotels. A hotel develops a mission statement to indicate the activities that it intends to undertake in the present and the near future. The statement should suggest the uniqueness of the hotel from its competitors and should address the interest of guests, owners, employees, and society. It should be clear in terms of its intention, and should be feasible and achievable. A well-designed mission statement offers guidance to managers to develop sharply focused, result-oriented objectives, goals, and strategies to achieve the organization's mission.

The mission and vision statements provide the framework for developing the strategies and operations of an organization.

An illustration of a mission statement is given below:

> The mission of our hotel is to provide outstanding lodging facilities and services to our guests. Our hotel focuses on individual business and leisure travel, as well as travel associated with group meetings. We emphasize high quality standards in our Rooms and Food & Beverage divisions. We provide a fair return on investment for our owners and recognize that this cannot be done without well-trained, motivated, and enthusiastic employees.

Objectives

As compared to mission, objectives are more precise—they are used to identify the end results that a hotel wants to achieve over varying periods of time. If objectives were general and non-quantifiable, then it would be impossible at the end of a certain period of time to see whether hotel's actual results match with the planned objectives or not. Objectives help to measure the progress of a hotel vis-à-vis its mission and vision.

Goals

Goals are those activities and standards an organization must successfully perform or achieve to effectively carry out its mission. Goals shall be:

- Specific and numerical
- Observable
- Measurable

Strategy

The methods employed to achieve goals are known as strategy.

The objectives, goals, and strategies should be planned in a way that they do not contradict and create conflict at departmental level.

Tactics

Day-to-day methods to reach the strategies is called tactics.

An illustration, to one of the front office department goal (a registration-related goal), a strategy to reach it and a related tactic is given below:

> **Goal:** Operate the front desk efficiently and courteously so that guests register within 2 minutes of arrival.
> **Strategy:** Pre-register guests with reservation guarantees as room become available from the housekeeping.
> **Tactic:** Pre-print registration cards for arriving guests and separate the cards of all guests with a reservation guarantee.

It is of extreme importance that managers shall continuously control and evaluate their strategies and tactics, and hence revise them (if necessary) so that department goals and objectives are reached fully at the end of the planned period.

WORK SHIFTS

As hotel industry functions round the clock, generally there are three shifts:

Straight shift: This type of shift extends for a period of 9 hours with a break of 1 hour.

Morning shift—7:00 AM to 03:30 PM

Afternoon shift—3:00 PM to 11:30 PM

Night shift—11:00 PM to 07:30 AM

Break shift/Split shift: This type of shift is split into two sessions that add up to a regular shift of 9–10 hours. This includes a break of approximately 3–6 hours.

Rotating shift: An employee may be given a particular shift for a week or two, and then changed over to the next shift. This rotation is done to ensure that all employees get a fair share of all the shifts.

Flexitime: In this kind of scheduling, an employee can work any time according to his convenience and is paid accordingly.

Each shift's duration is 9 hours and these shifts are normally scheduled to overlap by half an hour with the next shift to facilitate handovers and takeovers.

However, the workload on the shift depends on various factors like type of establishment and modes of transportation available. If the hotel is situated near the transit of a city then there will be a heavy load of traffic throughout check-in and checkout. It is therefore very important to get the overview of the work to be completed, right from the expected check-in, check-out, cash at the counter, complaints, proper work and briefed up from the housekeeping department to know the position of the rooms. During the change of shift care is taken that it does not hamper the smooth flow of operation.

To avoid conflicts and confusion between two shifts a log book is maintained in the respective department. It acts as a record if there is any carelessness or negligence being made by any individual and helps in tracing as it contains the column of signature. Before starting the shift it is wise to see that everything is in its place.

JOB DESCRIPTION

It is a written description of a job performed by a job holder. It specifies the parameter within which a job is done. Job description lists all tasks and subtasks that compose a work position. Moreover, it may outline reporting relationships, responsibilities, working conditions, equipment and materials to be used.

All job descriptions shall be tailored and customized to reflect the needs of each single hotel property, and work position. Moreover, job descriptions shall be task-oriented rather employee-oriented, which means that hotels shall try to search for employees who can fit their job descriptions, not design jobs to fit the skills of certain job applicants.

Job descriptions shall be revised periodically to cope with the ever changing demands and needs of the industry and to respond to the sophisticated needs of guests. While doing so, managers shall let their employees be involved in the revision process.

A well-defined job description brings about greater certainty of what is expected in terms of performance, and when actual results match expected ones, both morale and efficiency are raised.

Contents of Job Description

- Job title
- Reports to
- Level of job
- Coordinates with
- Supervises
- Duties and responsibilities

Advantages of Job Description

- Newly recruited employees know exactly what their job is all about.
- Job descriptions set a basic foundation for achieving the standards of performance.
- It serves as a legal document for any disputes arising from a lack of definition of roles.
- A job description may come to the employees' aid when dealing with an unreasonable superior who overburdens them with tasks not in their purview.
- The document ensures that the supervisor and the subordinate have a clear understanding of their role in the common work; else there may be a misinterpretation of the job, leading to friction.

JOB SPECIFICATION

A job specification is a document detailing the minimum qualities or traits required by an individual to perform a particular job. Job specifications list the personal qualities, skills, and traits a person needs to have in order to perform successfully the tasks outlined in a job description. That's why departments shall first design job descriptions, and later job specifications.

Job specification is generally used by HR department as tools for the selection of the right employee for a particular job, as defined in the job description.

Job specifications usually serve as a basis for advertising job vacancies, and as a tool to identify current employees for promotion purposes.

Contents of Job Specification

- Job title
- Gender
- Age limit
- Educational qualification: Bachelor degree in HM from a recognized institute
- Work experience: Minimum 3 years as assistant housekeeper in a reputed hotel
- Equipment skills: Sound knowledge of housekeeping equipment (laundry machines, carpet shampooing machine, scrubbing machine)
- Physical characteristics: Physically fit
- Mental traits: Stability under stress
- Personality: Leadership and motivation skills, excellent communication, administrative skills, eye for detail
- Language skills: Local language/foreign language
- Special requirements: Budget preparation.

HOTEL ORGANIZATION

To carry out its vision, mission, objectives, and goals, every hotel requires a formal structure, known as the organization structure. This structure defines the company's distribution of responsibility and authority among its management staff and employees. It establishes the manner and extent of roles, power, and responsibilities, and determines how information flows between different levels of management. This structure depends entirely on the organization's objectives and the strategies chosen to achieve them. The most common way to represent the organization structure is by an organization chart.

Organization Chart

An organization chart is a schematic representation of the various positions within an organization and also denotes the working relationships and the line of authority. An organization chart is a hierarchical, graphic representation of the structure of an organization—a list of all positions and the relationship between them. It shows where each position fits in the overall organization, as well as where divisions of responsibility and lines of authority lie. It is a visual representation of how a firm intends authority, responsibility, and information to flow within its formal organizational structure. It usually depicts different management functions (accounting, finance, human resources, marketing, production, research and development, etc.) and their sub-divisions as boxes, linked with lines along which decision-making power travels downwards and answerability travels upwards. The chart indicates direct reporting relationships as well as indirect relationships, which, though not connected directly, involve a high degree of cooperation and communication.

The organizational structure depends on various factors such as the size of the hotel, the availability of the various services and facilities in the hotel such as restaurant, bar, banquets, health club, night club, etc., type of clientele, location of hotel, level of automation—fully-automated, semi-automated or manual.

Every organization chart shall be flexible, to reflect the ever-changing environmental dynamics and, hence be able to survive. In accordance, organization charts shall be reviewed periodically in order to determine whether the actual organization still match the environment needs (i.e. guests, employees, technology, competitor's needs…) or not. A SWOT analysis (i.e. Strengths, Weaknesses, Opportunities, and Threats) shall be a good approach to initiate a change in the organization chart. Last but not the least, it is of extreme importance that there are no two hotels having exactly the same organization chart, and that a hotel might have an organization chart change over a period of time. Hence, organization charts shall be tailored to fit the needs of each individual property. Figure 12.1 depicts the organizational chart of a large hotel.

CORE AREAS OF A HOTEL

The organization of a hotel today is very complex and comprises various departments. The number of departments varies from one hotel to another. All departments may have their own managers, reporting to the general manager. Some of the departments are operated by the hotel, while some may be operated by an external agency. The departments like front office, housekeeping, kitchen, F&B service, engineering and maintenance, accounts, etc. are operated by the hotel, and the departments or sections like laundry, retail shops in shopping arcade, casino, and recreational services may be operated by an external agency on contract. Figure 12.2 depicts different departments of a hotel.

Classifying Functional Areas

There are two approaches to classify departments in typical hotels:

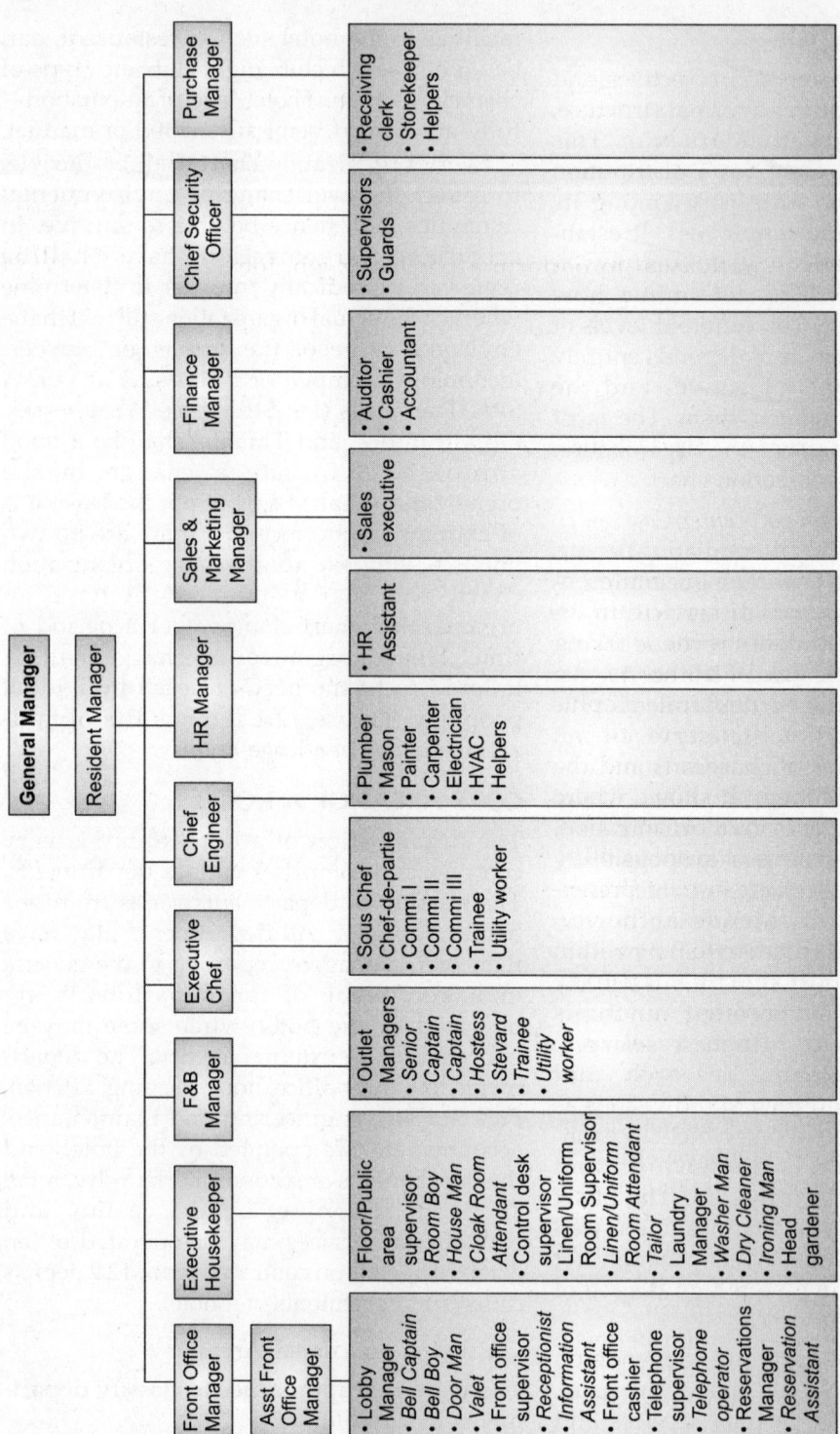

Fig. 12.1: Organizational chart of a large hotel

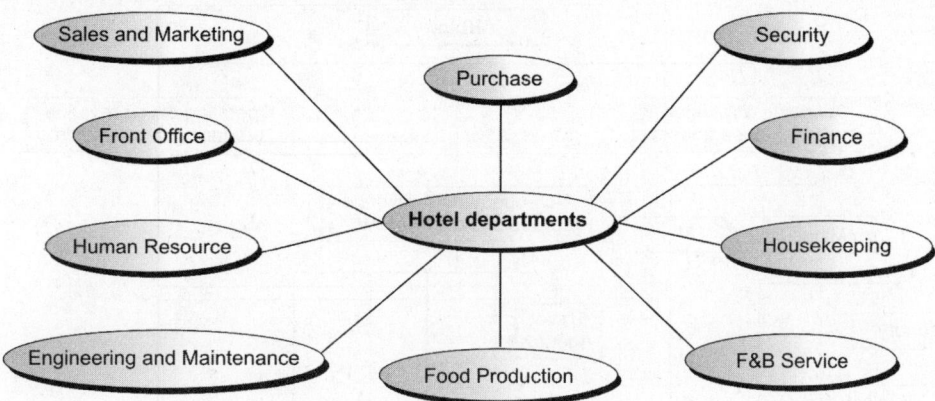

Fig. 12.2: Departments of a hotel

Revenue center versus Support center/Cost center: The first approach is by differentiating departments with respect to revenue-generation. A revenue center sells goods or services to the guests and thereby generates revenue for the hotels. The front office and F&B service are major revenue-producing departments. The minor revenue-producing services are telephones, membership charges from non-resident users of swimming pool and health club, and laundry services for resident guests. Support centers or cost centers do not generate revenue but play a supporting role to the revenue centers. The housekeeping department is major support center within the rooms division. Other support centers include the areas of accounting, engineering and maintenance, human resource, purchase, sales and marketing, security, etc.

Front of the house versus Back of the house: This approach classifies departments according to department staff's frequency of communication with guests. If communication between staff and guest is frequent like in front office and F&B service, then the department is said to be a front of the house department. On the other hand, if the communication between department staff and guests is non-existent or on occasions, then the department is said to be back of the house department, e.g.

accounting, engineering and maintenance, human resource. Although member of housekeeping department have some contacts with hotel guests, the department is generally considered as back of the house functional area (Fig. 12.3).

Various Departments and Sections in a Large Hotel

Front Office: Front office is the first department of the hotel with which guests come in contact at the time of their arrival and is also the last department they interact with when they depart from the hotel. Headed by the front office manager, this department performs various functions like reservation, reception, registration, room assignment, answering telephone calls and settlement of bills of a resident guest. Since the guests remain in contact with the front desk throughout their stay for all kinds of information and help, we can say that it is the 'nerve centre' of hotel operations. The various sections of the front office department are:

- Reservation
- Reception
- Information desk
- Cash and bills
- Communication
- Travel desk
- Uniformed services

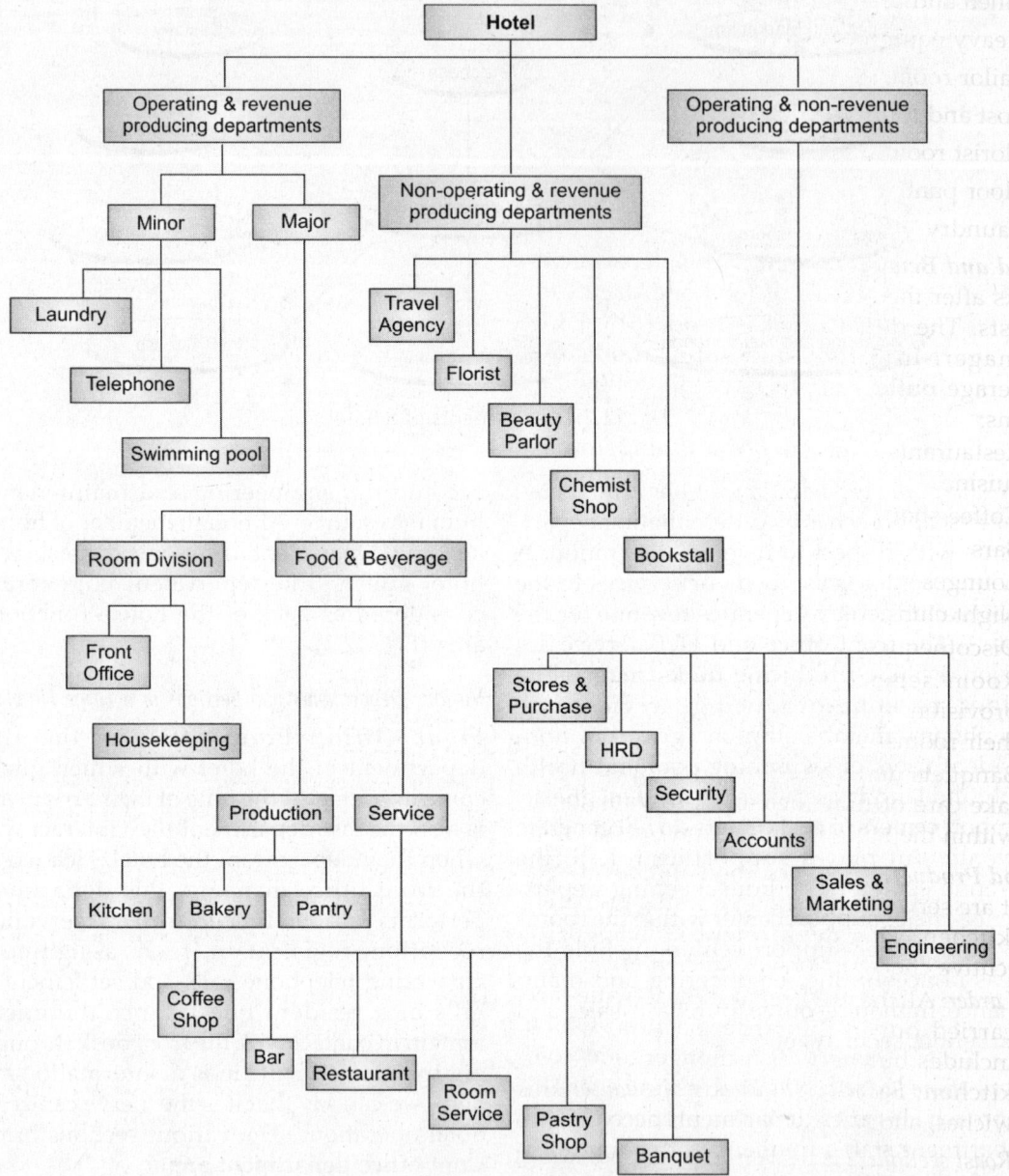

Fig. 12.3: Operating & revenue producing and operating & non-revenue producing departments

Housekeeping: This department is responsible for the cleanliness, maintenance and aesthetic upkeep of the front of the house as well as back of the house areas, so that they appear as fresh and aesthetically appealing as on the first day the hotel property opened for business. This department is headed by the executive housekeeper. The various sections of the housekeeping department are:

- Control desk
- Linen room
- Uniform room

- Linen and uniform store
- Heavy equipment store
- Tailor room
- Lost and found section
- Florist room
- Floor pantry
- Laundry

Food and Beverage Service: This department looks after the service of food and drinks to guests. The department is headed by F&B Manager. In a five-star hotel, food and beverage outlets might have the following forms:

- Restaurants—specialty/fine dining/multi-cuisine
- Coffee shops
- Bars
- Lounges
- Night clubs
- Discotheques
- Room service section looks after the provision of food and drinks to guests in their rooms.
- Banquets and Outdoor Catering sections take care of functions and programs both within the hotel premises and outside it.

Food Production: All the food and beverages that are served to the hotel guest is prepared in kitchen. This department is headed by executive chef. Sections of kitchen are:

- *Larder*: All the pre-preparation activities are carried out in the larder section, which includes butchery, fish monger, and cold kitchen. Salads, salad dressings, sandwiches, and juices are also prepared here.
- *Roast section*: It is responsible for providing all roast dishes of meat, poultry, and game.
- *Vegetable section*: It is responsible for the preparation of all vegetable dishes.
- *Sauce section*: It is responsible for preparing sauces required for all meat, poultry, game dishes, with the exception of those that are plain roasted or grilled.

- *Fish section*: It is responsible for supplying all fish dishes, with the exception of those that are plain grilled or deep fried.
- *Soup section*: It prepares all types of soups such as consommes, creams, veloute, purees, broths, bisques, and international soups.
- *Pastry section*: It prepares all hot and cold sweets, like breakfast rolls, cakes, pastries, and various desserts.
- *Specialty section*: Indian/Continental/Chinese/Mughlai, South Indian, etc.

Engineering and Maintenance: It is headed by chief engineer. The department is responsible for maintaining the property's structure and all kinds of maintenance and repair of equipment, furniture, fixture and fittings installed in hotel. The maintenance service is also referred to as facilities management, as it deals with the maintenance of ground, building, equipment, waste disposal system, store and sanitary, pollution control equipment, gas distribution system, electrical energy supply system, fuel supply system, water supply system, ventilation, refrigeration and air conditioning, fire-fighting, heating, telephone system, cable television, elevator, light, escalators, etc.

Human Resource: It is headed by HR Manager. Recruitment and selection, orientation/induction, training and development, employee welfare, salaries and compensation, labor laws, and employees' safety and working conditions come under the purview of HR department.

Sales and Marketing: Every product and services of the hotel needs to be utilized for maximizing towards revenue generation which is the responsibility of the sales and marketing department. Responsibilities of this department include—room sales, group sales, conference/convention sales, public relations and advertising. This department is headed by sales and marketing manager.

Purchase: The purchase department is responsible for procuring the inventories of all departments of a hotel. This department is led by the purchase manager. In most hotels, the central stores are a part of the purchase department. The requisitions from all departments are sent to the stores, on the basis of which a consolidated purchase order is made and goods are purchased in bulk.

Accounting/Finance: Every financial responsibility within the hotel lies with the accounting department. The accounts department monitors and records all the monetary transactions of the hotel. From the transaction at the front office, food and beverage to the daily review of transactions by the night auditor lie with the accounting department. This department is responsible for making payments against invoices, billing, collecting payments, and generating statements, handling bank transactions, processing employee payroll data and preparing the hotel's financial statement. In many hotels, the night audit and the F&B audit come under the purview of the accounting department.

Security: Headed by the chief security officer, the security department of a hotel is responsible for the overall security of the hotel building, in-house guests and their belongings, visitors, day users, and employees of the hotel. They are also responsible for conducting fire drills, monitoring surveillance equipment and patrolling the property. The security personnel should be trained to handle situations like vandalism, thefts, terrorist attacks, bomb threats, and also to prevent and fight fire.

REVIEW QUESTIONS

1. Discuss in detail the duties and responsibilities of various departments in a large hotel.
2. Write in detail about organizational mission, goals and strategies.
3. Draw the organizational chart of a large hotel depicting the lines of responsibility among positions, departments, and divisions.
4. Discuss briefly about minor operating department.
5. Write short notes on:
   - Organization
   - Mission statement
   - Goals and Strategies
6. Explain the need of organization in hotels.
7. What are the three traditional work shifts? What variations on the traditional work shift might a hotel adopt?
8. How do a hotel's goal relate to its mission statement and to departmental and divisional goals and strategies?
9. Differentiate between:
   - Revenue center and Cost center
   - Front-of-the-house areas and Back-of-the-house areas
   - Operating and revenue producing department and Operating and non-revenue producing department
   - Job description and Job specification.

Part III

Front Office Operations

13 | An Introduction to the Front Office

Learning Objectives

After reading this chapter, you will be able to understand the following:
- An overview of the front office department
- Layout of front office department
- Various sections of front office—reservation, reception, information, cash and bills, communication, travel desk, uniformed services

AN OVERVIEW OF THE FRONT OFFICE DEPARTMENT

Front office is the interface between a hotel and its guests. The front office department is the central point of the activities that take place between guests and a hotel. The employees of the department are among the first employees of the hotel to interact with the guest. This interaction starts with the processing of the reservation request and continues through the stages of arrival, stay, departure, and even after departure (when the hotel forwards mails received for the guest).

The front office may be regarded as the *show window* of the hotel and hence must be well designed and maintained in a well-organized and orderly manner. Regardless of the class or type of the hotel, front office is the most visible and essential focal point of a hotel.

Front office is the name given to offices situated in the front of house, that is, the lobby, such offices where the guest is received, provided information, his luggage handled, his account settled at departure, and his problems, complaints, and suggestions are

looked after. The front desk is the link between the guest and the hotel and represents the hotel to the guest and is a liaison between the hotel management and the coordination of all the guest services.

Front office can be defined as 'a front of the house department located around the foyer and the lobby area of a hospitality property'.

The reception desk is one small part of the front office and therefore in a world class type of hotel, forms a single section within the entire front office department.

Functions Performed by the Front Office Department

The front office of a hotel generally performs the following basic activities:
- Processing the reservation request of guests, this involves making room reservations, amendments, and cancellations.
- Receiving/welcoming guests at the time of their arrival.
- Making arrangements for the traditional welcome of guests.

- Registration of guests and the assignment of rooms.
- Handling guests' luggage from the guest vehicle to the assigned room on arrival and from the guest room to the vehicle at the time of departure.
- Accepting guests' valuables and cash for keeping in safety deposit lockers.
- Delivering messages and mails of resident guests.
- Handling guests' keys.
- Guest paging.
- Posting and verifying the room charges and any other credit charges in the guest folio.
- Providing information to guests about hotel products and services, and events or places of tourist interest.
- Arranging postage and courier of mails and other documents.
- Making travel arrangements like sight-seeing tours or intercity travel for guests.
- The parking of guests' own vehicles.
- Preparing, presenting, and settling guests' bills at the time of departure.
- Providing left luggage facility, etc.
- Handling guests' complaints.

Importance of Front Office Department

- The front office is one of the two major departments which produces revenue in a hotel, the other being food and beverage.
- The front office generates nearly 60–70% of the revenue for the hotel and is responsible for achieving a high occupancy of the hotel.
- One of the major functions of the front office is selling accommodation.
- The front office co-ordinates as a link between the hotel and the guest and at times is the only area which comes in contact with a guest and the guest knows the hotel by the front desk only.
- The front office is like a dispensing post to handle complaints and suggestions of the guest.
- Front office is indeed the nerve centre, the hub and the heart of the hotel.

- Front office co-ordinates with other departments of hotel in order to provide more efficient service to the guest by communicating with every other department with instructions and directions for services, care and relations with guests.
- Front office monitors the guest cycle and co-ordinates all guest services such as information, mail and message handling, guest accounting and bill settlement, telephone, telex and other communication services.
- Safety and security of guests and his belongings are also function of the front office.
- Front office maintains relations with travel agents, tour operators, airlines and other hotels.
- Front office has a complementary role of image-building, which is the first and last point of contact of every guest.

LAYOUT OF FRONT OFFICE DEPARTMENT

Layout is the physical demarcation of the sections of a department. A well-designed layout should involve proper space utilization, aimed at improving the efficiency and control of the staff. The front desk should be located at a prominent place in the lobby.

Many times in addition to the building ambience, the main entrance and approach play a very important role in the selection of hotel for a guest. When a guest enters a hotel, it is the entrance and thereafter the lobby, which is subjected to scrutiny. The main entrance must be identifiable and directly lead to lobby of the hotel. Care should be taken that the main entrance is able to accommodate the guests of the hotel who may come by a car or taxi or coach. To complicate the situation, a number of guests may arrive by several modes of transport at the same time. As a general rule the driveway in front of the lobby entrance should be at least 18 feet wide so as to allow at least two cars to pass easily. Sufficient height clearance to allow coaches loaded with luggage on top should also be given consideration. A minimum of 16 feet high clearance from road should be allowed. Another important point to be considered is the number of steps for entrance into the lobby area, as generally a raised entrance approach by steps is used. To avoid the problem of carrying of luggage through steps it is advisable to have separate luggage entrance in the form of a ramp starting from driveway to the main entrance. The ramp should not be very steep. The recommended inclination with road of the ramp for luggage trolleys and wheelchair is 1:10 as most convenient. Figure 13.1 shows the layout of the front office department.

Lobby

A lobby is an area furnished with seating arrangements and is a meeting place common to all hotel guests whether residents or not,

located immediately upon entry into the hotel building. The front office is strategically located in the lobby area.

Hotels spend considerable funds and efforts to make the lobby aesthetically appealing and rationally convenient. The atmosphere, décor, and the staff are suggestive of what the guest is going to get on his visit to the hotel. Size of the lobby depends upon the size and type of hotel. The systems used, such as manual, mechanical or automatic, will also be important while deciding the size of the lobby. Lobby shall be spaciously designed, of course without wasting any valuable space. The pillars should be avoided as they may obstruct the view and may create problem in the movement. If pillars are to be used to give support to superimposed floors, then they may be made multifunctional by arranging service ducts through them. Also they may be used for display of items such as jewelry and other small but expensive items.

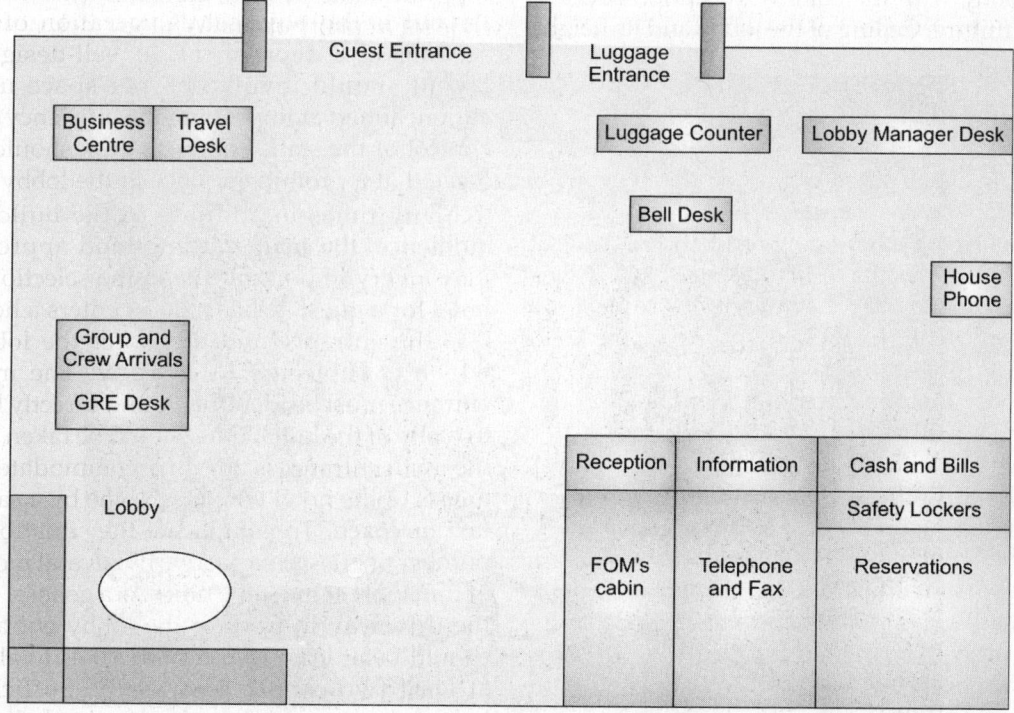

Fig. 13.1: Layout of the front office department

Lobby of the hotel includes the general circulation and waiting area which leads to check-in, information, and the cashiers' counter and also to the bell desk, travel desk, hospitality desk, GRE desk, lobby manager's desk, elevators, cloak rooms, etc. Within this area there also may be a shopping arcade, a coffee shop, etc. As the lobby usually serves as a meeting or gathering area for guests and their visitors and is the most common area it should be well planned, designed and furnished so as to give all a first and best last impression. The reception desk which is located in the lobby should be so located that it has open view of the entrance, exits, elevators, shops, etc. so as to oversee activities in the areas and if possible, the dining area. Another important point is that in addition to the entry from the porch or drive-in, there must be an entrance to lobby from the car parking area which is usually in the basement.

Color of distemper used must be in harmony with the color of reception counter or furniture. Ceiling of the lobby and its height is also important. It should give the impression of spaciousness. As far as possible, maximum utilization of natural light and air should be done for lobby planning. The ceiling of the lobby may be made the focal point of the lobby. The design and shape of lobby such as flat or dome shape should also match the architectural theme of the hotel. Figure 13.2 shows the lobby of a hotel.

The Reception Counter

The reception counter which is located in the lobby is where the various activities connected with guests such as check-in, information, check-out, mail handling, luggage handling, bill settlement, etc. are handled from; as such the reception counter must be fully functional and operational and well planned. Figure 13.3 shows reception counter of a hotel. The following points are important:

Size: Basically size depends upon the size and systems used by the hotel. For a large hotel using automatic system, the size may be small while for a small hotel on Whitney rack system

Fig. 13.2: Lobby of a hotel

Fig. 13.3: Reception counter

or manual system, the size may be comparatively big.

Shape: Another important factor is that the counter should be designed matching with the shape of the lobby. For example, L-shape, straight shape, curved shape or circular shape. Circular or semi-circular shape provides an effective service to more guests and appears more modern and innovative but since guests will approach the front desk from all angles, more staff is needed. In a straight shape, fewer staff is needed, but fewer guests can be served at the same time. Nowadays, the concept of deskless environment is prevalent in most of the hotels. Figure 13.4 shows different shapes of front desk.

Dimension: Usually the counter dimensions are: Height between 38 inches and 42 inches; width is 30 inches approximately and the length depending on various factors such as size of lobby, type of hotel, business profile of the hotel and the systems used, etc.

Material used for making the counter varies from wood and cement to concrete with finishing of laminated surfaces, stone, marble, granite, etc. depending upon various factors such as cost and design, etc.

Apart from the front desk, other counter and offices which are located in the lobby of a large hotel are lobby manager's desk, concierge desk, bell desk, travel desk, GRE desk, etc. and right behind the counter the reservation area, safe deposit lockers, bills section, telephone exchange may be situated.

Reservation Section

The location of the reservation section depends upon the size of hotel and the volume of business of the hotel. In a very small hotel, the functions of reservations can be performed by the front desk staff since the volume of reservation is less. In a large hotel, a separate section is needed since the job of reservation involves a lot of paperwork and generally does not involve direct contact with the guest. It should ideally be located behind the reception counter and there should be direct access of reception staff to this area through a door, preferably a swing door.

Curved front desk

Circular front desk

Straight front desk

L-shaped front desk

Fig. 13.4: Different shapes of front desk

The reservation section of a large hotel is headed by the reservation manager who is assisted by reservation assistants/clerks. The size of the reservation section shall vary according to the size of the hotel, and also the systems used. A large hotel with automated reservation system will require a smaller area as compared to a hotel operating on manual or semi-automated system. However, the basic requirements will remain the same. Proper storage place in the form of filing racks, cabinets, etc. should be there for storing stationery and reference materials such as reservation form, reservation slips, files for various categories of guests such as companies, travel agents, airlines, etc. who provide business to the hotel very frequently. Principle of motion study must be kept in mind while planning the layout of the reservation section. Fig. 13.5 shows the layout of the reservation section of a hotel.

Bell Desk

The bell desk is located very close to the main entrance of the hotel.

VARIOUS SECTIONS OF FRONT OFFICE DEPARTMENT

The front office department of a hotel comprises various sections. Depending on the size of the hotels, the sections may vary. In small or

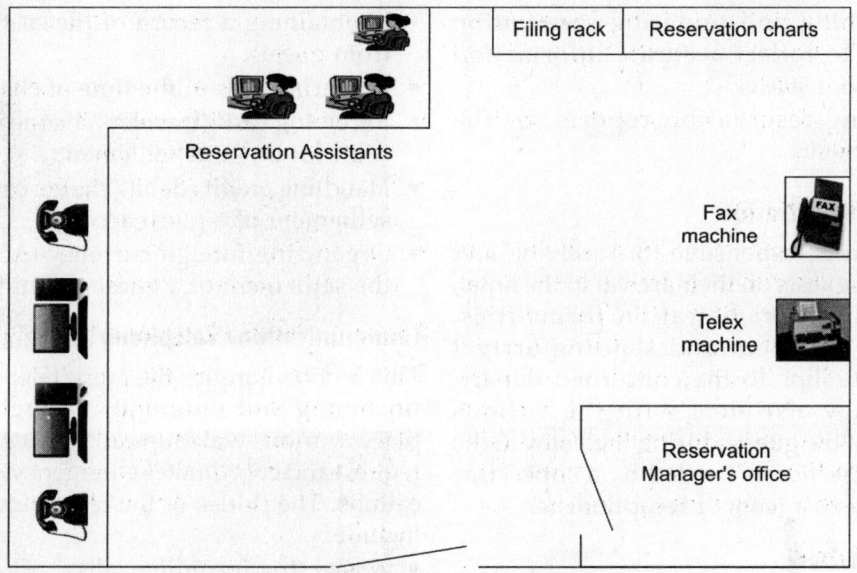

Fig. 13.5: Layout of the reservation section

medium-sized hotels the sections may be merged and handled accordingly. Figure 13.6 shows different sections of the front office department.

Reservation

This section is responsible for receiving and processing reservation queries, which involves making room reservations, amendments, and cancellations. The section is headed by a reservations manager, who is assisted by a reservations supervisor and a team of reservation clerks or assistants. The following functions are performed by the reservation section:

- Receiving reservation requests through various means like telephone, fax, e-mail, websites, sales representatives, or central reservations department.
- Processing reservation requests received from all means on the hotel property management system (PMS)
- Depending upon the availability of desired room type and projected sales during the requested dates, the reservation request may be confirmed, waitlisted or denied.
- Updating the room availability status after each reservation transaction, i.e. after each confirmation, amendment, and cancellation.

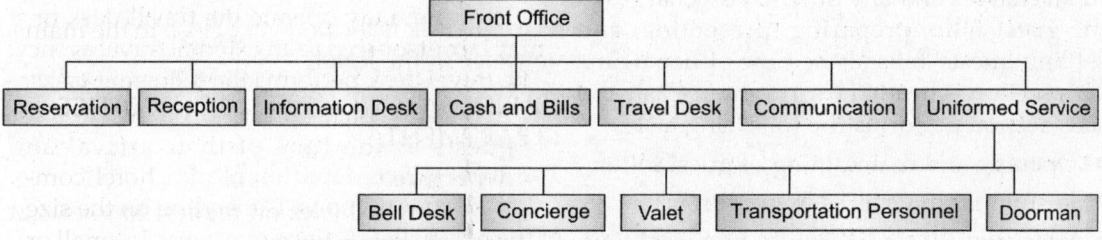

Fig.13.6: Sections of front office

- Maintaining and updating reservation records to reflect accurate information about room status.
- Preparing reservation reports for the management.

Reception / Registration

This section is responsible for receiving and welcoming guests on their arrival in the hotel, completing the registration formalities, assigning the room and sending arrival notification slips to the concerned departments. They also deal with the various problems of the guests during their stay at the hotel. This section is headed by a supervisor and comprises a team of receptionists.

Information Desk

This section is responsible for providing information to guests regarding hotel (facilities and services offered by hotel, rules and regulations applicable for guests, different F&B outlets—timings, location, etc.) and the city at large (railway timings, flight schedule, and historical, business, cultural places of interest, etc.). It is manned by an information assistant. In a small hotel, the same function may be performed by the receptionist. They are also responsible for delivering messages, mails, faxes, couriers, parcels, etc. of resident guests; handling guests' keys; paging guests.

Cash and Bills

This section records various monetary transactions between the guest and the hotel. It is responsible for posting and verifying the room charges and any other credit charges in the guest folio; preparing, presenting, and settling guests' bills at the time of departure. This section is headed by a cashier. Cash and Bills section performs the following tasks:

- Opening and maintaining of guest folios.
- Posting room charges in guest folios.
- Recording all credit charges in guest folios.

- Maintaining a record of the cash received from guests.
- Preparing bills at the time of check-out.
- Receiving cash/travelers cheques/demand draft for account settlement.
- Handling credit/debit/charge cards for the settlement of a guest account.
- Organizing foreign currency exchange for the settlement of a guest account.

Communication / Telephone Exchange

This section handles the guest telephone calls (incoming and outgoing) through EPABX, places various wake-up calls for the guests on request and coordinates emergency communications. The duties of the telephone operator include:

- Answering incoming calls.
- Directing calls to guest rooms through the switchboard/EPABX
- Providing information on guest services
- Processing guest wake-up calls
- Answering inquiries about hotel facilities and events
- Protecting guests' privacy
- Coordinating emergency communication

Travel Desk

The travel desk takes care of travel arrangements of guests. It assists in the booking of air tickets, railway reservations, airport or railway station pick-up or drop, providing vehicles on request to guest. It also arranges/organizes city tours, sightseeing tours to the guests on request and arranges for guides who can communicate in the guests' language.

The hotel may operate the travel desk or it may be outsourced to an external travel agency. The travel desk performs the following tasks:

- Arranging pick-up and drop services for guests at the time of their arrival and departure.
- Providing vehicles on request to guests at pre-determined rates

- Making travel arrangements like railway reservation/air-tickets
- Organizing half-day or full-day sightseeing
- Arranging for guides who can communicate in the guest's language

Uniformed Services

Employees who work in the uniformed service department generally provide the most personalized service in a hotel. Because of the high degree of attention awarded to the guests by this department, some hotels refer to uniformed service as *guest service*.

Often the first and last person a guest sees at the hotel is a uniformed service employee. The care and attention of bell boys, doorman, parking attendants and other uniformed service personnel convey a critical message to guests about the hotel's commitment to service.

Because of their direct contact with guests, uniformed service employees can be excellent salespersons for restaurants, health club and other hotel revenue outlets. Uniformed service staff also serves as the eyes and ears of the hotel since they are stationed at or make so many trips to various points in the hotel. It is their duty to report any irregularities or unusual circumstances to the front desk.

- Bell Attendants/Bell Hops/Porter/Bell Boys: Mainly responsible for luggage handling between the lobby and guest rooms during the arrival and departure of the guests. Also responsible for arranging postage and courier of mails and other documents and left luggage room. They escort guests to their rooms and familiarize them with hotel facilities, safety features, as well as in-room facilities. The bell desk is located very close to the main entrance of the hotel. This section is headed by bell captain.
- Door Attendants/Doorman: Ensure opening the door of the hotel, baggage service and traffic control at hotel entrance.

- Valet Parking Attendants: Ensure parking services for guest's vehicles.
- Transportation Personnel: Ensure transportation services for guests from and to the hotel.
- Concierge: The concierge service is relatively new to modern hotel-keeping though it is quite old in its concept. It is an extended arm of 'Information desk' and is located separately in the lobby.

Basically, a concierge serves as the guest's liaison with both hotel and non-hotel services. Many hotels actually enlist front desk staff to provide concierge-like services. Some hotels, however, find that front desk staffs are too busy to provide the personal services characteristic of a concierge. By adding a skilled concierge to the front desk staff, a more personal and specialized approach can be given to guests.

Concierge specializes in assisting the guest, regardless of whether inquiries concern in-hotel or off-premises attractions, facilities, services or activities. A concierge must be unusually resourceful and knowledgeable about the hotel and the surrounding community. They are also called *Man-about-town or Mister-know-it-all*.

Key to a concierge's list of achievements is the local contacts he has made at box office, restaurants, airlines, car rental offices, etc.

Some hotels actually encourage concierges to visit corporate houses and organizations go establish and strengthen relationship. Finally most successful concierges speak several languages.

In some hotels, the head concierge assumes additional responsibilities for supervising all uniformed service staff.

REVIEW QUESTIONS

1. 'Front Office is the nerve center of a hotel's operation'. Explain this statement with suitable examples.

2. 'First impression is the last impression which is created by front office department'. Justify this statement.

3. Discuss the importance and functions of the front office department.

4. Draw a layout plan of the lobby and 'back of office' of a front office department and discuss its various sections.

5. Draw a neat diagram of a 5 star hotel lobby, depicting the layout of different functional areas as well as entry and exit points. Label the diagram clearly.

6. Discuss in detail the functional areas within the front office department.

7. Discuss various functions performed by the information desk.

8. Discuss the duties performed by various sections of front office department.

9. Describe the role of concierge in a hotel.

10. Discuss the planning and layout of the reservation section.

Front Office Communication

Learning Objectives

After reading this chapter, you will be able to understand the following:
- Interdepartmental communication
- Intradepartmental communication

INTERDEPARTMENTAL COMMUNICATION

No individual department in any hotel can work in isolation. A willingness to cooperate and coordinate, with the assistance of efficient methods of communication, is essential if the establishment is to run smoothly. Proper and appropriate coordination between the departments and sections is essential for smooth operation and for superior service to the guests.

The front office department is just one of the departments in a hotel working towards the satisfaction of the guests, and each department is dependent on others for information and services if its work is to be accomplished efficiently. Frictions between departments must be kept to a minimum and there should be close interdepartmental liaison.

The front office department plays a pivotal role in delivering quality services to guests. It sets the stage for a pleasant or an unpleasant visit. The front office communicates guests' requirements to other departments, which work in close coordination and cooperation to deliver required products and services. In order to maintain the desired level of service,

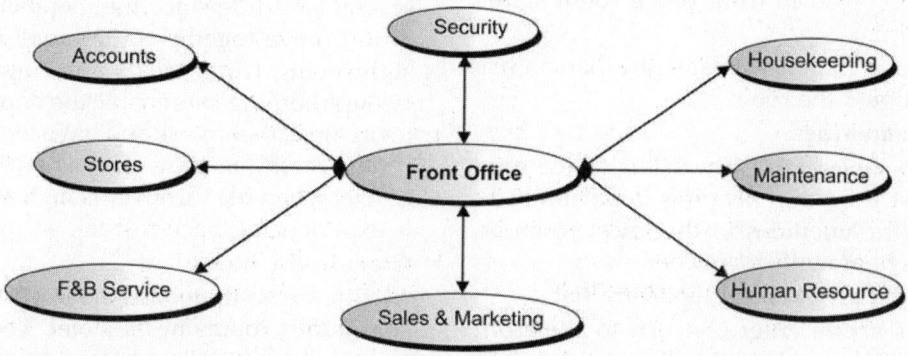

Fig. 14.1: Co-ordination of front office with other departments

the front office department communicates with the following departments of the hotel.

Co-ordination between Front Office and Housekeeping Department

The front office and housekeeping departments communicate with each other for the following information:

- **Room status**

  As rooms generate maximum revenue for hotels, the information about the room status should be updated correctly and frequently. The front office and housekeeping departments must coordinate on the room status. The housekeeping department prepares 'room status report' or the 'housekeeping status report' which indicates the physical count as seen by room boy. This report indicates the current housekeeping status of each room. This room status report is compared with the *front office occupancy report* or *night report*, and discrepancies are brought to the attention of the front office manager. A room status discrepancy is a situation in which the housekeeping department's description of a room's status differs from the room status information being used by the front office to assign guest rooms. This helps to:
  - Update room status
  - Find sleepers (a room from which the guest has checked out but it is showing as occupied in front office room status records)
  - Charge the guest if an extra person has occupied the room.

- **Extra amenities**

  For special guests, the front office may request the housekeeping department to put extra amenities in the guest room by means of amenities voucher.

- **Expected arrival and departure list**

  Rooms are of chief concern to the front office and housekeeping departments. The front office must provide lists for expected arrivals and departures for the day in advance, and notify housekeeping of actual arrivals and departures as and when they occur. Rooms received by housekeeping for cleaning are called 'departure rooms' while cleaned rooms handed to the front office for sale are called 'clear rooms'.

- **Room change slip**

  It is also important to intimate room changes by means of room change slip, so that items left behind by guests may be handed over, 'extras' retrieved, and laundry delivered.

- **Flower arrangements**

  Sometimes the management extends its compliments to a guest with a special gesture of flower arrangement in the room as recognition of the importance of a person. This requirement of flower arrangements for certain guests is conveyed to housekeeping by the front office on a daily basis.

- **VIPs in the house**

  The list of VIPs coming to stay at the hotel is passed on from the front office to the control desk. On receiving this information, the executive housekeeper personally supervises the servicing of the VIP rooms.

- **Groups in the house**

  It is vital that the control desk be informed about the groups being registered in the hotel, as servicing their rooms requires a special schedule since the members of the group move together in terms of arrivals, sightseeing trips, meals and departures. Group rooming lists enable the department to organize their work and have the group's rooms ready on time. This is particularly crucial when the turnover is high and hotel is experiencing back-to-back occupancy.

- **Crew in the house**

  Airline crew members are registered in contiguous rooms in the hotel. They often have odd sleeping hours because of international time differences. Because of odd timings for international flights, these crew rooms may display a DND Card at times when other guests are normally out. All these factors require that the house-keeping department be given advance notice of crew arrivals for special schedu-ling of GRAs cleaning these rooms.

- **Rooms under maintenance**
  Knowledge of rooms, which are out of order or under repair, is important for proper room management.

- **Security concerns**
  The housekeeping staff should inform the front office about any unusual circum-stances that may indicate a violation of security for the hotel guests.

- **Uniform issue**
  The front office needs to depend on housekeeping for the provision of clean uniforms to its staff.

Co-ordination between Front Office and Engineering and Maintenance

The engineering and maintenance department is responsible for the provision of engineering facilities that contribute to the comfort of guests and increase the efficiency of staff. The front office department depends on maint-enance to keep things in order. Any deficiency or fault should be immediately reported to maintenance over telephone and these req-uests are usually dealt with promptly if the rapport between the two departments is good. To be able to 'clear' a room for sale, it is necessary that all malfunctioning items in a guest room are attended promptly by maintenance. Hotels usually use work-order form to report maintenance problems.

Co-ordination between Front Office and HR Department

Front office coordinates with the human resource department for recruitment of its staff; managing their salaries and wages; addressing indiscipline; issuing identity cards

for employees; running induction programs; maintaining locker facilities; completing income tax formalities; effecting transfers, promotions, appraisals, and exit formalities; procuring trainees; and organizing training sessions to keep them up-to-date with the latest happenings in the hotel industry.

Co-ordination between Front Office and Sales and Marketing

The front office staff must take every effort to keep the information on room availability status and guest histories current and accurate. The sales and marketing executives may have to check the availability of rooms three, six or even twelve months in future to devise marketing strategies for off season period. This information helps the sales and marketing department to sell hotel products by bundling two or more hospitality products, like rooms with meals; rooms, meals and entertainment—all-in-one package. Thus a close cooperation and coordination between the front office and sales and marketing departments is important for hotel profita-bility.

Co-ordination between Front Office and F&B Department

The front office and F&B department communicate with each other for the fol-lowing information:

- The arrival and departure of guests.
- Special arrangements like cookies, fruit basket, soft drinks, etc. (by means of amenities voucher).
- In-house and expected VIPs
- In-house and expected groups
- In-house and expected crews
- Scanty baggage guests (all points of sales are notified to collect all payment in cash)
- The banquet manager should ensure that the intimation of forthcoming banquet functions is conveyed to front office well in advance.

Co-ordination between Front Office and Stores

Coordination with stores ensures the availability of day-to-day necessities of front office department. Communication with stores is by way of stores requisition form, which front office sends to stores when it requires certain items.

Co-ordination between Front Office and Accounting

The front desk provides a daily summary of the financial transactions after night auditing to the finance controller. The information provided by the front desk helps the finance controller to make budgets and to allocate resources for the current financial period. The front desk provides the controller the financial data for billing and maintenance of credit card ledgers. High balance reports, etc. enable the controller to formulate policy guidelines and strategies to recover the money from guests and companies.

Co-ordination between Front Office and Security

The coordination here is mainly concerned with the prevention of fire, prevention of thefts, safe keeping of keys and lost property. Front office staff should report anything of a suspicious nature immediately to the security staff. A guest may take advantage of his privacy and may be engaged in certain illegal activities such as gambling, smuggling and so on. Front office staff has to be alert to this risk and seek the security department's intervention, if necessary. The security department is responsible for conducting training sessions on handling emergency situations for the staff. For example, they conduct fire drills to train staff to gear up in a fire emergency.

INTRADEPARTMENTAL COMMUNICATION

As discussed earlier, the front office department itself comprises numerous sections:

- Reservation
- Reception/Registration
- Information Desk
- Cash and Bills
- Communication
- Travel Desk
- Uniformed Services

Again within each section there is a need for communication. The front desk may keep a **'Log Book'** so that all front office employees are aware of important events and decisions that occurred during previous work shifts. It is a very important register maintained by the front desk staff and serves as an important mode of written communication between two consecutive work shifts. A typical front office log book is a journal which chronicles unusual events, guest complaints or request or relevant information. Front desk agents record notes in the log book during their shift, which should be clearly written in prescribed form so that they serve as a reference material for the next shift. Before beginning their shift, front desk agents should review and initial log book, noting any current activities, situations that require follow up or any potential problems. The employee must also note what action was taken. These notations become an important link in the communication network. By reviewing these notes, the front desk agent on duty can respond intelligently if the guest contacts the front desk.

Doorman/Bell Boy: While the taxi stands at the main door of the lobby—the doorman should open the door of the car with proper care and wish the guest and then open the main lobby door. At this point the bell boy escorts the guest with his luggage to the front office. As soon as the registration is complete, the front desk personnel informs the bell boy of the room number and the bell boy escorts the guests up to the room and familiarizes him with the salient features of the room. Usually when the guest wishes to check-out calls the bell desk to get his luggage down. Then bell boy informs the front office cashier to prepare his bills. After settlement of guest's bill and having received the room key the front desk staff issues a checkout slip to the bell boy to carry all luggage of guest to his departing vehicle.

Room change: Changing of room often takes place in every hotel. Then with guest's consent the receptionist will issue a room change slip to the bell captain. The bell boy will shift all luggages to the new room and distribute the room change slip to all concerned departments so that all concerned personnel can be informed about the change of room.

Escape checkout: There are some guests (although not very much) who try to leave the hotel without paying the bill. This kind of leave may be possible because of his scanty luggage. In such cases, the guest himself can carry the baggage and secretly check-out without the notice of bell boy or the receptionist. In such situation the bell boy should inform the front office about the scanty baggage guest so that the front office personnel can take advance payment from the guest and minimize the chances of loss of revenue.

Mail delivery: As soon as the reception receives any message, fax, e-mail, postal mail or any kind of article for guest, the receptionist can take the help of the bell boy to send them to the guest room.

Guest paging: Paging takes place when a phone call is received demanding a certain in-house guest. In this system the name of the guest is displayed in a small board with a long handle and there is a small bell at the corner of the board. Now the page boy holds the handle and rings the bell to draw the attention of the guests to the board. The page boy does the paging in restaurants and in the public area. As soon as the concerned guest notices the board, he comes to receive the telephone.

All the above co-ordinations are needed in between reception, bell boy, page boy and front office cashiers.

Information directory: Front desk staff must be able to respond in a knowledgeable way when guests contact front office for information. Common guest queries involve:

- Local restaurant recommendations
- Directions to the nearest shopping center, place of worship, bank, theatre, travel agency or other points of interest.
- Information about hotel policies.

Front desk staff may require to access some obscure information to answer guest queries. Some front office accumulates such data in a bound guide called an information directory. The front office information directory may include simplified maps of the area; taxi and airline company telephone numbers; bank, theatre, etc.

Hotel XYZ
FOA's HANDING OVER AND TAKING OVER SHEET

Date: _____   Shift: _____

VIP's in the House: _____

VIP's Expected: _____

Rooms Blocked: _____

Messages, if any: _____

Room Change	From	To	Reason

Complaint for follow-up	Room No.	Time	Nature of Complaint	Complaint No.

Miscellaneous information: _____

Handed over by: _____   Taken over by: _____

Fig. 14.2: Handing and taking over sheet

Hotel XYZ

LOG BOOK
(Front Desk)

Date: _____
Day: _____
Shift: _____
FOM: _____

S. No.	Details	Remarks

Fig. 14.3: Front desk log book

REVIEW QUESTIONS

1. Write a note on the importance of 'Interdepartmental Communications' between front office and other departments.
2. Explain the co-ordination between front office and housekeeping in terms of room inventory control.
3. Intradepartmental communication is the backbone for efficient front office operations. Discuss.
4. Draw the format of:
   - Room status report
   - Log book
   - Discrepancy report
   - Occupancy report
   - Maintenance work order
5. Explain the use of a 'log book'
6. What is the purpose of a front desk information directory? What sort of information does it contain?

15 | Front Office Organization

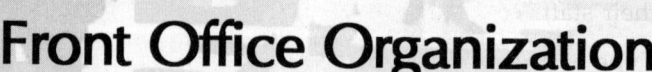

Learning Objectives

After reading this chapter, you will be able to understand the following:

- Organization of front office staff
- Duties and responsibilities of front office personnel
- Qualities of front office staff
- Rules of the house for the front office staff

ORGANIZATION OF FRONT OFFICE STAFF

The front office department is organized on the principles of division of labor and span of control.

Division of Labor

It refers to narrow specialization of tasks within a process so that each employee can become a specialist in doing one thing. The process is divided into several separate tasks, each performed by one person. The specialization and division of labor defines each employee's sphere of competence and ensures that employees perform their individual tasks without overlapping others.

The degree of division of labor depends on the degree to which the performance of particular task is measurable, the degree to which wages affect task performance, and the implementation of technology. Computerization has enabled organizations to increase the variety of tasks performed by employees, consequently reducing specialization and division of labor.

Span of Control

Span of control in an organization is defined as the number of employees reporting directly to one supervisor. Traditionally, the span of control has been defined as four to seven subordinates under one manager. The average size of the span of control, together with the total number of employees, determines the number of levels in an organization structure.

Organization Structure of Front Office Department

It is necessary to have a well-defined organizational structure for smooth operations of front office. The structure of the rooms division will vary from hotel to hotel. The organizational structure of the front office depends on many factors:

- **Size of the hotel**
  Bigger the hotel, the more specialized the staff is required to be. Whereas in the smaller hotels, one employee/staff member may perform a wide variety of duties. A large hotel will have a complex structure of hierarchies and positions, while a small hotel will have a simple structure.

- **Standard of service**

  High-class hotels usually provide more personal services for guests and, therefore, they expect greater specialization from their staff.

- **Type of guests**

  The needs of guests usually differ on the basis of their purpose of visit, i.e. business client prefers less time to be spent on checking-in and checking-out, and it is quite possible they would not mind carrying their own luggage. So the emphasis is more on staff in the front desk section than at the concierge.

- **Type of hotel**

  A hotel situated in the airport area knows that a guest may check-in or check-out at any time during the day. So more emphasis

needs to be there on front desk as a full team is needed to be on duty at all times.

The organization chart of front office of small, medium and large hotels is depicted in Figs 15.1 to 15.3 respectively.

DUTIES AND RESPONSIBILITIES OF FRONT OFFICE PERSONNEL (Fig. 15.4)

Front Office Manager

Front office manager is in charge of the front office department and allocates the available resources (men, machine, materials, and money) of the department to achieve the goals of the department and the organization. Responsible to the general manager for the operation of the front office, including reservations, front desk, uniformed service and telephone operation, through his

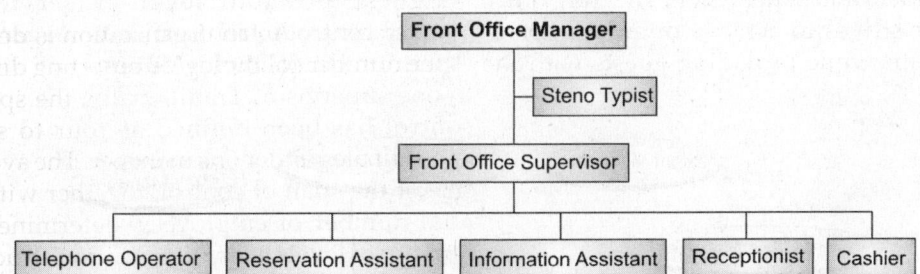

Fig. 15.1: Front office organization of a small hotel

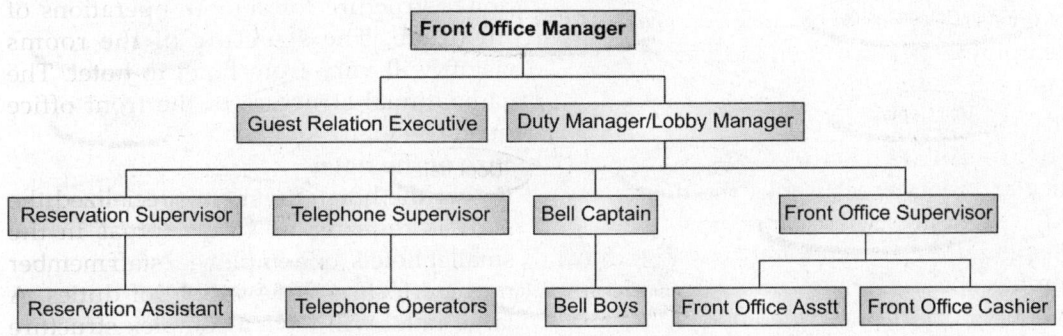

Fig. 15.2: Front office organization of a medium hotel

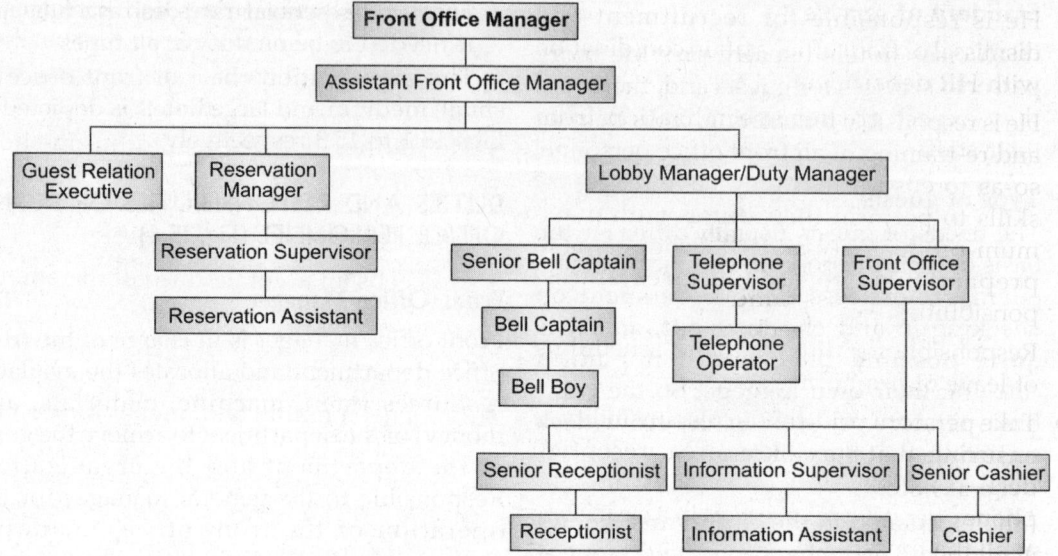

Fig. 15.3: Front office organization of a large hotel

management and supervision. The front office manager strives to achieve optimum operating results while providing guests with the highest possible level of service and satisfaction. A front office manager has to perform the following duties:

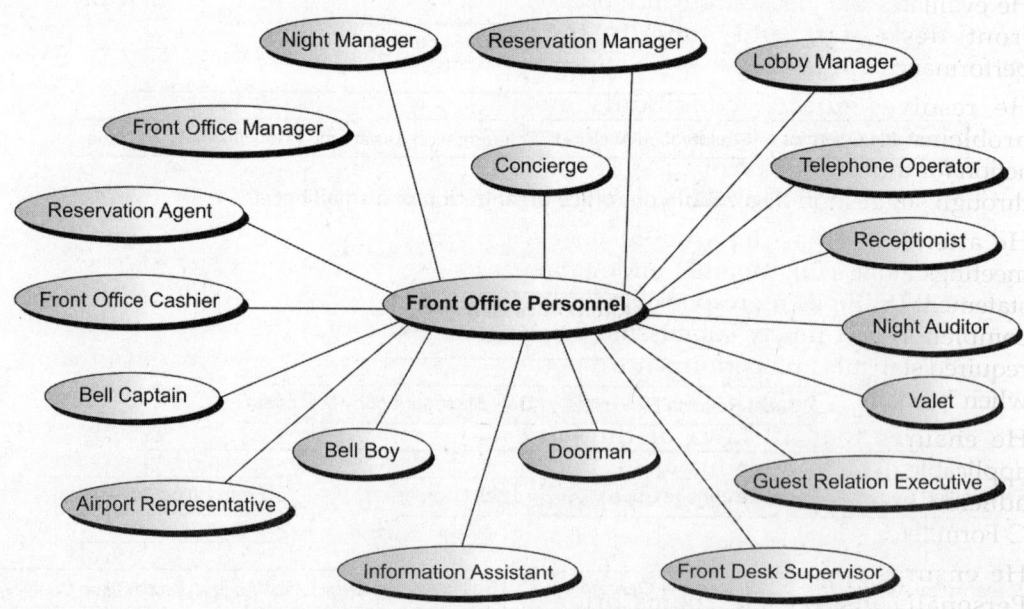

Fig. 15.4: Front office personnel

- He is responsible for recruitment and dismissal of front office staff in coordination with HR department.

- He is responsible for training, cross training and re-training of all front office personnel so as to ensure that they have necessary skills to perform their duties with maximum productivity and efficiency so as to prepare them to assume greater responsibilities.

- Responsible for duty rota and scheduling of leave of front office staff.

- Take personal pride in the responsibility of ensuring that the lobby area, reception desk, concierge desk, as well as the telephone operator area are kept in a clean state at all the times.

- He conducts regular meetings of front office staff.

- He develops SOP for guest arrival patterns, control of keys, receiving telephone calls, handling mails and messages to ensure efficiency and higher guest satisfaction.

- He evaluates the job performance of each front desk staff and submits their performance appraisal.

- He resolves guests' complaints and problems effectively to take corrective action for the satisfaction of guest. He goes through suggestion book daily.

- He attends regular department heads' meetings along with monthly information statement (MIS). Being responsible for the completion and timely submission of all required statistics and performance reports when needed.

- He ensures that all laws of the land applicable to the hotel industry are strictly adhered to, e.g. foreign exchange, taxes, C-Form, etc.

- He ensures special handling of VIPs. Personally inspect VIP rooms prior to guest's arrival. Welcome and escort VIP guests to their rooms.

- He decides special rates and packages to different companies on the basis of business that the hotel gets from a particular company.

- He defines policies regarding no-show, early arrival, overbooking, cancellation, etc. to ensure optimum occupancy with maximum average room rate.

- He keeps an accurate track of the amount of contribution of the front office department in the total revenue earned by the hotel.

- Prepares monthly and annual forecasts of the total revenue to be generated by the front office department from the sale of rooms. Ensures that the target/budget sales and occupancy are achieved.

- Ensures all allowances, paid-outs, registration cards, etc. are signed by him on daily basis.

- Reviews and complete credit limit report. Ensures that credit control procedures are strictly adhered to, that no bills exceed the stipulated limit without prior approval in order to minimize the unpaid city ledger accounts.

- To monitor trends within the industry and make suggestions how these could be implemented.

- To be fully aware of the market situation, as well as the competitors, to conduct at least twice a year a survey of market and competitors.

- To be fully aware of the booking situation together with the reservation manager to avoid any uncontrollable situation, with maximizing the rooms revenue.

- To maintain close working relationship with the hotel sales, to ensure that groups, conventions are properly handled and that sales is kept informed as to occupancy levels.

- To maintain a good relationship with potential tour operator/travel agents/corporates to increase their materialization with the hotel, and to solve any problem that could affect rapport with the hotel.

Lobby Manager

- Brief staff of uniformed services and ensures that the lobby is clean and tidy.
- Manages the discipline of the uniformed services.
- Manages all guest complaints and take action appropriately.
- Manages scanty baggage procedures.
- Monitors airline crew movements and coordinate with the respective airline control rooms for wake up calls and schedules.
- Oversees the left-luggage procedures and the safety of the left-luggage room.
- Attends to any disputes over guest billings.
- Updates on operational policies and procedures and inform the uniformed services.
- Trains staff of uniformed services.
- Appraises the performance of the uniformed staff and recommend rewards and recognition.
- Coordinates guest requests with concerned department of the hotel.
- Assists the security in lobby surveillance.
- Ensures that group and crew arrivals and baggage movement is conducted efficiently.
- Oversees the concierge, bell desk services, valet parking, doorman and transportation personnel.
- Acts on behalf of front office management at night.
- Custodian of the master key and all the keys to all stores (at night).
- Welcomes the VIP guests and escorts them to their room.

Night Manager

Responsible to the front office manager for the front office operation during the night hours, and for the general supervision of the entire hotel operations. Through his supervision, he should maintain and control the night staff performance with a special attention to the front office operation. The duties and responsibilities of a night manager are as under:

- Ensures that the front office staff is on duty, as well as the rest of the department is running the smooth operation.
- Reviews the rooms availability and to be familiar with the expected VIP arrivals or any other likely unusual situation to occur.
- Ensures that the lighting levels through the hotel are appropriate in view of power conservation and management policy.
- Ensures the security staff is on duty as scheduled.
- Ensures that all access to the hotel is secured as required in the hotel policy.
- Makes regular random patrols through the hotel including guest corridors and back of the house, and hotel outlets to ensure the good security and orderliness.
- Completely involved in any problem occur during the night hours, or any guest complaint, and to report the problems and action taken to the front office manager to be discussed next morning.
- Makes spot check on night cleaning staff to ensure their productivity, and to check the standard required.
- Makes frequent visits to the restaurants and bars to ensure service levels and that problems are dealt with promptly.
- Ensures that all hotel stores are locked.
- Ensures the night staff services are done efficiently, and up to the standard of the hotel policy.
- Keeps close to the front desk for giving any suggestion and supervision during the night hours.
- Supervises and follow up the night run and make sure of the accurate bucket check of all the in-house rooms.

- Ensures that the wake up calls is done efficiently, and as per the required standard.
- Be aware of the early arrivals and departures, especially for groups, and to supervise the front desk and bell staff preparation, luggage down, airport pick up, breakfast box, and wake up calls, etc.
- With the engineering on duty, check hotel refrigerators, boiler room and power station.
- Inspects staff cafeteria cleanliness and that night meal is served during the allowed time in a smooth atmosphere.
- To record any activities which may be of interest to the management in the night manager report, to be handed to the front office manager next morning.

Reservations Manager

Responsible to the front office manager for the maintenance of accurate and complete reservation records. The duties and responsibilities of a reservations manager are as under:

- Be fully aware of the booking situations at all times, and supervise the reservation clerks/supervisors in order to maximize the hotel sales with the highest possible average room rate.
- Following up on tentative booking, watching cut-off dates, and monitoring group tour business accounts blocks for productivity.
- Training reservation agents and setting up cross-training programs.
- Reviewing all VIP reservations and working with the rooms division manager and the front office manager on assignments
- Developing and maintaining a good working relationship with the central reservations office and travel agents
- Ensures that all reservations, both group and individual, are recorded on standard forms attached with the concerned correspondence, and they are filed by arrival date in easily accessible files.
- Prepares the 10 days and 3 months forecasts on timely basis for review by the front office manager.
- Ensures that incoming mails/faxes requiring a reply are answered with minimum delay.
- Bring to the attention of the front office manager dates when the hotel availability status should be changed, and update the sales office with the hotel availability accordingly.
- Fully responsible of the tour operators/travel agents correspondence regarding reservation decisions especially in the critical periods, allotment increase, stop sales, release period amendment, etc.
- Ensures that the reservations taken over the phone are answered promptly and politely, and according to the hotel answering script.
- To use and to train the reservation personnel using the up-selling techniques in order to maximize the rooms revenue.
- To ensure that all rooming lists for the group's reservations are received on time and accurately entered into the computer system.
- To ensure that all the definite reservations are guaranteed, to minimize the last minute non-guaranteed cancellations.
- To update the hotel data system with the updated contracted travel agents rates.
- In coordination with the front office manager, set up the over-booking margin for the normal and peak periods to ensure filling the house without making unnecessary turn away.
- Be fully aware of all the hotel facilities and services, and to train the reservation staff on using the hotel facilities during answering reservations.
- To prepare and submit all statistics and reservations reports requested by the front office manager or the general manager.

Reservations Agent/Reservations Assistant/ Reservations Clerk

Reservation assistant processes the reservation requests that reach the hotel by any mode. He should possess great salesmanship skills by suggesting higher room categories, and also selling other hotel services like spas, restaurants, etc. to the guest. As we know the reservation section generates the maximum revenue for the hotel, so reservation assistant should understand, anticipate, and influence consumer behavior in order to maximize the profits. The duties and responsibilities of a reservations assistant are as under:

• To receive and process the reservation requests that reaches the hotel by any mode—telephonic, written, or online.

• To process reservations from the sales office, other departments of the hotel, travel agents, tour operators, and corporate booking agents.

• Assists in pre-registration activities when appropriate.

• Processes cancellations and amendments. Promptly relay this information to the front desk.

• To maintain reservation records by completing reservation forms, sending reservation confirmation or amendment letters, and updating the status of rooms after processing each reservation request (i.e. confirmation, amendment, and cancellation).

• Knows the type of rooms available as well as their locations and layouts.

• Knows the selling status, rates and benefits of all package plans.

• Knows the credit policy of the hotel and how to code each reservation.

• Creates and maintain reservation records by date of arrival in alphabetical order.

• Determines room rates based on the selling tactics of the hotel.

• Prepares letter of confirmation.

• Understands the hotels policy on guaranteed reservations and no shows.

• Processes advance deposits on reservations.

• Tracks future room availability on the basis of reservations.

• Helps develop room revenue and occupancy report.

• Prepares expected arrival list for front office use.

• Handles daily correspondence: Respond to enquiries and make reservations as needed.

• Makes sure that files are kept up-to-date.

Guest Relations Executive (GRE)

Responsible to the front office manager for the efficient follow up on the hotel guest's special requirement, assisting the hotel guest through all the hotel departments, giving special care to the VIP and regular guests. The duties and responsibilities of a GRE are as under:

• Bids farewell to guests, noting their comments.

• Properly attend to guest's specific requests such as, assisting in booking/reconfirmation of the flight ticket, accepting reservations for the next destination, providing assistance with specific queries or problems.

• Completes guest questionnaires with departing guests.

• Prints special arrival list in morning and together with GM review arriving guests, identifying frequent guests and VIPs and return guests.

• Allocates rooms to VIP and return guest and ensure welcome letters, welcome back letters and gifts are place in room.

• Checks allocated rooms for cleanliness and that guest supplies are in room.

• Makes presence felt in restaurant, bar and banquets.

- Handles guests' problems and complaints.
- Escorts VIPs and frequent guests on arrival.
- Meets incoming guests, offer welcome drinks.
- Contacts selected guests by telephone to obtain information concerning customer satisfaction.
- Spends time on desk ensuring front desk staff are following procedures, train and guide where necessary.
  - Answering calls
  - Greeting guests
  - Knowledge of services provided by the hotel
  - Knowledge of all aspects reflected on guest services directory, e.g. shopping, doctor, etc.
  - CLS system
  - Messages procedure
  - Asking/communicating with guest as to achieve rapport and create ambience
  - Knowledge of company mission, vision and values
  - Recommends other hotels in group, etc.

Telephone Operator/Switchboard Operator

Responsible to the front office manager for the speedy and courteous answering of incoming telephone calls both internal and external in a warm and friendly tone. The duties and responsibilities of a telephone operator are as under:

- Answering the internal and the external calls with the hotel approved script
- Directs calls to guest rooms, staff or departments through a PBX or EPABX system.
- Places outgoing calls.
- Receives telephone charges from telephone companies and forwards charges to front desk for posting.
- Receives and distributes messages for the guest.

- Log all wake-up call requests and perform wake-up call service.
- Provides information on guest services.
- Answers questions about hotel facilities, events and activities.
- Be acutely aware of the procedures for handling fire alarm and other emergency procedures.
- To have available all important and emergency numbers, and also other numbers that might be requested by any of the hotel personnel.

Front Office Cashier

During the stay in a hotel, guests may perform various credit and debit transactions with the hotel. At the time of departure, guest settles his bill at cashier. It is essential for the front office cashier to keep the guest folio updated by posting all credit and debit transactions. The duties and responsibilities of a front office cashier are as under:

- He receives and records the advance payment made by the guest at the time of room reservation.
- He establishes the method of settlement of bill by the guest on his arrival at the hotel and processes the mode of payment.
- He opens the folio or account of the guest as soon as the guest registers himself in the hotel.
- He posts the various charges incurred by the guest in the various revenue centers of the hotel such as restaurant, coffee shop, laundry, etc. on the basis of vouchers.
- He disburses various petty amounts to the third party on the request of the guest to pay the taxi bill, medicine charges, etc.
- He prepares account allowance vouchers in order to give concessions to the guests due to poor services and products offered to him during his stay in the hotel.
- He prepares account correction vouchers in order to rectify any posting errors in the folio of guest, which may increase or

decrease the overall account balance of the guest.

- He exchanges the foreign currency given by the guest and also gives an encashment certificate to the guest as a proof of foreign currency exchanged by him.
- He performs the important function of processing the checkouts and settlement of the guests' bill at the time of departure of the guest from the hotel.
- He performs the important function of routine night audits to verify the accuracy of the account of the guest and to resolve any sort of discrepancies.
- Operates equipment used in front office cashiering, i.e. EFTPOS, credit card imprinter, electronic cash register (NCR).
- Obtains the house bank and keeps it balanced.
- Completes cashiers pre-shift supply checklist.
- Takes departmental reading at the beginning of the shift.
- Transfers folio charges to the non-guest ledger to each company's master file.
- Balances departmental totals at the close of shift.
- Manages safe deposit lockers.
- He is responsible for the safe keeping of payments made by the guest—cash, travelers' cheque, etc. under lock and key.

Receptionist

Responsible to the front office manager for checking in and out the hotel guests, assisting guest efficiently, courteously and professionally in all front office related functions, and to maintain a high standard of service and hospitality at all times. The duties and responsibilities of a receptionist are as under:

- Greets all guests in a friendly and helpful manner, and attempt to learn and use guest's names at every opportunity.
- Upon check in, ensures that the guest completes his registration card completely

and legibly, and that the guest is assigned a room of the type and the rate indicated on the reservation.
- Accommodates guest's special requests whenever possible, assists in preregistration and room blocking whenever necessary.
- Stay up-to-date on room rates, special packages, discounts and how to handle each.
- In the case of walk in, the guest should be sold a room with the highest possible room rate.
- To use the up-selling techniques in order to maximize the rooms revenue.
- Being knowledgeable of all the credit cards and cashing policies, and how to handle cash properly and efficiently.
- Develops detailed knowledge of the room locations, facilities and types.
- Develops detailed knowledge of the hotel's key personnel, service, outlets, and hours of operation for each.
- Handles the safe deposit boxes according to the hotel procedures.
- Prepares and reports guests with high balance to the attention of the front office manager.
- Be thoroughly aware of the hotel reservation system, and cancellation policy.
- Communicates with all other departments through the proper channels, and through the communication forms.
- Promptly notifies the housekeeping of all checkouts, early check-in, special requests in the rooms.
- Actions the housekeeping reports immediately upon receipt, record discrepancies and report to the shift leader.
- Develops a working knowledge of the reservation department, take same day reservations, and be aware of the cancellation procedures.
- Uses proper telephone manners.
- Understands and uses properly the mail, parcel, message delivery.

- Reports any unusual occurrence or request to the manager on duty or the front office manager.
- Maintains the cleanliness and neatness of the front desk area at all the times.
- Reads and initials log book and front office bulletin board to keep updated.

Bell Captain / Head Hall Porter

The bell captain is in charge of bell desk and is responsible for organizing, supervising and controlling all lobby services with the help of bell boys. The duties and responsibilities of a bell captain are as under:

- He controls the movement and activities of bell boys on the lobby control sheet which is a summary of the activities of all bell boys.
- He prepares duty rota and schedules leave for bell boys.
- Conducts daily briefing of bell boys.
- Handles left luggage room formalities and supervises left luggage register, baggage check and luggage inventory sheet.
- Trains bell boys to maximize their efficiency. Responsible for implementation of SOP as laid down by the management.
- Controls the sale of postage stamps to guests.
- He maintains record of all guests with scanty baggage and informs front desk staff.
- He reports irregularities of suspicious persons to the lobby manager.
- He prepares errand cards for the bellboys.
- Coordinates and controls the distribution of morning newspaper.
- Organizes and supervises check-in/out baggage, formalities of groups, crews, etc.

Bell Boy / Bell Hops / Porter

Responsible to the bell captain for welcoming and escorting guests with their luggage to and from their rooms during check-in and check-out, perform various other functions related to the guest needs. The duties and responsibilities of a bell boy are as under:

- He escorts the guest to his room, familiarizes guests with the guest room and in-room facilities and services.
- Checks the room for any damage and missing items at the time of departure of guest.
- Conducts light housekeeping services in the lobby.
- He carries luggage, parcels, etc. to guest room.
- He should possess thorough knowledge about the topography of the hotel.
- He distributes mails to various departments as well as to the guest.
- He distributes newspapers to the guest.
- He should be familiar with luggage storage procedures and luggage room.
- He keeps postage stamps available for guest request.
- He is responsible for paging of guest.
- He sees that the Form C is deposited to the FRRO/LIU or the nearest police station.

Night Auditor

A night auditor performs the following duties and responsibilities:

- Establishes the end of the day.
- Ensures the accuracy of front office accounting records and balances them.
- Reconciles all the financial transactions between a hotel and its guests.
- Calculates the total revenue generated during the day.
- Verifies and validates the cashier's posting of charges in the guest and non-guest accounts.
- Posts the room charges and taxes in the guest folio.
- Transfers the unpaid guest accounts, i.e. accounts of guests who have left the hotel without settling their bills—to city ledger.
- Monitors the credit limit of guests.

- Prepares a high balance report of guest accounts exceeding house limit.
- Monitors the current status of discounts, meal coupons, and other promotional activities that are carried out by the front desk employees.
- Prepares important reports which detail the results of operations for the management of the department and the hotel at large.
- Tracks room occupancy percentage and other front office statistics.
- Prepares a daily summary of all the cash and credit card settlements that took place at the front desk.

Front Desk Supervisor

- Trains and cross-trains front desk personnel in the tasks of registration, mail handling, information services, and check-in and checkout procedures
- Regulates the service given in the front desk
- Acts as a liaison between the guests and management, particularly with regard to problem-solving activities
- Is responsible for seeing that daily and hourly computer reports are run and distributed
- Assigns VIP rooms to ensure guest satisfaction
- Resolves room discrepancy report
- Prepares the weekly/monthly schedule of front desk employees
- Assists the group coordinator with all group arrivals, either directly or through the delegation of this duty to other front desk staff

Doorman/Door Attendant/Link Man/ Commissionaire/Carriage Attendant

His place of duty is outside the main entrance of the hotel. He must be well-informed about hotel features and the local community. Guests frequently ask door attendants for directions to restaurants or local attractions. He is the first staff to greet guests upon arrival and last to bid farewell upon departure. The duties and responsibilities of a doorman are as under:

- Opening car doors and assisting guests in and out of their vehicles with a smile.
- Brings umbrella, if needed.
- Helps bell boys load or unload guest luggage.
- Opening hotel entrance doors for guests.
- Checks taxis to ensure that the guest has not left any belongings.
- Ensures smooth traffic flow in the porch and ensures that the porch is always clear.
- Calls for cars parked in the parking area.
- Keeps the keys of the guests' car safely.
- Provides valet parking service, if required.

Information Assistant

Provides information to the guest about the hotel's products and services, nearby food and beverage outlets, places of tourist interest in the city and around, etc. He also handles guest mail, messages and keys. The duties and responsibilities of an information assistant are as under:

- Provides information to guests about the hotel's products and services, nearby food and beverage outlets, places of tourist interest in the city and around, etc.
- Keep information aids like time-table, road maps, hotel guide, etc.
- Handle guests mails and messages
- Assist in guest paging.
- Maintain information rack.
- Sign departure errand cards in acknowledgement of receipt or room keys from a departing guest.

Airport Representative

- Meets all booked guests at the airport, welcomes them and assists in arranging transport to the hotel.
- Meets and maintains rapport with airlines counter staff, airport staff and counter staff

of various travel agencies on a continuous basis to generate layover/FIT business for hotel.

- Receives, meets VIP guests on arrival and escorts them to the hotel.
- Sees off VIP guests at the airport.
- Maintains liaison and good rapport with custom authorities and helps in quick clearance of VIPs.
- Provides constant feedback to front office manager on airport happenings.
- Assists hotel VIP guests, in case of excess luggage, ticket reconfirmation and securing seats.
- Keeps record of all business generated by airport staff. Prepares weekly reports and attends sales meeting every week.

Valet Parking Attendant

Valet is generally available at hotels offering world class or luxury service for parking guest and visitor automobiles. Guests do not have to worry about finding a parking space, walking to the hotel in inclement weather or finding their vehicles in the parking lot.

Valet parking attendants are responsible for the security of guest vehicles. Security checks must be made and a proper system installed. The security measures must be taken for the safety of the parking area of the hotel. Particularly, when the parking is in the basement, the security becomes more important.

Valets do not take a car into their custody without issuing a ticket (usually in duplicate) to the guest. Cars are not issued to the guests unless he returns the ticket issued by the valet. Car keys must be kept in a secure area, because if lost or given to the wrong person, the hotel can be held financially responsible.

In some hotels, non-resident guest using parking facilities are levied some charges according to the duration while resident guests are usually provided free parking facility.

Valet parking attendants are also responsible for keeping the motor entrance area clear by providing quick service to incoming and outgoing guests and visitors.

In some hotels, the car parking is given to an outside agency on commission basis. Hotels which do not have in-house parking facility make arrangements with the nearby municipal or private parking lot on payment.

Concierge

- Providing directions and information about the hotel, its services and amenities, city, town, country, travel and transport, banks, etc.
- Booking flight and train tickets
- Reservations for dining
- Arranging for secretarial services
- Handling mail and parcel services
- Arranging sightseeing tours
- Transportation arrangement
- Limousine services
- Obtaining tickets for theatres, concerts, sporting, or any other special events
- They are involved in itinerary planning.
- Organize special functions such as VIP cocktail parties.
- Arrange for hotel doctor as and when required.

QUALITIES OF FRONT OFFICE STAFF

It is well known that the people who are on duty at the front desk are among the most important contacts between an arriving guest and the hotel. Guests often derive the initial information and sometimes lasting impression from the attitude of the front office staff. It is important that the front desk personnel greet every guest with a pleasant, personable smile and a warm greeting regardless of the time of day or the amount of activity. In view of the important role they play, the front office staff must possess the following essential attributes to discharge their duties effectively:

- **A high sense of personal grooming:** Uniforms must be clean and neatly pressed. Hair should be groomed well. Nails should be manicured. It is preferable for ladies to tie their hair up in a bun. A soft cologne is preferable to heavy perfumes. Jewelery should be restricted to a minimum. In short, the front desk staff must be seen at their best at all times.

- **Personal hygiene:** The front desk employees should follow the highest standards of personal hygiene because they are in direct contact with guests throughout the day. A good sense of personal hygiene is imperative for front desk employees as their appearance influences the image of the hotel in the eyes of the guests.

- **Self confidence:** This is necessary as the front office staff meet guests of different countries, statuses and cultures. They should be comfortable and feel at ease in dealing with these people.

- **Excellent communication skills:** The front office personnel must possess excellent communication skills as they interact with guests. The front desk staff should be proficient in English which is the official language of the world and should be also aware of the local language of the place where the hotel is located. It becomes important for the front desk staff to know a foreign language in order to deal with the guests coming from different countries.

- **Diplomacy:** Diplomacy is the quintessential characteristic needed in front desk personnel. Very often there are situations when a guest is irate over something or the other; a diplomatic dealing helps in diffusing the explosive moment. It is quite common for a busy hotel to have no room to offer a guest who has come with a confirmed booking. A diplomatic approach is the only way by which the guest can be pacified without upsetting or offending him.

- **Honesty:** The front office employees should be honest and trustworthy. They should not succumb to the temptations that may arise during the day-to-day working of the department.

- **Quick decision making ability:** Guests often approach the front desk with problems and requests. The front office staff must be able to decide quickly the course of action that satisfies the guest, at the same time keeping the interests of the hotel alive.

- **Ability to remember names and faces:** This single attribute distinguishes the good from the average amongst the front office staff. Every individual has an ego and his name is most precious and personal to him. If the front office staff can call most guests by their names, this immediately flatters them and personalizes the guest experience. The guest begins to feel he is welcome as staff recognizes him by name.

- **Good manners:** This is supposed to be the basic quality that should be present in all the staffs of the hotel industry. Good manners are the essence of hospitality and will thus attract more guests and will maintain the goodwill of the hotel and the industry at large. Thus, good gestures on behalf of the front desk staff help in creating a good image on the guests coming to the hotels.

- **Physical fitness:** Front office operations require the staff to stand for long hours at stretch. So the staff should be physically fit and active so that they are able to serve the guests with smile and pleasure.

- **Ready smile:** Guests like to be handled by a cheerful staff at the desk. Their smile exudes cheer to the guests and puts them at ease.

- **Tower of patience:** Being the nerve center of the hotel, the front office is constantly in touch with guests and therefore invariably comes under tremendous pressure. The front desk staff should thus have a high degree of tolerance for pressure of work and be calm and composed at all times. The calmness and patience of the front desk personnel in high-pressure situations will help to diffuse tension, resolve the problem, and win the guest's faith and loyalty.

- **Salesmanship:** Front desk personnel should possess the quality of salesmanship. There are many instances when they can push slow-moving services or products of the hotel. If the room category desired by a guest is not available, they can suggest the guest to book a room of higher rate category rather than rejecting the reservation request. They can motivate guests to increase their length of stay by informing them about the nearby places of interest or upcoming events and activities. They should be equipped with complete knowledge about the hotel and its facilities, as well as happenings in the city.

RULES OF THE HOUSE FOR THE FRONT OFFICE STAFF

- Do not lean on the reception counter. It is a natural tendency and easily developed into a habit. When you have the impulse, step back from the counter or if possible go into the back office for some time.
- Always address a guest standing and never while sitting and maintain eye contact with him.
- Do not turn your back to the guest.
- Do not gossip over telephone. No private conversation over telephone.
- Do not grumble about your work or discuss the hotel's policy to your fellow workers in front of the guests. Do not discuss personal matters within the hearing of the guest. Even low voice conversation within the hearing of the guest should be avoided as this may create suspicion.
- Do not run. Walk briskly, if required.
- Do not invite your friends and guests to visit you while you are working.

- Be oblivious to guest's mistakes. Do not say, 'You are wrong' or 'You misunderstood me'; rather say 'I am sorry I did not make it clear'.
- Do not criticize competing hotels. This never helps anyone.
- Avoid using slang terms and phrases.
- Speak in the positive; every negative reply can be rephrased.
- The front desk staffs meet many celebrities at close quarters. He should not show curiosity about them. Do not stare at them. Do not try to listen to their conversation or laugh on their jokes. Photographs and autographs should be avoided.
- Information regarding guest and his movements should not be given to out-siders or other guests. Do not reveal information about a guest to any visitor including the room number without the permission of the guest.
- Do not enter the guest room without his permission.
- Always refer to guest by his name and never by his room number. Address them as Sir and Ma'am.
- Many times guests ask unreasonable questions. In such cases be patient and diplomatic in approach.
- Any policy regarding the bill settlement such as 'No personal cheques will be accepted' or 'Advance payment from scanty baggage guests' should be communicated very tactfully and diplomatically to the guest.

REVIEW QUESTIONS

1. Draw an organisation chart of front office department of a large, medium and small hotel.
2. Explain the duties and responsibilities of front office manager.
3. Write a short note on functions and role of GRE in a hotel.

4. Explain the rules of the house to be followed by the staff at the front office.

5. What are the essential attributes of a front office staff?

6. Write the job description for the following job positions:
   - Front Office Manager
   - Bell Boy
   - Lobby Manager

- Front Office Cashier
- Front Office Assistant
- Concierge
- Receptionist
- Bell Captain
- Reservation Agent
- Telephone Operator
- Night Manager
- Doorman

Equipment used in Front Office Operations

Learning Objectives

After reading this chapter, you will be able to understand the following:
□ Various pieces of equipment used in front office operations

EQUIPMENT USED IN FRONT OFFICE

A large number of factors play a very important role in the choice of equipment to be kept or installed in the front office of a hotel to carry out the day-to-day administration and management of the department. These factors include the system of operations prevailing in front office, the size of the hotel, the level of automation and the budget of the hotel.

Various pieces of equipment usually used in front office department are discussed below.

Room Rack

The room rack is a wooden framework with an array of metallic pockets designed to hold room rack slips arranged by room number. The room rack summarizes the reservation and housekeeping status of each guest room in the hotel.

Mail, Message, and Key Rack

It is a wooden framework containing an array of pigeon holes with each pigeon hole used to store the various mails and messages received for in-house guest. This rack also contains the keys of guest room when he is not in the room. The rack is located underneath the counter of the front desk.

Fig. 16.1: Mail, message, and key rack

Information Rack

An alphabetical index of registered guests used in routing telephone calls, mail, messages, and visitor inquiries. The information rack normally consists of aluminium slots designed to hold information rack slips.

Reservation Rack

A part of Whitney System, where all the information of the prospective guest is recorded. The slips are arranged in alphabetical order. The racks are filed chronologically by the guests' scheduled dates of arrival.

Voucher Rack

A container for storing vouchers for future reference and verification during the night audit.

Fig. 16.2: Voucher rack

Folio Tray/Folio Well/Folio Bucket

This equipment contains a large number of slots where the folios are arranged sequentially according to the room numbers. It is used by the front office cashier to store and track the guest folios of various registered guests of the hotel and is used to maintain the folios safely for future use and reference.

Account Posting Machine

A device used to post, monitor and balance charges and credits to guest account. There is a keypad in the account posting machine which is used by the cashier to enter the room number of the guest, department key (i.e. room, tax, food) and also the type of transaction (i.e. debit, credit and transfer). A posting machine normally provides:

- A standardized means of recording transactions.
- Account statement.
- A basis for cash and default payments management.

- An analysis of departmental sales activity.
- An audit tray of charges, purchases transactions.

Wake-up Device/Reminder-O-Timer

This is usually a specially designed clock with multiple alarms setting to remind front desk agent or telephone operators to place wake-up calls.

Call Accounting System

It is used by the telephone exchange section to automatically trace and bill the outgoing calls made by the guests during their stay in the hotel.

Cash Register

Used to record cash transactions and maintain cash balances.

Fig. 16.3: Cash register

Credit Card Imprinter

An imprinter presses a credit card voucher against a guest credit card. The impact causes the raised credit card details like the card number, expiry date, name of the card holder etc. onto the credit card voucher for use in credit card billing and collection procedures.

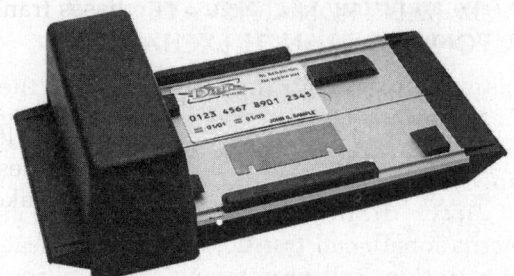

Fig. 16.4: Credit card imprinter

Credit Card Validator

The equipment is a computer terminal linked to a credit card data bank, which holds information concerning the validity of the credit card of the guest. The equipment assures the management that the guest has credit balance high enough to cover the projected charges and it also verifies that the card presented by the guest is not a stolen property.

Typewriter

The front office employees use typewriters to prepare guest reservation confirmation letter, to prepare the registration card of the guest and also to conduct the other word processing jobs of the department.

Fig. 16.5: Typewriter

Time Stamp

It is used to record the check-in and check-out time of the guests and delivery time of any mail or message for the in-house guest. Folios, mails and other front office paperwork are inserted into a time stand device to record the correct time and date.

Fig. 16.6: Time stamp

FAX (Facsimile) Machine

This is a facsimile reproduction machine that operates through telephone lines and is used to receive and send official documents. While sending a fax message, the operator dials the destination fax machine number and then sends the fax message by inserting the message page in the machine. The machine scans the page and makes an electronic representation of the text, sketches and graphics, compresses the data to save transmission time and transmits it to the dialled fax machine. The receiving machine decrypts the signals and uses its in-built printer to produce an exact photocopy of the original page. The fax machine can be run in auto mode, which makes it useful even if there is no one to answer the call.

Fig. 16.7: Fax machine

Computer, Printer, UPS and other related devices

These are used widely in the front office department for the purpose of reservation, registration, accounting and auditing. It helps to store and retrieve important data of the guest from time to time to carry out various guest services.

Fig. 16.8: Computer

EPABX (ELECTRONIC PRIVATE AUTOMATIC BRANCH EXCHANGE)

It is widely used in the hotels around the world nowadays for managing the incoming and outgoing calls of the hotel. This system is a modern system where many functions such as direct dialing (domestic as well as international), call transfer, call barring, call interruption, call monitoring, call waiting, speed dialing, hands free speakers, voice messaging service, conference facility do-not-disturb service, wake-up call facility, caller identification, etc. are available. The equipment has specialized software called the Call Accounting System which is interfaced with the computer system which automatically records, prices and posts the telephone charges in the respective guest folio.

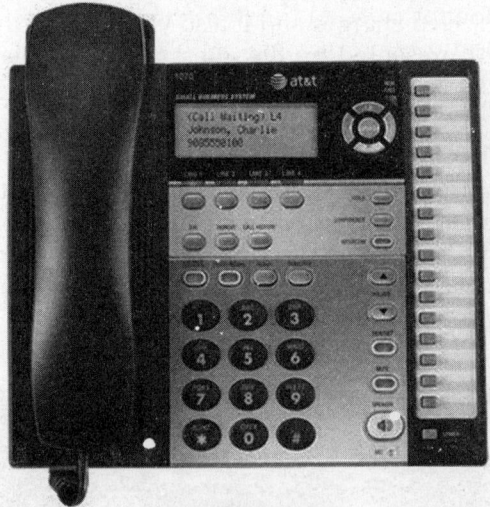

Fig. 16.9: EPABX

Magnetic Strip Reader

It reads data magnetically recorded and stored in the magnetic strip on the reverse of the credit card and transmits this data to a credit card verification service. On the basis of the credit card data and transaction data, the credit card verification service either approves or disapproves the transaction.

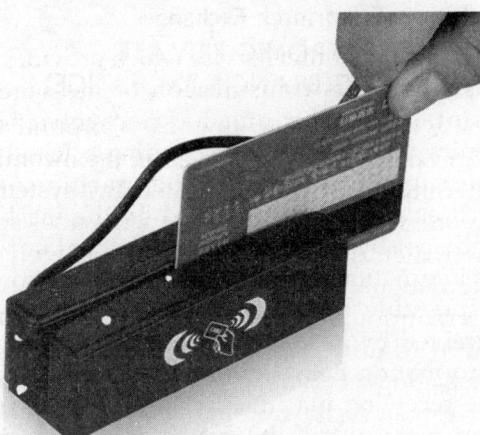

Fig. 16.10: Magnetic strip reader

Security Monitor

Close circuit television (CCTV) monitors allow front office or security personnel to monitor certain areas of hotel from a central location.

Fig. 16.11: Security monitor

Fig. 16.12: CCTV

Luggage Trolley/Bell Cart

Used at the bell desk to carry guest luggage to and from rooms.

Fig. 16.13: Luggage trolley

Franking Machine

Machine used for printing postage stamp value on the envelope.

Fig. 16.14: Franking machine

Voice Mail Box

This is an electronic device which is used for recording, storing, and playing back telephone messages for the guests in their absence.

Safe Deposit Box

Individual locker like bank vaults allotted to guest for keeping their valuables and documents.

Fig. 16.15: Safe deposit box

Self-registering Kiosk/Checkout Terminals

Fully automated hotels provide self-registration and checkout terminals for guests. These terminals do not eliminate front desk agents, but can free them to attend to other hotel duties, which can enhance guest service.

Fig. 16.16: Self-registering Kiosk

Telex or Teleprinter Exchange

24-hour teleprinter service which provides for instantaneous transmission of message in print. The teleprinter has a keyboard for typing messages. When sending a telex, the message is typed into the machine. The written data gets converted by the machine and sent by telephone link (national as well as international) to the chosen destination. It is possible to send a telex message to be received by an unstaffed telex machine, the information being automatically typed onto the receiving machine, where it can be read and print out of the message can be taken when the operator is on duty. Nowadays, this method is not used much because of the popularity of fax and computerized messaging or mailing system.

Fig. 16.17: Telex

Electronic Funds Transfer at Point of Sale (EFTPOS) machine

It is an electronic payment system involving electronic funds transfers based on the use of payment cards, such as debit or credit cards, at payment terminals located at points of sale.

EFTPOS technology originated in the United States in 1981 and was quickly adopted by other countries. Since 2002 the use of EFTPOS has grown significantly, and it has become the standard payment method, displacing the use of cash. Subsequently, networks facilitating the process of money transfer and payment settlement between the consumer and the merchant grew from a small number of nationwide systems to the majority of payment processing transactions.

On making a purchase, the EFTPOS customer gives his debit or credit card to the cashier who inserts it into an on-site EFTPOS machine. When the EFTPOS customer confirms the purchase, either by signature or security PIN, the EFTPOS equipment contacts the hotel's bank electronically about the transaction. A message is also sent to the customer's bank. Unless there is reason for the EFTPOS transaction not to be completed, the funds will then be transferred between the two accounts. The EFTPOS transaction takes a matter of only a few seconds. Before the EFTPOS customer has had the goods put into a bag, the EFTPOS transaction will be complete. Confirmation of the EFTPOS transaction is sent to the hotel and passed on to the customer in the form of a printed EFTPOS transaction record.

Fig 16.18: EFTPOS

Postage Scale

This scale is located in the concierge area and is used to determine the postage required for letters and small packages which a guest or hotel wishes to mail.

Fig. 16.19: Postage scale

Arrival Bucket

This bucket is located adjacent to the registration area and holds the pre-registration envelope for guests due to arrive today.

Folio Printer

These are laser printers dedicated to print guest folios and with multi-purpose trays attached to print system generated paid-out and foreign exchange vouchers.

Fig. 16.20: Folio printer

Key Encoder

Key card issuing machine which should be interfaced with PMS.

Fig. 16.21: Key encoder

Registration Card—Buckets

These buckets are used to store in-house guests' registration cards, supporting documents and credit card imprints. Each room number to be identified on a thin, plastic divider and placed inside.

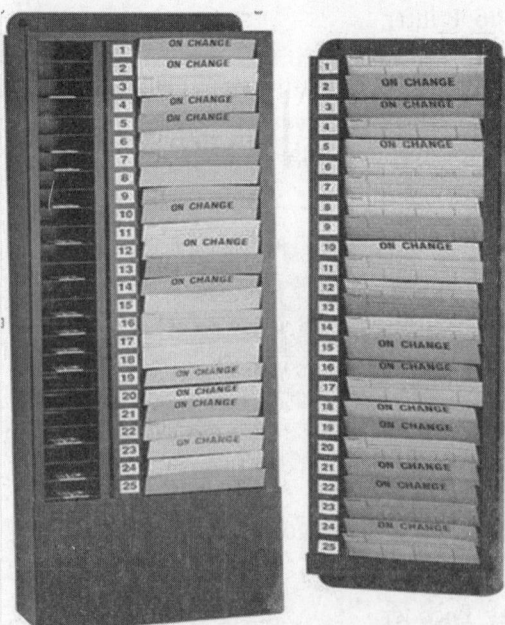

Fig. 16.22: Registration card—buckets

Support Devices

Fig. 16.23: Filing cabinet

Fig. 16.24: Money verifier

Fig. 16.25: Calculator

Fig. 16.26: Telephone

Fig. 16.28: Paging board

Fig. 16.27: Photocopying machine

Fig. 16.29: Function board

REVIEW QUESTIONS

1. Describe the various pieces of equipment used in front office operations.
2. Enlist and give uses/applications of automated, semi-automated and non-automated pieces of equipment used in the front office department.
3. To which extent are the operations of the front office enhanced with the use of the automated equipment.

Knowledge about Accommodation Products

Learning Objectives

After reading this chapter, you will be able to understand the following:
- Importance of product knowledge
- Various meal plans offered by hotels
- Different types of guest rooms in hotels
- Types of room rates
- Room status terminology

IMPORTANCE OF PRODUCT KNOWLEDGE

It is very important that the receptionist has an intimate knowledge of the product he is selling. It may appear that the product which the receptionist of a hotel has to sell is merely a room or a number of rooms, but it is much more than that in fact, and is not as simple as selling a room but far more complex. The product which he has to sell is whole range of services which the hotel has to offer to a guest depending upon its standards. This is particularly more important for a new entrant to this industry. It is suggested that he should be made to complete a checklist of the business for him, so that he is able to deal with the various possible questions which the guests may ask him from time to time during the process of selling the room to the guests.

This is further suggested that the checklist must be very detailed. Start with an analysis of location of business itself, its public areas so that he is able to show the guests the way or at least is able to direct him to what they want. The location of F&B outlets—their opening and closing hours, formal or informal dress requirements, etc. and other facilities and services like information about swimming pool, car hire, health club, details about rooms —their rates, locations, plans, etc. are equally important.

It is generally observed that in many hotels the front office staff which is directly responsible for the sale of the rooms has virtually no knowledge of the rooms in terms of its location and situation, the view from its window or balcony, etc. Many a times he may not have seen the room even which he is trying to sell. He may not know about the layout, its furniture, color scheme, type and size of bed and other such facilities and amenities available in the room. Some hotel management realize this problem and allow their staff to spend a night or two in their vacant rooms to give them a fairly good idea of the product which they are required to sell later. But in busy hotels this may not be possible. Some hotels keep color photograph of various types of rooms at the front desk. This helps the staff in equating the room requirement of the guest to the available room when shown. This would also help the potential guest in visualizing the room they are about to occupy. This is a useful aid in small hotels and where the rooms of the same type may vary in size, decoration and furnishing, etc.

Some small hotels use room-cards also at the reception counter for reference of the receptionist. This card contains information about decoration and furnishing. Another use to which the room card may be put is that it would tell the receptionist effectively as to the names of the guest occupying the room in recent weeks and any complaints that have been received. In large and medium-sized hotels this card is no more in use as this objective can be easily achieved by the use of visual display unit (VDU) very effectively.

MEAL PLANS

The room tariff of a hotel may be based on the choice meal plans offered to guests. Depending on the needs of their target audience, hotels offer a variety of meal plans or tariff plans (Fig. 17.1).

European Plan (EP)

This plan includes room rate only and the meals are charged separately as per actuals. It is generally preferred in a commercial hotel where business executives have to socialize with their clients and do not take meals at the hotel.

Continental Plan (CP)

This plan includes room rate and continental breakfast. A continental breakfast includes a choice of fresh or canned juices; breads like croissant, toast, brioche, rolls, etc with butter or preserves like jam, jellies and marmalade; beverages like tea or coffee, with or without milk.

American Plan (AP)

This plan includes room rate and all meals (i.e. breakfast, lunch and dinner). This plan is popular in resort hotels located at remote places where guests do not have a choice of food outside the hotel premises. This plan is also known as *Full Board Basis or Full-Pension Plan or En-pension.*

Plan	Plan includes				
	Room rent	*Morning tea*	*Breakfast*	*Lunch*	*Dinner*
European Plan (EP)	√	×	×	×	×
Continental Plan (CP)	√	√	√ Continental breakfast	×	×
American Plan (AP)	√	√	√	√	√
Modified American Plan (MAP)	√	√	√	Either lunch or dinner (one major meal)	
Bed and Breakfast (B&B) or Bermuda Plan	√	√	√ American breakfast	×	×
Note: √ *means included in plan*		× *means not included in plan*			

Fig. 17.1: Meal plans

Modified American Plan (MAP)

This plan includes room rate, breakfast and one major meal (either lunch or dinner). This plan is also known as *Half Board Basis or Demi-Pension Plan.*

Bermuda Plan (BP)

This plan includes room rate and American breakfast. American breakfast generally includes most or all of the following: Two eggs (fried or poached), sliced bacon or sausages, sliced bread or toast with jam/jelly/butter, pan cakes with syrup, cornflakes or other cereal, coffee/tea, orange/grapefruit juice. This plan is also known as *bed and breakfast plan (B and B).*

TYPES OF GUEST ROOMS

In order to suit the profile and pocket of various kinds of guests, hotels offer different types of rooms that cater to the specific needs of guests. The rooms may be categorized on the basis of the room size, layout, view, interior decoration, and services offered. The various types of rooms offered by a hotel is illustrated in Fig. 17.16.

Single Room

A room with a single bed to provide sleeping accommodation for one person. The size of a single bed is generally 3 feet by 6 feet. Nowadays single room does not exist. Mostly, hotels have twin or double rooms and charge for single room if occupied by one person.

Double Room

Meant for double occupancy and has one double bed. The size of a double bed is generally 6 feet by 6 feet. An extra bed may be added to this room on the request of a guest and charged accordingly.

Fig. 17.3: Double room

Twin Room

Meant for double occupancy and has two identical single beds separated by a bedside table and has a separate headboard for each bed. The bed size is normally 3 feet by 6 feet. An extra bed may be added to this room on the request of a guest and charged accordingly.

Fig. 17.2: Single room

Fig. 17.4: Twin room

Triple Room

This room has either a double bed or two twin beds and an extra bed to accommodate three persons. It is provided mostly for families.

Quad

A quad room has four separate single beds and can accommodate four persons together in the same room.

Hollywood Twin Room

Twin bedded room with a common head-board which is attached to the wall and not to the bed.

King Room

A king room is a room with a king-size bed.

Queen Room

A queen room is a room with a queen-size bed.

Interconnecting Room

Rooms with individual entrance doors from the outside and connecting door in between. Guest can move between rooms without going through the corridor. This room is ideal for families.

Fig. 17.5: Hollywood twin room

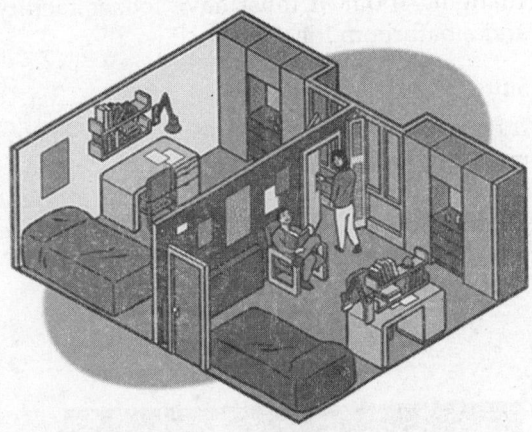

Fig. 17.6: Interconnecting room

Double-Double Room

A double-double room has two double beds and is normally preferred by a family.

Adjoining Room

Two rooms with a common wall, but no common door in between.

Fig. 17.7: Adjoining room

Parlor

It is a sitting room/living room. It does not have a bed.

Studio

A room with a studio bed (bed without headboard which can be used as sofa during the day and can be pulled out into a bed during the night). May also be called *multi-utility room or extension room*.

Fig. 17.8: Suite

Cabana

A cabana is situated in the vicinity of a swimming pool. It is generally used as a changing room. It must have locker facility and a bathroom for shower.

Suite

It consists of a parlor connected with one or more bedrooms. The décor of such rooms is of high standards, aimed to please the affluent guest who can afford the high tariffs of the room category.

Junior Suite

This is a large room with a wooden partition separating the living room from the bedroom.

Duplex Suite/Bi-level Suite

This type of suite has two rooms on two successive floors, which are connected by an internal staircase. This suite is generally used by business guests who wish to use the lower level as an office and meeting place and the upper level room as a bedroom. This type of room is quite expensive.

Fig. 17.9: Junior suite

Fig. 17.10: Duplex

Utility/Efficiency Room

An efficiency room has an attached kitchenette for guests preferring longer duration of stay.

Hospitality Room

It is a room which is used by the residential guests to entertain their guests and to hold parties and meetings outside their allotted rooms. It is charged on an hourly basis.

Penthouse

It is a room on the topmost floors, which has an open terrace or open sky space. The open part is covered with glass or bamboos. It has very opulent décor and furnishings, and is among the costliest rooms in the hotels, preferred by celebrities and political personalities.

Fig. 17.11: Efficiency room

Fig. 17.13: Penthouse

Fig. 17.12: Hospitality room

Presidential Suite

It is a suite with two or more bed rooms and a parlor. The best possible furnishings and decoration are used in these rooms. The bathroom will have the best possible sanitary fittings in it.

Lanai

It is type of room normally found at hill resorts and overlook a view of a waterfall, lake, mountain, forest or a garden from the balcony of the room.

Murphy Room

A room which has a murphy bed or sico bed (pull out or convertible or foldaway bed). A murphy bed is a bed that folds up into the walls and looks like a bookshelf when folded away.

Fig. 17.14: Presidential suite

Fig. 17.15: Lanai

Single Lady Room

The room which is designed with the safety and security features for single lady travelers. The features may include:

- In room check
- Room with a telephone for the guest to view who is at the door
- All telephone calls are screened

Non-smoking room

A room which is designed to cater to the non-smoking guest only. This room does not have an ashtray. There is usually a separate zone for non-smoking rooms.

Kids Room

A guest room attached with a small room to entertain kids. This room has games and toys to keep kids engaged

TYPES OF ROOM RATES

A hotel generally designates a standard rate for each category of rooms offered to guests. Apart from the standard rates, hotels also offer discounted rates to attract additional business from multiple market segments. Hotels may have various rate designations as illustrated in Fig. 17.17.

Rack Rate

It is the maximum possible room rate which is charged from the guest by the hotel for overnight accommodation. This is the normal published room rate mentioned on the tariff card. Rack rate is also called *top retail rate*. It is

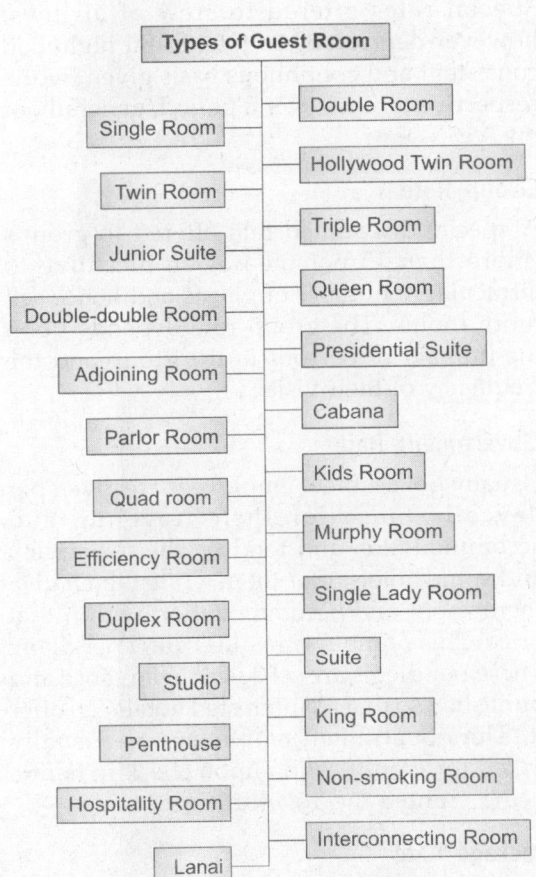

Fig. 17.16: Types of guest rooms

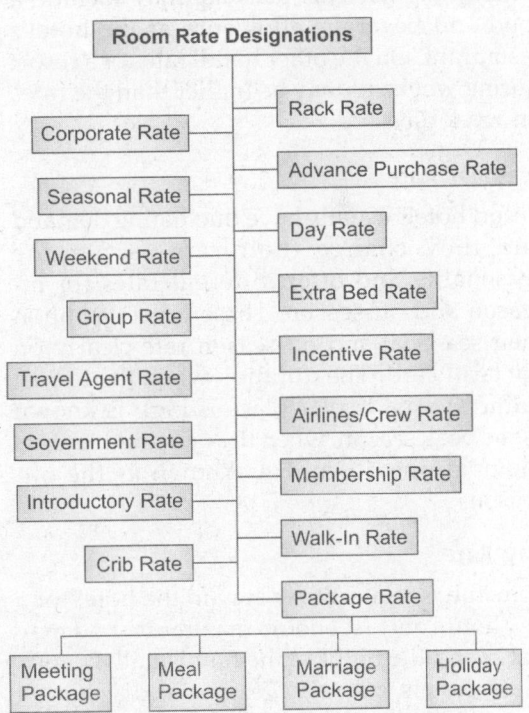

Fig. 17.17: Room rate designations

the highest published rate of the rooms. It is usually offered to walk-ins. Traditionally, a rate board was placed near the room rack, hence the name rack rate.

Corporate Rate/Commercial Rate

A special rate given to a company which signs a contract with the hotel guaranteeing a certain volume of business within a specific time frame and accepts financial responsibility for no-show. It is also known as *Corporate Volume Guaranteed Rate* (CVGR).

Weekend Rate

Some hotels observe a fluctuation in their occupancy levels with regard to the days of week. A commercial hotel, which relies heavily on the business travelers typically, has a low occupancy rate on weekends. A special weekend rate may apply for weekend packages—usually Friday, Saturday and Sunday nights. This package may include a food and beverage allowance at the hotel's restaurant. On the other hand, rates of a resort during weekend may be higher than the rates on week days.

Seasonal Rate

Resort hotels usually have fluctuating demand and they change their rate as per the seasonality and offer different rates for in-season and off-season. These hotels mention their seasonal and off-season rate clearly on the tariff card. The duration when the tourist traffic at a particular place is high is known as the peak season; when the demand for hotel rooms drops down, it is known as the off-season.

Day Rate

Sometimes a guest may stay in the hotel for a few hours only (six hours maximum) and may not spend the night in the hotel at all. In such cases, hotels may charge special discounted rate (which is usually 50% of the rack rate) and the rate is called day rate.

Advance Purchase Rate

The advance purchase rate concept was copied from the airlines industry. An advance purchase rate offers heavy discount on room rates based on the number of days in advance it is booked. A 7-, 10-, 21-day advance purchase rate will have a corresponding lower price. A hotel wishing to get more room reservations booked in advance will offer advance booking rates with heavy discounts. The advance booking of rooms ensures a certain amount of revenue at a given time and thus helps the management in planning revenue management strategy.

Airlines/Crew Rate

Special rates offered to crew of airlines; however dependent on total room nights on consistent and continuous basis given by the respective airlines over a period, generally of one year.

Group Rate

A special discounted rate offered to groups (more than 15 persons) as an incentive to attract large number of guests and hence sell more rooms. The group rate depends upon the number of persons in the group and the frequency of their visits.

Government Rate

Usually government employees are given per day allowance for their traveling and accommodation and food by their company and some hotels offer them a rate which give them room and accommodation within that price. This reimbursement is called per diem. These per diems are set a year in advance and published so that all interested hotels can offer it. The government employees are usually asked for identification upon check-in before being granted the government rate.

Package Rate

A package offered by the hotel is a combination of one or more hotel products or services (such as accommodation, meals, transfers, transport, A/V equipment, meeting room charges, sightseeing, etc.). Often this entails pricing the package below the cost of purchasing the items separately. These rates are tailor-made for specific guest require-ments. Hotel may offer the following packages:

- Meeting package
- Meal package
- Marriage package
- Holiday package

Crib Rate

A special rate applicable to children above five years and below 12 years of age accompanied by the parents.

Extra Bed Rate

A rate applicable to a third person occupying the room, and is provided with an extra bed in the room. Generally it is ⅓ or ¼ of the applicable tariff.

Incentive Rate

Some hotels give special rates to the members of various esteemed organizations such as FHRAI, TAAI, IATA, etc. This is also called *industry rate*. FHRAI cardholders gets 30% discount on room rent and F&B (soft). Those who work in the travel industry (travel agents, airlines, group leaders, meeting planners, tour operators and others capable of providing the hotel with additional room sales) are often extended the professional courtesy of incentive rate. These rates can vary from 30 to 50% off rack rates.

Membership Rate

Membership rates are offered to guests who are members of influential organizations that provide volumes of business to hotels. Organizations such as the American Automobile Association (AAA), FHRAI and the American Association of Retired Persons

(AARP) have a large constituency of members who enjoy travel. To promote patronage these organizations are offered membership rates which is typically 50% off rack rate. By offering these membership rates, the hotels are allowed to display the logos of the AAA and AARP. The members will then find it easier to seek out hotels that offer these membership rates. Another benefit of offering membership rates is the free advertising given to participating hotels by being listed in the organization (AAA or AARP) guidebooks. It is also called *promotional rate*.

Walk-In Rate

This type of rate may vary from night to night. It is set each night by FOM based on the remaining unoccupied rooms in the hotel. Literally 'walking in off the street' these guests can help fill any remaining hotel rooms. With a few remaining rooms, the walk-in rate may be set fairly high to maximize room revenue.

Complimentary Rate

When a hotel does not charge the room rent from a guest, it is known as complimentary rate. Hotels generally offer complimentary rooms to the tour/group leader. They may also offer complimentary rates to tour operators, travel agencies and local dignitaries who are vital to the public relations program of the hotel. Hotels also provide compli-mentary rooms along with marriage packages and bulk bookings.

Introductory Rate

The introductory rate is offered by a hotel on the opening of a new property in town. It is a part of a new hotel's marketing strategy to make inroads into the existing market by offering a price lower than what is offered by competitors with the same standards.

Travel Agent Rate

Travel agents sell travel products like hotel rooms, airlines bookings, etc. on a commission basis to end users. They provide a substantial business to hotels, hence hotels offer them special discounts and commissions. Some major travel agencies include Cox and Kings, Thomas Cook, etc.

ROOM STATUS TERMINOLOGY

During the guest's stay, the status of the guest room changes several times. The various terms defined are typical of the room status terminology of the lodging industry. Not every room status will occur for each guest room during every stay. It is very much crucial for front desk staff to have an accurate room status information so that they can sell the rooms.

Occupied

A guest is currently registered to the room.

Complimentary

The room is occupied, but the guest is assessed no charge for its use. This sort of offer is given to management staff from other hotels, regular customers, special guests, group leaders, etc.

Stay Over

The guest is not expected to check out today and will remain at least one more night.

On-change

The guest has departed, but the room has not yet been cleaned and readied for re-sale. Also known as *vacant and dirty*.

Do Not Disturb

The guest has requested not to be disturbed. This is usually indicated with a sign left hanging on the room's door knob or electronically by a red light indicator outside the guest room door.

Sleep-out

A guest is registered to the room, but the bed has not been used.

Skipper

The guest has left the hotel without settling his account.

Sleeper

The guest has settled his account and left the hotel, but the front office staff has failed to properly update the room's status.

Vacant and ready

The room has been cleaned and inspected and is ready for resale.

Out-of-order

The room cannot be assigned to a guest. A room may be out-of-order for a variety of reasons including the need for maintenance, refurbishing, repair and extensive cleaning.

Lock-out

The room has been locked so that the guest cannot re-enter until he is cleared by the hotel official.

Due out

The guest is expected to leave after the following day's checkout time.

Checkout

The guest has settled his or her account, returned the room keys, and left the hotel.

Late Checkout

The guest has requested and is being allowed to checkout later than the hotel's standard checkout time which is usually 12 noon. Normally one hour is allowed after checkout time, and after that it is charged.

DNCO (Did Not Checkout)

The guest made arrangements to settle his account (and thus is not a skipper), but has left without informing the front office.

Double Lock

The guest room door is locked from inside and outside (double-locked) so that no one can enter. No other key can open this room door except the grandmaster key.

Scanty Baggage

The guest is staying in the room but with very light luggage.

No Luggage

The guest is staying in the room but is without luggage.

REVIEW QUESTIONS

1. Explain the various types of meal plans offered by hotels and justify their suitability to different hotels.
2. Explain various types of room rates offered by a hotel.
3. Not knowing your product can be a disadvantage to sales. Explain.
4. Describe the various types of rooms available in a hotel.
5. Define the terms used to identify the status of guest rooms throughout the day.
6. Explain the following types of room:
   - Lanai
   - Cabana
   - Junior suite
   - Penthouse
   - Twin room
   - Studio room
   - Interconnected room
   - Duplex
   - Efficiency room
   - Hollywood room
   - Quad
   - Suite
7. Distinguish between:
   - Double room and Twin room
   - Interconnecting room and Adjoining room
   - Double-Double and Double room
   - Suite and Single room
   - Single room and Single lady room
   - Duplex and Penthouse
   - King room and Queen room
   - Studio and Cabana
   - Murphy room and Studio room
   - Efficiency room and Hospitality room
   - Twin room and Hollywood twin room
   - Advance Purchase rate and Package rate
   - Crib rate and Extra bed rate
   - Crew rate and Corporate rate
   - Tariff and Tariff Plan
   - Rack rate and Corporate rate
   - EP and BP
   - AP and MAP
8. Write short notes on:
   - Rack rate
   - Corporate rate
   - Promotional rate
   - Package rate
   - Meal plans
   - Stay-over
   - Sleep-out
   - Skipper
   - Lock-out
   - Sleeper
   - Scanty baggage
   - On-change
   - CVGR

Room Tariff

Learning Objectives

After reading this chapter, you will be able to understand the following:

□ Room tariff card
□ Basis for establishing room tariff—cost, level of services, competition, target market, location
□ Room tariff fixation—cost-based pricing and market-based pricing

ROOM TARIFF CARD

A hotel tariff card is a document developed by hotels for use of the travel trade (travel agencies, tour operators), companies, and individual guests called FITs (free individual travelers). The tariff card includes prices of the guest rooms classified into different categories based on types and meal plans, taxes, applicable policies, etc.

Hotel tariffs are sales and marketing tools which help to sell the accommodation, food and beverage, the facilities and services of a hotel.

Room Type	Period: Till 30th Sep 15 INR		Period: 1st Oct 15 – 31st Dec 15 INR	
	Single	Double	Single	Double
Executive Club	11,000	12,500	15,000	16,500
Executive Club Exclusive	13,000	14,500	18,000	19,500
The Towers	15,000	16,500	23,000	24,500
ITC One	20,000	21,500	25,000	26,500
Deluxe Suite*	30,000		50,000	
Luxury Suite*	60,000		100,000	
Presidential Suite*	125,000		250,000	
Grand Presidential Suite*	500,000		500,000	

Note:

1. 15 % Luxury tax will be levied on published FIT Tariff, 8.40 % service tax will be levied on charged rate and 1.25 % VAT if the rate includes breakfast.
2. Rates are subject to change without notice.
* Inclusive of buffet breakfast.

Source: http://www.itchotels.in/hotels/itcmaurya/published-tariff.aspx, last accessed on 25th Sep 2015

Fig. 18.1: Room tariff of ITC Maurya, New Delhi

The hotel tariff may also include the details about the hotel such as the location, distance from the airport, railway station, or bus terminus. It may also give a brief description of all the outlets such as coffee shop, restaurant, banquet halls, bar as well as services and facilities such as discotheque, swimming pool, beauty parlor, foreign exchange, baby sitting, doctor on call, etc.

Tariff cards have also the information like hotel name and address, website, e-mail, telephone, fax, corporate office address, etc.

Room Category	Single	Double
Deluxe	6,500	7,500
Club Royale	11,000	12,000
Club Premiere	9,000	10,000
Executive Suites	14,000	14,000
Club Suite	16,000	16,000

Note:
- The above rack/full published rates are exclusive of taxes.
- The above rates are valid form 1st April, 14 till 30th Sep, 15.
- The above rates are on CP basis.
- The published rates may change as per management's decision.
- 15% Luxury tax will be levied on the rack rates.
- 8.40% Service tax on the display rates.
- 1.25% VAT on display rates where breakfast is included.

Source: http://www.jaypeehotels.com/vasant-continental/tariff, last accessed on 25th Sep 2015

Fig. 18.2: Room tariff of Jaypee Vasant Continental, New Delhi

BASIS FOR ESTABLISHING ROOM TARIFF

Room rate, which is the daily rate charged for the usage of a hotel room and services, is among a traveler's basic criteria for choosing a particular hotel for stay. The basis for

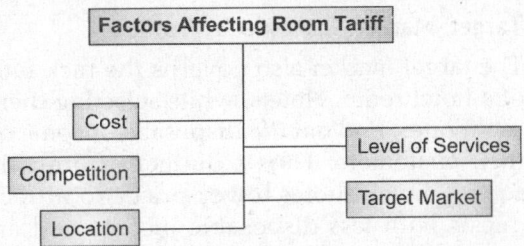

Fig. 18.3: Factors affecting room tariff

charging room rate differs from hotel to hotel. The rooms of a hotel generate the maximum revenue, so an accurate and competitive room rent is one of the prerequisites for running a successful hospitality business. The rate of a hotel room is decided by several factors that is illustrated in Fig. 18.3.

Cost

The total expenditure that is incurred in providing services and products to the ultimate consumer of the hotel services is the cost. The total cost can be divided into fixed cost, material cost, and labor cost. The higher the investment that has been made in a hotel property, the higher would be the room rent because the hotel proprietor expects a fair and equitable return on his investment.

Level of Services

The level of services offered by a hotel determines the room rent to a large extent. A hotel offering the best of services like spa, gym, banquet halls, speciality restaurants, etc. will charge a higher room rent in comparison to other hotels offering limited services.

Competition

Competition between similar hotels (i.e. hotels with similar standards and providing similar services and facilities in same vicinity of the city) in the market also plays an important role in determining the rack rate of the hotel. Rates must be competitive with other hotels of the same standard.

Target Market/Customer's Profile

The target market also governs the rack rate of a hotel room. Hotels, while selecting their room rates, find out the disposable income of their customers. Thus a budget or limited-service hotel quotes lower prices to attract guests with less disposable income and an upmarket hotel quotes higher prices for its products.

Locality of Hotel and Room Location

The locality in which the hotel is situated gains prominence while fixing room rates. Hotels in a city center or business centre, near places of tourist interest, or on scenic locales would have a higher tariff as compared to those located in far-off localities. The location of the room also determines the room rate. Rooms with a better view would have higher charges as compared to rooms facing a parking lot or a noisy commercial street.

Once the room tariff has been decided, every hotel has to decide about the criteria for establishing the 'end of the day' to post the room charges into guest accounts. The end of the day is an arbitrary time that is supposed to be the end of the financial transactions for a particular day. As hotels remain functional round the clock, it is very important to ascertain the time which will be treated as the end of the day and the beginning of a new day.

The hotel may choose to charge room rent on the following basis (Fig. 18.4):

Check-in and checkout basis: Here a day of stay is calculated from 12 noon to 12 noon of the following day. This means that a day begins in the hotel at 12 noon every day, regardless of the actual check-in time of the guest. Most of the hotels following this system may allow relaxation of a few hours before and after checkout time in charging the room rent. This system is normally adopted by commercial hotels.

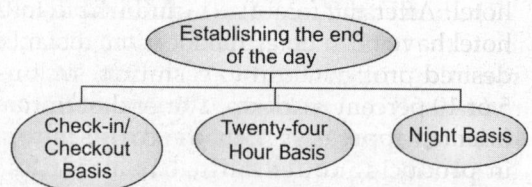

Fig. 18.4: Basis for establishing the end of the day

24-Hour basis: Here a day of stay is calculated on 24 hours basis, i.e. from the hour of check-in up to the hour of checkout, in a cycle of 24 hours. The room is assigned to a guest for twenty-four hours from the time of his arrival. This system is generally followed in transit hotels and hotels that are located in the vicinity of railway stations, where guests normally stay for a few hours.

Night basis: In this system the guest is charged on the basis of number of nights that he spends in the hotel irrespective of time of arrival. If a guest does not stay at night, a half day rent is charged from him. This system is now outdated and not so popular.

ROOM TARIFF FIXATION

Fixing of room tariff is a difficult task for the management. If the management fixes a low room rent, the hotel operations might not be economical. However, if the rate is too high, guests may not patronize the property. Hence, an accurate and competitive room rent is one of the prerequisites for running a successful hospitality business. A hotel fixes the room tariff on the following two approaches (Fig. 18.5).

Cost-based Pricing

Cost-based pricing is a room rent determination technique that covers the basic cost of operations at a given level of service, plus the predetermined percentage of return on investment. It involves the determination of all fixed and variable costs associated with a hotel. After the total costs attributable to the hotel have been determined, managers add a desired profit margin to each unit, such as a 5 or 10 percent mark-up. The goal of the cost-oriented approach is to cover all costs incurred in producing or delivering the products or services and to achieve a targeted level of profit.

The traditional pricing policy can be summarized by the formula:

Cost + Fixed profit percentage = Selling price

This method is simple, requiring only the study of the hotel's financial and accounting records to determine prices. It does not involve examining the market or considering the competition and other factors that might have an impact on pricing. Cost-oriented pricing is also popular as it uses internal information that managers can obtain easily.

The following are the two widely used cost-based pricing techniques:

- **Rule of Thumb Approach**: Rule of thumb is the oldest method of determining the room rent of any hotel. According to this approach, the room rent should be fixed at the rate of ₹ 1 for each ₹1,000 spent on the construction and furnishing of the room, assuming that the average occupancy is 70% for the year. If the hotel incurs a total expenditure of ₹15,00,000 on a room, the room rent will be ₹15,000 according to the rule of thumb approach. This is also called *Cost Rate Formula*.

There are several drawbacks associated with the rule of thumb approach; some of them are as under:

  - It only considers the cost incurred in constructing the room. It does not consider factors like inflation, competition, fixed expenses, etc.
  - It considers the average occupancy to be 70 percent, which might not be achieved by many hotels due to several reasons.

Therefore, hotels expecting lower occupancy should set a higher rate to attain the same revenue.

- The return on investment (ROI) is not considered. If the money invested in constructing and furnishing the hotel room had been invested in the market for the duration of one year, it would have generated income for the investor.
- It does not consider the depreciation of fixed assets or the elevation of land costs.
- Unexpected and unavoidable expenses and the provision for the same is not made in this approach.
- This approach also fails to consider the contribution of other facilities and services (laundry, food and beverage service, recreational services) to the profitability of the hotel while setting the price of a room.

- **Hubbart Formula:** The rule of thumb approach is an old traditional way of determining the room rent of a hotel. As seen, it is unscientific and suffers from many drawbacks. The Hubbart formula, which is a scientific way of determining the room rent, was developed by Roy Hubbart in America in the 1940s. It resolves all the problems of the rule of thumb approach. This approach is based on the principle of covering all the cost that is incurred in providing the accommodation plus a reasonable return on investment. The following steps are involved in calculating the room rent according to the Hubbart formula:

  - Calculate the total investment including the owner's capital and loans, both secured and unsecured. Once the total investment has been calculated, calculate the fair rate of return on investment (ROI). ROI is the amount that would have been generated if the money invested in the hotel business had been invested in the open market.

- Calculate the total expenses—like operating expenses, overheads, depreciation of fixed assets, interest paid, heating and lighting, etc.—that will be incurred during hotel operations.
- Combine steps 1 and 2 to find out the gross operating income that is necessary to cover the operating cost, investment, and return on investment.
- Calculate the income generated from other sources of income, like food and beverage sales, laundry, rent and lease of the hotel area, fitness center, etc. Subtract the same from the amount calculated in step 3 to find out how much profit is expected from the room sales. This will be the total revenue generation by the room sales.
- Calculate the total number of the guest rooms available for sale by multiplying the total number of rooms with the number of days in the year. Make the provision for expected average vacancy that is expected during the year. This step will provide the total number of rooms available for sale.
- Divide the revenue generation (result from step 4) by the total number of rooms (result from step 5); the result obtained will be the average daily rate, which will cover the cost of operations and fair return on investment.

Market-based Pricing

Market-based pricing is setting a price based on the value of the product in the perception of the customer. The concept is based on an idea of what the ultimate consumer of goods and services, i.e. the guest, is prepared to pay and then use this as a starting point. In this case, the hotel works backwards as it first makes an accommodation product available at a price that a guest is willing to pay, and then it tries to cut down on the cost to achieve a reasonable rate of return on that basis. Some common methods of market-based pricing are:

- As per competition: Arriving at a pricing based on competing hotels' rates.
- Market tolerance: Checking competing hotels' best available rates for a room. These rates can be found out by hotels by calling up the competing hotels without disclosing their identity.
- Rate cutting: Lowering of rates to increase occupancy levels, especially during off season.
- Inclusive and non-inclusive rates: Charging room rates on the basis of meals, provided on a CP/MAP/AP basis.
- Guest requirements: Varying room tariff as per guest requirements, e.g. early check-in on CP basis or late checkout on MAP basis.

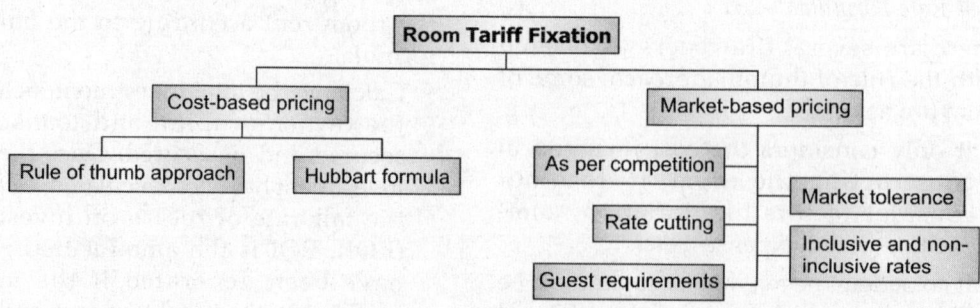

Fig. 18.5: Basis of tariff fixation

REVIEW QUESTIONS

1. What is tariff? Explain the various basis of charging the guest in hotel with suitable example.

2. Define a tariff card and draw out a typical tariff card of a 5 star hotel.

3. State the various basis of charging room rates.

4. What is rack rate? Explain different factors which affect the room tariff of a hotel.

5. What is 'Rule of thumb' method of pricing a room? Illustrate with a suitable numerical example. What are the major drawbacks of the rule of thumb approach?

6. What are the three popular approaches to pricing rooms? Explain

7. How is Hubbart formula used for calculating room rate?

8. Why does a hotel have different types of rates?

9. Explain the advantages and disadvantages of having different types of room rates.

10. Differentiate between:
    - Hubbart Formula and Rule of thumb approach
    - Cost-based pricing and Market-based pricing
    - Night basis and 24-hour basis

Guest Cycle

Learning Objectives

After reading this chapter, you will be able to understand the following:
- Concept of guest cycle
- The stages of a guest cycle—pre-arrival, arrival, stay, departure and post-departure—and the role of the front office in taking the guest through each of these stages.

GUEST CYCLE

In order to effectively manage front of the house activities, hotels have adopted the notion of a guest cycle. The guest cycle identifies the physical contacts and financial exchanges that occur between guests and various revenue centers within a lodging operation. The traditional hotel guest cycle was based on interactions in terms of a sequence beginning with the arrival of a guest, continuing through the guests' occupancy, and ending with the guest's departure. Many hotels have revised this traditional guest cycle into a sequence of phases beginning with the presale events, continuing through point of sale activities, and concluding with post-sale transactions. The primary reason for revising the traditional concept of the hotel guest cycle is the increasing need for improved co-ordination among various operating departments.

There are four main stages of interaction between a hotel and its guests—pre-arrival, arrival, stay, departure and post-departure. All guests go through the same procedure as they proceed from reservation to arrival and allotment of rooms, to their stay in the hotel, to the settlement of their bills and departure from the hotel. These various stages of activities constitute the guest cycle (Fig. 19.1). At each of these stages, guests are assisted by the front office department, which provides services like processing reservations, receiving guests and allotting rooms, preparing guest folios, looking after guests' requirements during their stay, and settling guests' accounts.

STAGES OF GUEST CYCLE

Pre-arrival

The guest chooses a hotel during the pre-arrival stage of the guest cycle. The guest's choice can be affected by many factors, including previous experiences with the hotel, advertisements, word of mouth referral by the travel agents, friends or business associates, location, hotel brand name, loyalty program, etc.

The guest's decision of making the reservation may also be influenced by the ease of making the reservations, the way reservation agent interacted and described the hotel and its facilities, room rates and amenities, recreational facilities and other attractions near the hotel, etc.

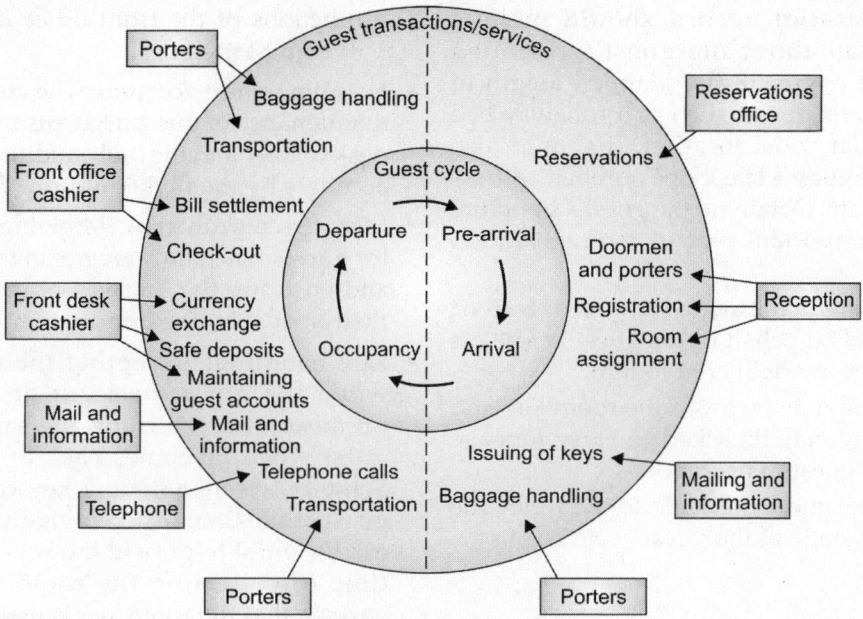

Fig. 19.1: Guest cycle

The attitude and efficiency of the front office staff may influence a caller's decision to stay at a particular hotel. A reservation agent must be able to respond quickly and accurately to requests for future accommodation. The efficiency and competence of reservation staff in handling the reservation request creates a good impression of the hotel in the mind of the guest.

If a reservation can be accepted, the reservation agent creates a reservation record. The creation of this reservation record initiates the hotel guest cycle. This record enables the hotel to personalize guest service and schedule needed staff and facilities. By confirming a reservation, the hotel verifies a guest's room request and personal information and assures the guest that his needs will be addressed. The detail which is collected during the reservation also helps the hotel to complete pre-registration activities like room assignment according to guest request and creation of guest folio (in case the hotel has received any advance payment). The hotel also makes any special arrangements like airport transfers, extra bed, etc.

Arrival

Guests have their first face-to-face interaction with the hotel staff on their arrival at the hotel. This is a very critical stage as guests form an opinion about the standards and services that the hotel can provide to them. During this phase, guests are greeted by the front desk staff and the registration process begins.

Front office staff should determine the guest's reservation status before beginning the registration process. Guest with reservation and guest without reservation commonly known as walk-ins also provide an opportunity of business for front desk staff. To sell successfully, the front desk staffs must be very familiar with the guest room types and guest services and be able to describe them in a positive way. A guest will not register if he is not convinced of the value of renting a particular hotel room.

A registration record should include information about the guest's intended method of payment, the planned length of stay, special requests such as a rollaway bed or particular room location. It should also include the guest's telephone number, address and signature. Obtaining the guest's signature is a very important part of the registration process.

If any mails or messages have been received by the hotel on behalf of the guest before his arrival, they are delivered to him.

The guest may be given the room key and the bell boy may be asked to carry luggage and escort guest to his room.

When the guest checks-in to the room, the occupancy stage of the guest cycle begins.

Stay

During this stage the guest gets a first-hand experience of the facilities and services offered by the hotel. These services and facilities are the most important part of a guest's overall experience at a hotel. An excellent level of services would lead to the satisfaction of the guest, which would make him come back to the hotel and give positive feedback to other potential customers. The front office is the interface between the guest and the other departments of the hotel, so it must coordinate well with other departments to ensure that the guest receives smooth and efficient services and facilities. The stay phase is the most important phase in the guest cycle for the hotel. During this phase, the guest would interact with the front office staff for various reasons, like asking for directions in and around hotel, arranging for inter-city travel, wanting to know about the history of the city or hotel. The front office staff should handle the guests' queries politely and patiently, and provide satisfactory answers. In case of any lapse in services, the front office staff must be courteous and use all possible resources to satisfy the guest.

The functions of the front office during the stay of a guest include:

- Handling guest accounts: The creation and maintenance of guest accounts by the front desk cashier and the daily auditing of guest accounts by the night auditor.

- Message coordination: Receiving messages for guests when they are not in their rooms and ensuring the delivery of the same on their arrival by the information assistant.

- Key handling: Accepting the room key when the guest goes out of the hotel premises and returning the same to the guest when he comes back to the hotel. Some hotels also issue a key card to the guest at the time of accepting the key, and ask the guest to present the key card at the time of collecting the room key. This ensures that the room key is delivered to a genuine person. A key card is either a card or a small booklet which has the guest's name, room number and room rate entered on it. On the reverse of the card, there are details of the services and facilities offered by the hotel.

- Guest mail delivery: Accepting mails of guests and delivering the same to them. When a guest is not in the room, the front desk receives his mail, keeps it in the room key rack, and delivers it when the guest comes back. If the guest is in his room, the bell boy delivers the mail to him. If the hotel receives mails for a guest who has a reservation in the future, the mails are sent to the reservation section and are kept in the reservation docket. The same is attached with the pre-filled guest registration card, which is given to the guest at the time of registration.

- Guest paging: Locating the guest in a specific area of the hotel when he is not in the room. If a guest is expecting a visitor, he may request for this service by filling a form.

- Safety deposit locker: Providing the locker facility to guests to keep their valuables like important documents and jewelery. The safety lockers are located in the back office of the front desk. Some hotels may have lockers in guest rooms.
- Guest room change: Changing the room of a guest, in case the guest's preference for view, type of room or location could not be fulfilled at the time of check-in due to non-availability of such a room. The room can also be changed if there is any defect in the room that requires excessive maintenance work.
- Handling guest queries and complaints: Responding to guests' queries and communicating guests' complaints to relevant departments.
- Information about the hotel: Providing information to guests about the products and services offered by the hotel.
- Information about the city: Providing information to guests about the city, like places of tourist interest, shopping malls, cinema halls, restaurants, bars, public offices, etc.
- Travel arrangements: Making intra- and inter-city travel arrangements for the guests, if required.

Departure and Post-departure

The maxim goes that 'all's well that ends well'. The front office should try to cover up any unpleasant episodes during a guest's stay by ensuring a smooth and hassle free departure of the guest.

At departure, the guest vacates the room, receives an accurate statement of account for settlement, returns the room keys and leaves the hotel. Once the guest has checked out, the front office updates the room availability status and notifies the housekeeping department.

In case a guest wishes the hotel to keep his luggage for a short duration of time after checking out of the hotel, the hotel keeps the same in the left luggage room. The front office makes a luggage tag and hands over the guest copy to the guest, which the guest has to produce when he comes back to claim the luggage.

At this stage front office also collects the feedback of the guest experience in the hotel by handing over the guest feedback form. The front office determines whether the guest was satisfied with the stay and encourages the guest to return to the hotel in the future.

The more information the hotel has about its guests, the better it can serve their needs and develop marketing strategies to increase business. Once the guest has checked out, the front office can analyze data related to the guest's stay. Front office reports can be used to review operations, isolate problem areas, indicate where corrective action may be needed and point out business trends. Analysis can help managers establish a standard of performance, which can be used to evaluate the effectiveness of the front office operations.

Hotels can still remain in contact with the guests even after a guest has departed from the hotel by means of guest history card wherein we can get his details like date of birth or marriage anniversary. Hotels usually wish their regular guests on special occasions like New Year, Deewali, Holi, Christmas, etc. This is generally taken care of by the sales and marketing team, which sends guests mailers or flyers with special offers or discounts, hotel updates like changes in room rates on a regular basis.

REVIEW QUESTIONS

1. What do you understand by guest cycle? Explain with the help of a diagram.
2. What activities are involved in the various stages of the guest cycle?
3. State the various functions carried out during the following stages:
   - Pre-arrival
   - Arrival
   - Occupancy
   - Departure

Handling Guest Complaints

Learning Objectives

After reading this chapter, you will be able to understand the following:

- Types of guest complaints
- Complaints are sales opportunities and not threats
- Resolving guest complaints

TYPES OF GUEST COMPLAINTS

Complaint is an expression of resentment or displeasure which results in protest.

Guests often express their displeasure when certain situations or services at the hotel are not up to their satisfaction. Many guests curb their tendency to complain when they are not pleased with one or two of the hotel's services, but when the displeasure builds up through a series of problems, the guest does complain.

When guests find it easy to express their discontent to hotel employees, both the hotel and the guests benefit. The hotel gets a feedback about its staff and services and can take corrective actions, while the guest can have a comfortable stay if his problems are addressed.

It may be that for every guest who complains there are 5–6 who keep quiet. Then again, hotels also have guests who complain just for the sake of complaining and like to find fault with everything. In hotel jargon, these guests are called 'handle with care' (HWC) guests.

The guests' complaints can be grouped into four major categories:

Mechanical Complaint

Mechanical complaints are related to the malfunctioning or non-functioning of systems and pieces of equipment installed in guest rooms, like climate control, lighting, geyser, room furnishings, door keys, plumbing, television sets, elevators, etc. Even an excellent preventive maintenance program cannot completely eliminate all potential equipment problems.

Attitudinal Complaint

Guests may make attitudinal complaints when they feel insulted by the rude or tactless staff members. Guests who overhear staff arguments or who receive complaints from staff members may also make attitudinal complaints. It also arises when a member of staff has misunderstood the guest or staff member has forgotten to pass on a message.

Service-related Complaint

Guests may make service-related complaints when they experience a problem with service. These complaints can be wide-ranging and can be made about such things as long waits for service, lack of assistance with luggage, untidy rooms, missed wake-up calls, cold or ill-prepared food or ignored requests for additional supplies, delay in the clearance of soiled crockery from the room after meals, etc. A hotel generally receives more service-related complaints when it is operating at or near full occupancy.

Unusual Complaint

Guests may also complain about the absence of swimming pool in the hotel, or discotheque in city, lack of public transportation, about bad weather (heavy rainfall or scorching heat). Hotels generally have a little or no control over the circumstances surrounding unusual complaints. Front office manager should alert his staff that some guests will complain about things they can do nothing about. Sometimes the complaint arises simply because the guest has had a bad day and what normally would have been a minor hiccup now becomes a major tragedy.

COMPLAINTS ARE SALES OPPORTUNITIES AND NOT THREATS

Complaints show both the good and the bad points of operation. Complaints must be used to have a better understanding of 'short-comings' and to correct them.

When guests find it easy to express their opinions, both the hotel and the guest benefit. The hotel learns of potential or actual problems and has the opportunity to resolve them. For a guest, this means a more satisfying stay, when problems are resolved. A guest often feels that the hotel cares about his needs. From this perspective, every complaint should be welcomed as an opportunity to enhance guest relations because guests who leave a hotel dissatisfied may never return.

A popular axiom in the hotel industry is that it takes ₹ 10,000 to attract a guest for the first time, but only ₹ 1,000 to keep the guest coming back.

Benefits derived from Guests' Complaints

- They highlight guests' views and needs.
- They reveal weakness in standards and systems.
- They enable hotel to prevent re-occurrence of the event.
- They provide information about competitors.
- They allow retaining a customer.
- They can be used in staff/management appraisals.
- It develops inter-departmental communication.

RESOLVING GUEST COMPLAINTS

The front office should handle guests' complaints tactfully, exercising patience, empathy, and decision-making skills. As hospitality is a service-oriented industry, the hotel staff should always try to resolve the customer's problems immediately and thus appease him. If a front office agent is unable to handle a guest's complaint, he should call his superior before the situation gets out of control or becomes worse.

The following guidelines may be followed while handling guest complaints:

- Listen attentively. Lend an ear to the customer's problem. If he is really angry, let him blow his top.
- Do not interrupt. An interruption will encourage the complainant to carry on louder and longer.
- Always act on the complaint. What seems to be a small thing to you is important to the guest.
- Do not ever take complaints personally.
- Take notes. Writing down the key facts saves time if someone else must get involved. Also, guests will slow down when they are speaking faster than you can

write. Most important, the fact that a hotel staff is concerned enough to write down what he is saying is reassuring to guests. If necessary, ask clarifying questions, without appearing to cross-examine the guest or doubt his judgement.

- Wait until the guest has completely finished. Before saying anything at all, be certain that the guest has completely finished talking, rather than just taking a pause for breath.
- Apologize for the inconvenience caused to the guest. Apology should be clear and concise and should not be qualified in any way by any excuse or explanation. Do not show disapproval or disbelief when a guest complains.
- Isolate the guest if possible, so that other guests do not overhear.
- Speak normally. A guest complaining is often further aggravated by a staff whose voice rises to match that of a irate guest. The result is an unseemly slanging match which may be watched by other guests or staff members.
- Summarize the complaint. Repeating the essence of a complaint serves two purposes. Firstly, it ensures that everything has been covered and that there has been no misunderstanding about the cause of the complaint. Secondly, a factual dispassionate summary helps to defuse the situation as it cannot produce further dispute.

- Thank the guest for bringing the matter to the attention of the management and show that the trouble he is taking in making the complaint is appreciated by you.
- Tell the guest what can be done. Offer choices. Do not promise the impossible and do not exceed your authority.
- Set an approximate time for the action. Be specific, but do not underestimate the time period it will take to solve the problem. By giving a definite time, the guest is encouraged that something will be done about the grievance.
- Monitor the progress of the corrective action.
- Follow up. Even if the complaint was resolved by someone else, contact the guest to see if the problem was satisfactorily solved. Report the entire event, the actions taken and the conclusion of the incident.

REVIEW QUESTIONS

1. What is guest complaint? What steps are to be followed in order to resolve the complaint?
2. Explain the various types of complaints in detail by giving suitable examples.
3. Why should guest complaints be welcomed by hotel staff? How may a hotel benefit from analyzing the complaints it receives?

Room Reservations

Learning Objectives

After reading this chapter, you will be able to understand the following:

- Reservations—concept, functions and importance
- Modes of reservation—written and verbal
- Various sources of reservations
- The types of reservations—tentative, waitlisted, and confirmed
- Systems of reservations—manual and automated
- Tools of room availability
- Processing individual reservation request
- Processing group reservation request
- Policies and procedures regarding amendment, cancellations, no-show and overbookings
- Typical reservation reports that can be generated from reservations data

RESERVATIONS—CONCEPT, FUNCTIONS AND IMPORTANCE

Due to globalization, advancement in the means of travel, and increase in the disposable income of people, more and more people are traveling to different locations. There has been a tremendous increase in the number of people traveling for business or pleasure. This has led to a rise in the demand for hotel accommodation at various destinations. To ensure a safe and secure place for stay during their visit to another town, people make advance reservations in hotels.

A reservation is an activity of booking room in advance for a prospective guest on his request and mutual agreement whereby the hotel is bound to provide the guest accommodation on the scheduled day of arrival of the guest and the guest is bound to take it.

A reservation in the context of a hotel means the booking of a room by a guest and involves a particular type of guest room being reserved for a particular person for a certain period of time.

Reservation in the hotel industry is defined as 'blocking a particular type of guest room (e.g. single room, double room, deluxe room, executive room, suite, etc.), for a definite duration of time (i.e. number of days of stay), for a particular guest'.

A reservation is a bilateral contract between the hotel and the guest, according to which the hotel must provide the specified room to the guest and the guest should bear all the relevant charges.

A reservation is the activity of booking the room in advance for a prospective guest on his request for future, which may be from a few days to months in advance. Below is the detailed process of reservation:

- Conduct the reservation enquiry
- Determine room and rate availability
- Create the reservation record
- Confirm the reservation record
- Maintain the reservation record
- Produce reservation reports

Functions of Reservation

- To help the hotel in generating revenue from future and prospective room sales.
- To receive reservation requests from prospective guests, check availability of rooms, to process the request, and either to accept, wait list or deny it, communicating it and then recording it, i.e. maintaining of reservation correspondence, files, charts and racks and computer records.
- To receive requests made by prospective guests for amendments or cancellations, processing them, communicating them and then recording them.
- To ensure a 100% or near 100% occupancy for the future.

Importance of Reservation

The role of the reservation section is not limited to making reservations. It maintains records of the hotel occupancy, which help in planning sales and marketing strategies and estimating manpower requirements. At the same time, properly executed reservations go a long way in ensuring a comfortable stay for

guests. Thus, the reservation section is important for the hotel as well as for the guest. The reservation process is of vital importance to a hotel because of the following reasons:

- Reservation gives the hotel a chance to equate the guest enquiry with the rooms' availability and thus, gives the hotel sufficient time to arrange and prepare for the most suitable accommodation for the guests.
- It gives good indication of the level of business that the hotel might get in future.
- To a great extent the hotel can forecast the future revenue generation keeping in mind the reservation business.
- It helps the hotel in scheduling and reorganizing the staff if the need be.
- The main importance of the reservation section is that it sells the main product of the hotel, i.e. accommodation and generates customers, thereby increasing the chances of revenue generation for the other departments in the hotel.

MODES OF RESERVATION

The reservation request may be made by a guest either verbally or in writing (Fig. 21.1). **Written Mode:** When a reservation request reaches the hotel in writing, the mode is classified as a written mode of reservation. The advantages of the written mode of reservations are that they are clear, unambiguous, and provide a written record for the hotel, which can be referred to in case of any miscommunication or confusion. The

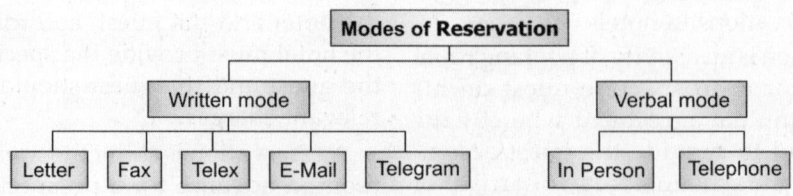

Fig. 21.1: Modes of reservation

correspondence with the guest is filed for future reference. The various written modes for reservation request are as under:

- **Fax**

  Advantage: On spot reservation and confirmation can be done.

  Disadvantage: It is expensive.

- **Telegram**

  Disadvantage: Very limited information can be sent.

- **Letter**

  Advantage: It is convenient, cheap and one could really make himself very clear.

  Disadvantage: Though so popular there is always a fear that it might not reach in time or might not reach at all. Nowadays, letter writing has largely gone out of fashion.

- **E-mail:** Electronic mail or e-mail as popularly known is a system that allows users to send and receive messages through the Internet instantaneously. The system requires a computer, a telephone line and a modem.

  Advantage: It is fastest (the delivery and response time is less) and convenient (guests can send an e-mail from the comforts of their home or office).

  Disadvantage: One requires a computer and a phone line. Phone line could be seen to be cheap but a computer could be said to be expensive. Even if cyber café is used to access Internet and emailing, not all could afford it.

- **Telex**

  Disadvantage: Nowadays, this method is not used much because of the popularity of fax and e-mail.

Verbal Mode: Reservation requests may also be made through oral communication known as verbal mode of reservation request. The advantage of oral communication is that it is fast, convenient, and generates immediate response or feedback; and one can get the complete information and clear any doubt through oral communication. The disadvantage is that it does not provide a permanent record. The various modes of verbal reservation request are as under:

- **Telephone**

  Advantage: It is the fastest and convenient way.

  Disadvantage: It is affordable if it is a local call, it is a bit expensive if it is a national call and it is very expensive if it is an international call.

- **In Person (either guest himself or someone on his behalf)**

  Advantage: It is face-to-face and straight. Person is available to consider the various options and suggestions, in case the room, or its availability, or its rate does not match the guest's expectation.

  Disadvantage: The words if misunderstood may put one in uncomfortable situation. Sometimes language also becomes a problem.

As each of the above-mentioned modes of reservation requests suffers from some drawbacks—either in terms of speed, or convenience, or written record—there are no ideal means of making reservation. Guests choose their modes according to their personal preferences, so any hotel receives reservation requests through all modes.

SOURCES OF RESERVATION

A hotel receives reservation requests from different sources/channels (Fig. 21.2).

Direct Reservation

A reservation request that a hotel receives directly from an individual (FIT) or a group (without a mediator) is known as a direct reservation. The direct reservation request is processed by the reservation manager and his team of reservation assistants in large hotel. In case of a small hotel the same may be processed even by a receptionist.

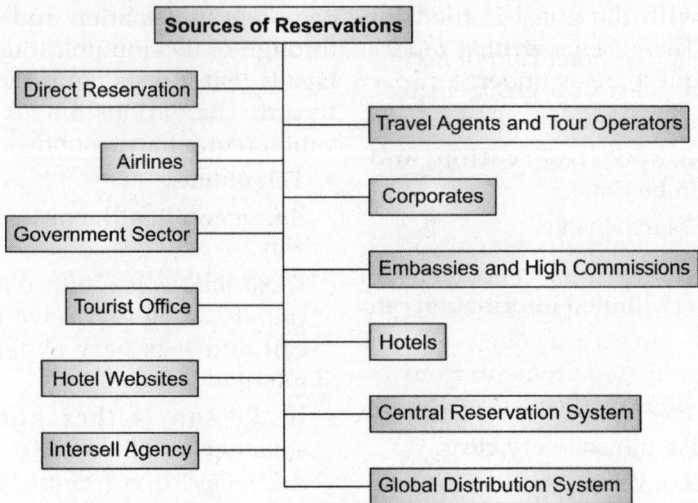

Fig. 21.2: Sources of reservations

Travel Agents and Tour Operators

Many guests make reservations through travel agents or tour operators. The agent will normally take a pre-payment from the guest, send a confirmation to the hotel, and issue travel agent voucher. Travel agencies receive a commission for their services from the guest, or the hotel, or both. As hotels receive bulk bookings and huge volume of business throughout the year from travel agents and tour operators, they offer very low rates to these agencies for various room categories.

Airlines

They make room reservations when they have to give accommodation to their crew and passengers when there are flight cancellations or long flight delays. The airline also makes reservation for their passengers during transit stay, as there is a long time (at least 6 hours) during the connected flights.

Corporates

They make reservations to accommodate their company delegates and business associates in a hotel. The company sends a letter of reservation to the hotel, the accounts

department at the hotel then verifies the credits of the company and accordingly confirms and sends the confirmation letter to the company. Then the company issues the letter to their delegates or associates. This letter is produced at the front desk when the guest of the company arrives at the hotel.

Government Sector

Hotels receive bookings from government sectors or public sector undertakings. As the government officials travel to different places on official work and need accommodation at the place of visit, they constitute a major source of reservation and revenue to the hotel.

Embassies and High Commissions

They make reservations when they have their delegates, counselors, ambassadors and head of state visiting.

Tourist Office

They make room reservation but only of hotels or inns, which belong to the government. Every state has a tourist office to make reservations convenient to the local people as well as tourist.

Hotels

This happens when the hotel cannot accommodate a walk-in guest or a group in their own property and refers them to another hotel in the vicinity.

Hotel Websites

A hotel's website is another potential source for receiving reservations. The website contains a link for reservation requests. By clicking the link, guests can make a hotel reservation as per their requirements from the comforts of their house/office/cyber café.

Central Reservation System

Central reservation system (CRS) is a computer-based reservation system, which enables guests to make reservations in any of the participating lodging properties (members) at any destination in a single call. The central reservation office typically deals with direct guests, travel agents, corporate bookers, etc. by means of toll-free telephone numbers. The central reservation offices operate twenty-four hours a day, all round the year.

The success of central reservation system depends mainly on communication network which these days is very strong through satellite links, internet, and conventional systems such as telephones and fax machines, etc. and upon the accuracy of transfer of information between the central reservation system and hotel members regarding room availability to avoid any double booking or rooms remaining unlisted for blocking.

The reservation agents can be creative and can upsell room categories (e.g. suites and other high-category, low-moving rooms) in case the requested rooms are not available, or suggest alternative destinations. They must be very efficient and well informed about the hotels and the details of the rooms which they are selling.

The central reservation offices help a guest get the information and room availability status for all the hotel members and to plan their itinerary in one toll-free call.

Hotels are required to provide accurate and current room availability data to the central reservation office to enable them to sell their rooms effectively. The members allocate a particular percentage of the total number of rooms to central reservation office. It becomes the responsibility of central reservation office to book these rooms for the hotel. Over a period of time a close watch on the performance of central reservation office is kept. If the central reservation office is not able to book the allocated number of rooms, then the member can withdraw rooms from CRO and also if the CRO is able to book more rooms than allocated to them, more rooms can be given to CRO for booking.

Members may pay a flat fee for obtaining the services of a CRS and an additional fee for each reservation received through the central reservation office. The CRS is of two types:

- **Affiliate Network:** The affiliate CRS is the central reservation network formed by the properties of a particular hotel chain, like Welcome net by Welcome group of Hotels, Holidex by Holiday Inn Hotels, Image by Hyatt Hotels, and ITT by Sheraton Hotels.

  Sometimes, independent hotels also participate in the affiliate reservation system of a particular chain in order to gain the benefit of central reservation network and so these independent hotels are called *overflow facility*. Overflow facilities receive reservation requests only after all room availabilities in chain hotels within a geographic area have been exhausted.

- **Non-affiliate Network:** A non-affiliate reservation system connects independent or non-chain hotels, like the Leading Hotels of the World (LHW), Small Luxury Hotels of the World (SLH). This enables non-chain properties to enjoy the benefits of CRS.

Intersell Agency

An intersell agency is an agency that deals with multiple products such as hotel reservations, car rentals, airline reservations, railway bookings, etc. This is also called *one call does it all*. Examples of such agencies are Expedia, Trevelocity, Travelguru, Make-Mytrip, etc.

Global Distribution System (GDS)

Global distribution system (GDS) is a world-wide computerized reservation network, which is used as a single point of access for reserving hotel rooms, airline seats, rental cars, and other travel-related items by travel agents, online reservation sites, and large corporations. GDS provides a bundle of products and services to the prospective user across geographical boundaries and is a link between the producers and end users of travel products and services. Some examples of GDS are: Amadeus IT, Galileo CRS, SABRE, Worldspan, etc.

- Amadeus IT: Owned by the Amadeus IT Group, it was formed in 1987 out of an alliance between Air France, Lufthansa, Uberia Airlines, and Scandinavian Airlines. It specializes in the bookings of hotels, airlines, cruises, travel services, and car rentals.

- Galileo CRS: This computer-based reservation system is owned by Travelport, and is used for the reservation of travel, tourism, and hospitality products and services.

- SABRE: Semi-automated business research environment (SABRE) is a computer based reservation system used by airlines, hotels, travel agents, railways and other travel-related companies for reservation of their products and services. It is a unit of Sabre Holding's Sabre Travel Network division and is one of the largest electronic travel reservation systems.

- Worldspan: This is a GDS used by travel agents and tour operators for travel and hospitality-related bookings. Owned by Travelport, it was created in 1990 by Delta

Airlines, Northwest Airlines, and Trans World Airlines to sell their GDS services to travel and hospitality operators worldwide.

TYPES OF RESERVATION

Reservations can be of the following types (Fig. 21.3):

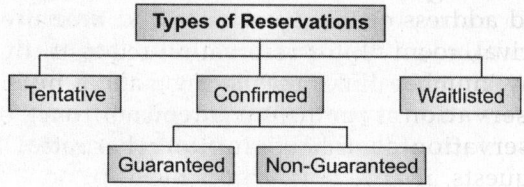

Fig. 21.3: Types of reservations

Tentative/Provisional Reservation

It is a reservation request that a prospective guest makes on a tentative basis for particular stay dates. The hotel holds the room for the guest till a cut-off date, by which the guest should confirm the reservation. Upon confirmation from the guest the hotel changes the tentative reservation to a confirmed reservation, otherwise it cancels the tentative reservation, and updates its records accordingly.

Waitlisted Reservation/Stand by Reservation

When a hotel does not have accommodation available for the requested date but foresee a possibility of accommodation being available, they may offer the guest a waitlisted reservation. The waitlisted reservation is confirmed when the hotel receives a cancellation request for a room of the same category. This way the hotel ensures that its rooms will not remain vacant in case of cancellations. The hotel does not guarantee a room for waitlisted reservations; it is understood that the guest will be assigned a room only in the case of a cancellation or a no show (the same is informed to the guest at the time of processing the reservation request).

Confirmed Reservation

A confirmed reservation is a reservation in which the hotel has assured the guest of the accommodation in the property on the requested date and in the requested room category. The confirmation of reservation is sent to the guest through letter or e-mail containing the following information: Name and address of the guest, date and time of arrival, room type and room rate, length of stay, number of persons in a group, if any, reservation type (guaranteed or not), reservation confirmation number, special requests, if any, like airport pick-up, non-smoking room, etc.

A confirmed reservation can be of the following two types:

• **Guaranteed Reservation:** A guaranteed reservation assures the prospective guest that the hotel will hold a room up to the checkout time following the guests scheduled arrival date. However, this requires that the guest make an advance payment which is not refundable even if the room is not used, unless the reservation is cancelled as per the hotel's cancellation policy. Guaranteed reservations ensure the reservation materialization and protect the hotel in case of a 'no-show'. The guaranteed reservation can be obtained through one of the following ways:

  – Pre-payment: This type requires that a prospective guest makes a full payment in advance for the period of stay, so as to confirm the reservation. Pre-payment can be made by sending demand draft or depositing cash at the hotel (or in its bank account). Guests can alternatively choose to pay the full amount in advance through their credit cards. A guest should send a letter authorizing the hotel to charge payment to their credit card account for obtaining guaranteed reservation, along with a copy of the front and back of the credit card (photocopy if sending by letter or fax, and scan if sending by e-mail).

  – Travel Agent Guaranteed: In this type of guaranteed reservation the travel agent makes the reservation for the guest and takes responsibility for charging the guest in case of a 'no show'. The hotels guarantee these reservations on the basis of vouchers issued by the travel agency.

  – Airlines Guarantee: In this case the airlines management takes the responsibility of payment of the accommodation booked by them for their passengers and crew and in case of their no show.

  – Company Guarantee: In this case a company, who makes a booking for their employees or sponsored guests in the hotel, holds the responsibility of paying the retention charges to the hotel in case of no show of the guest for whom the booking has been made. Hotels guarantee these reservations on the basis of a letter from the company, called a bill to company or BTC letter, acknowledging the guest as its employee or client and agreeing to pay his bills as per the contract.

• **Non-Guaranteed Reservation/6 PM Release Reservation:** This is usually when the prospective guest does not send any advance deposit in the form of guarantee but simply confirms his reservation at a hotel. In such cases the hotel holds the reservation till the cancellation hour (usually 6:00 PM) only on the day of arrival of the prospective guest and release the accommodation after that time and may sell it to a walk-in guest, and in case the guest who has made the reservation arrives after 6:00 PM, the hotel is not bound to provide him accommodation. If the guest arrives after the cut-off time, the hotel will accommodate the guest only if the room is available. This enables the hotel to cover the probable loss due to a no show. Cancellation hour is also known as *cut-off time/time limit/6 PM hold.* (It is the specified time at which the unclaimed reservations are released and can be sold to any walk-in guest, i.e. if the guest who had booked the room comes after this time, he may be refused accommodation. In the absence of any guarantee, the hotel usually holds the reservation until 6 PM and if the guest fails to check-in by that time, the reservation gets automatically cancelled.)

SYSTEMS OF RESERVATION

The hotels around the world have developed different systems of processing and recording reservation requests from the guests depending on their needs and volume of business. The following are the various systems of reservation in the hotels.

Manual System

In a manual system, all the reservation records are maintained manually. This old system of reservation is suitable for a small property, where the number of rooms is less and the volume of reservation requests is also low. The hotel may use one of the following systems of manual reservation:

- **Diary System of Reservation:** This system consists of a bound book called Booking Diary or Hotel Diary which is maintained by the reservation office. The booking diary has 365 pages. Each page is for each day of the year for which guest rooms of the hotel will be reserved. All the necessary information that is received from the prospective guest is recorded in the diary. The booking diary records the reservations according to the date of arrival of the guest (Fig. 21.4).

Advantages

- All the reservation records are available in one consolidated book, and chances of losing the records are very less and hence it is safe.

- It is useful for small hotels.

HOTEL XYZ
RESERVATION BOOKING DIARY   Date: 20th Oct, 2015

Sl.No.	Date of arrival	Date of departure	Guest name	PAX	Address	Expected time of arrival	Room type	Room rate and plan	Mode of payment	Reservation status	Reservation number	Special instructions	Booking made by
1.	22/11/15	25/11/15	Akshat	01	Pune	1230 Hrs	Deluxe	INR 8000 EP	Credit Card	Confirmed Guaranteed	1489	Room near elevator	Mr Rajesh
2.	15/12/15	21/12/15	Shreya	02	Goa	1400 Hrs	Suite	INR 25000 CP	BTC	Confirmed Guaranteed	1490	Corner Room	Mr Rahul

Fig. 21.4: Reservations booking diary

Disadvantages

- Cancellations and amendments can create problems and disturb the sequence of the diary.
- Since the diary is bulky, its movement for reference from reservation section to front desk is difficult.
- Diary can be maintained in date sequence only and it is not possible to arrange it in alphabetical order.

- **Whitney System of Reservation:** The Whitney system of reservation, developed by the American Whitney Duplicating Check Company of New York in 1970, is suitable for large hotels. It is based on the use of standard size slips, known as Whitney slips or Shannon slips (Fig. 21.5). To indicate and identify the status of the guest at a glance, hotels may use color coded slips. For example, white slips for FIT, blue slips for travel agent booking, pink for airline booking and yellow for group booking and so on.

  The Whitney system uses wooden racks called Whitney rack that are vertically mounted on walls. It requires a total of 43 racks, out of which 31 racks are kept for the current month (one for each day, arranged as per the date of the month), 11 racks for the next eleven months of the year, and 1 rack for the next year.

Date of arrival	Name of guest	Room type	Rate	Date of departure
Mode of Reservation	Reserved by	Date received		
Agency (if any)				
Billing instruction		Confirmation date		

Fig. 21.5: Whitney slip

Advantages

- It is useful for taking reservations in large hotels.
- Cancellations and amendments are very easy as only the removal and readjustment of slip is needed.
- The status of the guest can be seen at a glance as colored slips are used.
- Bookings can be kept in order of the date of arrival.

Disadvantages

- The Whitney slips may be lost and the reservation information can be lost.

Automated System

Automated reservation systems are computerized reservation systems that are used to store and retrieve room status information and conduct transactions. The information stored in the automatic system is the same as in a manual system. However, the processing of reservation request does not require manual study of density control charts or conventional charts. The reservation assistant can check the availability of rooms by clicking on a link on the computer. In this system, the reservation information is keyed into the electronic format of the reservation form, and this information is transferred to the central server where the room status is updated automatically. The automated system saves the trouble of manually updating the records. It also generates electronic confirmation letters that are sent to the guests' e-mail addresses. The system is also equipped to automatically generate reports like occupancy list, expected arrival/departure lists, etc. Central Reservation System, Global Distribution System (*we have discussed earlier*) and Instant Reservation System are examples of automated systems.

- **Instant Reservation System (IRS):** The Instant Reservation System is a fully computerized reservation network and is based on the principle of Wide Area

Network. The Instant Reservation System operates through the Instant Reservation Office (IRO). The Instant Reservation Office is a separate reservation section other than the hotel's own reservation section and the IRO accepts the reservation request of all other participating properties of the reservation network except the property in which it is located; as the reservation section of the particular property takes its reservation requests. IRS is useful for guests who are staying in the hotel while CRS can be approached by guests through toll free number from the comfort of their homes/offices. The IRS is used by hotels which cannot afford to establish a CRS due to heavy investment.

TOOLS OF ROOM AVAILABILITY

Before accepting the request made by a prospective guest for booking a room, it is important to know the rooms' availability position. Hotels use various tools to know the room availability position which are as follows.

Room Status Board/Perpetual Year Planner/ Stop and Go Board

As the name suggests, this chart shows the rooms booking position for one year on continuous basis. The status is shown under three categories: Sold out, on request, and free sales by three different colored plastic discs. 'Sold out' means no rooms are available for booking for that period or date. 'On request' means rooms can be blocked subject to cancellation and the guest is given the status of waiting list. 'Free sale' means that the rooms are freely available for booking for the requested dates. The bookings keep coming. The free sale status changes to on request and further to sold out and with cancellation the status changes from sold out to on request and further to free sales.

Advance Letting Chart (ALC)/Conventional Booking Chart (CBC)

It is an old system of recording reservations in small hotels which do not have computerized system of reservation. The chart has 31 columns in which the dates of the month are arranged horizontally while the guest room number and types are arranged vertically for reservation (Fig. 21.6). Each individual reservation is completed by showing line with arrows with the name of the guest written on the line with block letters while each group reservation is shown by boxes with the name of the group written in block letters. The small arrow head is placed at the end of each reservation to indicate when a particular reservation starts and finishes. All housekeeping status such as OOO, UR, etc. for that period is also recorded in the chart so as to avoid booking of such rooms. Many hotels use this chart in conjunction with the hotel diary, as it is not possible to record all the details of reservation in the chart.

Advantage

- The chart is useful for small hotels with a limited number of rooms where the guest stays for a longer duration.

Disadvantages

- It is not easy to find out how many rooms are booked and how many are available for sale at a glance from the chart.
- The chart becomes untidy in the case of cancellations and amendments.
- It does not have space for doing over-booking and this limits the chart from being used in big hotels.

Density Control Chart

It is an old system of recording reservations in large hotels which do not have computerized system of reservation. The chart has 31 columns in which the dates of the month

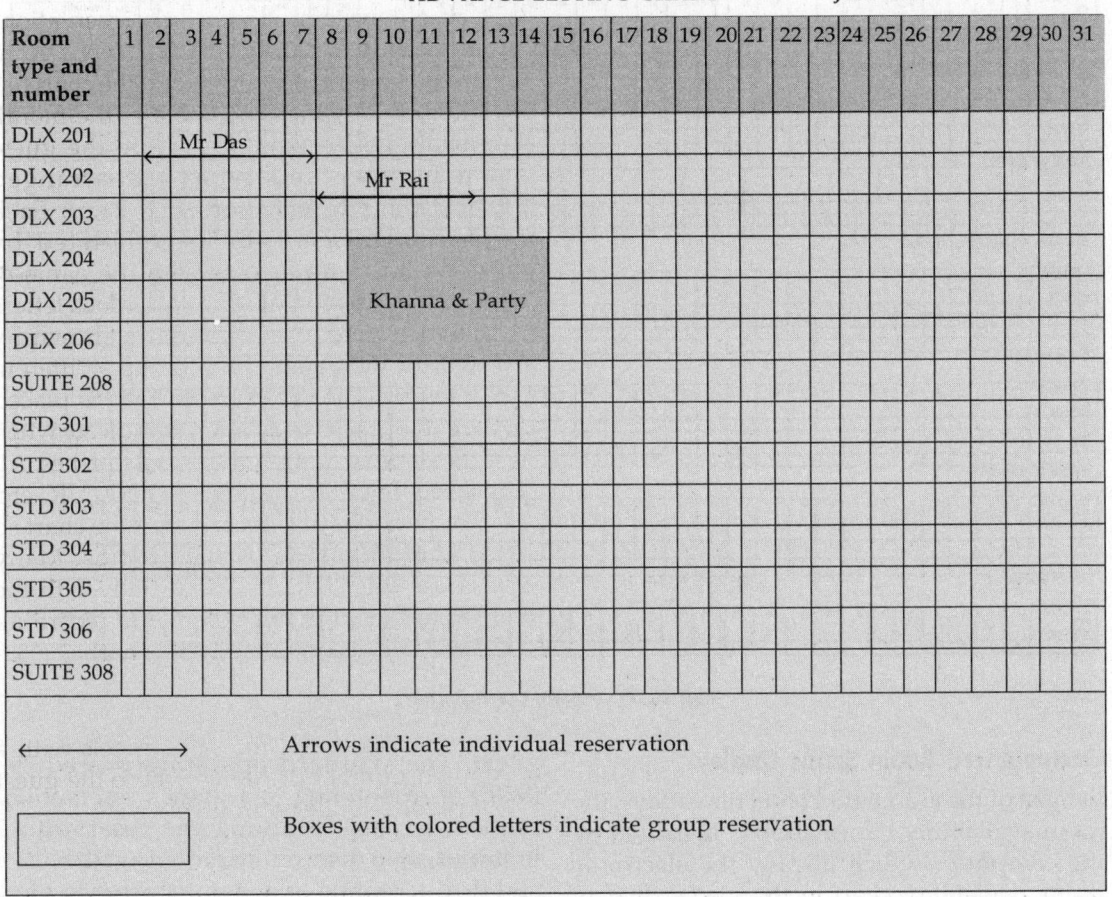

Hotel XYZ

ADVANCE LETTING CHART   June 2016

Room type and number	1	2	3	4	5	6	7	8	9	10	11	12	13	14	15	16	17	18	19	20	21	22	23	24	25	26	27	28	29	30	31
DLX 201		←	Mr Das		→																										
DLX 202									Mr Rai		←			→																	
DLX 203																															
DLX 204																															
DLX 205										Khanna & Party																					
DLX 206																															
SUITE 208																															
STD 301																															
STD 302																															
STD 303																															
STD 304																															
STD 305																															
STD 306																															
SUITE 308																															

←————→   Arrows indicate individual reservation

[ ]   Boxes with colored letters indicate group reservation

Fig. 21.6: Advance letting chart

are arranged horizontally while the number of rooms of specific type is written in descending order (Fig. 21.7). This chart is based on the principle that each reservation reduces the availability of the rooms and each cancellation increases the availability of rooms. In this chart, the various types of guest rooms available in the hotel are grouped together separately and shown with separate table of overbookings for each of these categories. Strokes (/) are put in the appropriate boxes for indicating a reservation.

Advantages

- It allows one to see at a glance how many rooms of a particular type have left.
- It has got space for doing overbookings.

Disadvantages

- While completing the density chart with strokes, it is easier to make mistakes, and more difficult to check them, since one has no idea which stroke represents which guest.

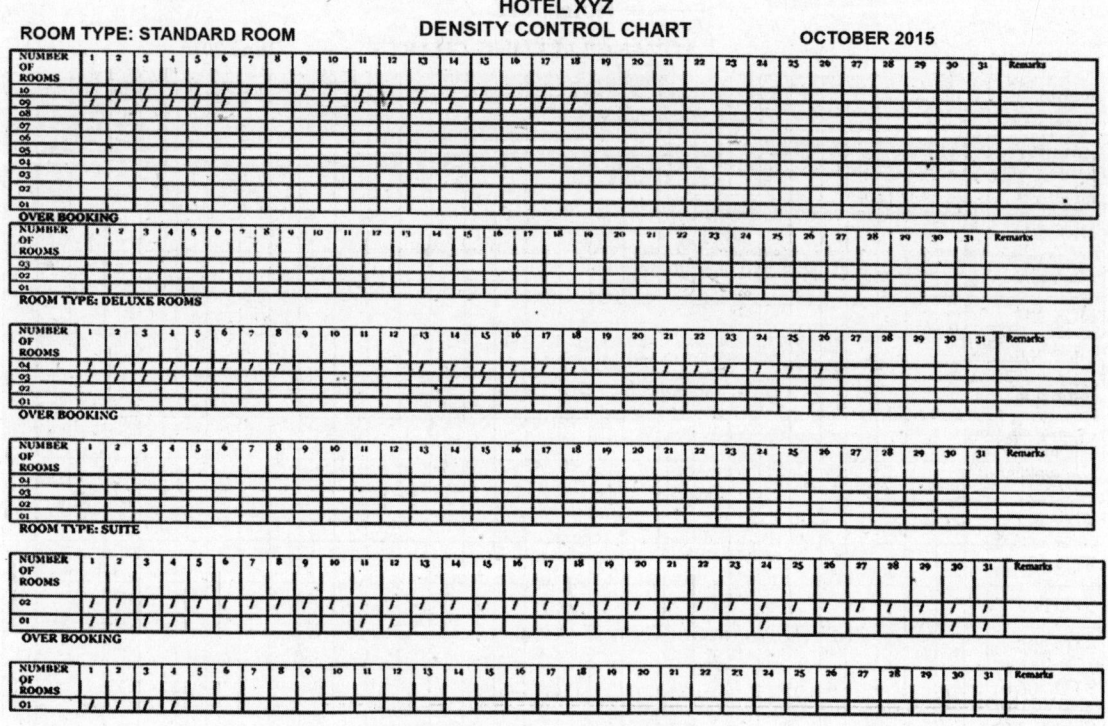

Fig. 21.7: Density control chart

Computerized Room Status Display

In most of the automated hotel nowadays, the room availability information is managed by the computers which display the electronic room availability chart in the visual display units or the monitors. The reservation assistant can check the availability of rooms by clicking on a link on the computer.

Advantages

• Efficient, effective and time saving, saves manpower, little paperwork.

Disadvantages

• Expensive to install in small hotels.

PROCESSING INDIVIDUAL RESERVATION REQUESTS

Every hotel follows a systematic procedure for dealing with a reservation request from a guest. The standard operating procedure (SOP) of responding to a guest's reservation request is first receiving the reservation inquiries, then determining room availability, and then accepting or denying the request for reservation (Fig. 21.8).

Steps of Processing Individual Reservation

• The reservation request may be made by a guest either verbally or in writing (telephone, in person, fax, letter, e-mail, etc.)

• The reservation clerk then enquires about the date and time of arrival, date and time of departure, number and type of rooms required.

• Then the reservation clerk matches the reservation request of the guest with the availability of the accommodation product and this is done with the help of tools of

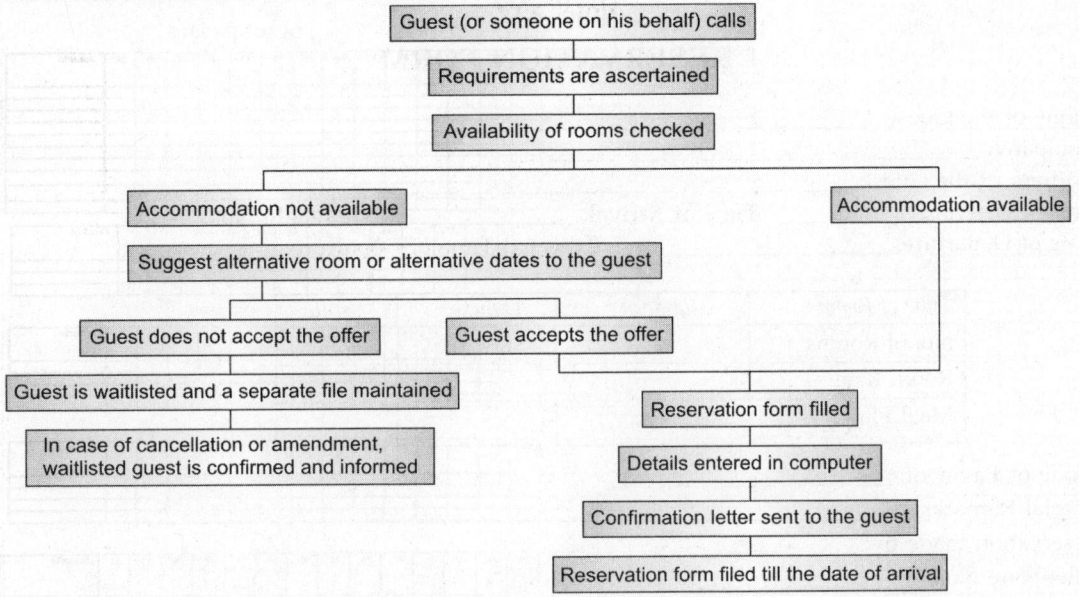

Fig. 21.8: Diagrammatic representation of a telephone request process

availability such as advance letting chart or density control chart or room status board or computerized system.

- If the reservation request does not match with the room availability, the reservation clerk suggests alternative dates or room type to the guest. The clerk may suggest alternative hotel to the guest.
- If the reservation request matches with the room availability, the reservation clerk proceeds further with the processing of the reservation requests and takes other details of the reservation enquiry such as the name of the guest, his company name and address, billing instructions, person requesting the reservation for the guest and his contact number, and any other information relevant from the point of view of the reservation (inter-connecting rooms, non-smoking room, river facing room, airport pick-up, etc.). All the above information about the reservation is recorded in reservation form or slip (Fig. 21.9).
- Finally, the reservation clerk gives a confirmation number (alphanumeric code)

to the guest, which is the proof of the confirmation of the reservation by the hotel. Reservation clerk may send a confirmation letter to the guest's mailing address or e-mail (Fig. 21.10).

- All documents (reservation form along with correspondence) are filed till the arrival of the guest.

PROCESSING GROUP RESERVATION REQUESTS

A group, as the name suggests, is a number of people traveling together and booking a number of rooms in hotels in advance. Normally a hotel gets group business from travel agents, tour operators, housing or convention bureaus, meeting planners, embassies, etc. Most of the hotels around the world have a large number of guests coming in groups and these groups are of major importance to the hotels from the point of view of the revenue generation. The various types of groups visiting hotels are airline crew members, corporate groups for seminar,

Hotel XYZ

RESERVATION FORM

Name of the Guest: _____

Company: _____

Address of the Guest: _____

Date of Arrival: _____ Time of Arrival: _____ Mode/Flight No.: _____

Date of Departure: _____ Expected Time of Departure: _____

Type of Room	Standard	Deluxe	Suite
No. of Rooms			
Room Rate			
Meal Plan			

Mode of Payment: _____

Special Remarks: _____

Reservation made by: _____

Telephone No.: _____

Reservation Status: Confirmed/Waitlisted

Reservation Number

Name and Signature of Reservation Agent

Date: _____

Fig. 21.9: Reservation form

conference or convention, educational groups of students, travel groups of vacationers or holiday makers or religious travelers, sports teams, businessmen attending trade fairs, etc. Thus, the hotels take great care while handling the reservation requests of the groups as the groups give a huge magnitude of revenue to the hotels.

Because it involves volume business, management offers special discount to the travel agents. The discount depends on the size of the group and number of room nights, meal plans and the frequency of visits organized by the travel agent. Generally following discounted rates are given:

10 to 14 full paying pax—50 % discount on 1 pax on plan

15 to 30 full paying pax—Complimentary 1 pax on plan

31 to 45 full paying pax—Complimentary 2 pax on plan

However, for the accompanying Indian Escort, hotels generally give 30% discount on published room tariff. An escort is the person who accompanies the tour from the beginning to the end and looks after the arrangement.

Steps of Processing Group Reservation

- The reservation request may be made by a group (through group leader or travel agent) either verbally or in writing (telephone, in person, fax, letter, e-mail, etc.)
- The reservation clerk then enquires about the date and time of arrival, date and time of departure, number and type of rooms required.
- Then the reservation clerk matches the reservation request of the group with the

---

HOTEL XYZ

RESERVATION CONFIRMATION

Dear **Mr Akshat**,

Thank you for your reservation at the Hotel XYZ. Should you require any further information, please do not hesitate to contact us at any time.

YOUR RESERVATION DETAILS

Guest Name:	Akshat Mishra
Company Name:	Samsung
Confirmation Number:	4579
Arrival Date:	Tuesday, 15 December, 2015
Departure Date:	Thursday, 17 December, 2015
Room Category:	Suite
Number of Rooms:	01
Number of Adults:	2 Adults
Number of Nights:	2 Nights
Room Rate:	INR 20,000
Rate Information:	Including breakfast
Preferences:	Room on top floor
Guaranteed by:	American Express Credit Card
Cancellation Policy:	The room is reserved on a guaranteed basis. Cancellations less than 72 hours prior to arrival will be forfeited with a one night charge.
Check-in/Checkout Time:	12 Noon

We look forward to welcoming you to Hotel XYZ.

With best regards,

Yours sincerely,

Reservation agent

Fig. 21.10: Reservation confirmation letter

---

availability of the accommodation product and this is done with the help of tools of availability such as advance letting chart or density control chart or room status board or computerized system.

- If the reservation request does not match with the room availability, the reservation clerk suggests alternative dates or room type to the group. The clerk may suggest alternative hotel to the group.
- If the reservation request matches with the room availability, the reservation clerk

discusses with the group leader or the travel agent about the number of rooms to be kept aside for the group as Block and then decides the cut-off date. Cut-off date or Option date (generally 30 days before the arrival of group) is an agreed upon date between a group and the hotel after which all the unclaimed reservation in the group block are released into the general room inventory pool for resale.

- After receiving the confirmation, the reservation department shall change the desired number of rooms' status from blocked to booked (or reserved) rooms, and release the remaining rooms (if any left) as vacant for sale.

- The reservation clerk then proceeds further with the processing of the reservation requests and takes other details of the reservation enquiry such as the name of the group, name of the group leader and the group members, meal plans accepted by the group, mode of settlement by the group, and special instructions by the group. All the above information about the reservation is recorded in Bulk Reservation Form (Fig. 21.11).
- After receiving the final list, a group folder needs to be prepared for the group that contains all details of group, with the name and code number of the group on top:
  - Name of the group
  - Name of all the group members along with the nationality and passport details of each member
  - Room break-up: Type and number of rooms required.
  - Billing instructions
  - Meal plan

- Special instructions
- Name of the group leader so that all matters pertaining to the group may be referred to him
- Name, address, and telephone number of the travel agent

- The rooming list is sent by the travel agent in advance. Along with the rooming list the travel agent will also send Travel Agent Voucher. The voucher gives details about facilities and services included in the package. Any service or facility not mentioned in the voucher shall have to be paid extra by the guest availing that service. A duplicate copy of the same voucher will be brought by the group leader on the day of arrival. The two will be tallied by the front desk staff.
- Finally, the reservation clerk gives a confirmation number (alphanumeric code) to the group leader or travel agent, which is the proof of the confirmation of the reservation by the hotel. Reservation clerk may send a confirmation letter to the group leader or travel agent's mailing address or e-mail.
- All documents (reservation form along with correspondence) are filed till the arrival of the group.

				Sl No.: _____
Group No. _____			Tour Operator: _____	
Group Leader: _____			Telephone No.: _____	
PAX: _____			E-Mail: _____	
Arrival date: _____			Arrival time: _____	
Departure date: _____			Departure time: _____	

Accommodation				Meals Break up		
Rooms	No.	Rate	No. of Days	Breakfast	Lunch	Dinner
Single						
Double						
Special requirements:						
Date: _____				Signature of Reservation Agent		

Fig. 21.11: Bulk reservation form

Group Meal Information Sheet

The hotel sends this to the group leader or the travel agent to be completed and sent back. This sheet is circulated to all the coordinating departments (F&B and Kitchen) at least a week in advance of the arrival date to prepare themselves for group arrival (Figs 21.12 and 21.13).

Group Amendments

Amendments of a group are a continuous process till the group finally checks in. These amendments are made through the travel agent. After receiving the amendment request, the hotel confirms whether the request for amendment can be met with or not. In case it can be accepted, proper entries are made on computer and the group folder and the same is intimated to all concerned including the travel agent.

Group Cancellation

Cancellation policy for the group varies from hotel to hotel and also depends on the size of the group. Generally the cancellation should be received 30 days in advance. The

Hotel XYZ

GROUP MEAL INFORMATION SHEET   **Date**: 10 Aug, 2015

Dear **Mr. David** (Group Leader),

We have made the following meal arrangements for your group. The venue has been shown against the dates and meals. We would appreciate if you could mention the approximate timings for the meals to help us serve you better. Also if you could indicate any special preference of food you would like us to serve for your group.

Date	Plan	Meal	Venue	Time	PAX
20 Aug, 2015	AP	Breakfast	Mehfil		
		Lunch	Sammelan		
		Dinner	Rang Mahal		
21 Aug, 2015	AP	Breakfast	Mehfil		
		Lunch	Sammelan		
		Dinner	Rang Mahal		
22 Aug, 2015	AP	Breakfast	Mehfil		
		Lunch	Sammelan		
		Dinner	Rang Mahal		
23 Aug, 2015	AP	Breakfast	Mehfil		
		Lunch	Sammelan		
		Dinner	Rang Mahal		
24 Aug, 2015	AP	Breakfast	Mehfil		
		Lunch	Sammelan		
		Dinner	Rang Mahal		

Remarks: _____

Front Office Manager

Note: Groups on AP and MAP plan may please collect meal coupons from front desk and present them prior to meals at venues marked above.

CC: GM, F&B Manager, Ex Chef, EHK, GRE

Fig. 21.12: Group meal information sheet

```
                          Hotel XYZ

                 MEAL COUPON              Sl.No.1442
                                         Date: _____
Guest Name: _____
Room No.: _____
No. of Guest: _____
Signature of FOA

         (This coupon is non-transferable and non-refundable)
                      Have a Nice Day!

   ┌──────────────────────────────────────────────────────┐
   │  Buffet Lunch and Dinner only at 'Mehak' 10th Floor   │
   │            (Valid for this day only)                  │
   └──────────────────────────────────────────────────────┘
```

Fig. 21.13: Meal coupon

relationship between the hotel and the travel agent is also important. The inconvenience caused to hotel and possible loss of revenue is also considered while considering the cancellation request. Depending up on the management policy, retention charges may or may not be charged.

AMENDMENTS, CANCELLATIONS, NO-SHOW AND OVERBOOKINGS

Amending Reservation

Any alteration or change in original reservation in terms of guest name, date of arrival or departure, type of room, number of rooms, billing instructions, the type of reservation (guaranteed or non-guaranteed), is termed amendment. In case of amendments, the hotel has to check the availability of rooms again as per the fresh details given by the guest. The reservation agent should ascertain that the person requesting the amendment is the same as the one who has made the original booking. This is done to avoid any problem or confusion that may arise at the time of the arrival of the guest. The changes are recorded in a specialized form known as the reservation revision or cancellation form (Fig. 21.14).

Cancellation of Reservation

This is a request which is made by the guest to void his reservation previously made. When a cancellation request is received, the reservation staff must issue a cancellation number as proof of the cancellation along with the name of the person who has cancelled the reservation. As cancellation might lead to the loss of room revenue, hotels discourage cancellations by imposing retention charges (which may be equal to the rent of one night or more).

Reservation clerks accepting a reservation cancellation shall behave in a polite, courteous and effective manner even though that reservation might make the hotel faced with unsold room(s). The main reason is that guests are doing the hotel a favor, especially under the non-guaranteed type of reservation, to communicate the hotel their cancellation to let you adjust your room availability, and try to find alternative potential guests beforehand.

No-Show

A guest who does not arrive on the scheduled date of arrival after making a room reservation is called a no-show. No-show guest is also referred to as DNA (Did Not Arrive). In case any advance money has been deposited by the guest (guaranteed reservation), the same may be forfeited. This amount is called retention charge. This is the charge levied on individual guest or group for default in reservation (no-show or cancellation). The principle behind a retention charge is to offset the inability to sell the booked rooms again. Retention charge

<table>
<tr><td colspan="3" align="center">**Hotel XYZ**
REVISION/CANCELLATION FORM
Guest Name: _____
Address: _____</td></tr>
</table>

Hotel XYZ **REVISION/CANCELLATION FORM** Guest Name: _____ Address: _____
Original Reservation
Room Type ☐ Single ☐ Double ☐ Suite Arrival Date: _____   Departure Date: _____
Cancellation
Reason:
Revision
Room Type ☐ Single ☐ Double ☐ Suite Arrival Date: _____   Departure Date: _____
Requested by: _____   Telephone No.: _____ Received by: _____   Date: _____

Fig. 21.14: Revision/cancellation form

may be waived by the hotel depending upon the relation between hotel and the guest or the group.

Overbooking

It is the process of accepting more booking than the actual number of rooms available in the hotel. This is a situation which arises not due to an error but is a deliberate act by the hotel (reservation section) to achieve 100% occupancy or maximum possible occupancy, because if there are reservations, there are certainly going to be some cancellations and no-shows. So to avoid loss of revenue and to meet the financial obligation a certain percentage of overbooking is done. Factors to be taken into consideration before deciding the % of overbooking:

- No-show and cancellation%
- Understay and overstay%
- Source of booking and its past record
- Profession and purpose of visit of the prospective guest such as conference, trade fair, holiday, etc.
- Period of business—lean or peak
- Lead time, i.e. the time between when a reservation is made and when the guest is due to arrive.

Overbooking ratio: To find the overbooking ratio, just divide the cost of walking a guest by the sum of the cost of walking a guest plus the cost of an empty room:

$$\text{Overbooking ratio} = \frac{\text{Walk}}{(\text{Walk} + \text{Empty room})}$$

Cost of an empty room = ADR – Variable cost

RESERVATION REPORTS

The reservation section compiles many reports for the use of all departments. Some of the most commonly used reservation reports include:

- **Reservation transaction report**: The reservation transaction report is the summary of the daily activities of the reservation department in terms of:
  - Creation of reservation records
  - Amendment request
  - Cancellation of reservation request
- **Commission agent report**: This report includes the amount payable by the hotel to the different commission agents (like travel agents, tour operators, CRS, GDS, etc.)
- **Turn away or refusal report**: At times hotels have to 'turn away' guests due to unavailability of rooms. The reservation section compiles a report of the total 'turn away' during a period of time. This report aids the management in planning the expansion and developing new properties in the city.
- **Revenue forecast report**: The revenue forecast report is a projection of the volume of business that the hotel will be generating in a specified duration. This is calculated by multiplying the reservations and the room rent offered to guests.
- **Expected arrivals and departure lists:** A list of guests, along with their respective room numbers who are due to arrive or to depart on a particular day.
- **Stayover list**: The list of names and surnames, along with the respective room numbers, of the guests who are expected to continue to occupy their rooms the next day.
- **Room availability report**: A list showing the number of rooms sold/available on a daily basis.

REVIEW QUESTIONS

1. List the six activities during a reservation cycle.
2. What information does a reservation agent require to create a reservation record? Explain with the help of reservation form.
3. Give the use of a room status board.
4. What is the use of a density control chart?
5. Explain the use of a whitney system.
6. Briefly explain the CRS. Give its benefits.
7. Write in detail various types of reservation.
8. How are reservations taken by the aid of density chart?
9. Explain the importance of reservation process in front office operations.
10. Briefly describe the various sources and modes of reservation.
11. Explain the process of reservation alongwith a flow diagram.
12. What are the different types of reservations? Explain the process of cancellation and amendment of reservation and draw necessary form.
13. How does a hotel handle group reservation? Draw the format of Bulk Reservation Form.
14. Explain the function of typical management reports and reservation records that can be generated from reservations data.
15. What is the main purpose of a reservation confirmation letter? Draw its format.
16. Draw the formats of the following:
    - Booking dairy
    - Conventional chart
    - Reservation form
    - Advance letting chart
    - Density chart
    - Reservation form
17. Write short notes on:
    - CRS
    - Cancellation
    - No-show
    - Cut-off date
    - Retention charge
    - Time limit
    - Overbooking
    - Intersell
    - FIT
    - GIT
18. Differentiate between:
    - Confirmed reservation and Guaranteed reservation
    - Density control chart and Advance letting chart
    - IRS and CRS
    - Cancellation and Amendment
    - Affiliate reservation network and Non-affiliate reservation network

Registration

Learning Objectives

After reading this chapter, you will be able to understand the following:

- Pre-registration activity
- Registration process
- Various methods of registration—hard-bound register, loose-leaf register, guest registration cards and self-registration terminal
- The check-in procedures—manual, semi-automated and fully automated systems
- Room selling techniques

PRE-REGISTRATION ACTIVITY

The activities that are carried out by the front desk agents before the arrival of guests and are primarily aimed at saving their time during the registration process are termed pre-registration activities. The pre-registration activities are only performed for guests having a confirmed reservation with the hotel and not for guests coming to the hotel without reservation (walk-in). The various activities which are performed in the pre-registration process are given below.

Preparation of the Expected Arrival and Departure List

The expected arrival list is prepared on a daily basis to indicate the number and names of guests expected to arrive the next day, along with their time of arrival, date of departure, rooms requested, reservation status, special requests, etc. (Fig. 22.1).

Calculation of the Room Position

The front desk then calculates the room availability status for the next day after going through the expected arrivals and departures list and equating the other factors affecting the room position such as out-of-order rooms, overstays, understays and no-shows. Room position refers to the number of rooms available for sale. It can be calculated as

Room Position = Vacant Rooms + Expected Departures – Expected Arrivals

Plus position implies that rooms available for sale exceed expected arrivals. *Minus position* indicates that expected arrivals exceed the number of rooms available for sale.

Preparation of the Amenities Voucher

Amenities vouchers are prepared for the arriving guests and sent to concerned departments like housekeeping and food and beverage. These vouchers instruct the departments to provide the mentioned amenities in the guest rooms, like cookies, fruit basket, soft drinks, dry fruits, cake, flowers, etc. prior to the arrival of the guest (Fig. 22.2).

Preparation of the Guest Registration Card

Pre-filling of guest registration card (GRC) based on the information gathered from reservation form and guest history card. Guests can experience a quick check-in when they arrive at the registration desk, as they only have to verify the information already entered in the registration card and provide a valid signature in the appropriate place on the registration card.

Room and Rate Assignment and the Creation of Guest Folios

In case, advance payment has been received by the hotel.

Review of the Condition of the Vacant Guest rooms

Once the room allocations have been completed, the front desk staff checks the condition of the vacant rooms to make sure that the rooms are ready to move in.

Expected Arrival List for 25 September, 2015										
Reservation no.	Guest name	Company name	Arrival date	Arrival time	Departure date	PAX	Room type	Billing instructions	Status

Fig. 22.1: Expected arrival list

Hotel XYZ
AMENITIES VOUCHER

From: Front Office

Date: _____

To: Room service/Housekeeping/Pantry

Please provide

- ▢ Fruit basket
- ▢ Flowers
- ▢ Soft drinks
- ▢ Dry fruits
- ▢ Cookies
- ▢ Mineral water
- ▢ Beer
- ▢ Toiletries

To: _____

Room No.: _____

(Name of Guest)

Authorized by: _____

Signature of FOA

Fig. 22.2: Amenities voucher

Arrangement for Welcoming Guest

Aarti, Tilak and Garlanding

REGISTRATION PROCESS

Registration begins when the front desk staff extends a sincere welcome to the guest.

Registration is a term given to the entire procedure followed by the hotel authorities on the arrival of the guests to confirm their stay in the hotel and includes the process of signing of the guests on the registration card.

The process of assigning a guest room to a guest and checking him in after the completion of other legal formalities (establishing the method of payment and creating a registration card) on his arrival is called registration.

Registration is the process of gathering information from the guest that is mandatory as per the laws prevailing in the country. According to the Foreigners' Act, 1946 and the Registration of Foreigners' Rules, 1992, the innkeeper should keep the records of the guests staying in his premises as per Form F (of the Registration of Foreigners' Rules, 1992).

Registration is the formalization of a valid contract between the guest and the hotel, in which the hotel offers safe and secure boarding and lodging facilities to the guest and the guest accepts to pay for the services and facilities received.

Importance of Registration

Registration satisfies Legal Requirements

It is a mandatory requirement that all guests over the age of 16 years—whether ordinary or VIP, Indian or foreigner, confirmed or walk-in, arriving in any hotel which may be small or large, categorized or uncategorized, one star or 5-star deluxe—must give basic information about them and fill up either a visitors' register/hotel register (also called Red Book/F Form) or guest registration card. The guest is required to fill the document in his own handwriting and sign it. It is mandatory to store this document for a minimum of three years or as required by the law prevailing in the state. The same records can be accessed by a competent local authority as and when required.

Registration provides a Record of Actual Arrivals Against Bookings

The process of registration gives the information about the number of guests who have actually arrived in the hotel. Thus, the hotel is also able to know about the actual number of no-shows (guests who did not arrive even after having a confirmed reservation and neither did they cancel their reservation). Thus, the hotel charges retention charge from the guests who were no-shows.

Registration helps to confirm the Guest's Acceptance of Hotel Rules and Regulations

Registration cards contain an acknowledgement on the part of the guest that he will abide by the rules of the hotel regarding payment and other aspects for a comfortable stay and it also serves as a proof of his stay in the hotel.

Creates Impression on the Guest

Registration provides first face-to-face contact of the guest with the hotel and establishes the front desk as focal point for guest services.

Steps of Registration Process

The registration process involves many stages (Fig. 22.3). The different stages of the registration process are discussed below.

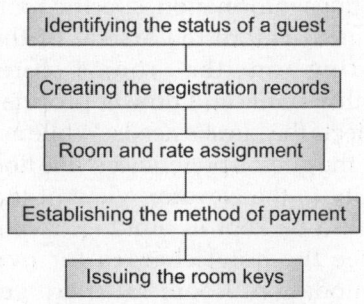

Fig. 22.3: Registration process

Identifying the Status of a Guest

The identification of a guest's status is important as the hotel processes the registration of guests with confirmed reservations and walk-ins in a slightly different way. The front desk agents verify the status of guests with confirmed reservations by referring to the day's arrival list. As the pre-registration activity has been carried for these guests, they have a speedy check-in. For walk-in guests, the front desk first refers to the room availability status. If the room is available, the front office staff collects the relevant information from guests for filling the guest registration card. The check-in of a walk-in guest usually takes longer than the guests with confirmed reservations.

Creating the Registration Records

After the arrival of the guest in the hotel, the front office staff creates a registration record, which is a collection of important guest information. The registration process requires the guest to provide his name, company name and address, date of arrival and departure and other personal data (or he verifies the information in the pre-filled GRC). The guest's signature completes and formalizes the registration record.

Room and Rate Assignment

Room assignment is an important part of the registration process and involves identifying and allocating an available room in a specific room rate category to a guest. On the basis of reservations information, specific rooms may be assigned before the arrival of the guest depending on the rooms forecasted availability status and how appropriately the room meets the guest's needs. While assigning a room, the guest's preferences, like floor level, proximity to the elevator, view of the room, etc. should be kept in mind. A room rate is the price the hotel charges for overnight accommodation. Room rates are generally assigned during the reservation process except for walk-ins who are assigned the room rates only during the registration process. Front desk staff are often allowed to offer discounts to the guests during reservation or registration and thus they can offer a room at a lower price than its rack rate when the management thinks it is appropriate.

Establishing the Method of Payment

The establishment of the method of settlement of the bill by the guest during the stay is an important part of the registration process. Regardless of whether the guest intends to pay by cash, cheque, credit card or any other acceptable method, the hotel should take precautionary measures to ensure payment.

Issuing the Room Keys

The front office staff completes the registration process by issuing the room key to the guest. Some hotels also issue a key card and ask the guest to present the key card at the time of collecting the room key. This ensures that the room key is delivered to a genuine person. The guest is finally escorted with his luggage to his room by a bell boy, who also explains the salient features of the room to the guest. The front desk sends arrival notification slips (ANS) to all concerned departments (Fig. 22.4).

VARIOUS METHODS OF REGISTRATION

The hotels around the world have been using different methods of registration depending on their number of rooms, system of operation and their budget.

Bound Book

The bound book is a big register placed on the reception counter and is used for the purpose of registration in small hotels. Thus, all the arriving guests are required to fill in their details in the bound book as a step towards the completion of the registration formalities. Format is given in Fig. 22.5.

```
HOTEL  XYZ
ARRIVAL NOTIFICATION SLIP

                                              Date: _____
Guest name: _____                           Room no.: _____
Nationality: _____                          PAX: _____
Date of arrival: _____                      Time of arrival: _____
Date of departure: _____
Billing instructions: _____                 Status of the guest: _____
Room rate: _____
                                              Sign of Receptionist
CC: Guest Services (Telephone)/HK/F&B/Kitchen
```

Fig. 22.4: Arrival notification slip

S. No.	Guest Name	Address	PAX	Nationality	IN CASE OF FOREIGNER GUESTS ONLY					Date and Time of Arrival in Hotel	Purpose of Visit	Date and Time of Departure	Sign of Guest
					Passport No.	Date of Arrival in India	Whether Employed In India	Registration Details	Proposed Duration of Stay in India				
1	Akshat	Delhi	02	Indian						20.08.15	Business	25.08.15	
2	Adam	London	01	British	BM6579JK	15.08.15	No		30 days	20.08.15	Business	27.0.15	

Fig. 22.5: Format of a hotel register (bound book and loose-leaf)

Advantages

- Since the bound book is very heavy, there are remote chances of the book getting lost.
- The records about all the guests are available in a single book. It provides in a compact form of chronological record of all guests in their order of arrival, and if date of arrival is known, a guest's registration can be easily located.
- No filing is required.
- It is a relatively cheap method as each entry occupies only one line of space.

Disadvantages

- Since the bound book is bulky and kept on the counter, with frequent usage it becomes loose and it looks shabby.
- The biggest disadvantage is that the information provided by the guest cannot be kept confidential as it can be seen by the next arriving guest. Privacy of guests cannot be maintained.
- If the book gets misplaced, although it is very unlikely, all the records for that duration are lost.

- The pre-registration activities cannot be performed.
- Only one guest can register at a time. During peak hours of guest arrival, guests will have to form a queue and wait for their turn for registration.

Loose Leaf Register

This system is almost similar to bound book system with the difference that the pages are not bound. One new page is used every day. Format is given in Fig. 22.5.

Advantages

- The loose leaf register has more confidentiality than the bound book as the sheets of the previous days are removed as soon as the registration for that particular day is over. Thus, to some extent the privacy of the guest can be maintained.
- If a sheet is lost, only one day's records are lost.
- It is convenient to handover to guests to fill their details.

Disadvantages

- The disadvantage is that the sheet can be easily misplaced if the front desk staff are careless.
- The sheet may not be fully filled on days when there are very few arrivals and rest of it may be a waste.
- Filing is required on a daily basis.
- Only one guest can register at a time.
- The pre-registration activities cannot be performed.

Guest Registration Card

This is the most prevalent system these days. In this system, one separate card is used for each guest. It may be made in duplicate or triplicate as per the policy of the hotel (Fig. 22.6).

Advantages

- The advantage of this system is that complete privacy of the guest information can be maintained.

- Another advantage is that at rush hours many guests can be registered at the same time using different cards.
- Another very important benefit of the system is that the guest can be pre-registered which will give more time to the receptionist to give personal attention and concentrate on providing services to the guest at the time of arrival.
- The cards can be stored more systematically and arranged alphabetically or in order of date of arrival.
- The card is more mobile and at the time of group arrival a separate group arrival counter can be opened for registration.

Disadvantages

- The individual card system is quite expensive and hence this system is prevalent in the large hotels only.
- If not stored properly, they can be lost or misplaced.

Self-registration Terminal

Self-registration terminals are an outcome of the advancement of technology and 'do-it-yourself' competency in guests. This system of registration is used in fully automated and world-class hotels (Fig. 22.7).

A self-registration terminal is like an interactive ATM machine, which may be located at the airport or at a convenient place in the hotel lobby. Guests with confirmed reservations can check into the hotel without any assistance from the hotel staff, simply by operating the self check-in terminal and using their credit cards. The machine requests the guests to enter their reservation numbers and then their credit card number. As the self-registration terminal is electronically linked to the reservation management and rooms' management system, the terminal is easily able to track the reservation information of the guest in the hotel. They are assigned a room by the terminal and the room keys are also

HOTEL XYZ
GUEST REGISTRATION CARD

Sl. No. 1560

Surname: _____ First name: _____

Designation: _____

Company's name and address: _____

_____

Permanent address: _____

_____

Date of arrival: _____ Time: _____

Date of departure: _____ Time: _____

Purpose of visit: _____

Arrived from: _____ Proceeding to: _____

Mode of payment: Cash-Credit card/BTC TA

Credit card no.: _____

Remarks: _____

Date of birth: _____ Marriage anniversary: _____

Passport no.: _____ Nationality: _____

Date of issue: _____ Validity: _____

Place of issue: _____

Date of arrival in India: _____ Place: _____

Proposed duration of stay in India: _____

Whether employed in India: _____

Registration Certificate No.: _____

Date of issue: _____

Place of issue: _____

I agree to abide by the hotel rules and regulations.

I declare that information furnished herein is correct and true.

Terms and Conditions:

- Check-in/Checkout time: 12 Noon
- Hotel is not responsible for the safety of any valuables left in the room.
- Safe deposit lockers available with cashier, free of charge.
- Visitors not permitted in the guest room after midnight.

Manager's Signature   **Guest's Signature**

Room no	Persons		Room type and room rate	Billing instructions	Booked by	Initials of FOA
	Adult	Children				

Fig. 22.6: Guest registration card

dispensed by the machine. Room status gets automatically updated at the front desk.

Advantages

- These terminals are intended to reduce the check-in time at the front desk.
- The efficiency of the front desk is increased when such terminals are used as it reduces the load on front office agents. This gives the front desk staff more opportunity to look after the other hospitality needs of guests.
- It reduces manpower requirements.

Disadvantages

- However, some customers may feel the lack of human touch in using self check-in terminals.

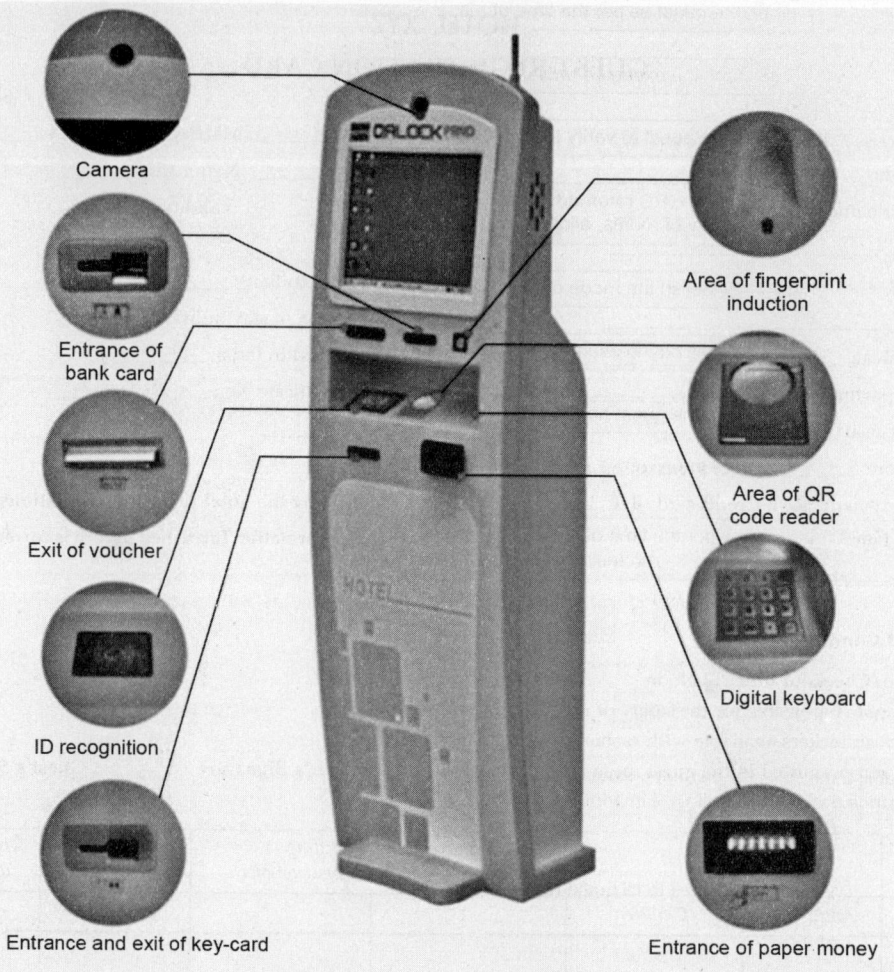

Camera

Entrance of
bank card

Exit of voucher

ID recognition

Entrance and exit of key-card

Area of fingerprint
induction

Area of QR
code reader

Digital keyboard

Entrance of paper money

Fig. 22.7: Self-registration kiosk

CHECK-IN PROCEDURES—MANUAL, SEMI-AUTOMATED AND FULLY AUTOMATED SYSTEMS

The check-in procedure involves all stages from the arrival of a guest to the issuance of the room key to the guest. Check-in process is critical, since a lot of vital information is exchanged between the guest and the hotel staff during this process, irrespective of the fact whether check-in process is manual or fully automated. In case of manual or semi-automated systems, the check-in process begins as the guests arrive at a hotel and are greeted by the front office staff, subsequent to which they complete the registration formalities, and finally the guests are assigned rooms and issued the room key by the front office personnel. In a fully automated system, the same activity is performed automatically by self-registration terminals located at convenient places like the hotel lobby or the airport.

The check-in procedures of guests with different status are discussed as follows.

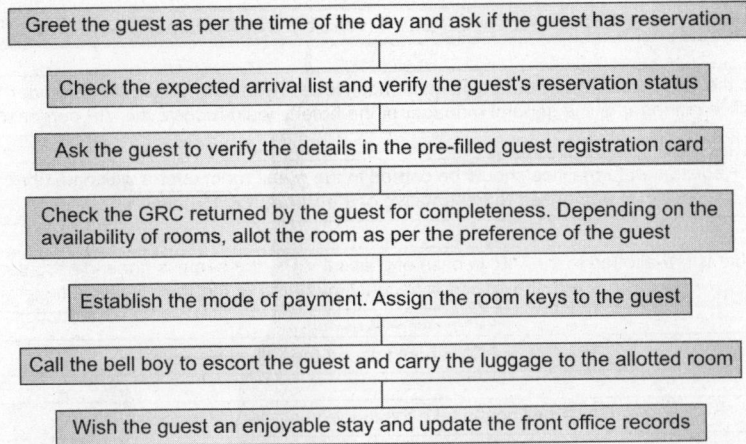

Fig. 22.8: Processing registration for guests with confirmed reservation

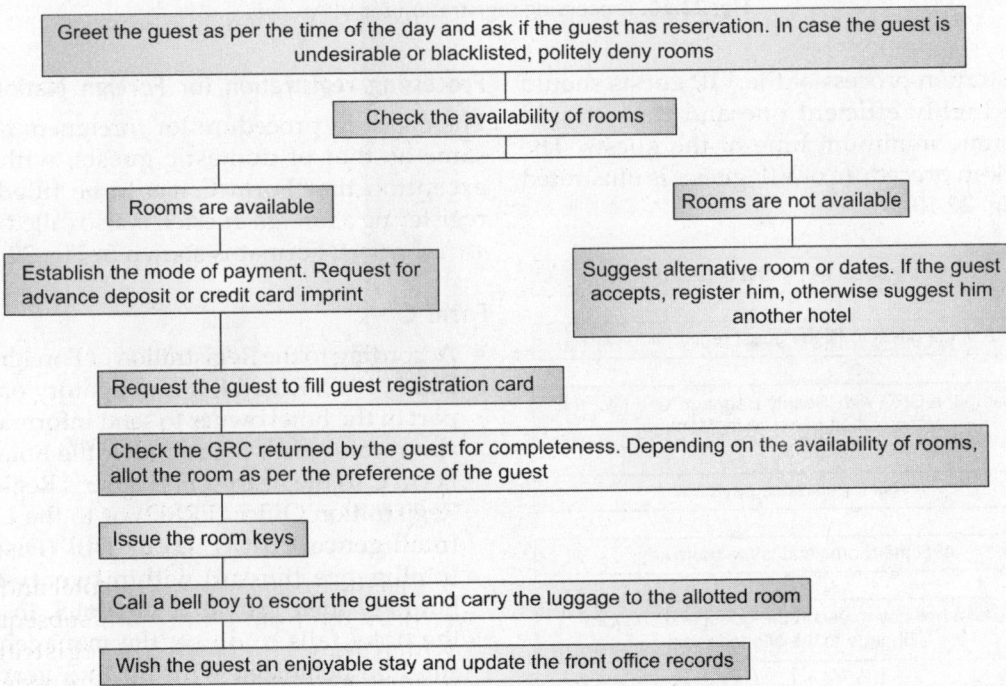

Fig. 22.9: Processing registration for walk-in guests

Processing registration for VIP guests

The VIP guests generally include heads of states, ministers, senior media personnel, dignitaries, commercially important persons (CIP), politically important persons (PIP), distinguished guests (DG) and well-known personalities from the field of entertainment, sports, academics, etc. and thus are huge source of revenue for the hotels besides publicity. It is extremely important that the

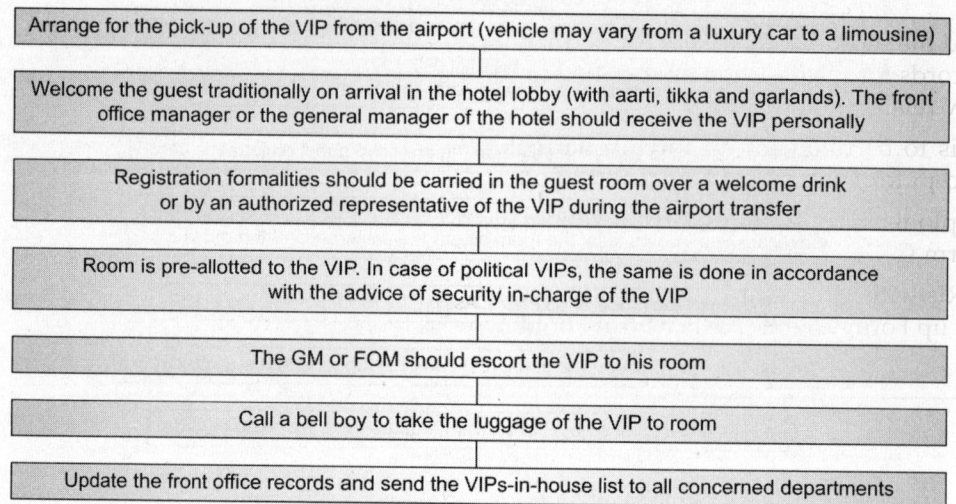

Fig. 22.10: Processing registration for VIP guests

registration process of the VIP guests should be a highly efficient one and thus should consume minimum time of the guests. The check-in procedure of VIP guests is illustrated in Fig. 22.10.

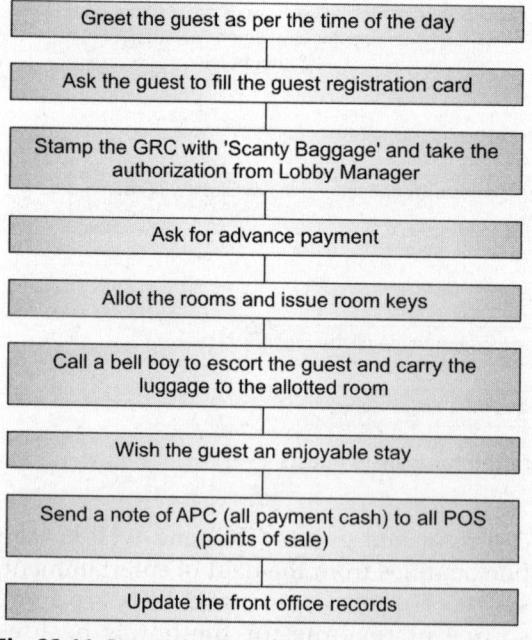

Fig. 22.11: Processing registration for scanty baggage guests

Processing registration for Foreign Nationals

The check-in procedure for foreigners is the same as that of domestic guests, with the exception that Form C has to be filled for registering a foreign guest. It is also called *hotel arrival report*. Format is shown in Fig. 22.12.

Form C

- According to the Registration of Foreigners' Rules, 1992, Rule 14, it is obligatory on the part of the hotel owner to send information about foreigners registered at the hotel on Form C to the nearest Foreigner's Regional Registration Office (FRRO) or to the Local Intelligence Unit (LIU) or DIB (District Intelligence Bureau) within twenty-four hours of their arrival at the hotel. In case the hotel fails to do so, the management shall be penalized with up to 5 years of imprisonment and a fine of ₹ 2 lacs.
- In the case of Pakistani, Bangladeshi, and Chinese nationals, this information should reach within 12 hours to the nearest FRRO or LIU, and also to the local police station.
- Form C should be prepared in duplicate and serial numbered—the top copy is sent to the competent authority (FRRO or LIU)

and the second copy is kept for permanent records for the duration as specified in the law related to the same.

- It is to be filled for all foreign nationals except for nationals of Nepal and Bhutan.
- Diplomats are exempted from filling the Form C.
- NRIs with 'green card' are not required to fill up Form C while NRIs who are holding passport of other countries have to fill up Form C.
- In case of a foreigner group registering in the hotel, only one Form C is filled and signed by the group leader (group leader provides the passport and visa details of all the group members). But in case if the group members are of different nationalities, then separate Form C as per the number of different nationalities shall be filled up.

HOTEL XYZ

FORM C

(Under Rule 14 of the Registration of Foreigners Rules, 1939)

Hotel Arrival Report

(To be completed in duplicate)

Sl. No.: _____

Date: _____

Name of the hotel: _____

Name of the foreign visitor: _____

(In full in block capitals, surname first)

Nationality: _____

Passport no.: _____

Date of issue: _____

Place of issue: _____

Address in India: _____

Date of arrival in India: _____

Date and time of arrival in hotel: _____

Arrived from: _____

Whether employed in India: Yes/No

Proposed duration of stay in India: _____

Proposed duration of stay at hotel: _____

Proceeding to: _____

Registration Certificate No.: _____

Date of issue: _____

Place of issue: _____

Prepared by: _____

Manager's Signature

Fig. 22.12: Form C

Processing registration for Groups/Crews (domestic and international)

Activities performed by the Front Desk

Pre-arrival

- The evening before the date of arrival of the group, folder containing details like name of the group, name of travel agent, date of arrival and departure, accommodation required, meal plans and rooming list are sent to reception from reservation to prepare for group arrival.
- After ascertaining the type and number of rooms required, the front office should block rooms for the group, preferably on the same floor.`
- Room service is informed for the arrangement of welcome drinks on arrival and housekeeping department is informed to clean the rooms blocked for the group.
- Bell desk is informed to organize for arrival.
- GRE is informed to arrange for ATG.
- Usually a blank itinerary is prepared which has information about wake-up calls, breakfast, lunch and dinner, baggage down and departure date.
- Meal coupons are made and keys of the rooms to be kept in key cards.

During Arrival

- The group is welcomed at the special group arrival counter by GRE in traditional Indian style. The group members are offered welcome drinks on arrival by the room service staff.
- The front desk then fills one registration card for the group under group leader's name. In case the group is a foreigner group, only one Form C is filled and a copy of the rooming list is attached with it.
- At the cashier desk, the front office cashier prepares a master folio for the group and individual folios for each member of the group after the mode of payment is settled.
- Rooms are allotted to the group members as per the group leader's instructions and the final rooming list handed over to him. Rooming list (Fig. 22.13) contains the names of all group members along with their nationality, passport details, visa details and the corresponding room numbers.
- The front desk then finalizes the meal timings and venue of meal for the group and then hands over the meal coupons to the group leader.

HOTEL XYZ

GROUP ROOMING LIST   Date: _____

Flight no: _____

Arrival date: _____   ETA: _____

Departure date: _____   ETD: _____

S. No.	Group code	Room no.	Guest name	Nationality	Passport no.	Date of issue	Place of issue	Signature

Fig. 22.13: Group rooming list

- Once the group leader hands over the signed registration card, room keys (enclosed in key cards placed in a tray) are handed over to the group leader for distributing the same to the group members.
- Wake-up call, baggage down and departure date and timing are noted down on itinerary in consultation with the group leader.

Post-arrival

- The front desk finally distributes the copies of the rooming list of the group to different departments—telephone exchange, room service, housekeeping, etc. in order to notify them about the arrival of the group.
- Passport details copy is sent to FRRO along with the Form C.

Activities performed by the Bell Desk

- The bell captain makes sure that he has sufficient bell boys to assist him during the arrival of the group.
- The moment the group arrives in the hotel, the bell boys help in the transportation of the luggage of the guest from the coach to the lobby of the hotel.
- Then the bell captain counts the total number of baggage of the members of the group and attaches luggage tags on each of the luggage.
- Finally, the baggages are loaded on the luggage trolley and transported to the guest rooms.

ROOM SELLING TECHNIQUES

Selling skills of a receptionist come to real test at the time of room assignment. He has not to merely dispense the rooms but at the same time provide the guest with the fulfillment of their needs and management satisfaction by selling the house to its advantage.

A good sale is really dealing with the requirements of the guests and trying to accommodate them wherever possible in courteous manner. In order to sell accommodation successfully the receptionist must learn how to use eyes, ears, and his intelligence in assessing of judging his potential customer. All guests differ emotionally, physically and temperamentally and will have a wide variance of their dislikes, taste, etc. Many guests may require detailed guidance as to the best type of accommodation according to their needs, whereas others will have fairly strong predetermined ideas of what they want. A tactful and good receptionist will be reacting accordingly.

Offering Alternatives

Guests are not always aware of the range of services available in a hotel. Front desk staff can promote the sales of the services by suggesting the most appropriate service to the guest or offering alternatives for the guest to choose from. Two strategies that can be used when offering alternatives of accommodation to guests are the top-down and bottom-up approaches.

The *top-down technique* requires the receptionist to start from the most expensive option and then offer progressively cheaper ones if the guest does not intend to take the most expensive offer. This method is most appropriate with guest whose prime concern is comfort and service, rather than cost.

The *bottom-up technique*, on the other hand, requires the receptionist to start with the cheapest option and then persuade the guest to take progressively more expensive packages. This method is most appropriate with guests whose prime consideration is the cost of the service. By starting with the cheapest option and then suggesting that for a small amount more the guest could have much better accommodation, the receptionist may be able to persuade the guest to accept services of the medium or higher price ranges.

Front desk staff may have to know more about the guest, such as the purpose of his visit in order to establish a guest's needs and to be able to recommend the most appropriate service. In general, it can be said that:

- Well-dressed, affluent guests are less likely to be on a tight budget and may be more concerned with the quality of service than cost.
- Guests whose full accounts are settled by their companies tend to spend more than guests paying for themselves.
- Guests who want to impress business clients tend to spend more on high-quality services.
- Guests who desire comfort are more likely to treat themselves to expensive services.

Up-selling/Sell High/Suggestive Selling

Up-selling is the process of offering a better accommodation to the guest than what he had asked for during the time of reservation or registration and thus encourages the guest to buy a higher priced room.

It is a sales technique whereby a guest is offered a more expensive room than what he originally requested, at an extra cost, by highlighting the added features, benefits and services of the room.

A good receptionist with the ability of being a good salesman can persuade a prospective guest to buy a higher priced room by projecting the features of the room in such a way that the guest is allured to buy the proposed accommodation without raising many queries.

Up-selling gives direct and immediate results, a more satisfied customer, a better and more profitable sales to the hotel and quite possibly, a better tip to the receptionist.

Up-selling is an art and skill of good salesmanship. For a walk-in guest the up-selling is easy as compared to a guest who has made an advance reservation. In case a guest has made an advance booking, the guest has already decided as what he wants, i.e. what type of room and at what rate he wants and

up-selling may not be possible in such case, but if the guest is a walk-in guest, then a smart receptionist can definitely, by his sales skill convert an ordinary sales into a more lucrative business.

Some techniques like highlighting the unique sales proposition (USP) and also like sandwiching the price of the higher room between two plus points of room may prove to be quite beneficial.

'Sir I have an ordinary room for ₹ 6,000 and also I have a room with a view of sea on top floor for ₹ 10,000 and the room has antique furniture'. In this case the receptionist is very cleverly putting the price factor between two positive features (room facing sea and the antique furniture) and is highlighting the features of the room.

Up-selling is a sales skill which should be used with great skill and care, because if the receptionist is not a careful salesman then in his enthusiasm to sell higher accommodation to a prospective guest he may spoil the sale completely. He should make efforts to sell high but he should never be aggressive.

Downselling

It means selling a product or service at a little bit less than the actual price. Downselling may appear to be a process of losing revenue initially, but carefully planned downselling program may prove to be quite beneficial for the hotel particularly when dealing with group sales, as they will make it a point to patronize the hotel in future also.

REVIEW QUESTIONS

1. What is guest registration? Describe various methods of registration used in hotel with necessary format.
2. What is the importance of registration of a guest in hotels? Discuss the legal aspects of registration.
3. Describe the different types of registration records.

4. Explain the procedure for basic check-in activities.

5. What is the importance of guest registration card? Explain GRC with format.

6. Describe in detail the step-by-step procedure for a group arrival process.

7. What is the importance of pre-registration?

8. How can pre-registration assist in front office operations?

9. Why is assessing the guests requirement important?

10. Explain the pre-arrival procedure for a VIP guest.

11 Explain the arrival procedure for a guest with guaranteed reservation.

12. Explain the arrival procedure for a scanty baggage guest.

13. How is check-in procedure different for a guest with guaranteed reservation and a walk-in guest?

14. Write a short note on 'C-form'. Draw its format.

15. What are the steps of the registration process?

16. Which documents are generated during the registration process?

17. What are the regulations involving a C-form?

18. Discuss briefly about arrival and departure register.

19. Explain the importance and purpose of maintaining the guest history in a five star hotel.

20. What is room position? Explain the different types of room position with examples.

21. Differentiate between Form F and C-Form.

22. What is up-selling? When is it appropriate? What are the things a hotel can do to up-sell a guest?

23. Write a short note on offering alternatives.

24. Discuss the details of room selling techniques used in front office.

25. Draw the formats of:
   - Notification slip (Arrival and Departure)
   - Rooming list

26. Write short notes on:
   - Walk-in
   - Walk-out
   - Registration
   - FRRO
   - PIA
   - Rooming the guest
   - Pre-registration
   - Plus position

Handling Situations

Learning Objectives

After reading this chapter, you will be able to understand the following:

- Dealing with guests of different personalities: Drunken guest, inquisitive guest, conceited guest, chaotic guest, mysterious guest, guests who are quiet and hate to talk, etc.
- Dealing with blacklisted guest
- Situations when guests cannot be accommodated
- Handling overbooking situation
- Any other situation pertaining to front office

DEALING WITH GUESTS OF DIFFERENT PERSONALITIES

Drunken Guest

A guest under the influence of alcohol may disturb or trouble other guests and could be a cause of embarrassment for the hotel. He needs patient handling. Such a guest is emotionally sensitive and could take offence of your behaviour. Do not laugh at his actions or comments. To avoid problems, the hotel staff should politely remove the drunken guest from the hotel lobby or public area at the earliest and escort him to an isolated place, like the back office. If the guest is overdrunk and requires any medical assistance, inform the doctor. If the guest acts in an unruly manner and does not cooperate, the hotel security must be called.

Inquisitive Guest

While dealing with an inquisitive guest, who refuses to accept your words or demands proof for any policy, try the following:

- Do not feel hurt, when questioned or go on an ego trip, the guest is not testing your job knowledge or authority for such implementation.
- Be patient and calm but at the same time be firm.
- Answer all his questions truthfully.
- Provide him with all the facts, as to why the hotel has adopted any particular system or procedure.
- Assure him that, he has not been singled out for such treatment.

Conceited Guest

One who tries to boast off his position and acquaintances, needs special handling in the society.

- Do not prick his inflated ego.
- Give him a patient hearing.
- Try to praise and comment on his ideas, if he is giving one.
- Show sign of acceptance of his comments and give him the chance to satisfy his ego.

- Do not get angry or frustrated, if he has exceeded your patience.
- Do not be obstinate.

Chaotic Guest

A guest who does not follow the rules and regulations laid by the management. One may wonder why a well placed guest, who has attained a level of understanding, breaks the rules. To seek attention he violates law that gives him subtle pleasure or satisfaction. A security staff or as a matter of fact, every hotel staff, should remember that for every instance of such behavior, there is always a cause, try to understand this, a guest may not like being dictated or told, what to do, so do not tell him, what he should be doing and what not. Try to convince him that rules and procedures are to be followed and not broken and that, by doing so, he is only helping management. Such a guest needs to be tactfully and efficiently dealt with.

Mysterious Guest

A person who is not clear about his room requirements.

- It is very difficult to understand the possible reasons behind guest, being unable to express his exact requirements. He needs patient and calm handling.
- Encourage to clarify his needs or requirements by positive questioning.
- Ask him direct questions and suggest alternative, when he does begin to speak, listen carefully and give him impression that you understand him.

Guests Who are Quiet and Hate to Talk

- The best way of meeting the requirements of guest, falling under this category is by being very brief in putting across the facts.
- It is very important to read and understand his body language and facial expressions while dealing with him.
- Make him talk, explain advantages and benefits of services, you are offering.

- Treat him, with as much respect and dignity as you would treat a friendly guest, who loves to talk.

DEALING WITH BLACKLISTED GUEST

The blacklisted guests are a list of unwanted guests who make a lot of fuss and create problems for the hotel. The list includes the names of persons whom the hotel would not like to accommodate as guests due to their bad reputation in terms of general conduct, payment of bills and other factors important from the point of view of hotel. A habitual drunkard guest may also be blacklisted.

It is a common practice in most hotels to have such a list which records the name of all persons who are not welcome at the hotel.

- An updated list of blacklisted guests should be circulated to all front desk staff.
- This is a highly confidential record and under no circumstances its contents be divulged to common public as it can place the hotel in legal problems.
- The list should be pasted on the inside of cupboard or drawer, so that it can be consulted unobtrusively by the front desk staff without the guest realising what is being done.
- A great deal of tact is required to cope with a person whose name is on the blacklist. It is essential that he never knows the real reason for the refusal. The most tactful thing to do is to tell the guest that there is no accommodation available.
- Sometimes a fake name is given by the guest. An enterprising receptionist should therefore make it a habit to go through the list often and try to remember the names of undesirable or unwanted guests along with their description, habits and modus operandi.

SITUATIONS WHEN GUESTS CANNOT BE ACCOMMODATED

In general, a hotel is obligated to accommodate guests. Discrimination is prohibited in places of public accommodation on the basis of race, sex, religion, or national origin.

Of course, denying of accommodation to guests means loss of business which means loss of revenue. At times the way a guest has been refused accommodation may lead to a permanent loss of business from that particular client. Denying accommodation is a decision taken by the management in the following circumstances:

- **If the hotel is booked to its capacity:** In case the hotel is fully booked you cannot help it but can only refuse the reservation politely or gently. In such cases, if possible, alternate hotels in the area can be suggested or a different property of the same group of hotel, if not available, any other hotel of the same level can be suggested.

- **If the requested category of accommodation is not available:** In case the requested type of room is unavailable, you may suggest alternate category of room available and always try to sell upper class rooms which, however, should not seem to be unethical.

- **The guest or travel agent is blacklisted:** At times some guests are blacklisted because of their previous record of drunk or disorderly behavior with hotel staff or other guests and non-payment or delayed payments of bills. The same holds true for the travel agents. In such a situation if any doubt or complication crops up, it is always advised to ask for the reservation supervisor/manager's help.

HANDLING OVERBOOKING SITUATION

Sometimes in spite of all care and calculations, the overbooking may cause problem for the hotel and inconvenience to guests because of overstays and complete arrivals. This creates poor guest relations and discourages repeat business. In addition, hotels may be subject to lawsuits if they fail to provide agreed-upon accommodation. To meet such exigencies hotels should keep good relations with other

hotels for 'referrals'. Overbooked guests should be sent to hotels which are at least of the same standard and the same tariff. Taxis should be arranged and paid for by the hotel to take the guest to the new hotel. A letter of apology written to the guest by the front office manager may help to offset loss of goodwill.

Also to avoid the problem of overbooking, guests may be asked to pay advance deposit. Fix cut-off time and inform the guest about the same. After this time rooms may be allotted on 'first come first serve' basis.

ANY OTHER SITUATION PERTAINING TO FRONT OFFICE

DNS (Did Not Stay)/Walk-out guests

The guest sometimes wants to move out of the hotel because of improper services or standard. If the room is not satisfactory to the guest, the receptionist must try to provide alternative accommodation (even in another hotel). If the guest departs for reason that is beyond the hotel's control, the receptionist should express his regrets and assist the guest in departure. If the room has not been used, no charge will be made to the guest. All forms and records should be marked 'DNS'. The management normally wants to be informed about all the DNS cases.

DNA (Did Not Arrive)/No-show

A guest who makes a confirmed room reservation but fails to cancel the reservation or arrive at the hotel on the date of the reservation.

At the end of the day, the clerk or receptionist should take the following steps:

- If there is reservation slip in the reservation rack.
- Check the room information rack that the guest did not check-in already.
- Double check the arrival date.
- Check with the airlines, if the airline numbers are given.
- Attach the time-stamped reservation to the folio and mark DNA if guaranteed or deposit is paid.

• Note DNA on reservation form and place it with next days arrival (often a guest arrives a day or two later).

RNA (Registered but Not Assigned)

A guest arriving early in the morning when rooms are not available may be asked to register, deposit luggage with the bell desk and return to the hotel for the room assignment later in the day. The registration card is kept at the desk with the notation RNA. As soon as the room will be available, the assignment is made. Usually happens with early arrival guests.

PIA (Paid In Advance)

If the guest does not have luggage (or scanty baggage), payment in advance is normally requested. This situation should be handled with extreme care and tact. If the guest holds a credit card, an imprint can be taken and an advance payment need not be requested.

If it is the policy of the hotel to collect the room rate in advance for 'No Luggage' guests, the receptionist should inform the guest politely and collect the charges for one night accommodation. PIA guests are often denied in-house credit or charge privileges due to their chance of being skipper. A note of APC (All Payments Cash) to all POS (Point of Sale) is sent.

NI (No Information)

The guest may request that no information regarding his presence in the hotel be given to the visitors. The 'No Information' status should be clearly marked so that the front desk staff can respond appropriately.

Do Not Disturb (DND)

Any guest who has displayed DND card outside his door should not be disturbed till 2 PM. After 2 PM if the DND sign is still displayed, following procedure should be followed:

• Front desk staff should check what time the guest checked-in.

• In case the guest checked-in sometime during the late night or early morning, the guest should not be disturbed till 4 PM as he might be sleeping.

• In case if he is a stay-over guest, the front desk staff should check with the telephone operator if the guest has left any message about not being disturbed.

• In case if the guest has not given any instructions to the telephone operator, the front desk staff should call the guest and speak to him.

• Apologize for disturbing the guest and ask him if he wishes to have his room cleaned. If the guest refuses any service, immediately offer him some fresh towels so that someone gets a peek into the room to verify that nothing illegal is happening in the room. (Most illegal acts being performed in hotel rooms is covered by a DND sign).

• In case if the guest does not pick the call, the guest room should be opened in the presence of security staff. In most of the cases in which a phone call has not been answered, entering the room will reveal that the guest is out of the room and has forgotten to remove the DND sign from the door.

REVIEW QUESTIONS

1. In the case of overbooking, how would you handle a guest with confirmed reservation?

2. How would you handle a blacklisted guest?

3. Explain the following:
   • DNS
   • DNA
   • RNA
   • PIA
   • NI

4. What can be the reasons for guests to be refused accommodation? How can the situation be handled?

5. Write short notes on:
   • Walking the guest
   • Dealing with drunken guest

24 | Bell Desk

Learning Objectives

After reading this chapter, you will be able to understand the following:

- □ Functions performed by bell desk
- □ Duties and responsibilities of bell desk staff
- □ Equipment needed
- □ Left luggage handling procedure
- □ Forms, formats and records maintained
- □ Guest arrival and departure procedure at bell desk

FUNCTIONS PERFORMED BY BELL DESK

Bell desk may be defined as a section of the front office department which helps in the transportation of the guest luggage at the time of his arrival and departure. It is also known as *Bell Stand or Porter's Lodge*. Bell captain is in charge of the bell desk who is assisted by the team of bell boys.

This is supposed to be the first department to come in contact with the guests on their arrival and hence is located near the main entrance of the hotel on the left hand side of the lobby. The bell desk should be situated in clear view of the front desk and particularly the doorman, so that the doorman may signal for a bell boy at the time of arrival of a guest. Further, it is also important that the bell desk is situated near the left luggage room and luggage entrance.

Because of their direct contact with guests, bell desk can communicate vital information to guests and make them feel welcome at the hotel. One of a hotel's best marketing opportunities occurs when the bell boy escorts the

guest to his room. Providing information on the surrounding area and the hotel's restaurants, recreational activities, swimming pool and health club facilities is part of every bell boy's job. By doing so, bell boys become key players in a hotel's sales and marketing efforts.

Some of the important functions performed by the bell desk are:

- Responsible for *luggage handling* of the guest at the time of arrival, departure and during stay. At the time of arrival when the luggage of the guest is moved from taxi to the lobby and further to the allotted room, the activity is called '*Up bell activity*'. When the luggage of the guest is moved from room to lobby and further to the taxi at the time of departure, the activity is called '*Down bell activity*'.

- Responsible for *left luggage room*. The term left luggage is attributed to luggage left by a guest who checks out of the hotel but wishes to collect his luggage later.

Guests leave their luggage in hotel premises under the guarantee by the management that the luggage would be safe. Some hotels may charge a fee for this facility but most hotels do not.

- Responsible for *paging services* for the guest in order to locate him in the hotel in case of any mail or message to be delivered to him.
- Handles and *distributes mails, letters, faxes and messages* received by the front desk in the absence of the guests to their respective rooms.
- Responsible for *delivering the newspaper* to all occupied rooms.
- Responsible for *collecting room keys* from the guest at the time of checkout and depositing the same at front desk.
- In hotels, the *wake-up call* to groups and crews is coordinated by bell desk. In such cases, it is the responsibility of the bell captain on duty in the night shift to prepare the wake-up call sheets of all the groups and crews in-house.
- The bell desk also acts as a *mini post office* for the guests of the hotels and provides them with the various types of postal services. The bell desk sells postal stamps, postcards, inlands and envelopes to the guests on their request. The bell desk also arranges for various courier services for the guests in order to send their urgent letters. The hotel maintains a postal letter box in which the guest can drop their outgoing letters. The clearance is done normally twice a day and the collected items are forwarded to the nearest post office immediately.
- *Miscellaneous jobs* such as carrying out outside errands for the guest and hotel like buying of cinema tickets, moving of files and documents as well as going to banks, post office and FRRO/LIU for delivering of Form C are done by the bell desk. At times when there is a room discrepancy, the bell desk staffs helps the front desk in checking and sorting out of the status of guest room.
- The bell boys also assist the housekeeping staff in *conducting light housekeeping services* in the lobby and the reception area. These services include light dusting of the front desk, bell desk and furniture of the lobby as well as wiping the main entrance of the hotel.

DUTIES AND RESPONSIBILITIES OF BELL DESK STAFF

Bell Captain/Head Hall Porter

The bell captain is in charge of bell desk and is responsible for organizing, supervising and controlling all lobby services with the help of bell boys. The duties and responsibilities of a bell captain are as under:

- He controls the movements and activities of bell boys on the lobby control sheet which is a summary of the activities of all bell boys.
- He prepares duty rota and schedules leave for bell boys.
- Conducts daily briefing of bell boys.
- Handles left luggage room formalities and supervises left luggage register, baggage check and luggage inventory sheet.
- Trains bell boys to maximize their efficiency. Responsible for implementation of SOP as laid down by the management.
- Controls the sale of postage stamps to guests.
- He maintains record of all guests with scanty baggage and informs front desk staff.
- He reports irregularities of suspicious persons to the lobby manager.
- He prepares errand cards for the bellboys.
- Coordinates and controls the distribution of morning newspaper.
- Organizes and supervises check-in/out baggage, formalities of groups, crews, etc.

Bell Boy/Bell Hops/Porter

Responsible to the bell captain for welcoming and escorting guests with their luggage to and from their rooms during check-in and check-out, perform various other functions related to the guest needs. The duties and responsibilities of a bell boy are as under:

- He escorts the guest to his room, familiarizes guests with the guest room and in-room facilities and services.
- Checks the room for any damage and missing items at the time of departure of guest.
- Conducts light housekeeping services in the lobby.
- He carries luggage, parcels, etc. to guest room.
- He should possess thorough knowledge about the topography of the hotel.
- He distributes mails to various departments as well as to the guest.
- He distributes newspapers to the guest.
- He should be familiar with luggage storage procedures and luggage room.
- He keeps postage stamps available for guest request.
- He is responsible for paging of guest.
- He sees that the Form C is deposited to the FRRO/LIU or the nearest police station.

EQUIPMENT NEEDED AT BELL DESK

Luggage Trolley/Bell Cart

Used at the bell desk to carry guest luggage to and from rooms.

Franking Machine

Machine used for printing postage stamp value on the envelope.

Postage Scale

This scale is located in the concierge area and is used to determine the postage required for letters and small packages which a guest or hotel wishes to mail.

Time Stamp

It is used to record the delivery time of any mail or message for the in-house guest.

Luggage Net

It is used for identifying and separating the luggage of groups.

Telephone

Used for proper and effective intra-communication with the front desk areas.

Stamp folder, glue, twines, scissors, etc. for packaging.

Umbrellas

LEFT LUGGAGE HANDLING PROCEDURE

The term left luggage is attributed to luggage left by a guest who checks out of the hotel but wishes to collect his luggage later. The hotels provide the facility of left luggage to the guests who would return to the hotel after a short visit to another city and may find it inconvenient to carry their entire luggage with them or may find it uneconomical to retain a room where they can keep their luggage. This facility is also provided to guests who check-out early from the hotel but eventually depart late from the hotel as they might be busy in official work or sightseeing and thus, may find it difficult to carry their luggage. Thus, the guests leave their luggage in hotel premises under the assurance by the management that the luggage would be safe and secure. Some hotels may charge for the left luggage facility but most hotels do not.

Hotels normally follow the following procedure while accepting the luggage to be stored in the left luggage room:

- Ascertain if the guest who is intending to keep his luggage in the left luggage room has cleared his hotel bill.
- String or tie the luggage tag/baggage check on each piece of luggage separately. The

baggage check has a number, which is also printed on the counterfoil.

- Enter details in the left luggage register.
- Tear off the counterfoil of each luggage tag and hand it over to the guest. The guest is supposed to present the same at the time of collection of his luggage.
- Keep the luggage in the left luggage room.

While delivering the luggage to the guest, the following procedure is followed:

- Take the counterfoil of the luggage tag from the guest.
- Tally the same with the luggage tag

attached to the luggage in the left luggage room.

- Enter the date of delivery in the left luggage register.
- The bell boy hands over the luggage to the guest.

FORMS, FORMATS AND RECORDS MAINTAINED AT BELL DESK

Bell Desk Log Book

It is an important register maintained by the bell desk personnel for leaving any important instructions for the staff of the next shift (Fig. 24.1).

	HOTEL XYZ	
	LOG BOOK	
	(Bell Desk)	
		Date: _____
		Day: _____
		Shift: _____
		Lobby Mgr/GRE: _____

S. No.	Details	Remarks

Fig. 24.1: Bell desk log book

Date	Room number	Guest name	Bell boy	Luggage tag no.	Description of the luggage	Date of delivery	Remarks

HOTEL XYZ
LEFT LUGGAGE REGISTER

Fig. 24.2: Left luggage register

Left Luggage Register

The register details out records of the entire luggage left by the guest at the bell desk at the time of departure from the hotel (Fig. 24.2).

Luggage Tag

It is a very important format used by the bell staff while handling the luggage of the guest during their departure from the hotel. The luggage tag has two parts and is printed on both front as well as back side. The first part (of front side) is filled by the bell desk at the time of the receipt of the luggage from the guest and is attached to the luggage, while the second part/counterfoil (of front side) is also filled by the bell desk and is torn off and given to the guest as a proof of receipt of luggage from the guest which is produced by the guest when he would like to collect his luggage. Both the parts (of front side) of the luggage tag have a number imprinted which is tallied when the guest produces the counterfoil to collect his luggage back from the bell desk. On the back side of the luggage tag, there is a declaration from the management of the hotel that the guest will assume all the risk of damage caused to the left luggage from fire, water and theft and the hotel will only take care of the luggage. (Format is shown in Fig. 24.3.)

Scanty Baggage Register

It is maintained by the bell desk staff to keep a record of all the guests who have registered in the hotel with very few luggage or without any luggage and are a matter of concern to the hotel as they may leave the hotel anytime without settling their bills. (Format is shown in Fig. 24.4.)

LEFT LUGGAGE TAG

(Front side of the card)

Date _____   Room No. _____

Baggage Check

No – 312

The Management is not responsible for goods left for over 30 days

_____ Suitcase   _____ Briefcase

_____ Handbag   _____ Golf Bag

_____ Umbrella   _____ Overcoat

_____ Package   _____ Others

No - 312

The Management is not responsible for goods left for over 30 days

Date   Room No.

HOTEL XYZ

(Back side of the card)

Storage Record

Name _____

Date _____

....

The owner of the baggage or personal property indicated in this receipt assumes all risk of damage to the goods by fire, water, theft or from other causes.

The Management only undertakes to exercise that it will take proper care of the baggage left by the guest.

The left luggage will be delivered only on showing the tag to the concerned authority.

MANAGEMENT

Fig. 24.3: Luggage tag

Lobby Control Sheet

It is the summary of all the activities performed by the bell boys during a particular shift. (Format is shown in Fig. 24.5.)

Bell Boy Errand Card

It is maintained by the bell captain for controlling the activities of the bell boys when they leave the bell desk for the jobs associated with the arrival and departure of the guests in the hotel. The arrival errand card (Fig. 24.6) is used during the arrival and up-belling of the guests as well as during the other services requested by the guests while the departure errand card (Fig. 24.7) is filled during room change or during the departure of the guest from the hotel.

HOTEL XYZ
SCANTY BAGGAGE RECORD

Date: _____

Serial No: _____

Shift: _____

S.No	Guest Name	Room No	Arrival Time	Bell Boy	Baggage Description	Remarks	Lobby Manager/GRE

Fig. 24.4: Scanty baggage register

HOTEL XYZ
LOBBY CONTROL SHEET

Bell Captain: _____

Sheet No: _____

Shift: _____

Date: _____

Room No	Bell Boy	Arrival Time	Departure Time	Service Call Time		Remarks
				From	To	

Bell Boy's Signature: 1. _____   2. _____   3. _____   4. _____

Bell Captain's Signature: _____

Fig. 24.5: Lobby control sheet

HOTEL XYZ				
ARRIVAL ERRAND CARD				
Bell Boy Name:				Call Time:
Name of Guest:				Room No:
Articles				
Suitcase	Handbag	Package	Briefcase	Overcoat
Others				
	Signature (Bell Captain)			Signature (Receptionist)

Fig. 24.6: Bell boy arrival errand card

HOTEL XYZ				
DEPARTURE ERRAND CARD				
Reception	*Cashier*	*Information*	*Departure date and time*	*Room no*
Bell Boy Name:				Call Time:
Name of Guest:				Room No.:
Articles				
Suitcase	Handbag	Package	Briefcase	Overcoat
Others				
Baggage brought down by:			Baggage loaded by:	
	Signature			Signature

Fig. 24.7: Bell boy departure errand card

Stamp Register

The stamp register is maintained by the bell desk to keep a record of all the stamps that have been purchased by the guests.

GUEST ARRIVAL PROCEDURE AT BELL DESK

- When a guest arrives, the doorman buzzes the bell desk for a bell boy.

- The bell boy should wish the guest and collect his luggage and bring it to the lobby via the luggage entrance and place luggage at the bell desk. The bell boy should wait for the guest to register at the front desk. Luggage tag should be attached to the luggage. Inform scanty baggage to the front desk.

- The information desk will indicate that the guest has been allotted a room by handing over the arrival errand card which mentions the room number. The room key is also handed over with the errand card.

- Escort the guest to his room along with his luggage. Some hotels have separate baggage elevators.

- The bell boy should open the door of the guest room and let the guest enter first. After positioning the luggage at the luggage rack in the room, the bell boy explains the salient features of the room to the guest.

- Offer any other help and if not required, wish the guest a pleasant stay. Do not solicit for tips.

- Report back to the bell desk.

GUEST DEPARTURE PROCEDURE AT BELL DESK

- The bell desk receives call from the guest about his intention to checkout. Write the room number carefully on the departure errand card. Inform the bell captain and proceed to the room. The cashier is also informed so that he takes the necessary action needed for departure.

- The bell boy goes to the guest room, knocks on the door, and announces himself. After getting due permission from the guest, he enters the room. Give a cursory look to the room, look for any guest articles left by mistake, any damaged hotel property and switch-off the AC or lights. Check the refrigerator for the consumption of

---

HOTEL XYZ

Luggage Out Pass

No. 3112

Name of Guest: _____   Room No.: _____

Date of Departure: _____   Time: _____

Bill No.: _____

Billing Settlement

☐ Complete   ☐ Partial   ☐ Corporate Settlement   ☐ Others_____

_____
Authorized Signatory

_____
Date

Fig. 24.8: Luggage out pass

minibar items (if chargeable). Collect the room key, guest luggage and depart from the room letting the guest lead the way. Ensure that the guest room is locked. Hang 'clean my room' card on the door knob, so that the housekeeping department can quickly clean the room for the next occupant.

- Place the luggage at the bell desk. Stick any hotel stickers or publicity tags. Handover the room key to the information desk and take his signature on the departure errand card. Handover the departure errand card to the cashier. Wait for the guest to settle his bills.

- The bell boy will receive an authorization to take the luggage out of the hotel only after the cashier has signed on the errand card.

- Take the luggage to the porch and load it in car after obtaining luggage out pass/cashier clearance card (Fig. 24.8) from the cashier.

- Report back to the bell desk.

REVIEW QUESTIONS

1. Give step-by-step procedure for baggage handling on FIT arrival. Support your answer with any one document used during the process.

2. Discuss the procedure for handling left luggage.

3. Give step-by-step procedure for baggage handling on FIT departure.

4. Differentiate between a FIT and group baggage handling procedure during arrival.

5. Discuss the functions performed by bell desk.

6. Discuss the forms and formats maintained at bell desk.

7. Draw the formats of:
   - Left Luggage Register
   - Scanty Baggage Register
   - Luggage Out Pass
   - Luggage Tag
   - Errand Card
   - Lobby Control Sheet

Learning Objectives

After reading this chapter, you will be able to understand the following:

- Guest services and its importance
- Handling guest mails
- Message handling procedure
- Procedure for guest paging
- Safe deposit facility
- Procedure for guest room change
- Handling wake-up calls
- Procedure for hiring a car

GUEST SERVICES AND ITS IMPORTANCE

Once guests check into a hotel, they avail the various services and facilities offered by the hotel during their stay. This is the third phase of the guest cycle. This stage is very important for the hotel as the guest's experience is crucial in generating repeat business and positive word-of-mouth publicity. So the hotel staff should provide various services to guests in a caring and personalized manner to ensure that they come back to the hotel and recommend it to their colleagues and friends.

During a guest's stay in a hotel, the front office staff provides various kinds of guest services. These services may vary from hotel to hotel and from guest to guest. Some of the services provided by front office of the hotel to the guest are:

- Handling guest mail
- Message handling
- Guest paging
- Safe deposit facility
- Guest room change
- Wake-up call

HANDLING GUEST MAIL

When guests are away from their homes, they need a contact address where they can receive any urgent mails, letters, parcels, packets, or faxes. During their stay in a hotel, guests may provide their family and clients the contact details of their hotel for any urgent communication.

Mail handling is a very important activity of the front desk and the way the mail is handled shows the efficiency and attitude of the hotel staff. Any delay and carelessness shown by the staff may result in great dissatisfaction.

Mails can be divided into two categories.

Incoming Mails

All the mails addressed to the hotel are received by the front desk, information desk or bell desk as per the house policy and are recorded in the mail log book (Fig. 25.1) and the guest's signature is taken at the time of delivery. The person who receives the in coming mails at the hotel stamps them with the date and time of receipt. Doing so provides evidence of when the mail was received, in case any question arises on how promptly the mail was delivered.

Incoming mails can be further classified as shown in Fig. 25.2.

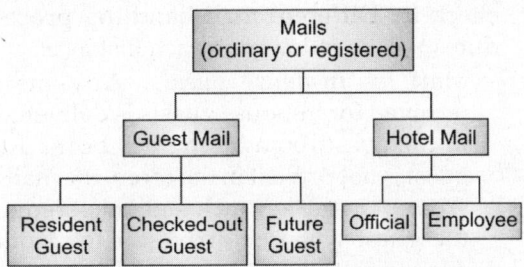

Fig. 25.2: Types of mails received by the hotel

- **Mail for guests:** When mail arrives, front office records should always be checked immediately to see if the guest is either registered, has checked-out or due to

HOTEL XYZ							
INCOMING MAIL LOG BOOK							
S.No.	Date and time of receipt	Name of addressee	Type of mail	Received by	Delivered to	Signature	Remarks
13008	15/09/1 12:10 p.r.	Mr Deepak Dixit	Registered	Bell desk	Time office		
13009	15/09/15 1:10 p.m	General Manager	Ordinary	Bell desk	GM office		
13010	15/09/15 02:05 p.m	Mr Pawan Kumar	Parcel	Bell desk	Front desk		
13011	16/09/15 10:10 a.m	Mrs Priyanka	Insured Mail	Bell desk	Front desk		
13012	16/09/15 11:15 a.m	Mrs Preeti	Ordinary	Bell desk	Front desk		
13013	16/09/15 11:50 a.m	Mr Akshat Mishra	Courier	Bell desk	Front desk		
13014	17/09/15 09:20 a.m	Mrs Sudha	Ordinary	Bell desk	Front desk		
13015	17/09/15 09:50 a.m	Mr Bid	Registered	Bell desk	Front desk		

Fig. 25.1: Incoming mail log book

check-in. Different mail handling procedures will be followed in each instance:

– *Mail for in-house guests*: Any mail
received for in-house guests is delivered
in the guest room by the bell boys. If a
guest is not present in the room, the mails
are kept in the key rack and delivered to
the guest when he arrives at the front
desk to collect his room key. Sometimes
the guest may go out of his room without
leaving the key at the front desk. In such
cases a mail advise slip (Fig. 25.3) is put
on the door knob so that if he goes to the
room directly he would know about the
mail which the hotel has received in his
absence.

– *Mail for guests who have checked-out:* The
mails of checked-out guests are sent to
the back office, from where the mail
forwarding address (Fig. 25.4) is taken
and mails are re-directed to that address.
In case there is no forwarding address,
the mails are sent back to the sender.

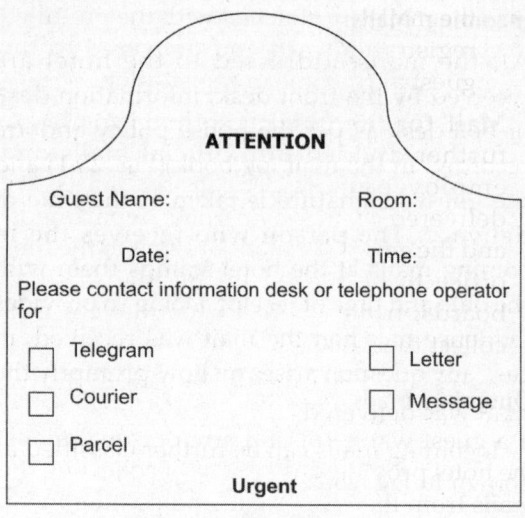

Fig. 25.3: Mail advise slip

– *Mail for guests expected to arrive:* The mails
of future guests are sent to the
reservation section, where they are
placed along with the reservation record.
On the date of the arrival of the guest,

HOTEL XYZ
MAIL FORWARDING ADDRESS SLIP

This address will be in file for 15 days, unless otherwise requested. Please fill the details and handover at the reception.

Mail forwarding instructions: Forward until: _____

                                                 Hold until: _____

Forwarding Address: _____

_____

_____

Date: _____   Name and Signature of Guest

Record of Forwarded Mails

S. No	Date	Type of mail (Letter, Fax, Parcel, Telegram)	Forwarding address	Forwarded by

Fig. 25.4: Mail forwarding address slip

the mails are attached with the pre-filled registration card and delivered to the guest at the time of registration.

- **Mail for the hotel:** The hotel mails are further divided into official mails and employee mails. The hotel mails are delivered to the concerned departments and the employee mails are sent to the time office to be placed in the mail display boards, from where the employees may collect their mails.

Outgoing Mails

If a guest wants to send any personal mails, the hotel provides the service of collecting the mails from the guest room and posting them. The charges for the service are added to the guest account through a miscellaneous charge voucher. The miscellaneous charge voucher is authenticated by a competent authority and sent to the front desk cashier for posting into the guest folio. A record of the same is maintained in the outgoing mail register (Fig. 25.5).

MESSAGE HA. 'DLING PROCEDURE

At times, there a e telephone calls or visitors for a resident gue t when he is not present in the hotel. In suct situations, the front desk take the message for the guest and deliver the same as soon as the guest comes back. The process of receiving and delivering messages to resident guests is known as message handling. The prompt and timely delivery of messages to guests reflects the degree of professionalism of the front desk staffs.

If a resident guest is expecting a call or a visitor during his absence, he may leave a location slip at the front desk. In such a case, the front desk staff follows the instructions of the guest on receiving the telephone call or visitor for that guest.

Every hotel has its own standard operating procedure for handling guest messages. Most hotels follow the given procedure, with some variations:

- When there is a visitor or a telephone call for a guest, the front desk staff should look at the information rack to see whether the guest is a resident guest, future guest, or checked-out guest.
- In case of a resident guest, the staff must check whether he is present in the room or not. If the guest is present in his room, connect the call to the guest room or inform the guest about the visitor.
- If the guest is not present in the room, then the staff must check the key rack for the

S.No	Room No	Name of guest	Addressed to	Description of mail	Charges	Received by	Date and time of receiving	Posted by	Date and time of posting	Remarks

HOTEL XYZ
OUTGOING MAIL REGISTER

Fig. 25.5: Outgoing mail register

location form. If the same is found, then act according to the instructions of the guest. If guest has not left any instructions or the location form at the front desk, the front desk staff should take down the message for the guest on a message slip (Fig. 25.6).

- The message slip is prepared in triplicate— the first copy is placed in the key rack, the second copy is placed in a message slip envelop and slipped through the door of the guest room by a bell boy and the third copy is retained at front desk for records. The purpose of preparing the message slip in triplicate is to ensure the delivery of the message to the guest.

- If there is a visitor or a call for a guest who has checked out of the hotel, then the front office staff should give the information as per the instructions left by the guest.

- If there is a call for a future guest, then the agent should note the message on a message slip and send the slip to the back office, where it would be placed along with the reservation record. On the date of the arrival of the guest, the message slip is attached with the pre-filled registration card and delivered to the guest at the time of registration register.

- Some hotels have automated systems for delivering messages to guests. The telephone in the guest room has a message indicator that can be switched on by the front desk agent in case any message is waiting for a guest. This prompts the guest that there is a message for him and he may call the front desk to receive it. In some hotels, guests can read messages on the television screen by dialing a number.

GUEST PAGING

Paging is the process of locating guests in a specified area of the hotel. Many times the in-house guest expects a phone call or a visitor but decides not to wait in his room and might

---

HOTEL XYZ

MESSAGE SLIP

Guest Name: _____   Room No.: _____

Date: _____   Time: _____

DURING YOUR ABSENCE

Mr./Mrs./Miss _____

of _____

Called by telephone	Please call back
Came to see you	Will return
Wants to see you	Urgent

Message _____

_____

Received by: _____

Fig. 25.6: Message slip

decide to go to a public area of the hotel. In such cases the hotel requests the guest to tell about his whereabouts through a location form. Location form is usually kept in the stationery folder in the guest room or the guest may request it from the information section of the front desk (Fig. 25.7). Generally it is filled in by the guest but many times it may be filled in by front desk staff on the request of guest. The completed location form is kept in the key rack. The purpose of the location form is to earmark the area of paging and save time.

HOTEL XYZ

LOCATION FORM

Name of Guest: _____

Room No.: _____

I am expecting:

* Mr/Ms _____ to visit

* Telephone Call

  If I am not in my room kindly locate me at :

* Coffee shop

* Gym

* Swimming pool area

* Night club

* Restaurant

* Other (Specify) _____

Between _____ and _____

Or convey my message to caller/visitor

Message _____

_____

                                    Signature of Guest

Fig. 25.7: Location form

In hotels different systems of paging are used:

Page Board System

It is the traditional system of paging in hotels. When the visitor comes to meet the guest, the front desk staff writes the guest's name and room number on the paging board (Fig. 25.8) and sends a bell boy to the area mentioned by the guest on the location form. The bell boy holds the page board above his head and shakes it so that the bells attached to it ring and attract people's attention. The guest contacts the bell boy, who escorts him to the front desk to meet the visitor. To avoid any embarrassment to the guest, no message should be written on the paging board. This method of paging may create disturbance to other guests. Moreover, if the guest has not left any whereabouts, this method is time consuming as the paging might have to be done in various public areas one by one.

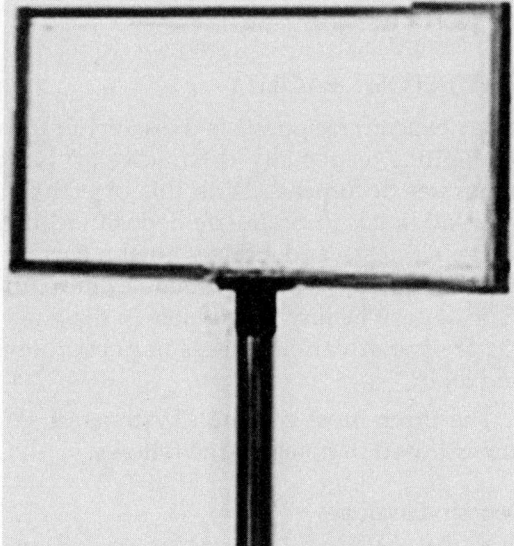

Fig. 25.8: Paging board

Public Address System (PA System)

It is a new concept used by many hotels these days. In hotels, channeled music is played from a central room. One switch of the same is with the front desk staff who on receiving the phone call or a visitor for a guest switches off the music from all public areas and announces the name and room number of the guest to be paged. This is transmitted to all public areas instantaneously. The communication skill of the person announcing is very important in this system. His voice, manner of speech, modulation of voice, correct use of phrases and words and tone (friendly, interesting and helpful) are very important. It is a quick method of paging as it saves time as well as efforts of bell boys.

Pager/Beeper System

With the development of electronic communication equipment some hotels/resorts which are spread in a vast area, use pagers/beepers for paging guests who are given pagers where they can be immediately reached if there is a phone call or visitor for them. A pager is a telecommunication device for short messages.

SAFE DEPOSIT FACILITY

A key concern for guests is the safety of their belongings, especially cash, jewelery, and important documents. With this objective in view, all hotels provide safe deposit facilities to their guests and notices in the form of 'Please do not leave your valuables in room', 'Safe deposit facility is available in the hotel', etc. are put at various places in guest rooms and lobby.

The three most common systems of safe deposit used in hotels are as follows.

Deposit Envelopes

This system is used mainly in small hotels. The procedure is as follows:

- The guest is offered a strong envelope into which he can seal his valuables and sign over the seal.
- The cashier marks the envelope with the guest's name and room number and issues a receipt for the same.

- The receipt is countersigned by the guest and the receipt number is noted on the envelope, which is kept in a safe at front desk with the envelopes of other guests.
- When the guest wants his valuables back, he gives the receipt to the cashier and signs a register.
- The envelope is returned with the seal intact after the guest's signature is verified.
- The returned receipt is attached to the register as a permanent record.

In-Room Safe

The in-room safe (Fig. 25.9) is another option for storing guest valuables. These safes are usually located in the guest room closet; most are larger than the typical safe deposit box. Convenience is the main advantage of in-room safes as the guest does not have to come again and again to the front desk for operating the locker. These safes have combination locks so the guest can use a personal code which only he knows, to access the safe. The combination can be changed as often as the guest wants.

In most states, in-room safes are not considered by law to offer the same level of protection for guest valuables as safe deposit boxes. This means that if a valuable is lost after being placed in an in-room safe, the hotel cannot be held liable since the item was not in the care, custody, and control of the hotel. Claims against hotels for articles that are stolen from these safes are very rare.

Fig. 25.9: In-room safe

Safe Deposit Lockers

These lockers are located at the front desk (usually in the back office) and are managed by the front office cashiers. Safe deposit boxes (Fig. 25.10) should be located in a limited-access area. Unauthorized persons, whether guest or employee should not be permitted in the area.

These lockers can be opened by using two keys simultaneously—one is issued to the guest and the other is with the front office cashier called the master key. This means that the locker can only be opened when both the keys are used. Whenever a guest wishes to operate the locker, the front office cashier and the guest use their respective keys to open the lock. Usually this facility is given without any charges to the guest. Guests who wish to use this facility have to sign in the safe deposit locker register (Fig. 25.11) to get the keys of the safe deposit box.

Procedure for using Safe Deposit Locker

Every hotel has its own operating procedure for the allotment of safe deposit lockers. The standard procedure has two stages:

Issue of Locker

When a guest wishes to use the locker facility extended by the hotel, the following procedure is followed:

- An empty safe deposit locker is allocated to the guest with the locker number.
- A safe deposit locker registration card (Fig. 25.12) is handed over to the guest and the guest is requested to fill the necessary information.
- The locker is assigned and the locker key is handed over to the guest.
- The guest keeps his valuables and documents in the locker, locks the box, and carries the key.
- Each time the locker is operated by the guest, an entry is made in the safe deposit box registration card.

Surrender of Locker

When the guest surrenders the safe deposit box, the following procedure is followed:

Fig. 25.10: Safe deposit locker

HOTEL XYZ

SAFE DEPOSIT LOCKER REGISTER

S. No	Date	Name of guest	Room No	Locker No	Key issued	Guest signature

Fig. 25.11: Safe deposit locker register

- The guest is requested to withdraw the articles placed in the locker.
- The guest is requested to sign an acknowledgement (in the safe deposit box registration card) that he has received all the articles that had been placed in the safe deposit box.
- The guest surrenders the locker key to the front office cashier.

Guiding Principles in Key Control System

- The master key must be accounted for at each shift change. Only those front office staff authorized to provide access to safe deposit boxes should ever have possession of the master key.
- Under no circumstances should there be more than one guest key for each safe deposit box, even when more than one guest is using, the same box.
- If a guest key is lost, the safe deposit box should be drilled open in the presence of the guest and security staffs.
- The identity of the guest must be verified before granting access to a safe deposit box. Some hotels ask guests to include a piece

---

HOTEL XYZ
SAFE DEPOSIT LOCKER

Locker No.	Date issued	Issued by	Room No

Terms and Conditions:

1. I/We shall not hold the hotel liable for any loss of, theft, or shortage in the contents of the safe deposit locker which is being used by me/us exclusively.
2. In the event of the loss of the keys of the safe deposit locker, I/We shall reimburse the hotel ₹ 3,000 only towards replacement.

Signature: _____

Address: _____

SURRENDER OF BOX

The undersigned hereby surrenders above numbered box and certifies that all property placed therein has been lawfully withdrawn and is now in possession of the owner(s); all claims against and liabilities of the custodian are hereby released and discharged.

Signature: _____ Date: _____ Time: _____ Cashier: _____

Date	Time	Signature of the Guest(s)	Cashier

Fig. 25.12: Safe deposit locker registration card

of personal information (for example, mother's maiden name) on the agreement as an additional safeguard. If there is some doubt about the identity of the person requesting access, the cashier can ask for the additional personal information, which an imposter would be unlikely to answer.

GUEST ROOM CHANGE

A room change is a process when a guest is moved from one room to another for various reasons which can be due to request by a guest or requirement by the hotel. In case a room does not match the guest's expectations, the guest may want to change the room. There are times when the hotel may wish to change the room of a resident guest. If the change of room is done in the presence of the guest, it is called a *live move*, and if it is carried out in the absence of the guest, it is known as a *dead move*. It is important for the hotel and the guest to mutually agree on the change of room to avoid any unpleasantness.

Possible Reasons for Change of Room

A guest may want to change his room in the following circumstances:

- If the room was not assigned to the guest as per his choice probably due to non-availability.
- If one or more pieces of equipment or facilities in the room are not working satisfactorily.
- If the number of occupants in the room changes.
- The guest does not like the view, location, etc. of the room.

The hotel may wish to change the guest's room for the following reasons:

- If the guest was upgraded due to the non-availability of the requested category of rooms.
- If the guest has overstayed in a particular room which has been blocked for some other guest.

- If the hotel has scheduled a spring cleaning for the room.
- If the room requires maintenance work.

Procedure for Changing the Guest Room

To change the room of a resident guest, the following procedure is followed:

- The front office informs the guest about the room change in advance so that the guest packs his luggage properly.
- The front office agent fills six copies of the guest room change slip (Fig. 25.13) for reception, bell captain, front desk cashier, telephone exchange, housekeeping, and room service and takes authorization from a competent authority.
- A bell boy is called and given the keys of the new room. He proceeds to the guest room to shift the guest's luggage.
- In case of dead move, the bell boy asks the room boy/floor boy to open the guest room. If it is a live move, he goes to the room and requests the guest to allow him to shift the luggage.
- The bell boy removes all the guest's belongings from the room and locks the room. He then carries all the belongings to the new room and hands over the new room keys to the guest. He collects the keys of the room being vacated from the guest and deposits the same at the front desk. (Flow chart for change of room is shown in Fig. 25.14.)

WAKE-UP CALL

Hotels offer wake-up call services, wherein the hotel staffs makes a telephone call at a requested time to awaken a guest. It is generally done in the morning, but a guest may require anytime of the day. The guest who wishes to be given a wake-up call by the hotel personnel may place a request with the front desk, bell desk, information section, or telephone operator. The wake-up call request is entered in the wake-up call sheet (Fig. 25.15). The telephone operator gives the

HOTEL XYZ

ROOM CHANGE SLIP

Sl. No.: _____

Date: _____

Time: _____

From	To
Room No.: _____	Room No.: _____

Guest Name: _____

Reason for change: _____

Authorized by:	Signature

Copy to:

Reception, Bell Captain, Front Desk Cashier, Telephone, Housekeeping and Room Service

Fig. 25.13: Room change slip

wake-up call to the guest at the time specified by the guest.

Since a guest may miss an important appointment, a flight, or simply a head start on a vacation by oversleeping, front desk staff must pay special attention to wake-up call requests.

Front office computer systems can be used to remind front desk staff to place wake-up calls, or the EPABX can be programmed to place the calls and play a recorded wake-up message. This feature is of a great help when many guests have to be wakening up at the same time in the morning. Despite advances in technology, many hotels still prefer that front desk staff place wake-up calls because guest appreciates personal touch.

Hotels also give wake-up calls to groups or crews staying in the hotel at the time registered by the group leader or by the airline operation office.

Here are some tips that telephone operator should follow:

- Give full attention to write proper room number, name and time to wake-up guest to avoid any mistake.

- Always ensure that guest really wakes-up after your call. If required, telephone operator can politely ask the guest if he wants to have a second wake-up call or not.

- If no reply is received by the guest while you are calling or guest just hang up the phone and hardly give any reply, then you should call him again.

- If after several attempts guest does not respond at all, then send the bell boy to knock his door and wake him up.

- While calling you should start this way, 'Good morning, Mr. X. It is 6 am in the morning, which is your wake-up call, have a nice day'.

HIRING A CAR

At times the guest staying in a hotel may require a car or a cab. The facility is provided to the guest in close coordination with the travel desk. The following steps are taken in this regard:

- The requirements of cab must be noted down in front office log book.

Request received from guest for change of room

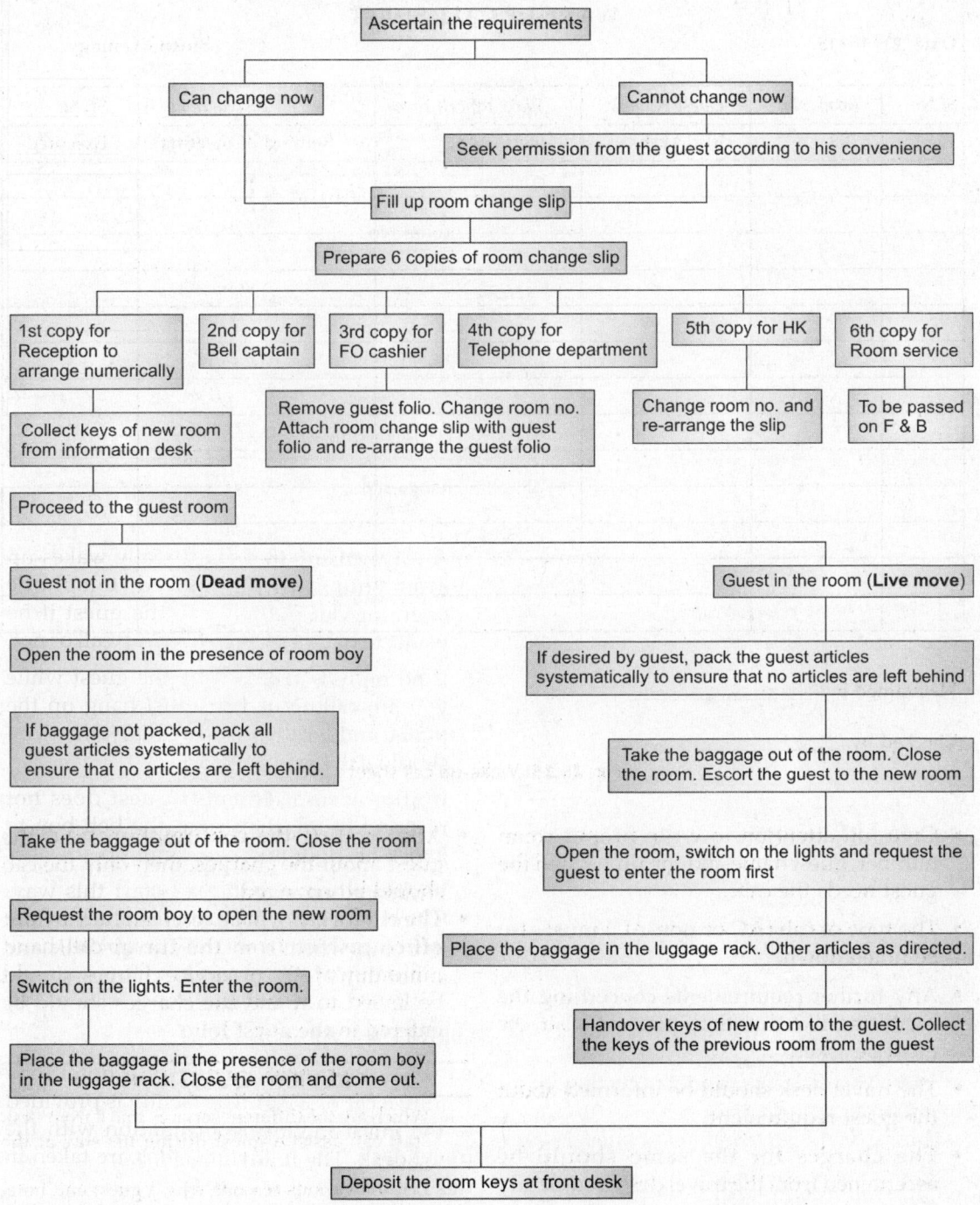

Fig. 25.14: Change of room flow chart

HOTEL XYZ

WAKE-UP CALL SHEET

Date: 21/10/15   **Shift**: Evening

Sl.No.	Room No.	Guest Name	Wake-up call time	Special Instructions	Status
1.	308	Mr. Akshat	08:00 AM	Remind of meeting	Executed

Maintained by: _____

Checked by: _____

Fig. 25.15: Wake-up call sheet

- Give full attention to write proper room number, guest name and the time when the guest needs the cab.

- The type of cab (AC or non-AC) must also be noted down.

- Any further requirements concerning the car hiring like duration, place of visit, etc must be asked.

- The travel desk should be informed about the guest requirement.

- The charges for the same should be ascertained from the travel desk and clearly informed to the guest.

- After getting the confirmation from the guest about the charges, then only the cab should be arranged.

- The bill for cab will be received by the front office cashier from the travel desk and minimum of 10% of service charges, should be levied to it and the charges should be entered in the guest folio.

REVIEW QUESTIONS

1. Which are the different services that a front desk staff can assist a guest during his stay at the hotel?
2. List the various reasons why a guest and hotel could require a room change.

3. Differentiate between a 'live move' and 'dead move'.

4. List the procedure for changing the room of a guest.

5. Explain the procedure for wake-up calls.

6. What is paging? Discuss three methods of paging.

7. Explain procedure for handling guest messages with the help of a flow chart. Draw the format of message slip.

8. List the various types of mail received by the front desk. Write in detail about the procedure of handling guest mails.

9. Explain the use of location form in handling telephone calls in the absence of guests.

10. Explain step-by-step procedure for the following:

   • Room change

   • Safe deposit

11. Write a note on house rules for safety lockers.

Front Office and Guest Safety and Security

Learning Objectives

After reading this chapter, you will be able to understand the following:

- Security of guests, staff and the hotel
- Safety and security measures
- Role of front office
- Keys and their control
- Handling unusual events and emergency situations—dealing with fire, terrorist activities and bomb threat, theft of guest property, theft by employee of the hotel, misconduct, situation of illness and epidemics, death of a guest in the hotel, robbery
- Fire prevention and fire fighting
- Safety awareness and accident prevention
- First aid

INTRODUCTION

Safety and security are concepts often used interchangeably, and it should be understood that both are means of safeguarding human and physical assets. The term 'safety' is used with reference to such things as disasters, emergencies, fire prevention and protection, and conditions that provide for freedom from injury and prevent damage to property. The term 'security' is used with reference to freedom from fear, anxiety, and doubts concerning humans as well as protection against thefts of guest, employee, or hotel property.

SECURITY OF GUESTS, STAFF AND THE HOTEL

Security has always been a concern for hotels worldwide. The recent increase in terrorist acts has had its toll on travel and tourism worldwide. Whilst there is no indication that hotels are a primary target for the perpetration of terrorist acts, hoteliers must ensure that their properties are secure—if anything to give a sense of security to guests and staff whilst at the same time protecting their investment. Today, guest safety is a 'luxury'.

Security System

The hotel should have a proper security system to protect the human beings (guests and the staffs), physical resources and assets (buildings, pieces of equipment, appliances) and also the belongings of the guests, i.e. his luggage. There are two types of security threats hotels should be concerned with:

- Threats that might affect a guest's health, comfort or well-being.

- Threats that affect the hotel directly, in particular, its fixtures and fittings, its revenue and its reputation.

Therefore, it is important for the management to select reputed and reliable security system and agencies which will provide protection against all such threats which will create problem for the hotel.

Types of Security

Security can be classified under following aspects (Fig. 26.1):

Physical aspects (internal and external security)

- **Internal Security**
  - Against theft (close circuit camera and burglars' alarms).
  - Fire security (smoke detectors, fire alarm, water sprinklers, fire extinguishers).
  - Proper lighting of corridors, fire escape, basement and other areas to facilitate detection of suspicious or unusual activities.
  - Safeguarding assets (proper inventory, regular physical checks, etc.)
  - Keeping track of unwanted guests.

- **External Security**
  - Proper lighting of boundary and outside of the building.
  - Proper fencing of the building.
  - Fencing of pool area to avoid accidents at night.

- Planting of shrubbery can also help in restricting perimeter access into the property.
- Avoid poisonous and thorny bushes, barbed wire and electric fencing.
- Manning of service gates to restrict entry.
- Fixing of closed circuit TV cameras.

Security aspect of persons (guest and staff)

- **Guests**
  - Guests suspected of taking away hotel property should be charged according to hotel policy.
  - Scanty baggage guest should be carefully watched.
  - Employee should be instructed not to divulge information about resident guests to outsiders.
  - Any suspicious person roaming in corridor must be immediately reported.
  - Housekeeping staff should never leave keys lying exposed on unattended carts in corridors.
  - While issuing the room key to a guest, ask for the key card.
  - Guest room security—wide angle door viewer, dead bolt locks, night torch, chain on door, etc. should be provided.

- **Staff**
  - Rigorous recruitment and selection procedure: References checked, proper screening, etc.

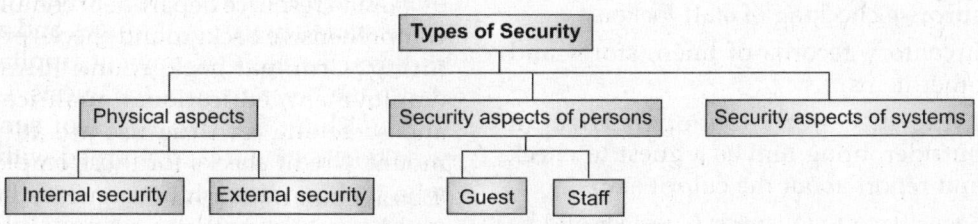

Fig. 26.1: Types of security

– Identification of staff: Issue identity badges (with photograph and employee code) and use distinct uniform for easy identification. Collect the uniforms, name tag, or identification badge when a staff member is no longer associated with the business.

– Restricted access: Identify staff requiring unlimited access to all areas of the hotel. Limit staff access to non-public areas so that staff enter only those areas necessary for their job functions and only during appropriate work hours.

– Prevent staff from bringing personal items (for example, lunch containers and bags) into food areas or storage areas.

– Key control: Issuance and return of room keys by the room attendants should be properly recorded.

– Red tag system: Proper system for hotel property being taken out.

– Training: Proper training to employees to note unusual things, safety drills and fire fighting skills.

– Adherence to management policy of security: An employee disregarding company policy should not be confirmed.

– Trash handling: Trash should be checked to see if employees are smuggling out things out of hotel with trash.

– Employees' parking: Proper checking of employees' vehicles at the time of going off duty. If possible it should be isolated from the main parking/ guest parking.

– Locker inspection: From time to time surprise checking of staff lockers.

– Inventory records of linen, stores and other items.

– Bring in experts (snoops): Hire an outsider, bring him as a guest to check and report about the culprit staffs.

– Security consciousness: Guest should be told to be careful about his property.

– Entrust employees: Employees should be asked to report about suspected persons, guests with scanty baggage, etc.

– Set example: Management should follow these rules and should encourage employees to follow the same.

– Unusual behavior: Watch for unusual or suspicious behavior by staff. This could be staff who, without an identifiable purpose, stay unusually late after the end of their shift, arrive unusually early, access files/information/areas of the facility outside of the areas of their responsibility, remove documents from the facility, ask questions on sensitive subjects, or bring cameras to work.

Security Aspect of Systems

Security aspect of systems in hotel is equally important to physical and persons' security. The objective of such security is to safeguard the assets of the hotel. Systems, procedures and the policies followed properly shall safeguard the assets and shall increase the life span of equipment.

• Record all losses and missing items immediately.

• Inventory control should be proper.

• Auditing should be done on a regular basis.

• Proper system for cash receipts and disbursements should be followed.

SAFETY AND SECURITY MEASURES

Some of the security measures adopted by hotels are:

• Employees should be thoroughly screened by human resource department completing comprehensive background checks prior to hiring. Criminal background, previous employment, educational qualifications and screening for drug use are all paramount. Credit checks for those employees who will handle cash or monetary instruments or who will be responsible for financial data is appropriate. These background checks apply to permanent, temporary and outsourced employees.

- Identity card has to be provided to the employees and insisted to use them regularly at all times during work.

- Temporary staff/casual workers should be hired from reputed contractors who maintain up-to-date database.

- The hotel's reservations and reception desks must obtain enough information from the guest at the time of booking to determine, before the guest arrives, that they pose no threat to the hotel.

- Ask for the guests' identity, especially in case of foreigners and walk-ins.

- Heavy drapes to be drawn during night on windows and exposed glass panels to cut out external light.

- Hotel should preferably use electronic locking system. Key control system should be implemented.

- Double lock system, magic eye and door chain system to be installed in guest rooms.

- Proper left luggage system to be followed.

- Safety lockers for guest valuables should be provided.

- Smoke detectors to be installed. Install modern and efficient fire fighting system.

- Fire escape route must be designed and highlighted.

- The staff of the hotel should be given the training for emergencies like earthquakes, tornadoes, terrorist attacks, fire, etc. Many hotels maintain an emergency manual. Hotels can also organize periodic drills for emergency procedures.

- The practice of people entering the lobby and taking the lift to any floor must be stopped. Elevators may be interfaced with the room electronic locking system so that the in-guests are able to go only to the guest floor they are on by swiping their room card in the elevator.

- Hotels must inform guests about the various security measures put in place for everyone's safety. This can be accomplished by placing literature prominently in the guest room or including it in the key card and distributing to each guest during check-in.

- Close circuit TVs to be installed in all public areas of the hotel, especially at the alighting point and entrance door for scanning people coming into the hotel. Parking area and guest corridors should also be under CCTV surveillance.

- The number of access points to the hotel needs to be restricted to a minimum.

- Frequent patrolling by the security staff must be made.

- Security frisking (body check) if needed (without offending the guest).

- Computer and data processing security installed (for safeguarding of computer information, so that it does not reach the competitors and protection against virus in the program). Eliminate computer access when a staff member is no longer associated with the establishment. Establish a system of traceability of computer transactions.

- At the perimeter, on the outside of the hotel itself, cars must be stopped, boots checked and mirrors put under the car chassis to detect explosives.

- Unless a hotel guest is being driven in a hotel automobile, all other vehicles such as taxis and private cars must be halted at a reasonable distance from the hotel and not allowed to roll into the hotel porte-cochere or stand in close proximity to the hotel. Some guests will unfortunately have to walk a longer distance, or the hotel should provide its shuttle service to bring in guests.

- Heavy shrubs and vines should preferably be kept away from the hotel building and trimmed close to the ground to reduce their potential to conceal criminals or bombs.

- Guest baggage should be passed through baggage scanners which can detect the shape and density of the objects with a high degree of clarity.
- Supplies should be sourced from approved and established vendors only. All supplies should be put through scanners. Establish delivery schedules; do not accept unexplained or unscheduled deliveries.

Safety and Security Measures for a Single Lady Traveler in the Hotels

In the 21st century women is taking over various roles in the business world as chief executive officers, entrepreneurs, sales representatives. The female customers are slowly replacing the valued male guest and this can be seen in the hotels globally. With more and more women pursuing careers, this segment is seeing a growth of 15–20 percent (Hotel and Food Service, 2004) every year. Indian Tobacco Company (ITC Hotels) has 10 percent of their clientele as domestic or foreign women travelers. According to Travel Industry Association, the female clientele values luxury and security above all the other factors in the hotel. The single lady traveler is a fast growing, niche market and has tremendous potential in India. Today, women are as big spenders as men have been and are willing to pay the money as long as they can see the value. Reaching out to them and satisfying their needs will be an important factor in translating *marketing into sales* for the hotels. Hotels that often perceive their customers as only men could be losing a lot of clientele if their practices or facilities are not tailored to suit the female customers. Women are now traveling without being accompanied by their male partner and are taking more trips on their own to visit relatives, friends with female friends or within a tour group. In the past, women do not seem to constitute a discrete market segment for hoteliers. Marketing practices accepted women as part of the family market segment

and were believed to gain access to hotels in their role as wives and mothers. Today, the women market is expanding and substantial where their purchasing power is large and profitable enough to be served by the hotel. The services could be tangible like price, physical appearance, and location and intangible like security, reputation, staff behavior.

Women are perceived to be especially interested in factors which include cleanliness, well maintained furnishings, comfortable bed and pillows, good service by the staff, ambience of the guest rooms, placement of amenities in the rooms, comfortable access. The other related factors include the location of the hotel, the convenience of meeting site, the hotel's reputation. By understanding how women customers rate these services, the hotel attributes can be changed and practiced to become more competitive within the market. The hotels apart from including the full scale facilities like 24-hour internet access, e-mail, voice-mail, secretarial assistance, etc. also offer women friendly facilities like lady butlers, lady attendants in housekeeping, screened telephone calls, entry allowed through swiping cards in elevators, video phone in the room to the extent that assistance in draping a sari is given.

Some of the facilities and amenities offered for a Single Lady Traveler are as follows.

Special Attention During the Allotment of Rooms

When rooms are allotted by the front desk staff to the single lady guest, special confidentiality is maintained. The information about the room number, personal details, and the room change details are kept confidential. They are allotted rooms near the elevator, so they are secluded in a corner and are not inconvenienced by having to walk through a long corridor alone. The staff will avoid giving the women customers, adjacent rooms to single male guests. The room to room dialing service is not given when the hotel has a single lady staying with them. The hotel offers in-room check-in with bouquet on arrival as most ladies prefer flowers. The guest relations executives keep in touch with them during their stay to have it as per their likings and preferences.

Safety and Security Services

Security is the main concern for these customers, so most of the hotels have free pick-up and drop services from the airport. When entire floors are dedicated to single women travelers (women-only floors) in the hotel, entry is allowed only through swiping the room key card in the elevator to direct it to the women's floor. The entry is allowed only to the resident women guest giving them the optimal security. The floor is guarded by women security guards and is maintained by a team of all-women staff all 24 hours in the hotel. The corridors are well lit during night. There are video phones to check the identity of the person seeking entry into the guest room or an 'interactive doorbell' that allows room guests to see who is at the door, and the screening of telephone calls is done at the desk by the front desk staff of the hotel. The electronic safe box is placed in the rooms, for the convenience and ease of the lady customers.

In-room Dining Services

The women's floor may have the service of a cocktail and hors d'oeuvres trolley in the room by a woman (room service staff) in the evenings. The female customers may feel a little awkward to have a drink all alone at the bar in the hotel, and hence such services might actually help to have repeat customers in the hotel.

Special Room Amenities and Facilities

The guest room has feminine needs like full-length mirrors, make-up mirrors, iron and ironing board, skirt hangers, appropriate-sized bath robes for females, and an assortment of extra organic cosmetics. The dressing preferences of the business women clients at the age of forty today are different from those in 1990's and hence the amenities also need to be upgraded all the time by the hotels. Most of the hotels claim that these are the part of their standard facilities. Some hotels place special amenities like face wash, sanitary napkins, a cane basket containing foot and bath salts, emery board, nail-cutter, comb, bindis, a manicure set, rose-scented soaps, and headband which is elegantly adorned with a flower for a lady guest in the room. The number and the logical placement of electrical plugs in the guest rooms is also an important aspect for the guest.

Services by Female Attendants

The servicing of the guest room is done by the female attendants for a lady guest in the hotel. The women may feel uncomfortable with male attendants serving them as they need to be well dressed all the time in their rooms. A single window service is given to the guests which includes all room related services to be given at one time only. The guest is not disturbed at different times to provide services to the room.

Check-in, Checkout and Travel Facilities

A desk may be provided on the Women's Floor with check-in and checkout facility on the floor without having to go to the lobby. The hotels provide airport assistance on the arrival of the guest; the guidance may be given in terms of traveling within the city by the guest relations staff to the guest. They help with the hiring of reliable transportation for the female guests. The travel desk also arranges female guides for sightseeing, excursions, shopping in the local market and tries to help them in all aspects of travel.

Bathroom Facilities and Amenities

The bathroom amenities include spa style showers, enlarged bath spaces with more color, shallow vanity counters with raised edges and good illumination for makeup, phone in the bathroom, good quality bathrobes and bath towels, silky robes and slippers, hair curlers, aroma oils, potpourri in the bathrooms, with special emphasis given on the sanitary items. Some hotels have introduced curved shower rods in their bathrooms, which creates more space and the shower curtain is less likely to touch the guests which gives them more sense of cleanliness.

The other complementary facilities and services provided by the hotel include the following:

- The hotel arranges for baby-sitters for the lady customers traveling with their children.

- The hotel staffs help the lady customers to drape the perfect sari and many hotels do the stitching of blouses in a day's time for their customers.

- The hotels provide feminine touch— upholstery in soft shades of pink, purple in the interiors of the guest room for the ladies special floors. An exquisite flower arrangement, women's magazines, cookery books, in-room exercise DVDs may be placed in the rooms.

- The lady guests may be given complimentary ayurvedic body massages. The hotel spa and beauty salons have lady masseurs' for the women clientele.

- The foreign guests are offered cultural experiences through complimentary yoga and cookery sessions.

- The hotel has a 'Dial a Chef' program where a lady guest can ask for a special diet meal, according to her preference.

- The restaurant has the 'Business Women's Networking Table' for those touring alone. It also has a special lounge where female guests can unwind after a long day's work. The tables are better attended than the other tables and the lady hostess interacts with the guests more often.

- The mini-bar in the rooms has champagne and smoothies instead of beer in the rooms.

- The hotel has a private women-only fitness room with lady trainers at convenient timings.

- The hotel may offer a pillow menu, with extra pillows being placed on the bed for the women guest.

ROLE OF FRONT OFFICE

A front desk professional is required to assume a variety of roles during a workday— chief among them being a gatekeeper, communications expert, phone whiz, troubleshooter, and receptionist. As a hotel's first contact with guests, vendors, visitors, and delivery people, the front office staff is the eyes and ears of the organization. It is up to them to spot potential troublemakers, identify red-flag behaviors, prevent breaches of security, and act with confidence and authority when situations threaten to get out of hand.

The front office staff plays a vital role in assisting the hotel's security personnel to carry out their tasks efficiently. The front office employees like door attendants, parking attendants, bell boys, and receptionists should observe all the people who are entering or leaving the hotel premises, and if they notice any suspicious activity or circumstances, they should immediately report the same to the security personnel.

KEYS AND THEIR CONTROL

Hotel Proprietors Act states that hotel management is liable for the safety of guests' property while they are resident. Depositing guests' valuables into safe deposit boxes is one method of security but the security of guests' luggage and other belongings in the guest rooms is also of great importance. Therefore, a well-controlled system of handling guests' keys is essential for ensuring the security of guest rooms. Moreover, it also prevents unauthorized access to the guest room.

Hotels may use the following two systems for securing the guest rooms.

The Conventional Key System (Metal Key/Hard Key)

Fig. 26.2: Metal key/hard key

The Electronic Key System

Nowadays, metal room keys are being replaced by electronically coded key-cards (plastic, metallic or hard-pressed paper). These cards look like credit cards with holes punched in them. Some have a magnetic strip instead of the holes (Fig. 26.3).

The system uses a master control console which is wired to every guest room door that codes the cards to lock and unlock the doors. At the time of check-in, the front desk staff inserts the card into the appropriate slot on the console to transmit its code to the guest room door lock.

An electronic locking system allows the hotel to issue a fresh key to each guest. This means that whenever the next guest checks into that particular room, a new door lock code will be re-programed to that room. Thus the old card is automatically rendered useless.

Advantages of Electronic Key Card

- These are expensive but effective.
- It ensures complete security since the loss of the key has no repercussions on the room security. No room number is printed on the card—hence if someone finds the key, it is of no use to him. The lock combination too can be changed with ease by issuing a new key in the eventuality of the loss of a key.
- At the time of issue, more than one key can be given to a guest if there is double occupancy in the room.
- The system also monitors the number of times the guest exits and re-enters the room, including the number of times the room boy

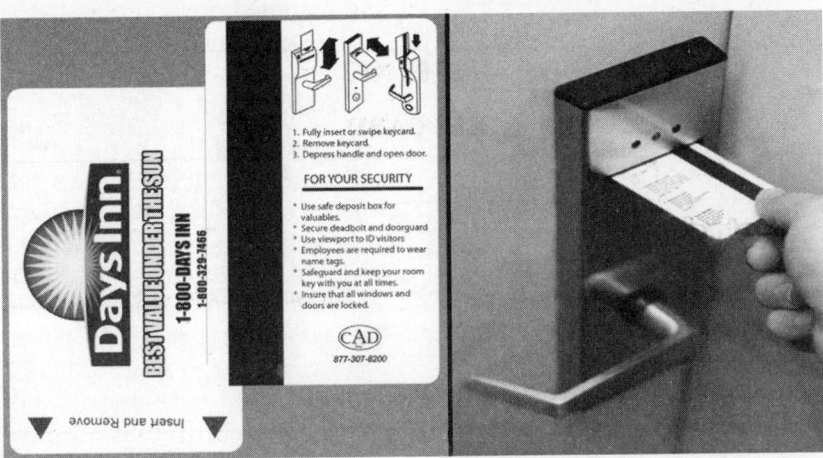

Fig. 26.3: Electronic locking system

enters the room, when using their floor key. This gives an additional dimension to the security of a guest room. If a person tries to enter using an incorrect key, then the system alerts the security staff.

- Some hotels also use the key as a trigger for the lighting and air-conditioning, so that without the key neither of these services will operate. Clearly this is a cost saving for the hotel, since many guests are careless about the costs of services.

Guiding Principles in Key Control System (In Case of Metal Key/Hard Key)

- Discourage guests to carry room keys with them while going out of the hotel premises. This is mostly done by putting heavy and large key tags, which are inconvenient to carry.

- Keys must be inserted into the correct room slot of the key rack (key rack should be out of view of passersby as another security precaution). If this is not followed accurately, an unnecessary chaos could result. Guest would get wrong keys and discover this on reaching their rooms.

- A hotel will very often issue only one key for each room, even if two guests are sharing.

- Duplicate keys should be kept on a separate board and should only be handed over when absolutely necessary.

- To ensure security for guests, issue of room keys should be made against key cards (Fig. 26.4) only.

- Guests should be reminded to return their room keys at the time of departure.

- Key control sheets are maintained by the front desk staff in the night to check that all keys are accounted for and that none are missing.

- When keys are lost or stolen, the locks should be replaced immediately.

- Many hotels do not list their name, address, logo, or room number on guest room keys. That way, if a guest room key is lost or misplaced and falls into the wrong hands, it cannot be traced back easily to the hotel for criminal use. A code number is typically stamped on the key in place of the room number and a master code list is maintained at the front desk.

- Keys must be signed in and out in a key control register (Fig. 26.5) by the housekeeping staff at the beginning and the end of a shift so as to ensure proper check and control over keys.

HOTEL XYZ
KEY CARD
Guest Name: _____
Room No: _____
Arrival Date and Time: _____   Departure Date: _____
Signature of Guest: _____
Note: Please produce the key card to receive your room keys.

Fig. 26.4: Key card

Date	Area	No. of key	Staff name	Time-out	Sign	Time-in	Sign

Fig. 26.5: Key control register

Types of Hotel Keys

Guest Room Key

These keys are under the control of the front desk and are issued to guests who have registered in the hotel. It opens the lock of an individual guest room so long as it is not double-locked.

Section Master Key or Sub–Master Key

This key opens all rooms in a section of the hotel that are not double-locked. This key is issued to room boy and floor supervisor. It is normal for hotels to insist that room boys wear their keys on a key belt around their waist, so that the key is not accidentally lost.

Floor Master Key

This key opens all rooms on a particular floor that are not double-locked. A room boy is given this key to open the rooms he is assigned to clean on a floor. Floor pantry can also be opened with this key.

Master Key

A master key opens all those guest room door locks that are not double-locked. The master key is under the control of the executive housekeeper of the hotel.

Emergency Key/Grand Master Key

This key opens all the doors in the hotel, even those that the guests have double-locked. (Double-lock is an internal safety locking device, in which, if the door is locked from inside the room, it cannot be opened from the outside by its own keys or master key). The emergency key should be highly protected and should only be used in the event of an emergency. It should not be taken out of the premises, and a strict key control should be maintained for the same. Generally, the emergency key is under the control of the head of the property. In addition, it may also double-lock a room if access to it has to be denied.

HANDLING UNUSUAL EVENTS AND EMERGENCY SITUATIONS

The nature of all emergencies is the same: They are uncontrollable and unforeseen. Apart from fire and bomb threats, the hotel staff at time has to handle a lot of other unusual situations also. Some such situations may be death and illness of guest, theft in the hotel, etc. and many others. Thus all hotels must be prepared for them and have emergency plans put in writing. Emergency plans must be a part of the SOPs.

Dealing with Fire

In case a fire breaks out, follow the guidelines given below:

- Immediately switch on the nearest fire alarm. But alarm should be used keeping in mind the gravity and intensity of fire and should be used only when needed to avoid any chaos amongst the guest and other staff.
- If possible, attack the fire with appropriate fire extinguisher, remembering to direct the extinguishers at the base of the flames. Do not attempt to fight a fire if there is any danger of personal risk.
- Close the windows and doors and switch off all electrical appliances, including fans and lights to prevent the rapid spread of fire.
- Report to your immediate supervisor for instructions.
- Remain at the assembly point until instructed to do otherwise.
- Do not use the lifts.
- Carry out instructions—for instance, rouse guests in the section and direct them to the nearest fire-escape route. Each guest room should have the route to the nearest fire escape drawn out and displayed in a place where it is most likely to be seen by the guests.
- Try to help the guests to get out of the building with the help of other hotel staff.

Terrorist Activities and Bomb Threat

Hotels that cater to VIPs like celebrities, politicians, etc. are potential targets of terrorist attacks and should be well equipped to handle terrorist threats. In case of a bomb threat, the hotel should liaise with the local police authorities and follow their instructions. If the bomb threat comes over the telephone, the person receiving such a call should follow the given procedure:

- Do not interrupt the caller.
- Write down the exact words of the caller.

- If possible, try to gather the following information and make a note of the same:
  - Where the bomb is located.
  - When it is supposed to go off.
  - The motive of the attackers.
  - The identity of caller.
  - Any background noise, etc.
  - Caller's voice and accent
  - Mannerism
  - Age and sex
- Inform the competent authority immediately. Do not alter the exact conversation between you and the caller while narrating the incident to the authorities.
- Do not spread any rumor.
- Do not attempt to defuse the bomb if you are able to locate the same. Contact the local police authority or bomb disposal squad for defusing the device.
- Whether or not the hotel should be evacuated will be the decision of the management.
- In most cases selected personnel who thoroughly know the hotel will be part of search teams.

Theft of Guest Property

When a theft of guest property is reported by any guest, the following points should be taken into consideration:

- Seal all entrances and exits.
- Restrict movements of everybody in the hotel.
- Go to the site or talk to the guest for the details.
- If the article is costly, like jewelery or huge amount of cash, the hotel security staffs may be allowed to make a search in the hotel. If required, the police should be informed. When the police arrives, assist them in the search and interrogation procedure. Nobody should be allowed to leave the hotel until the article is found or police allows them to go. Everybody

concerned should report to front office and leave their forwarding address.

Theft by Employee of the Hotel

The management should detail explicit regulations concerning employee theft. The employee handbook should spell out the consequences of stealing hotel property. It is important that the management not discriminate against any employee when enforcing these rules:

- While screening applicants for the job, a thorough check of the background, including a check for any criminal convictions, should be carried out. Reference checks over the telephone are therefore a good practice for managers when they are hiring new employees.

- Color-coded uniforms and identification badges with the employees' photographs and signatures discourage people bent on thievery from trying to pass themselves off as employees.

- Orientation and training programs should emphasize the value of honesty.

- Supervisors should closely monitor behavior and adherence to company policies and procedures during the employee training and probationary period.

- All storeroom doors should be kept locked and these locks should be changed periodically to reduce the opportunity for theft.

- An effective key-control program, lost-and-found procedure and gate pass system should be in place and enforced at all times.

- A detailed record of all room boys who enter the guest room should be maintained.

- Good inventory control procedures should be followed. Conduct a monthly inventory of all housekeeping supplies such as toilet paper, amenities and linen. If the items in storage do not match the usage rate or if too little stock is on the shelves, it may be an indication of employee theft.

- Regular locker inspections also discourage employees from stealing for lack of a hiding place for articles.

- Employee entrances should have a security staff office that monitors arriving and departing employees.

- Employee parking should be well lighted and sufficiently far from the hotel building.

Misconduct

Most misconduct cases in hotels fall under one or more of the following:

- Intoxication: Appearing obviously to be under the influence of alcohol. In some cases, these situations will have to be substantiated by a witness. In the case of a guest being under the influence of alcohol or drugs, one should escort the guest to his room and send for security and the hotel nurse or a responsible person.

- Dishonesty: This involves illegal possession of company property or that of fellow employees or guests.

- Excessive unexcused absence or tardiness: Every case of unexcused absence and tardiness must be documented.

- Conduct contributing to moral delinquency of the job: For example, selling narcotics or engaging in prostitution.

- Quarreling: Sound and abusive language and/or physical fighting in guest areas. Quarreling if by guest or staff, the parties should be separated and escorted to the back areas or an office so as not to disturb others and then resolve the conflict. Staff is normally dismissed for public quarreling and guests asked to check out.

- Refusal to perform duties: Be sure duties requested are reasonable, that they do not conflict with the union contract.

- Insubordination: Refusal to follow instructions or obey orders, excessive talking that interferes with his or others' job performance, abuse of privileges and lack of co-operation. All these reasons for termination must be documented with a record of each incident in the employee's file. Inefficiency or inability to do assigned work in a satisfactory manner is not considered willful misconduct. However, if inability to do the job is combined with any of the above misconduct charges, then there is a sufficient reason to terminate employment.

Situation of Illness and Epidemics

- On many occasions, hotels find a sick guest on their hands.
- If the guest is too ill to travel home or it is inconvenient for him to do so, as in the case of an overseas traveler, he should be seen by the doctor on call at the hotel or by a local doctor.
- If medical aid is on the way, the front desk may have to administer first-aid to the ailing guest.
- Hotel guests who are ill should be regularly visited by the front desk staff.
- In case of a contagious illness, it is advisable that the guest should be moved to a nursing home.

Death of a Guest in the Hotel

- Once the information comes to the front desk it should directly be reported to the front office manager. The front office manager will then report it to the GM and security manager.
- Do not panic and also do not inform other guests and other staff who are not involved. Go to the room or area and check the situation.
- The room should be locked and care should be taken to not touch anything in the room as it might be useful in establishing the cause of death. The room should be double-locked until the police arrive.
- GM should inform police. All the staff must cooperate with the police.

- Meanwhile the hotel will locate the residential address of the deceased and will inform the relatives.
- Once the police complete their investigation and gives permission, the dead body is fully covered and then removed from the room on a stretcher. For this purpose the service elevator is used.
- If the death is natural, the staff and other guests are allowed to move freely; if it is a murder then the movement of everybody in the hotel is restricted and nobody is allowed to leave the hotel as well as the city without informing the front desk or the police. If any witness is available, all his details are noted with front desk along with his forwarding address.
- In case if there is luggage and other belongings of the deceased, then they all must be collected, a list prepared, kept in luggage room or handed over to police.
- After obtaining clearance from the police the room is opened and thoroughly disinfected and then it should be sold.

Robbery

There is always a possibility of robbery in hotels as the front desk cash and bills section and the points of sale usually have large sums of money. Also, the valuables in the possession of guests may invite burglars. To discourage robbers, the guests should be asked to leave their valuables in the front office safety deposit locker or in the in-room locker. In the event of an armed robbery, the hotel employees should follow the below-mentioned procedures:

- Comply with the robbers' demand.
- Do not make any sudden movement
- Remain quiet, unless directed to talk by the robbers.
- Do not attempt to disarm the robbers, as this may jeopardize many lives.
- The cashier may switch on the secret alarm that might be installed in the cash drawer.

- Observe the robbers carefully, noting the physical characteristics like height, build, eye color, hair color, mannerisms, complexion, clothing, scar marks, or anything that can be helpful in their identification.
- Note the direction of escape, and the type and registration number of the vehicle used by the robbers.
- Do not touch any object that might have been touched by the robbers.

FIRE PREVENTION AND FIRE FIGHTING

Fire is among the major potential hazards associated with hotels. It could take place due to cigarette smoking in rooms, defective electrical wiring, faulty equipment, worn out insulation or gas leakage. Hotels must be equipped to safeguard guests and their property from fire by installing smoke detectors and by conducting routine checks to see that the wiring and equipments are not defective. All the employees must be aware of specific procedures laid down by the establishment and must be ready to comply with them at any time. There should be fire drills during low occupancy periods to check how the employees put their knowledge of laid down procedures into practice. Fire extinguishers should be installed in hotels at strategic locations and fire exits are earmarked for use in the event of fire.

Extinguishing Fire

The presence of three basic components—fuel (a combustible substance), oxygen (necessary as fire is an oxidation reaction), and heat (ignition temperature)—results in the outbreak of fire. If any one of them is absent, fire cannot break out. Fire can be extinguished by:

Starving: Starving is the removal of fuel from the vicinity of fire so that it does not spread.

Smothering: Smothering is the removal of air. Fire can be extinguished by cutting off the supply of air (oxygen), which is necessary for the existence of fire.

Cooling: Heat (ignition temperature) is also essential for the existence of fire. Cooling—by adding water—brings down the temperature and puts out fire.

Classification of Fire

Fire is classified depending upon the type of fuel/combustibles involved and the source of heat. Each of the different types needs separate and specific means of control to extinguish it.

Class A: Fire in ordinary combustible materials such as wood, paper, textile, grass, garbage, and materials composed of cellulose. They are extinguished by quenching and cooling effects of water or of solutions containing large percentage of water.

Class B: Fire in flammable liquids, which must be vaporized for combustion. These include oil, gasoline, petroleum products, varnishes, and paints. These fires are extinguished by blanketing the source of burning substances and eliminating the supply of O_2.

Class C: These are electrical fire. Electrical fire is usually caused by a part of circuit overheating or by short circuit. Controlling the sizes of electrical fuses and circuit breaker will often minimize this class of fire.

Class D: Fire that occurs in combustible metals such as magnesium, aluminium, zinc, potassium, etc. Special extinguishing agents and techniques are needed for fires of this type.

Class E: These are the fires of pressurized gases such as liquefied petroleum gas (LPG), methane, compressed natural gases (CNG), etc.

Fire Fighting Equipment

Portable fire extinguisher

A fire extinguisher is an active fire-protection device. It is normally a manually operated and typically a hand-held device (a cylindrical or conical pressure vessel), which produces fire-resistant agents and can be discharged to extinguish a fire (Fig. 26.6). These are to be used to extinguish or control small fires only, often in emergency situations. It is not designed for use on fire apparently going out-of-control, such as one which has reached the ceiling or puts the incumbents and the user of the extinguishers in danger, for example, there is no escape route or huge smoke is billowing out or there is a risk of explosion, etc. Such large-scale fire requires the expertise of a fire department to control.

Different types of fire extinguishers are designed for use in specific classes of fire.

- **Soda acid fire extinguisher:** It is used for class A fire. The extinguishing agent is H_2O. The fire extinguisher is a cylinder type of pan in which a rubber or flexible hose is attached to the top. When it is desired to use the extinguisher, it is carried to the fire and inverted. A small bottle of acid usually H_2SO_4 is spilled when the cylinder is inverted or turned upside down. Powdered sodas, bicarbonate of soda (baking soda) is mixed with H_2O when the tank is charged or filled with water. The chemical reaction of acid and soda water creates a pressure

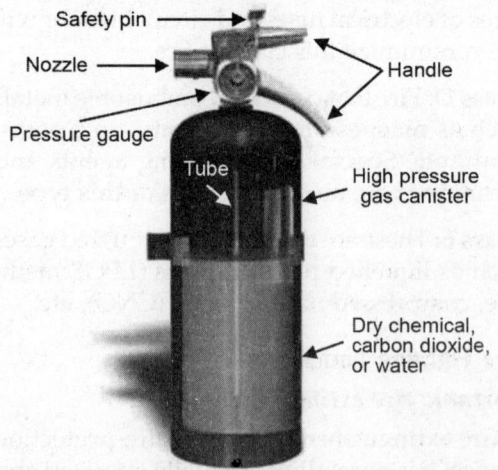

Safety pin

Nozzle

Handle

Pressure gauge

Tube

High pressure gas canister

Dry chemical, carbon dioxide, or water

Fig. 26.6: Fire extinguisher

which forces the water out of the cylinder or tank. The hose is used to direct the flow of water to the fire.

- **$CaCl_2$ fire extinguisher:** It is also used on class A fire. $CaCl_2$ is a salt which when added to water form brine which has very low freezing temperature. CO_2 cartridge is used as pressure agent to force H_2O and $CaCl_2$ out of the cylinder to the fire. These extinguishers are used where freezing is a potential hazard.

- **Foam type extinguisher:** It is used on class B type of fire. The extinguisher is charged with special chemical (Al_2SiO_4), the chemical spread on the burning material and the solution, blanket the fire by excluding O_2.

- **CO_2 fire extinguisher:** It is used on C, D and E class of fire. The CO_2 fire extinguisher spray a chemical fog towards the fire. The fog quickly excludes the O_2 from the burning material and blanket the combustible material.

- **DCP (dry chemical powder) extinguisher:** It can be used on C, D and E class of fire. The most common extinguishing agent is sodium bicarbonate or plain baking soda. The extinguisher is charged with the dry chemical and a small tank of CO_2 gas. The CO_2 gas exerts pressure on dry chemical and forces it out of a nozzle directly to the fire. The powder restricts the fire; and the heat from the fire breaks down the chemical which releases CO_2 gas on a large scale which helps in extinguishing the fire.

Stationary Fire Fighting System

- **Automatic sprinklers:** It is generally mounted just below the ceiling height with a temperature detector or smoke detector, attached with each sprinkler. The temperature from the fire melts the fusible link on the detector, which opens a water valve. The water is then sprayed on the ceiling and falls on the floor, extinguishing the fire. If the fire area should spread, more sprinklers are automatically opened, thus confining the fire to a small area. The temperature detector can be purchased for different activating temperature. The high temperature detectors are often used in kitchens.

- **Fire hose system:** It is a semi-portable system. In this system the fire hose box is permanently located but the flexible hose can be moved to various distances throughout the building. The hose used to fight fire within a building should be of linen type. The linen allows some water seepage through it which will prevent its burning when in use.
- **Fire alarm:** These are red in color, with a glass panel that needs to be broken to set off the alarm.
- **Smoke detector:** These are alarms triggered off when smoke blocks a beam of light emanating from the detector.

SAFETY AWARENESS AND ACCIDENT PREVENTION

According to Oxford Advanced Learner's Dictionary, an accident is 'an unpleasant event that happens unexpectedly and causes injury or damage'.

Safety awareness should be an ongoing program at all establishments. The management of all establishments should be aware of the laws concerning safe work environments and should be concerned about the safety of their employees. Periodic training should be provided to all staff in order to raise awareness about safety. All employees should be aware of the potential hazards in their respective departments. All head of the departments must ensure that employees follow safe job procedures, correct unsafe conditions immediately, and take adequate time to do the job so that accidents are not caused due to haste.

Specific Causes of Accidents

Accidents may occur due to any one of the following reasons:

Excessive Haste: Excessive haste is among the prime causes of accident because a person in haste may overlook the safety rules or obstacles in the way. For example, bell boys, when in a hurry to carry the luggage of a group, may overlook objects in blind corners and may meet with an accident.

Carelessness: Carelessness is also a common cause of accidents. The careless handling of goods and equipment may lead to accidents.

Anxiety: Anxiety is a feeling of worry or fear about something. An anxious person will not be able to concentrate on the task at hand and this might lead to accidents.

Lack of Interest: If a person lacks interest in what he is doing, he becomes careless and does not follow the correct procedure for carrying out a task, leading to accidents.

Lack of Concentration: A person may not be able to concentrate on work due to personal problems, lack of interest, distractions, etc. The lack of concentration may lead to problems and accidents.

Failure to Apply Safety Rules: Safety rules, if followed, prevent accidents. Before operating any equipment, one should read and follow the operating instructions given in the product manual in order to eliminate the chances of accidents.

Spills on the Floor: Floors should be kept clean and dry.

Badly Maintained Areas: For example, back areas, service lift, staircase, etc.

Procedure to Follow in case of an Accident

When a guest or an employee has met with an accident at the hotel, the following procedure should be followed:

- With the help of another person, check if the victim requires any assistance.
- Remove the person who has met with accident from the site of accident (as early as possible and take him to a more comfortable area, use a stretcher in case the need be)

- Report the matter immediately to the manager concerned.
- Either administer first aid or get help from trained personnel.
- If the injury is serious, an ambulance may be summoned and the victim taken to nearby hospital. Follow all necessary first-aid measures until the ambulance arrives.
- Fill in the accident report form (Fig. 26.7) and hand it over to the manager concerned. Also make your comments as to the reason

of the accident and how could it have been prevented and what action is to be taken to avoid the same in the future.

FIRST AID

According to the Oxford Advanced Learner's Dictionary, first aid is 'the simple medical treatment that is given to somebody before a doctor comes or before the person can be taken to a hospital'.

First aid is the provision of initial care for an illness or injury.

HOTEL XYZ

ACCIDENT REPORT FORM

Sl. No. _____

Name of the injured person: _____

Section: _____

Supervisor:_____

Date of accident: _____ Time of accident: _____

Date of report: _____ Time of report: _____

Extent of injury: _____

_____

Was hospitalization required?   Yes _____   No _____

Nature of the accident: _____

Place of accident: _____

What happened/cause: _____

_____

Witness 1: _____

Witness 2: _____

Supervisor's Remarks: _____

_____

_____

_____

Date: _____   Supervisor's signature

Fig. 26.7: Accident report form

It is mandatory for an establishment to have adequate first aid equipment, facilities, and trained personnel to provide first aid in the work area.

First aid is usually performed by non-expert, but trained personnel to a sick or injured person until definitive medical treatment can be accessed. Certain self-limiting illnesses or minor injuries may not require further medical care past the first aid intervention. It generally consists of a series of simple and in some cases, potentially life-saving techniques that an individual can be trained to perform with minimal equipment.

If the injury is serious, the injured person should be treated by a doctor or a qualified nurse as soon as possible.

Aims of First Aid

The key aims of first aid can be summarized in three key points:

Preserve life: The overriding aim of all medical care, including first aid, is to save lives.

Prevent further harm: Also sometimes called prevent the condition from worsening, or danger of further injury, this covers both external factors, such as moving a patient away from any cause of harm, and applying first aid techniques to prevent worsening of the condition, such as applying pressure to stop a bleed becoming dangerous.

Promote recovery: First aid also involves trying to start the recovery process from the illness or injury, and in some cases might involve completing a treatment, such as in the case of applying a plaster to a small wound.

First aid training also involves the prevention of initial injury and responder safety, and the treatment phases.

First Aid Box

There should be a first aid box in the work area and it should be easily identifiable and accessible. It should be in the charge of a responsible person, who should ensure that the consumed medicines are replenished and the expired medicines are replaced regularly.

A first aid box must contain the following things:

- Plastic bandages
- Transpore tape
- Alcohol preps
- Adhesive bandages
- Micropore tape
- Gauze
- Extra large plastic
- Bandages
- Iodine prep pads
- Fingertip bandages
- Sterile pads
- Antiseptic towelettes
- Knuckle bandages
- Antiseptic ointment
- Ammonia inhalant
- Sponge packs
- Instant ice packs
- Sterile eye wash
- Eye pads
- Safety pins
- First aid cream
- Bandage scissors
- Tweezers
- Butterfly bandages
- Water tight utility box for contents
- Burn gel to treat burns
- Burn bandages
- Adhesive spots
- Extra large strips
- Surgical tape
- Sponges
- Pain reliever
- Elastic bandages

Principles of First Aid

- Keep the injured people lying down.
- Check for bleeding, stoppage of breathing, poisoning, wounds, fractures and dislocations.
- Keep the injured person warm.
- Call ambulance
- Keep calm
- Never give water to an unconscious person
- Keep on-lookers away.
- Make the patient comfortable.
- Do not let the patient see his injury.
- Notify the patient's family.

First Aid for Some Common Problems

Burns and Scalds

Burns may be caused by dry heat source like flame or hot articles. For minor burns on the limbs, immediately hold the injury under cold running water for five minutes. A small burn needs no further treatment. It should simply be left exposed to air. Do not apply any oil or ointment to the burn and do not prick or remove blisters.

Large and deep burns (covering more than 3 sq inches) need medical attention. If possible, relieve pain by immersing the area in cold water or applying cold wet clothes. Wrap or cover the injury with a clean cloth and a light bandage. If it is a major burn, seek the assistance of doctor.

Scalds are caused by a wet heat source like steam or boiling liquids. In case of scalds, remove very hot clothing from the skin immediately and pour plenty of cold water over the burnt area.

Fainting

Fainting may be caused by a sudden reduction in blood flow or oxygen to the head. It may be the result of a slowing down of the heartbeat from shock, anxiety, or even hormonal changes in early pregnancy. If someone feels faint, get the person to lie down with the feet raised above the level of head. Alternatively, sit the person in a chair with the head between the knees. If someone has already fainted, loosen any tight clothing around the neck, chest and waist. When he regains consciousness, he should be kept in fresh air for a while and it should be ensured that he did not injure himself when he fainted.

Fractures

The signs of a bone fracture are pain and tenderness even at a gentle touch, swelling and bruising or loss of control or deformity of the affected limb. In case of a fracture, the first step should be to remove any pressure from the affected part and to make it immobile. One should not try to realign the bone and seek immediate medical assistance. If there is bleeding, apply pressure to the wound with a sterile bandage or a clean cloth. To prevent swelling and to relieve pain till the medical assistance arrives, apply ice packs wrapped in a towel or piece of cloth. If the person feels faint or is breathing in short, rapid breaths, he should be made to lie down with the head slightly lower than the trunk, and, if possible, his legs should be elevated.

Shock

The signs of shock are faintness, sickness, clammy skin, and pale face. A person who has suffered a shock should be made to lie down. To make him comfortable and warm, he should be covered with a blanket or additional clothes.

Cuts

All cuts should be washed with an antiseptic lotion and covered with a waterproof dressing. In case of considerable bleeding, it should be stopped as soon as possible by bandaging firmly or by pressing the artery with thumb. If bleeding does not stop, immediate medical assistance should be sought.

Nose Bleeding

In case of nose bleeding, the person should be asked to sit down with the head bent forward. His clothing should be loosened round the neck, chest, and waist. To stop bleeding, apply pressure and pinch the nostrils closed for about five minutes. Nose bleeding occurs due to the rupture of the fine capillary of artery; applying pressure and keeping the nostril closed provides sufficient time for blood coagulation. If bleeding persists, seek medical assistance.

Muscle Strain

In case of a sprain, stop the activity that caused the injury and apply ice to the injured area to reduce swelling. Use a bandage to restrict the movement of the injured area, and, if possible, elevate the injured part to make the fluids drain away.

Heart Attack/Stroke

A stroke may be caused due to insufficient blood supply to the heart or a clot of blood in the heart or a major blood vessel or in the brain. The symptoms may include chest pain (angina), breathlessness and feeling faint. The person should be taken to the coolest possible area in the hotel. The patient should be propped up or allowed to sit forward on a chair and on no account moved until the doctor or ambulance arrives.

Fits or Convulsions or Epileptic Fits

This can be a major or minor attack, but it is not unusual for a major attack to follow a minor one. In a major attack, the person will suddenly lose consciousness and fall to the ground and will have a series of convulsions which may be violent. The place where the person has fallen should be cleared of obstacles so that the patient does not hurt himself. If possible the clothing should be carefully loosened and something soft placed under the head. When the convulsions have ceased, the person should be put under observation until he is fully recovered and a doctor should be informed of the attack.

Artificial Respiration

The most usual methods of artificial respiration are:

- *Mouth-to-mouth respiration or kiss of life:* The mouth-to-mouth method is easiest when the patient is on his back but it should be started immediately whatever the patient's position. The person's head should be tilted backward, putting a hand under his neck and pulling the chin upwards in order to get a clear passage way to the lung. The person should pinch the patient's nose and blow hard into the lungs, through the patient's mouth until his chest expands to the maximum possible. Once it falls, the procedure should be repeated. After two inflations the pulse should be checked to be sure the heart is beating, if it is then one should continue to inflate the chest 12–16 times per minute, until the doctor or the ambulance arrives.

- *Holger and Nielson method:* The patient is turned downward with the head turned to one side. The person administering the artificial respiration should kneel at the patient's head and place his hands at the patient's shoulder blades. Pressure is exerted by slowly rocking forwards, and the pressure is released by rocking backward. The patient's arms are raised by

Fig. 26.8: First aid

the elbows to expand the chest. The process is repeated until the doctor or the ambulance arrives. Each phase of expansion and compression should last about two and a half seconds and the complete cycle repeated twelve times per minute.

REVIEW QUESTIONS

1. Differentiate between the terms 'safety' and 'security'.

2. What safety measures need to be adopted for prevention of deaths due to fire and drowning inside the premises of a hotel?

3. What measures can be taken to ensure safety of women executives at the hotel?

4. How hotels provide safety and security to guest and guest valuables in current scenario?

5. List the do's and don'ts to be followed by the front office staff in case of a bomb threat.

6. What are the measures being taken by hotel to minimize guest and employee theft?

7. How will you handle death of a guest in hotel room?

8. What are the causes of fire? Explain classification of fire.

9. Write a note on fire extinguishers.

10. Explain different types of keys and key control system followed in hotels.

11. How do electronic locking systems change the nature of key control?

12. What are the causes of accidents? What are the advantages of reporting accidents?

13. What is first aid? Enlist the articles that a first aid box must contain.

14. Differentiate between:
   • Key card and Card key

15. Write short notes on:
   • Electronic locking systems
   • Role of the front office in protection of hotel funds

16. What is the Holger Nielson method?

17. Explain the mouth-to-mouth method of administering artificial respiration.

18. Elaborate on the commonly used fire-warning systems and fire-fighting equipment found in hotels.

19. Mention the first-aid procedure for the following:
   • Burns and scalds
   • Shock
   • Stroke
   • Fracture

Checkout and Account Settlement

Learning Objectives

After reading this chapter, you will be able to understand the following:

- The importance of departure procedure
- Handling FIT departure in manual, semi-automated and fully-automated systems
- Handling group departure
- Modes of settlement of bills—cash settlement and credit settlement
- Potential checkout problems and solutions—late checkout, improper posting of charges in the guest folio, unpaid account balances (late charges and skipper)
- Innovative checkout options—express checkout, self checkout, in-room folio review and checkout

IMPORTANCE OF DEPARTURE PROCEDURE

The last interact on of the guest with the hotel staff takes place uring the final phase of the guest cycle, i.e. checkout. During checkout, guests formally vacate their rooms, settle their bills, and leave the hotel. This phase is very crucial as guests settle their financial transactions with the hotel, and any dispute at this stage can ruin their entire experience. On the other hand, a smooth settlement of bills and checkout would enhance the guests' experience. As every hotel endeavors to achieve high levels of guest satisfaction, the activities of the various departments should be coordinated to ensure a smooth checkout. The speed and accuracy in the preparation and presentation of bills will lead to the maximization of guest satisfaction. Error-free billing and speedy processing of checkout requests reflects the professionalism of the hotel and imparts a lasting good impression to the guest. Remember, for repeat business,

it is very important that the guest departs at a good note. Many guests will forget all the previous courtesies and hard work of the hotel staff if checkout and settlement do not go smoothly.

The departure procedure may vary slightly from hotel to hotel according to the level of service and degree of automation of the organization. The amount of face-to-face contact between the guest and front desk personnel may also vary since some hotel offer special automated or express checkout service. Despite such variations checkout and settlement represents an essential front office responsibility. Like check-in, checkout gives the hotel an opportunity to make a positive impression on the guest.

HANDLING FIT DEPARTURE

The following steps are involved in the departure procedure in *manual or semi-automated systems*:

- The checkout request is received at the front desk or bell desk.
- The front desk sends a bell boy to transfer the guest's luggage from the room to the lobby.
- The bell boy fills the departure errand card.
- The front desk sends departure notification slips (Fig. 27.1) to all concerned departments.
- The front desk alerts all points of sale to rush last-minute credit transactions to the front desk, so that the cashier can add them to the guest account.
- Cashier updates the guest folio on the basis of recent bills received from the point of sales. He should also check if any late checkout charges are applicable (if the checkout time is 12 noon and the guest checks out at 6 pm). Cashier should pay special care to late charges (A late charge is an outstanding payment from the guest, the bill for which reaches the front desk cashier after the bill has been prepared).
- The guest arrives at the front desk and hands over the room keys.
- The cashier should greet the guest checking out with the same amount of enthusiasm and warmth that is used to welcome a guest

checking in. This is very important as the guests' departing experience may form a lasting impression that will form the basis of repeat visits.

- The cashier prepares the bill and presents it to the guest, along with supporting vouchers, for review. In manual hotel operation, the front office follows the given procedure to prepare a guest's bill:
  - Prepare bills in duplicate.
  - Check the room number.
  - Take out the guest folio.
  - Calculate the correct number of room nights and establish whether a late checkout charge is to be added.
  - Enter the guest's credit transactions in the master bill in the order of their occurrence.
- The payment is received from the guest as per the pre-determined mode of payment. The cashier should then bring the guest account balance to zero. This is typically called 'Zeroing out the account'.
- The original copy of bill is handed over to the guest in an envelope along with supporting vouchers.
- The front desk makes the 'luggage out pass'.

HOTEL XYZ

DEPARTURE NOTIFICATION SLIP

To: HK/Room Service/ All F&B outlets

This is to inform you that the following guest is departing from the hotel. Kindly rush the credit charges to the front desk.

Guest Name: _____   Room No.: _____

Date of departure: _____   Expected time of departure: _____

                    Front Office Cashier

Fig. 27.1: Departure notification slip

- As the guest and the hotel staff come face-to-face for the last time at the checkout stage, the front desk should use this opportunity for marketing efforts in the following ways:
  - The front desk should ask guests about their experience at the hotel and ask them to fill a feedback form.
  - If guests have any complaints, the front desk should note the same and assure the guests of a quick resolution.
  - The front desk should inform the guest about upcoming special offers.
  - The front desk may suggest making future reservation for the guest's return trip or for a hotel in the same chain at the guest's next destination.
- The front desk informs all departments about the departure of the guest to ensure smooth operation of the hotel. To ensure proper room management, front desk must inform housekeeping about the departure of the guest, so that the latter cleans the room and makes it available for sale.
- Duplicate copy of the guest bill along with his registration card is sent to accounts department for their records.
- The front office assistant updates the following front office records:
  - Room status records: The front desk changes the guest room status from occupied to an 'on-change' on the room status report. 'On change' is a housekeeping term that means that the guest has left the hotel and that the room he occupied needs to be cleaned and readied for the next guest.
  - Guest history card: The guest history card (Fig. 27.2) is a very important document prepared by the registration

Hotel XYZ

GUEST HISTORY CARD

Name: _____   Designation: _____

Company: _____

Official Address: _____   Nationality: _____ Passport No.: _____

Residential Address: _____   Date of Birth: _____

Marriage Anniversary: _____

Visit No.	Room No.	Room Rate	Date of		Special Remarks	Billing	Remarks
			Arrival	Departure			
1							
2							
3							
4							
5							
6							
7							
8							

Fig. 27.2: Guest history card

section of the front office department in conjunction with the cashier and reservation section and is a collection of the personal and financial information of the guest recorded by the hotel during his stay at the hotel.

It serves as an important tool of information for the front office staff to draft a profile of the guest based on his previous stays.

Hotels that believe in improving their relations with the guest often maintain guest history card which enables the front desk staff to know which guest is a regular client and therefore deserves a personal attention because of his patronage. Often these cards list the guest's personal likes and dislikes so as to serve them in a better way should they return.

Consider a situation where a guest is checking-in a hotel, and he is assigned the same room in which he stayed and enjoyed his previous stay because he highly appreciated the view of sea from his room. Entering the room the bell boy turns on the television showing his favorite channel already set in. The room air-conditioning set to the temperature most comfortable for him. The fruit basket full of fresh seasonal fruits but with no bananas as he dislikes bananas and a flower arrangement of only roses because he likes roses only. The wardrobe has much needed extra hangers. His favorite newspaper and magazines are on the table. A hard bed is already prepared for him that he had to request on his last visit, to top all, a ticket to the night's performance at the same famous theatre of the town, which he frequently visited during his last stay at the hotel.

Now imagine how happy and satisfied the guest would be when he will get all that he wanted even without asking for it. All the services and amenities provided and care shown by the hotel in the above case is possible because the hotel is well aware of the needs and wants of the service and facilities of the guest in advance, and which the hotel has come to know because its staff was taking interest in knowing his likes and dislikes on the previous visits and were recorded down for future reference.

– Arrival/departure register: It gives details about all the arrivals and departures for the day (Figs 27.3 and 27.4).

The departure procedure in a hotel running on a *fully-automated system* is smoother and more efficient. It involves the following steps:

• The checkout request is received at the front desk or bell desk.

• The front desk sends a bell boy to transfer luggage from the guest room to the lobby.

• The front desk informs all points of sale and other departments of the hotel about the

S. No.	GR Card No	Arrival Time	Guest Name	PAX	Nationality	Date of Departure	Room No.

Fig. 27.3: Arrival side of arrival/departure register

S. No.	GR Card No.	Departure Time	Guest Name	PAX	Nationality	Date of Arrival	Room No.	Folio No.

Fig. 27.4: Departure side of arrival/departure register

departing guest through the interlinked computer network.

- Since all the points of sale terminals are interlinked, any credit transaction of the guest will instantaneously get added in the guest folio.

- The front desk prepares the bill by selecting the bill option of the cashier module.

- The front office presents the bill, along with supporting vouchers, to the guest for review.

- The payment is received from the guest as per the pre-determined mode of payment.

- The front office makes the 'luggage out pass'.

- The front desk communicates the departure of the guest to housekeeping and all the other concerned departments.

- The front office records are updated automatically. These include:

  – The auto removal of the name of the departed guest from the in-house guest name list.

  – The automatic updating of the current room status—from occupied to on change.

  – The automatic updating of the guest history card. In fully automated systems, the property management system automatically directs the guest's information into a guest history database.

HANDLING GROUP DEPARTURE

At the Bell Desk

- The bell captain ensures that sufficient number of bell boys is available in the shift to handle luggage of the group.

- The baggage down time and wake-up call of the group are checked and followed strictly.

- The bell captain allocates the floors and rooms to the bell boys for bringing the luggage down to the lobby.

- The bell boys collects the baggage from rooms mentioned in their list. Generally, the group members are advised by the group leader to keep their baggage packed and ready for pick-up outside their room at baggage down time.

- If on the day of departure the guests are not in the room the bell boys go to the rooms of each group member's rooms and 'pull' each group member's baggage out of the room and bring it down to the lobby until the group is ready to leave. This process is called 'bag pull'.

- Baggage is brought down to the lobby and counted. After the tour leader settles the bill, the bell captain ensures that all the keys from the group members are retrieved.

- The room keys are handed over to the reception.

- After all the keys are received, the receptionist gives the clearance on the errand card.
- The bell captain then obtains a baggage out pass from the cashier. Finally the baggage is loaded onto the vehicle by the bell boys.
- Only one errand card is made for group departure; however, the number of all those bell boys who were engaged in the departure procedure is mentioned.
- An entry is made in the lobby control sheet.

At the Reception

- Departure notification slips are issued about half an hour prior to actual departure by the receptionist to telephones, house-keeping, room service, and coffee shop, etc. to avoid any late charges.
- On presentation of the departure errand card of the group, the receptionist checks the receipt of the room keys by scoring out the room numbers on the rooming list, the keys of which have been received.

At the Cashier's Desk

- The cashier prints out the master folio and individual folios (if any).
- A summary is made roomwise to facilitate easy collection of extra charges.
- The cashier presents the master folio to the tour leader for approval and the extra charges are collected with the assistance of tour leader.

- After ratifying the bill, the tour leader is requested to sign on it. The travel agent voucher is stapled with the master folio noting down the exchange voucher number, travel agent's name and travel agent's reference number.
- The cashier then issues a baggage out pass to the bell captain

MODE OF SETTLEMENT OF BILLS

A guest account can be brought to a zero balance in several ways. There are several acceptable modes of payment (Fig. 27.5).

Cash Settlement

The cash payment option is one of the most preferred modes of settlement of guest accounts.

Currency Notes

Guests can settle their accounts by paying through currency notes. While accepting currency notes, the cashier should check if the currency notes are genuine.

- **Local Currency:** At the time of settlement, the cashier zeroes the balance in the guest account.

If there is a debit balance in the guest folio, the cashier collects money from the guest against the bill and issues him with a cash receipt (Fig. 27.6) as a proof of receipt of payment from the guest and marks his folio

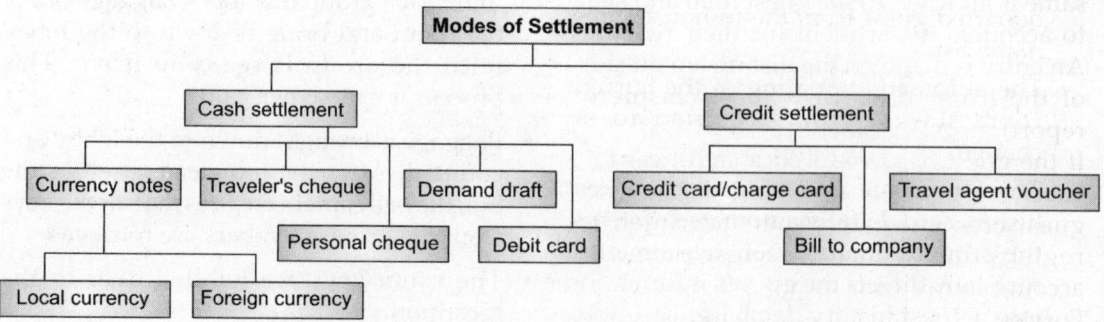

Fig. 27.5: Modes of settlement

HOTEL XYZ

CASH RECEIPT

Sl. No.: _____

Date: _____

Received with thanks from _____

the sum of Rupees _____

_____

Cash/Cheque No. _____ Date _____ Bank _____

on account of Advance/Bill No. _____

₹ _____

Receipt valid subject to encashment of cheque

Signature of Cashier

Fig. 27.6: Cash receipt

with 'paid' stamp. The second copy of the receipt is attached to the guest folio and sent to accounts department for their records. The third copy of the receipt is retained by the front office for future reference.

If there is a credit balance in the guest folio, the hotel will pay back the balance amount to the guest to make the balance zero. The guest is asked to sign a refund voucher as a proof that he received the money. The same is attached to the guest folio and sent to accounts department for their records. An entry is made on the disbursement side of the front office cash sheet (cashiers' report).

If the guest has produced a credit card at check in, the cashier should destroy the guest credit card voucher imprinted at registration when the guest pays the account in full with cash.

- **Foreign Exchange:** The hotel industry is a prime source for the generation of foreign currency (Fig. 27.7) exchange for any country. In a hotel the front office cashier is the authorized person on behalf of the management for foreign exchange dealing. To deal in foreign exchange, a hotel has to take a valid license from the Reserve Bank of India (RBI). RBI issues two types of licenses to deal with foreign exchange—one for the purchase and one for the sale of the

Fig. 27.7: Foreign currency

foreign currency. A hotel with the purchase license can only purchase foreign currency, which means that the hotel can accept foreign currency from the guest, but the refund amount exceeding the billing amount will only be made in Indian currency. Hotels generally obtain the purchase license; but if a hotel has obtained the license for buying and selling foreign currency, it can buy and sell foreign currency, i.e. accept foreign currency and give the balance amount also in foreign currency.

If a foreigner guest pays in Indian rupees, as an advance or in settlement of his account at the time of checkout, the local currency can be accepted if and only if the guest produces encashment certificate to the cashier for his reference.

Procedure for Accepting Foreign Currency

When guests wish to exchange the currency of their country into Indian currency, the following procedure is followed:

- The cashier requests the guest to produce his passport to determine his credentials.
- Confirm that the guest is a resident of the hotel by asking his room no. If the guest is a non-resident, the permission of the lobby manager is obtained.
- Find out the type of currency to be exchanged and determine whether it is exchangeable as per RBI guidelines.
- The cashier checks the exchange rate from RBI or a leading nationalized bank of the town. (Of course, the daily exchange rate should be prominently displayed for guests to see).
- Fill in all the details in the encashment certificate (Fig. 27.8).
- The cashier requests the guest to sign the exchange certificate.
- The guest gives the foreign currency to the cashier in notes or as traveler's cheques.
- If traveler's cheque (Fig. 27.9) is given, ask for guest signature and tally the signatures.

- Calculate the total amount to be paid in local currency (INR) by multiplying the foreign currency amount by the rate of exchange.
- The cashier dispenses the amount to the guest along with the original encashment certificate (as a proof of foreign currency exchange).
- Attach the second copy of the encashment certificate with the notes or traveler's cheque. (The same is sent to RBI.)
- Leave the third copy of the encashment certificate in the book for future reference.
- Fill in the details in the foreign currency control sheet (Fig. 27.10).

Traveler's Cheque

A traveler's cheque is an internationally accepted cheque for a sum in a specific currency that can be exchanged elsewhere for local currency or goods. Issued by a financial institution in various denominations, it functions as cash but is protected against loss or theft.

The purchaser of the traveler's cheque puts two signatures—one in front of the issuing authority, i.e. the bank and the second in front of the encashing authority, i.e. the hotel cashier. There is no danger of them being stolen as they can be encashed only when both the signature tallies.

While accepting a traveler's cheque from a guest for the settlement of bills, the cashier should proceed as follows:

- The cashier asks the guest to countersign the cheque in front of him. A traveler's cheque cannot be enchased if the signature does not tally.
- The cashier requests the guest to produce his passport to determine his credentials.
- A foreign traveler's cheque should be treated as foreign currency and the necessary records, statements and certificates must be maintained like in the case of foreign currency and should be sent to the Reserve Bank of India.

HOTEL XYZ

ECR

(To be issued by Restricted Money-
changer on official letterhead
indicating name and address)

(Valid for three months from the date
of purchase of foreign currency)
RBI Licence No.

Encashment Certificate

Serial No. _____   Date: _____

We hereby certify that we have purchased today foreign currency from Mr/Ms. _____ holder of

Passport No. _____ Nationality _____ and paid net amount in Indian currency after adjusting

the amount towards settlement of bills for goods supplied/services rendered as per details given below:

A. Details of Foreign Currency Notes/Coins/Traveler's Cheques Purchased

Currency purchased (indicating clearly Notes and Traveler's Cheques separately)	Amount	Exchange Rate	Rupee Equivalent
(1)	(2)	(3)	(4)

B. Details of adjustments made towards settlement of bills for goods supplied/services rendered

Bill No./s	Date/s	Amount
(1)	(2)	(3)

C. Net amount paid in ₹ : _____ Amount in words _____ (Total under A minus Total under B)
_____

_____
(Signature of Authorized Official)

┌─────────┐
│ Stamp   │
└─────────┘

Name: _____

Designation: _____

Registration No: _____

Note: This certificate should be preserved by the holder to facilitate the re-conversion of the rupee balance,
(out of the amount stated at C) if any, into foreign currency at the time of departure from India
and/or for payment of passage/freight cost in rupees, if necessary.

Fig. 27.8: Exchange certificate

Fig. 27.9: Traveler's cheque

HOTEL XYZ
SUMMARY OF FOREIGN CURRENCY EXCHANGED

Date: _____
Shift: _____
Cashier: _____

Foreign Exchange Certificate Number	Type of Currency	Amount	@TC	Notes	Indian ₹

Fig. 27.10: Foreign currency control sheet

Personal Cheque

Nowadays, almost no hotels still accept personal cheques (Fig. 27.11) as a method of payment due to the high probability of fraud associated with this method. However, hotels may accept personal cheques from frequent guests that give a high volume of business to the hotel. In case a guest insists on settling his bill by personal cheque, the cashier may request him to get an authorization from the lobby manager. On receiving authorization from the lobby manager, the guest has to

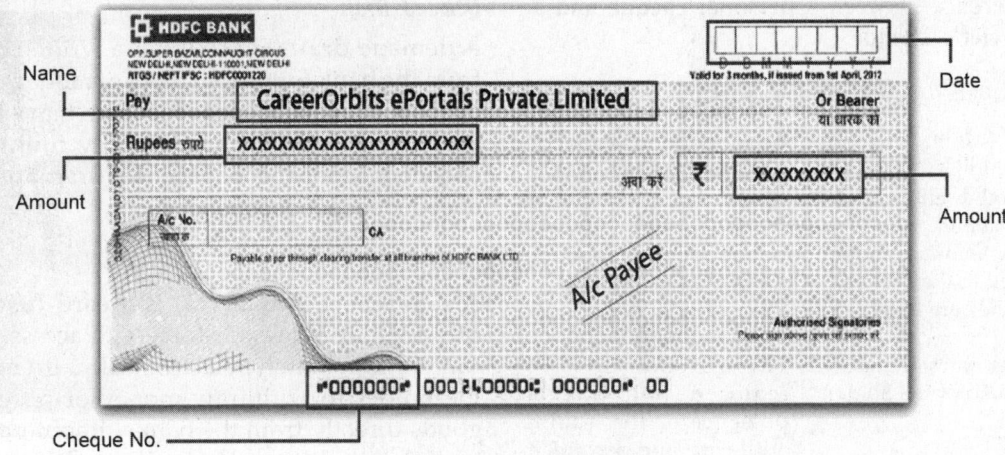

Fig. 27.11: Personal cheque

submit an application for making payment by personal application (Fig. 27.12). The cashier needs to check the following details before accepting personal cheque:

- Cashier must ensure that the date on the cheque is valid.
- The cheque has been marked 'A/c Payee only'.

- The name of the payee should be spelt correctly.
- The amount to be paid must be entered on the cheque; both in words and figures, and the two must tally.
- The drawer must sign the cheque.
- Accept only MICR (Magnetic Ink Character Reader) cheques.

APPLICATION FOR PAYMENT BY PERSONAL CHEQUE

Date: _____

Dear Sir,

I handover to you at Faridabad the Cheque No. _____ dated _____ ₹ _____ (Rupees _____) drawn on (name of bank and station) for value received in cash and/or for the bills submitted to me for my room, meals and other expenses provided in Hotel XYZ.

I further confirm that I have sufficient fund in the above bank and that Cheque No. _____ will be duly honored by the bank when presented for payment.
Yours faithfully,
Signature of Guest

Name: _____

(in block letters)

Room No.: _____

Full Address: _____

Fig. 27.12: Application for payment by personal cheque

Difference between a personal cheque and a traveler's cheque

Personal cheque	Traveler's cheque
For issuing, a person should have a bank account (either current or saving).	No need of any bank account for purchasing and encashing of traveler's cheque.
Any amount can be filled in the cheque as they are blank.	Have a fixed amount printed on its face and available in different denominations.
Only one signature is needed of the holder.	Two signatures are required (one in the presence of the issuing authority and second in the presence of encashing authority).
Ordinary cheques are valid only for 3–6 months.	Valid for indefinite period of time unless dated.
These cheques can be crossed for account payee.	No such provision.
Not safe as someone might force the owner to sign the cheque.	Quite safe because the second signature have to be put in front of the en-cashing authority.
Cheque may bounce as the balance in the account may be less than the cheque	No such possibility as the amount is already printed on the face of the cheque.
No slip/list of lost, damaged or stolen cheques is issued by the bank.	Many banks issue a stop list for stolen and damaged cheques.

Demand Draft

A demand draft (Fig. 27.13) is a written order from the bank for the payment of money upon presentation of the same. A person may obtîin a demand draft from a bank by filling the required form and paying the draft amount and the bank's commission.

Debit Card

A debit card (Fig. 27.14) is a card (usually plastic) that allows customers to access their funds immediately, electronically. It enables the holder to withdraw money or purchase goods directly from the bank without paying by cash or writing a cheque. It can also be used as ATM card, which is used to withdraw money from ATMs (automated teller machines). When the card is swiped, the electronic fund transfer point of sale (EFTPOS) terminal (Fig. 27.15) contacts the computer

Fig. 27.14: Debit card

Fig. 27.13: Demand draft

Fig. 27.15: EFTPOS machine

network of the bank to verify and authorize the transaction. The guest is supposed to enter his PIN (Personal Identification Number) and the amount is debited from the guest's bank account and instantly credited to the hotel's bank account. The guest and the hotel receive a slip notifying them of the details of the transaction.

Credit Settlement

A credit settlement is an arrangement for the deferred payment of goods or services, i.e. a settlement in which the hotel does not receive any payment on the day of departure of the guest but would receive it later.

Credit Card

Credit cards, nowadays, become a preferable method of payment by the guests and hotels worldwide. It is a payment card (usually plastic) that enables the holder to purchase goods and services on credit terms, without feeling the pinch of paying for it immediately (buy now, pay later concept). In addition to obtaining goods, credit cards (Fig. 27.16) can also be used to obtain cash. The account statement is sent to the card holder every month and must be settled in full.

Diners Club, a finance company, can be given the credit of creating the modern credit card system. The most commonly used credit cards in India are Master Card and Visa issued through various banks such as Citibank, Bank of Baroda, SBI, HDFC, HSBC, Standard Chartered, etc. Other than these, organizations such as American Express, Diners Club, etc. also issue credit cards.

The basic concept behind credit card operation is that the member pays an annual fee for membership. The financial institution that floats the credit card gets discounts from the establishment or merchant from which a customer has acquired a sale. It is from these discounts and membership fees that the credit card companies get their revenue. On receiving the bills of customers, the full value is recovered from the customer while what is paid to the establishment is the discounted value.

While processing payment through credit cards (by using EFTPOS machines), the cashier takes care of the following:

• Checks if the credit card is accepted by the hotel.

Fig. 27.16: Credit card

- Swipes the card through the EFTPOS terminal for verification and authorization from the bank. The card number may be manually entered via a numeric keyboard.
- If the terminal displays 'pick-up' card on the screen—the cashier should take the card in his custody and cut it with a scissor and then send to the credit card company to avoid any further misuse of the card. For this, he may be rewarded by the credit card company.
- If the terminal displays 'decline' on the screen—the cashier requests the guest to pay the amount by another card or by cash.
- If the transaction takes place smoothly, the cashier asks the guest to sign on the transaction slip (transaction slip indicates the purchase amount, date, time and other transaction details).
- Verifies the guest's signature on the transaction slip (Fig. 27.17) with the signature on the signature panel at the reverse of the card.
- Returns the credit card and the customer copy of the transaction slip to the guest.
- Towards the end of each month, the hotel send a list to the credit card company/bank of the amounts spent with them by the guests, together with the corresponding credit card numbers. The company/bank pays out the hotel the amount due less a commission which they retain to cover expense, risk and profit. Banks, in turn, collect the debts from the card holders.

Smaller hospitality businesses may still use an imprinter (Fig. 27.18) for credit card transactions. The procedure followed when using an imprinter is as follows:

- Check if the card is accepted in your business and is signed
- Check the validity of the card.
- Refer the warning bulletin for cancelled or stolen cards
- After imprinting the card, check the card numbers are legible on the payment slip
- Write the date and amount
- Check the expiry date and circle it
- If the amount is above the 'floor limit', ring for authorization. (The floor limit or merchant limit is the maximum credit card amount that a business can accept for an individual credit card transaction, above which the business must obtain phone authorization).
- Retain the card until authorization has been granted or declined
- Check the signature after the customer has signed the payment slip
- Return the card and the customer copy of the payment slip to the customer.

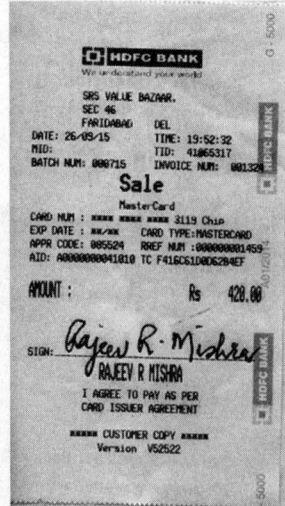

Fig. 27.17: Transaction slip

Fig. 27.18: Credit card imprinter

Corporate Billing/Bill to Company (BTC) Letter

Most of the companies around the world have contractual agreements with the hotels for allowing their executives and clients to register and stay on company letter settlement. The terms and conditions of payment are pre-determined. The reservations are made by the company on behalf of the traveling executives. The executives carry a letter from their company on the company letterhead, which is called a bill to company (BTC) letter, as a proof of their identity (Fig. 27.19).

The cashier proceeds as follows while processing BTC letters:

- At the time of registration, the guest is requested to produce the original copy of the BTC letter and his identity card issued by the company.

- Checks and verifies that the company is listed in the company volume guarantee rate (CVGR) list of the hotel.
- Checks the services that are included in the BTC letter; any service or facility utilized by the guest that is not covered in the BTC letter should be charged separately from the guest.
- At the time of departure, the guest verifies and signs the bill along with all the supporting vouchers and checks out.
- Finally the original copy of the bill duly signed by the guest is sent to the company along with the original copy of the BTC letter for settlement.

Travel Agent Voucher

Tour operators and travel agencies make bookings for a guest's accommodation, food and beverage and other services. Travel

---

Dated: 25th Aug, 2015

Reservations Manager,
Hotel XYZ
ABC Road

New Delhi – 110001

Dear Sir,

Mr Kapil Sharma is an honored guest of our corporation, and during his one night stay at your hotel on 27th Aug, 2015, you are hereby authorized to forward all bills for his stay there to our accounting office. They have been pre-authorized to immediately approve and pay any invoices from your hotel. We have placed a credit limit on the account of Rs 50,000 only. Should this account exceed the billing amount, please contact this office for approval in advance of any additional charges. Thank you in advance for your special consideration of his needs. We hope to make his visit to our company and your hotel most enjoyable and memorable. If we may be of any assistance in this matter, please contact our office as necessary.

Yours truly,

Deepak Dixit

Finance Manager

Dell India Pvt Ltd

Mumbai

Fig. 27.19: BTC letter

agencies receive advance payments from guests and in turn they send travel agent vouchers (Fig. 27.20) to the service provider (hotel), with the details of the billing procedure and the services (meal and accommodation) to be provided to the guest. Similarly airlines (that have contracts with hotels) also send Meals and Accommodation Order (MAO) and Passenger Service Order (PSO) to layover passengers in case of flight delays and cancellations. In such cases, the hotel obtains payments from the travel agencies or the airline.

The cashier proceeds as follows while processing travel agent vouchers/MAO/PSO:

- At the time of registration, the guest is requested to produce the original copy of the travel agent voucher.

- Checks and verifies that the voucher has been issued from a travel agency duly approved by the hotel.

- Checks the services that are included in the voucher; any service or facility utilized by the guest that is not covered in the voucher should be charged separately from the guest.

- At the time of departure, the guest verifies and signs the bill along with all the supporting vouchers and checks out.

- Finally the original copy of the bill duly signed by the guest is sent to the travel agent along with the original copy of the voucher for settlement.

Lotus Trans Travel Private Ltd.
15/28, Okhla Road, New Delhi

Voucher No.: _____
Date: 20.10.15

To

**Reservation Manager
Hotel XYZ**

Dear Sir,

Please provide the following services to our valued client

Accommodation	:	01 Double Room
Guest Name	:	Pawan Kumar
PAX	:	02
Arrival Date	:	15.11.15 Time: 14:00 Hrs
Departure Date	:	16.11.15 Time:10:00 Hrs
Plan	:	MAP
Meal Details	:	15.11.15 (Dinner)
		16.11.15 (Breakfast)

Note: The bills for the above services may be:

☐ Presented to guest for direct settlement

☐ Forwarded to us for settlement

Signature of Issuing Authority
Seal

Fig. 27.20: Travel agent voucher

POTENTIAL CHECKOUT PROBLEMS AND SOLUTIONS

A speedy checkout procedure and an error-free billing leave a lasting good impression on the guest. A guest who leaves the hotel with a good impression will return to the hotel and will also recommend the hotel/chain to his friends and relatives. Despite measures taken by hotels to avoid problems during the checkout process, the following problems may occur during the departure of a guest.

Late checkout

Most hotels have a fixed checkout time—generally 12 noon—at which the departing guests must vacate their rooms and settle their bills. If a guest vacates his room after the checkout time, it is considered late checkout. This may create a problem, especially during high occupancy periods, as the guests with confirmed reservations will be required to wait for the room to be vacated and cleaned.

To minimize late checkouts, the front office should post notice regarding checkout time in conspicuous places such as back of the guest room door, reception, etc. Information regarding the same can be printed on the tariff card, guest registration card, key card, etc. Guests should also be informed about the checkout time and late checkout charges at the time of reservation and also at registration. The practice can be discouraged by adding the late checkout charges in the guest bill.

Improper posting of charges in the guest folio

There are occasions when a guest's financial transactions are not properly posted in the guest folio and the final bill is inaccurate. This might be due to human error or system error. The front desk is flooded with checkout requests at peak checkout times and the cashier might make some mistake in the posting of charges or the calculation of bills. This could lead to a dispute with the guest

and delay other guests in the queue. The whole experience can be quite damaging for the hotel. To avoid this, hotels should install guest accounting systems, which are more accurate and faster, leading to higher guest satisfaction.

Unpaid Account Balances

No matter how carefully the front office monitors the guest's stay, there is always the possibility that a guest will leave without settling his account. Some guests may honestly forget to checkout. The front office may also discover late charges after a guest has checked out. Regardless of the reason, after-departure charges or balances represent unpaid account balances.

- **Skippers:** Guests may leave the hotel without settling their account. These guests are commonly referred to as skippers.

- **Late charges:** This is an outstanding charge which has not been posted into the guest folio and reaches the front desk after the guest has already checked out of the hotel. Restaurant, telephone, room service charges, etc. are examples of potential late charges. Since the guest may not pay for these purchases before leaving, the hotel may never collect for the transactions. Reducing late charges is important to maximizing profitability.

Late charges may be major concern in guest account settlement. Front office cashiers can take several steps to help reduce the occurrence of late charges.

 – Ask departing guests whether they have made any charge purchases or placed long-distance telephone calls, which do not appear on their folio. While most guests will respond honestly to a direct question, many guests may not feel obligated to volunteer information about charges not posted to the folio. These guests will simply pay the outstanding balance on the folio and disregard unposted charges. Guests are frequently unaware that they are responsible for paying unposted charges.

– A front office computer system that interfaces with POS (revenue center outlets) is often the most effective means of reducing or even eliminating late charges. A restaurant point-of-sale interface can instantly verify room account status, check credit authorization, and post charges to the guest's folio—before the guest leaves the restaurant. Similarly, a call accounting system interface can help eliminate telephone late charges. Guests who make telephone calls from their guest room and then go directly to the front desk to checkout will find all their telephone charges listed on their folio. Call accounting system interface will instantly post a telephone charge as soon as the call is completed.

INNOVATIVE CHECKOUT OPTIONS

As more and more guests now want a speedy and queue-free checkout, hotels are coming up with alternatives to the standard checkout. These alternative methods are:

Express Checkout

Guests may encounter line at front desk when checking out during the peak hours. To ease front desk volume, some hotels initiate checkout activity before the guest is actually ready to leave.

A common pre-departure activity involves producing and distributing guest folios to the guests expected to checkout in the morning (usually done before 6 AM). Normally the front office will distribute an express checkout (ECO) form along with the pre-departure folio.

The ECO form (Fig. 27.21) is an authorization by the guest to the hotel authorities to charge the outstanding balance to his credit

card. By signing the ECO form, the guest agrees to pay the amount finalized by the front desk cashier after his departure. Express checkout form may include a note requesting the guest to notify the front desk if departure plans change. Otherwise the front office will assume the guest is leaving by the hotel posted checkout time.

After signing the ECO form, the guest can leave the hotel at his convenience without having to go to the front desk for the standard checkout procedure.

Once the guest has departed from the hotel, the front office must complete the guest checkout by transferring the outstanding guest folio balance to a previously authorized method of settlement.

Because of late charges, the amount due on the guests' copy of the express checkout folio may not match the charges posted to his credit card. This possibility should be clearly stated on the express checkout form to avoid any confusion later on. The hotel mails a copy of the final bill to the guest's mailing address.

Self Checkout

Only fully-automated hotels are equipped with self-service terminals (Fig. 27.22), which allow guests to check-in/checkout promptly by operating these interactive machines.

Generally located in the hotel lobby, the self checkout terminals are used in fully automated and world-class hotels. These terminals are interfaced with the front office computer and are intended to reduce the checkout time at the front desk.

The self checkout terminal looks like an ATM machine. The machine requests the guests to enter their room number and then their credit card number. The guest can access and review their accounts. The checkout is completed when the guest's outstanding balance is posted to his credit card account and an itemized bill is printed and dispensed

EXPRESS CHECKOUT

Thank you for choosing to stay at our hotel.

Enclosed is a copy of your account for your perusal. Any charges incurred after this time will be added to your bill.

Should you wish to utilize our express checkout service, please fill in your details and simply leave this envelope and your room key at the front desk on your departure. A full receipt will be sent to your account address in the next mail.

☐   Check here if you would like your account emailed to you

Email _____

☐   Check here if you do not require a receipt.

Name _____

Room No. _____

Account Address _____

_____

I authorize for all costs incurred during my stay to be charged to my credit card.

Card No. (not necessary if provided on check in)

| | | | | | | | | | | | | |

Card Type: ☐ VISA   ☐ MC   ☐ Amex   ☐ Diners   ☐ JCB

Expiry Date   ☐☐   ☐☐

Signature _____

We look forward to seeing you again.

Fig. 27.21: Express checkout form

to the guest. Room status gets automatically updated at the front desk.

In-Room Folio Review and Checkout

This is an exclusive feature provided by the property management system in fully automated hotels which allows the in-house guest to checkout of the hotel by reviewing their folio and settling their bills through an in-room television set with a remote control device or guest room telephone. The guest can confirm a previously approved method of settlement for the account. The front office then prints a copy of the folio at the front desk for the guest to authenticate and sign at the time of departure. Another advantage is guests can look at their folios at any time during their stay.

Fig. 27.22: Self checkout kiosk

REVIEW QUESTIONS

1. What are the important functions of the checkout and account settlement process?

2. Describe in detail the step-by-step procedure for a group departure process.

3. Express checkout is an excellent way to ensure guest convenience as well as efficient settlement process. Discuss.

4. Give the benefits of self checkout to a guest, explaining the process.

5. Explain the different methods of bill settlement of a guest.

6. Explain the rules followed for foreign exchange.

7. How would you settle guest's bill using traveler's cheque?

8. 'Credit card is the safest method of bill settlement'. Explain.

9. What are the precautions to be taken while accepting traveler's cheque at the time of checkout?

10. Explain the activities carried out at cashier's desk during guest checkout.

11. Explain the procedure of departure and settlement of guest bills.

12. Late checkout and late charges are two of the most common problems during checkout and settlement. Describe why these two happen and how a hotel can control them.

13. Define guest history. How does it help in generating repeat business? Draw the format of guest history card.

14. Updating front office records after a guest's departure is very important. Discuss.

15. Differentiate between:
    - Traveler's cheque and Personal cheque
    - Self checkout and Express checkout

16. Write the precautions to be taken for handling:
    - BTC (Corporate Guest)
    - Credit cards
    - Foreign currency
    - Travel agent voucher

17. Draw the formats of:
    - Express checkout form
    - Guest history card

18. Write short notes on:
    - ECO
    - Encashment certificate
    - Late checkout
    - Late charges
    - Self checkout
    - Zero-out
    - Skipper

Front Office Accounting System

Learning Objectives

After reading this chapter, you will be able to understand the following:

- Front office accounting and its functions
- Types of accounts maintained by the front desk—guest account and non-guest account
- Various folios—master folio, guest folio, non-guest/city folio, and employee folio
- Different kinds of vouchers—visitor's paid-out, miscellaneous charge voucher, restaurant or bar check, cash voucher, telephone call voucher, travel agent voucher, commission voucher, allowance voucher, correction voucher, credit card voucher, transfer voucher
- Types of ledgers—guest ledger and city ledger
- Front office accounting cycle—creation, maintenance, and settlement of accounts
- Accounting systems—manual, semi-automated and fully-automated
- Credit control
- Internal control in the front office
- Terms and Terminology related to front office accounting system

FRONT OFFICE ACCOUNTING AND ITS FUNCTIONS

Accounting may be defined as the process of collecting, recording, summarizing, and analyzing financial transactions of a business. According to the American Institute of Certified Public Accountants (AICPA), 'Accounting is an art of recording, classifying, and summarizing in a significant manner and in terms of money, transactions, and events which are, in part at least, of a financial character, and interpreting the results thereof.'

The resident or in-house guests of a hotel are those who are staying in the hotel. They seldom pay on the spot for the use of hotel services and facilities like restaurants, laundry, spa, barber shop, etc. They sign the

bills to verify their use of the services and the amount incurred. The bills are posted in their folio, and a final consolidated bill is made at the time of their departure. In other words, the hotel provides them charge privileges (credit facility) at the time of the utilization of the hotel services. An accurate posting of the guest accounts is very essential to ensure the guest's goodwill and the recovery of all the charges incurred by the guest.

In order to maximize revenue, hotels offer some services and facilities to non-resident guests as well. These may include the use of health club, swimming pool, food and beverage outlets, games and entertainment, etc. When these services and facilities are offered to non-resident guests on credit, the account of the same is maintained by the front office cashier. As the number of financial transactions that take place between a hotel and its guests (resident and non-resident) is very high, the front desk cashier should maintain the guest accounts accurately and properly.

However, the collection of a non-resident account is not the responsibility of the front desk; it is collected by the accounts department.

The major functions of front office accounting system are as under:

- Creation and maintenance of guest and non-guest accounts accurately
- Tracking financial transactions of guests throughout the guest cycle
- Monitoring the credit limit of guests, and asking for a deposit from guests in case of high outstanding balance
- Preparing a high balance report for collection and informing the management about the same
- Providing an efficient management information system (MIS) to the management for departmental revenue generation
- Maintaining effective control over cash and credit transactions

Need of Front Office Accounting

The aim of the hotel business is to generate profit by providing services like accommodation, food and beverage, and the use of facilities such as health club, swimming pool, beauty parlor, etc. The timely and accurate posting of a guest's financial transactions in the guest account is very important for the successful running of the hotel business as it prevents the loss of revenue for the hotel and thus leads to higher guest satisfaction. To ensure an efficient and error-free billing, the front desk cashier should record the financial transactions properly and accurately in different types of accounts, vouchers, folios, and ledgers and keep them up to date.

An effective accounting system consists of tasks performed during each stage of the guest cycle. During the pre-arrival stage, a guest accounting system captures data related to the type of reservation guarantee and tracks payments made earlier and advance deposits. When a guest arrives at the front desk, the guest accounting system documents the application of room rate and tax at registration. During occupancy, the accounting system tracks authorized guest purchases. Finally, a guest accounting system ensures payment for outstanding goods and services at the time of checkout.

ACCOUNTS

An account is a form on which financial data are accumulated and summarized. It is a record of charges and payments. Adding a charge or payment is called *posting* to the account. A charge that is posted to a customer is called a debit, and a payment is called a credit. When a debit is posted, the amount of the debit is added to the account. When a credit is posted, the amount is subtracted. The increases and decreases in an account are calculated and the resulting monetary amount is the account balance. The value of debits and credits results from the use of double-entry bookkeeping, which is the basis for accounting in all modern businesses.

The most widely used representation of accounts is the T-Account, which summarizes debit entries on the left-hand side and credit entries on the right-hand side (Fig. 28.1).

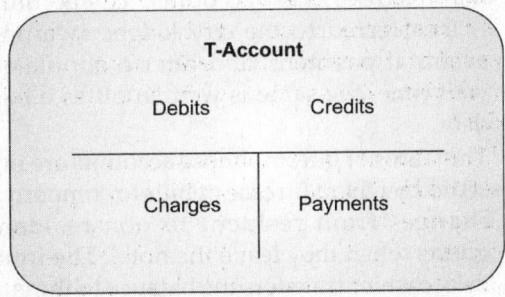

Fig. 28.1: T-Account

Hotels operating under the manual system get use of journal forms to account for different front office accounting transactions.

Types of accounts

In front office, accounts are segregated into two categories:

Guest Account

Guest account is a record of financial transactions, which occur between a resident guest and the hotel. Guest accounts are created when guests register at the front desk or during reservation, when they guarantee their reservations. The front office creates an individual folio for each guest for maintaining a record of all the financial transactions that take place during the stay of the guest. The front office usually seeks payment for any outstanding balance at the time of settlement of final bills or when the account exceeds credit limit.

Non-Guest Account

Non-guest account is a record of financial transactions, which occur between a non-resident guest and the hotel. These are also known as *house accounts or city accounts*. A hotel may extend in-house charge privileges to local residents in order to maximize revenue. *Besides local guests, the front desk cashier also maintains other types of non-guest accounts, which include:*

- Guests who leave the hotel without the settlement of their accounts are known as skippers; their accounts are also treated as non-resident guest accounts. The account is transferred to the city ledger awaiting eventual payment, and after a stipulated wait time, the same is written off as a bad debt.

- The status of guests, whose accounts are not settled by them (in case of bills to company), changes from resident to non-resident guests when they leave the hotel. The front desk cashier transfers the balance to the city ledger and the payment is collected by the accounts department.

- When advance payment has been received for a guaranteed reservation and it is subsequently a no-show, the account is normally recorded in the city sales ledger.

- Hotel employees with charge privileges shall also be opened non-guest accounts.

FOLIOS

A folio is a statement of all transactions (debits and credits) affecting the balance of a single account. When an account is created, it is assigned a folio with an initial balance of zero. All debits and credits are recorded on the folio, and at the time of settlement, a guest folio should be returned to zero balance by cash payment or by transfer to an approved credit card or direct billing account.

All folios shall have a unique serial number for internal control and storing purposes.

Types of Folios

There are basically five types of folios used in front office accounting.

Guest Folio

A guest folio (Fig. 28.2) is created for each guest as soon as the first financial transaction takes place between the hotel and the guest. Generally, a folio is created at the time of the guest registration. In case a guest makes an advance payment at the time of reservation, his folio is created at that time, with the posting of the advance amount in the credit side.

Master Folio

A master folio contains accounts of more than one guest. In small hotels, a master folio contains the record of all the guests staying in the hotel. By glancing through the master folio, one can find the net credit or debit balance on a particular day.

HOTEL XYZ

GUEST FOLIO

Folio No.: _____

Sl. No.: _____

Guest Name: _____

Company: _____

Address: _____

Room No.: _____

Rack rate: _____

Arrival date: _____

Departure date: _____

PAX: _____

Arrival time: _____

Departure time: _____

Date	Ref No.	Particulars	Debit	Credit
20.08.15		Cash advance		20,000
20.08.15		Room rate	10,000	
20.08.15		Ghazal bar	2,000	
20.08.15		Laundry	1,000	

Fig. 28.2: Guest folio

Non-Guest/City Folio

A city folio contains the financial transactions between a hotel and its non-resident guests. A hotel may offer credit facility to local businessmen to attract additional business for the hotel. A city folio is created for credit services, like the use of club facilities, fitness center, health center, sports facilities, etc. offered to guests who are not staying at the hotel.

Employee Folio

An employee folio contains the credit transactions between a hotel and its employees. This folio is created and maintained for employees to whom the hotel has permitted credit/charge purchases. The amount is later collected from the employees or deducted from their salaries, depending upon the hotel policy.

Split Folios

Accounts assigned to a guest on his request to split his charges and payments between two personal folios—one to record expenses to be paid by the sponsoring business company, and the other to record personal expenses to be paid by the guest. In this case, two folios are created for the same guest.

VOUCHERS

A voucher is a documentary evidence of a financial transaction. It depicts the details of the transaction information gathered at the source of the transaction. The voucher is then sent to the front desk for posting it into the guest account. *Some of the commonly used vouchers in the front office accounting are discussed below.*

Visitors Paid-Out (VPO) Voucher

It is a voucher used to support the cash disbursed by the hotel on behalf of a guest and charged to the guest's account as a cash advance. A VPO (Fig. 28.3) is generally made for the following charges:

- Payment for taxi fare
- Movie ticket charges

HOTEL XYZ

VISITORS PAID-OUT VOUCHER

Sl. No.: _____

Date: _____

Room No.: _____

Guest Name: _____

Room Account No.: _____

Detailed Explanation	Amount	
	₹	Paisa

Rupees (in words):   Total:

Signature of Recipient   Approved by: _____   Signature of Cashier

Fig. 28.3: Visitors paid-out voucher

- Florist charges
- Postage and courier charges
- Payment for medical bills

The paid-outs are made from the cash bank that is maintained by the front office cashier. A proper authorization should be taken from the lobby manager before making the VPO.

The VPO voucher has three copies—the first copy is attached to the folio of the guest and given to him at the time of departure, the second copy is sent to the accounts department for verification and the third copy is retained in the books for future reference.

Miscellaneous Charge Voucher

It is a voucher used to support a charge purchase transaction that takes place somewhere other than the front desk, like, laundry, health club, barber shop, etc. The guest verifies and signs the miscellaneous charge voucher, which is sent to the front desk cashier for posting the charges into the guest folio.

Restaurant/Bar Check

Resident guests may enjoy their meals in any of the food and beverage outlets in a hotel. Whenever a guest consumes food or beverage in a restaurant, a bill is raised; in case a resident guest wishes to utilize the credit facility offered by the hotel, he should sign the bill (Fig. 28.4). The signed bills serve as the proof of financial transactions at the food and beverage outlets and are treated as vouchers for posting the charges to the guest folio.

Cash Voucher

It is a voucher used to support a cash payment transaction at the front desk. It is prepared and issued to the person depositing cash as a proof of remittance of the deposited cash (Fig. 28.5)

Telephone Call Voucher

Nowadays, a lot of hotels use computerized systems, where, whenever a guest makes a

HOTEL XYZ
RESTAURANT CHECK

Sl. No.: _____
Bill No.: _____

Guest Name: _____ Room No.: _____ PAX: _____

Date: _____ Time: _____ Table No.: _____

Served by _____

Sl. No.	Particulars	Quantity	Rate/unit	Amount
1.	Yellow Dal	2	150	300
2.	Butter Naan	4	50	200
3.	Shahi Paneer	1	250	250
			Subtotal	750
			Service charge	37.50
			Net amount	787.50

Please do not sign if paying by cash or credit card.   Signature of Cashier

Fig. 28.4: Restaurant check

HOTEL XYZ
CASH VOUCHER

Sl. No.: _____
Date: _____

Received with thanks from _____

the sum of Rupees _____

_____

Cash/Cheque No. _____ Date _____ Bank _____

on account of Advance/Bill No. _____

₹ _____

Receipt valid subject to encashment of cheque

Signature of Cashier

Fig. 28.5: Cash voucher

call, the call accounting module automatically transfers the call charges to the guest folio.

In small hotels, where outgoing calls are routed through the operator, the responsibility of billing the call charges lies with the telephone operator, who puts down the call details on a telephone call voucher (Fig. 28.6) and sends it to the front desk cashier for posting into the guest account.

Travel Agent Voucher

Tour operators and travel agencies make bookings for a guest's accommodation, food

passengers in case of flight delays and cancellations. In such cases, the hotel obtains payments from the travel agencies or the airline.

Commission Voucher

Hotels offer commission to persons who provide regular business to them. Whenever a commission is paid by the cashier, a commission voucher (Fig. 28.7) is made. The commission voucher should be authorized by a competent authority of the hotel. Generally, it is authorized by the lobby manager. More

HOTEL XYZ		
TELEPHONE CALL VOUCHER		

Sl. No.: _____

Date: _____

Room No.: _____

Guest Name: _____   Room Account No.: _____

Detailed Explanation	Amount	
	₹	Paisa
Rupees (in words):	Total:	
	Signature of Telephone operator	

Fig. 28.6: Telephone call voucher

and beverage and other services. Travel agencies receive advance payments from guests and in turn they send travel agent vouchers to the service provider (hotel), with the details of the billing procedure and the services (meal and accommodation) to be provided to the guest. Similarly airlines (that have contracts with hotels) also send Meals and Accommodation Order (MAO) and Passenger Service Order (PSO) to layover

commonly, commission vouchers are made for the following:

- A taxi driver who brings a walk-in guest to the hotel. In case the guest stays at the hotel, the hotel pays a commission to the taxi driver.
- A travel agent/tour operator working on commission basis (generally, the commission is 10 percent on the room rates, excluding taxes).

HOTEL XYZ

COMMISSION VOUCHER

Sl. No.: _____

Date: _____

Name of Recipient: _____

Detailed Explanation	Amount	
	₹	Paisa

Rupees (in words): _____   Total: _____

Signature of Recipient   Approved by: _____   Signature of Cashier

Fig. 28.7: Commission voucher

Allowance Voucher

It is a voucher used to support an account allowance. The cashier prepares allowance voucher (Fig. 28.8) in order to give rebate to the guest due to poor services and products offered to him or to rectify any posting error in the guest folio (of course, after night audit process) which may decrease the overall account balance of the guest. Once a mistake has been overlooked and is posted in the folio, it cannot be corrected, because the revenue has been counted and posted to the revenue journal. In such cases the only way is to adjust off the amount by giving allowance. The front office cashier requires an authorization from a competent authority before giving any allowance.

Allowances can cause irreparable damage to the reputation of the hotel as it may shake the confidence of the guests and create distrust towards the hotel staff.

There are basically four accounts for allowance/rebate granting:

- Room rate: By mistake wrong tariff posted in guest folio (double room charges instead of single or discount not adjusted)
- Telephone charges: When a guest is wrongly charged for a telephone call which he has not made or the call did not mature.
- Unsatisfactory service: When the guest is not satisfied with the service, he will be annoyed and may refuse to pay. Hotels are liberal in giving such allowance but take drastic action against the employees who are found guilty.
- Wrong billing/posting: When the guest is charged for what he has not availed (like laundry, tea, parlor, etc).

In case if the allowance is given on taxable charges, suitable adjustments of taxes is required to be effected by way of inclusion of the proportionate amount of tax on such charges. For example, if the hotel levies 10% sales tax on food and ₹ 50 is incorrectly charged for food, which the guest did not consume, then an allowance for ₹ 50 plus 10%

HOTEL XYZ
ALLOWANCE VOUCHER

Sl. No.: _____

Date: _____

Room No.: _____

Guest Name: _____   Room Account No.: _____

Detailed Explanation	Amount	
	₹	Paisa

Rupees (in words):   Total:

Prepared by: _____   Approved by: _____   Checked by: _____

Fig. 28.8: Allowance voucher

tax, i.e. a total of ₹ 55 will have to be credited to the guest account by way of allowance. Such adjustment does not arise in case of non-taxable charges.

Correction Voucher

It is a voucher used to support the correction of a posting error which is rectified before the close of business on the day the error was made (Fig. 28.9). An account correction can either increase or decrease an account balance, depending on the error. For instance, an account would need to be adjusted if the front desk inadvertently posted a lower than normal room rate for a particular guest room. In this instance, the account correction would increase the guest's folio balance. If higher than normal room rate had been accidently posted, then the account correction would decrease the guest's folio balance.

Credit Card Voucher

It is the form designated by a credit card company to be used for imprinting a credit card and recording the amount charged.

Transfer Voucher

It is a voucher used to support a reduction in balance on one guest folio and an equal increase in balance on another. For example, when one guest offers to pay a charge posted to another guest's folio, the charge will need to be transferred from the first account to a second account.

LEDGER

A ledger is a summary grouping of accounts. A front office ledger is a collection of front office account folios. The folios represented in the front office are a part of the front office accounts receivable ledger. An accounts receivable represents money owed to the hotel. *There are normally two ledgers maintained by the front office.*

Guest Ledger

A guest ledger contains the details of all the financial transactions between a resident guest and the hotel, including charge purchases and the payments received from the guest. It has

```
┌─────────────────────────────────────────────────────────────────────┐
│                          HOTEL XYZ                                    │
│                     CORRECTION VOUCHER                                │
│                                          Sl. No.: _____      │
│                                          Date: _____         │
│                                          Room No.: _____     │
│                                          Room Account No.: _____  │
│  Guest Name: _____                                           │
├──────────────────────────────────┬────────────────────────────────────┤
│      Detailed Explanation         │             Amount                │
│                                   ├──────────────────┬─────────────────┤
│                                   │        ₹         │      Paisa       │
│                                   │                  │                  │
├──────────────────────────────────┴──────────────────┴─────────────────┤
│  Rupees (in words):                        Total:                     │
├─────────────────────────────────────────────────────────────────────┤
│  Prepared by: _____   Approved by: _____   Checked by: _____  │
└─────────────────────────────────────────────────────────────────────┘
```

Fig. 28.9: Correction voucher

two parts—debit and credit. The guest ledger is also known as *transient ledger*, or *front office ledger* or *rooms' edger*. In a manual system, the financial transactions are recorded in a tabular ledger, or tab ledger, which is of two types:

- **Horizontal Tabular Ledger**

  In a horizontal tabular ledger (Fig. 28.10), all the credit expenses of the guest are recorded in one horizontal row, and at the end of the row, the guests' credit or debit balance is shown. The vertical row of the table contains the room numbers. At the end of vertical column, the daily sales balance can be seen.

- **Visitors/Vertical Tabular Ledger (VTL)**

  The visitors' tabular ledger (Fig. 28.11) also known as the 'tab' has been accepted throughout the world as the most

Day: _____							Date: _____						
Room No.	Guest Name	PAX	Room Rate	B/fwd	Room Rent	B/Fast	Lunch	Dinner	Phone	Misc. Expenses	VPOs Total	Credit	C/fwd
101	Akshat	02	2000										

Fig. 28.10: Horizontal tabular ledger

satisfactory method of keeping the accounts of the hotel guests. It is an accounting system usually used in small hotels. As the name implies, this is a ledger in tabular form, usually loose leaf sheets, recording the daily transactions the hotel has with its guests. The entries are made from checks sent from the various department and they should be made as soon as they are received. This is to ensure that the visitors' accounts are always kept up-to-the-minute. It is from these entries on the 'tab' that the personal bill is prepared for presentation to the guests, and can be produced upon demand with the minimum delay.

The main features of the VTL are:

– A number of personal accounts of guests are arranged on the same sheet in tabular form for easy reference.

– Narration is reduced to the minimum through the use of printed description applicable to all accounts on the sheet.

As the cycle of the hotel operations is repeated daily, the guests accounts are balanced daily and one or more sheets of the ledger relate to a day's business.

Writing of VTL

Write the guest name in capitals, legibly in ink. Record all details of the transactions immediately. VTL has two sides—Dr and Cr. All guest charges are written on debit side and cash deposits on credit side. Service charges and sales tax should be written in appropriate column. At the end of the day the total for that day must be made and to this any brought over of previous day must be added to get the grand total. This amount is carried over to next day (if not paid by the guest).

Recording transactions in the VTL is as simple and similar as recording in GWB. Each guest weekly bill must tally with each guest account appearing in the VTL on a particular day. The VTL can either be in a vertical form or in a horizontal form. In case of vertical form, all the charges to the guest are recorded vertically, whereas in case of horizontal form, all charges are recorded horizontally.

Advantages

• Each guest name and room number appears in the sheet in one column and all the transaction made by him is recorded in the same line.

• At any time his account can be noted.

• Since personal account is debited directly, fewer mistakes are made and it is a fast system.

• In case of any controversy cross-checking can be done through check/voucher.

• Income can be bifurcated department-wise.

• Credit limit can be checked.

Disadvantages:

• Size is too big.

• Since columns are many and short, possibilities of wrong posting are there and the location of mistake becomes difficult.

• Not useful for bigger hotels.

City Ledger

It is also called the *non-guest ledger*. A city ledger contains the collective accounts of all the non-resident individuals/agencies to whom the hotel extends credit facility.

Typical city ledger account includes:

• All non-cash settlements (credit card payment accounts, bill to company accounts, travel agency accounts, etc.)

• Bad cheques account (bounced cheques of guest)

• Skippers' account—guests who left the hotel without settling their account.

• Disputed bills account—for guests who refuse to settle their account (in part or in full) because of a discrepancy.

HOTEL XYZ

VISITORS' TABULAR LEDGER

Date: 10 Nov 2015

Registration No.		1820	1821				TOTAL
Guest Name		Mr. Akshat	Mr. Rishav				
Arrival date		9 Nov 2015	9 Nov 2015				
Arrival time		11:00	13:00				
Departure date		11 Nov 2015					
Departure time		9:00	11:00				
Plan	PAX	CP 2	MAP 1				3
Room rate		8000	6000				
Room No.		**1101**	**1102**	**1104**	**1201**	**1202**	
Opening balance Dr/Cr							
DEBIT ENTRIES							
Apartment		8000	6000				14000
Guest charge							
Breakfast		650					650
Lunch		1500					1500
Dinner		1500					1500
Coffee/Tea		100					100
Room service		425					425
Restaurant							
Bar							
Coffee shop							
News stand							
Telephones		200					200
Laundry		350					350
Other charges							
Visitor paid out		250					250
Luxury tax							
VAT		175					175
Service charge		250					250
(a) TOTAL: Op balance+ Debit charges		13400	6000				19400
CREDIT ENTRIES							
Allowance/Discount							
Advance /Deposit		10000					10000
Transfer to Guest A/c							
Transfer to City Ledger							
(b) TOTAL: Credit entries		10000					10000
CLOSING BALANCE: (a) – (b)		3400	6000				9400

Fig. 28.11: Visitors' tabular ledger

- Retention charges account from guaranteed reservation, from DNA guests
- Late charges account (the amount that could not be posted before the check out of the guest).

An efficient method for keeping track of post due accounts should be used. For example, age analysis method. It provides a quick glance how old is the outstanding account.

In most hotels, accounts that are less than 30 days old are considered *current*. Accounts which are older than 30 days are considered *overdue*. In some cases, accounts which are older than 90 days are considered *delinquent*.

Analysis of the outstanding bills according to the length of time, for which they are outstanding, is called *account ageing*. A statement which shows how old the unpaid account has become is *ageing statement* (Fig. 28.12).

FRONT OFFICE ACCOUNTING CYCLE

An important function of the front office accounting system is to maintain an accurate and up-to-date record of all the financial transactions (credit and debit) between the hotel and each guest, so that all the outstanding accounts are settled and the hotel does not lose any revenue. *The front office accounting cycle has three distinct phases:*

Creation of Accounts

A guest account is created when the first financial transaction between the hotel and a guest takes place. It may happen at one of the following stages:

- At the time of reservation, if the guest pays an advance amount (may be part or full).
- At the time the hotel receives the advance payment for a booking—after the reservation has been made and before the arrival of the guest.
- At the time of guest registration, when a room is allotted to the guest.

Maintenance of Accounts

All the monetary transactions that take place between the hotel and a guest are recorded in the guest folio in the order of their occurrence. For creating the folio, the necessary information is taken from reservation and registration records.

Manual and semi-automated systems commonly use pre-numbered folios for internal control purposes, and the folio number is usually entered onto the guest registration card for cross-referencing. In a computerized system, guest information is automatically transferred from an electronic reservation or registration record and entered onto an electronic folio, which is cross-referenced automatically with computer-based records within the front office system.

Aged Accounts Receivable as on 20 Aug, 2015

Name	Balance	Current	Outstanding (in days)			
			30–60	60–90	90–120	120+
Akshat	51,000					51,000
Pawan Kumar	65,000			65,000		
Sambhavi	40,000	12,000	28,000			
Shreya	45,000				45,000	

Fig. 28.12: Aged accounts receivable report

An entry in the guest folio may be either debit or credit.

The most common debit entries in a guest account include the following:

- Room charges
- Food and beverage charges (restaurant, bar, coffee shop, room service, etc.)
- Telephone and fax charges
- Health center, business center, fitness center charges
- Laundry charges
- Postage charges
- Transportation charges
- Visitors paid-out

Credit entries in a guest account may include the following:

- Pre-payment, in part or in full (at the time of reservation or between reservation and arrival).
- Part payment during the stay.
- Allowances given to the guest.
- Adjustments made in case of any error in posting in the guest folio.
- Final payment for the settlement of accounts at the time of checkout.

Settlement of Accounts

This is the final and concluding phase of the front office accounting cycle. The settlement of account means zeroing the balance in a guest folio. The formula for calculating the outstanding balance is:

Opening balance + Debit entries – Credit entries = Outstanding amount

At the time of departure, the final bill of the guest is prepared and settled in such a way that the outstanding balance is brought to zero. The settlement of the guest account may be by cash or credit. In case of credit settlement, the account balance is transferred to the city ledger and the responsibility of collecting the balance is transferred to the accounts department.

ACCOUNTING SYSTEMS

Non-automated system

Guest folio in a manual system contains a series of columns listing individual debit/credit entries accumulated during the occupancy.

At the end of the business day, each column is totaled and the closing balance is carried forward as the opening folio balance for the following day.

Semi-automated system

Guest transactions are printed sequentially on a machine posted folio. The information recorded for each transaction includes the date, department code/cheque no., amount of the transaction and the new balance of the account. A column labeled 'previous balance pick-up' provides an audit trial within the posting machine framework that helps prove the current outstanding balance is correct.

Automated system

Electronic folios are maintained on the computers. POS transactions may be automatically posted to an electronic folio. Computer-based systems maintain accurate current balances for all the folios.

CREDIT CONTROL

Credit control refers to the various measures taken by a hotel to ensure that guests settle their accounts in full at an agreed time. For a hotel to be successful and profitable, it is necessary for it not only to be busy, but also to receive all the revenue owed to it by resident guests. To accomplish this, a hotel requires a series of stringent credit control measures.

The front office must monitor guest and non-guest accounts to ensure that they remain within acceptable credit limits. Carelessness on the part of the front office cashier may not only make it difficult to recover unlimited unpaid accumulated charges from the guest but may also let off the guest without payment of such charges which may be beyond his capacity to pay. *The hotel may adopt following credit control measures*:

- **House limit:** This is a credit limit established by the hotel for the guest and thus is the maximum limit up to which the hotel can accept charge privileges for the guests. Credit limit of a guest account depends upon the credibility of the guest and the floor limit of the guest credit card.
- **Charge privilege/credit facility:** Facility given to a guest to charge expenses to his account during his stay. Guests who present an acceptable credit card at registration may be extended credit facility equal to the floor limit authorized by the issuing credit card company.
- **APC (All Payment Cash):** For a walk-in guest with light or scanty baggage, payment in advance is normally requested, which may be equal to one or more days room charges, depending upon the house rules. These guests are often denied in-house credit or charge privileges due to their chances of being skipper. A note of APC (All Payments Cash) to all POS (Point of Sale) is sent.
- **High balance account/High risk account:** This is an account of the guest whose charge purchases are either approaching the credit limit or have exceeded the credit limit decided for the guest by the hotel at the time of registration.
- **High balance report:** This is a report which summarizes all the high balance guests and the amount due on them for settlement (Fig. 28.13). It is prepared by night auditor after screening all guest folios. The front office cashier can ask the guest for partial or full settlement of the outstanding balance in order to re-establish charge privileges (Fig. 28.14).

INTERNAL CONTROL IN THE FRONT OFFICE

In front office, the main purpose of internal control is to track transaction documentation, verify account entries and account balances, and to identify vulnerabilities in the accounting system. The keyword to internal control is auditing.

Auditing is the process of verifying front office accounting records for accuracy and completeness. Each financial interaction produces paperwork, which documents the nature and amount of the transaction, and these documents should be checked to ensure that proper postings have been made to the correct accounts. *Certain instruments used to*

HOTEL XYZ			
GUEST LEDGER HIGH-BALANCE REPORT			
Auditor: _____		Date: _____ Reviewed by: _____	
Room No.	Guest Name	Amount	Action Taken
1101	Akshat	25,000	Left message for the guest to contact
1204	Rishav	29,000	Contacted and received a draft of ₹ 22,000
1408	Shreya	35,000	Guest will be settling his bills partially today evening
1506	Sambhavi	31,000	Left four messages for the guest but could not be contacted. Action necessary

Fig. 28.13: High balance report

HOTEL XYZ

NOTIFICATION OF HIGH BALANCE

Room No. _____   Date _____

Dear _____

I would like to notify you that expense up until _____ are _____

This amount is in excess of the level of credit the hotel normally extends to its guests. We would, therefore, be grateful if you would contact our duty manager on _____ or the front office manager to establish how you wish to settle your account.

Yours sincerely,

_____

Front Office Manager

Fig. 28.14: Notification of high balance

exercise control in front office cash are described below:

- **Front office cash sheet/Cashier's report:** The front office is responsible for a variety of cash transactions affecting both guest and non-guest accounts, and front office cashier may be required to complete a front office cash sheet (Fig. 28.15) that lists the receipt and disbursement of cash to and from the front office cash section. The information contained on a front office cash sheet is used to reconcile cash on hand at the end of a cashier shift with the documented transactions, which occurred during the shift.
- **Cash bank/Cash float/House bank:** A second set of front office accounting control procedures involves the use of front office cashier banks. A cash bank is the amount of cash assigned to the front office cashier by the accounts department to carry out the various transactions smoothly during a particular work shift. Front office cashier requires cash for encashment of foreign currencies, giving change to guests, VPO, cash refund against deposit made at the time of check-in. Cashiers should sign for

their bank at the beginning of their shift, and only the person who signs for the bank should have access to it. During the day, the front office cashier collects cash and cheques for various transactions from different guests and also pays cash for various reasons (mentioned above). At the end of the day, front office cashier separates all excess money from the cash float and hands it over to either the night auditor or to the accounts department, duly sealed in envelope (Fig. 28.16), by recording it in the log book in the presence of a witness and then placing it in the front office vault. This amount is called *turn in*. Cash float requires periodic check to ensure that the cash is not being misused by the staff and is properly accounted for. Many times the front office cashier finds cash in excess of his float, which is referred to *overage* and sometimes he may find shortage in his cash float, which referred to as *shortage*. Neither overage nor shortage is considered good.

Net Cash Receipt = Amount of all cash, cheques, and other negotiable instruments in cashier's drawer – amount of the initial cash bank + all paid outs

HOTEL XYZ
FRONT OFFICE CASH SHEET

Shift: _____

Cashier: _____

Cash Receipts				Indian Residents	Foreign Nationals and Non-Residents			Exemption Under Clause No.	Room No.	Name	Expla-nation	Cash Disbursements		
Bill No	Room		Name		Received in Indian Currency	Received in Foreign Currency	Received in Indian Currency Under Exempted Category					Guest Ledger	City or Misc Ledger	Paid Out Ledger no
	In Advance	On Account	On Departure											

Fig. 28.15: Front office cash sheet

- **Due back:** This is a typical situation which occurs when the front office cashier is not able to restore his cash float because he pays out more than that he actually receives in cash. As soon as due backs are noticed, the cash float is replaced to its original limit by accounts department. Such a situation may arise when a cashier encashes large amount of foreign exchange offered by guests during a shift, whereby, a large

amount of outflow of cash from the bank takes place.

- **Audit control:** Apart from the above mentioned measures to verify correct proceedings in the front office cash, internal auditors may make unannounced visits to the cashier's desk for the purpose of auditing accounting records, as well as conducting spot-checks of the cash bank of the cashier on duty. A report is duly completed for management and ownership review.

TERMS AND TERMINOLOGY

Point of Sale

The term point of sale describes the physical location at which goods or services are purchased. Any hotel department that collect revenues for its good or services is considered a revenue center and, thus a point of sale. Large hotels typically support many points of sale (e.g. restaurants, bar, laundry, gym, etc.).

Some hotels offer guest operated devices that also function as points of sale, e.g. in-room movie and in-room vending systems.

The volume of goods and services purchased at scattered points of sale within the hotel requires a complex internal accounting system to ensure proper posting and documentation of sales transactions. An electronic transfer ensures this, under the fully automated system. Under manual and semi-automated systems, posting shall be done by a physical submission of different vouchers to the front office department. When posting charges, the following items shall be considered:

- Voucher transaction number
- Amount of the charge
- Name of the point of sales outlet
- Room number and name of the guest
- Brief description of the charge
- Guest signature and employee identification

CASHIER'S REPORT ENVELOPE			
Day_____ Date _____			
Cashier_____			
Deptt. _____			
Shift_____AM/PM TO _____AM/PM			
CURRENCY	5.00		
	10.00		
	50.00		
	100.00		
	500.00		
	1000.00		
COINS			
PAID OUTS :			
VOUCHERS AND CHECKS			
TOTAL AMOUNT ENCLOSED			
NET RECEIPTS			
DIFFERENCE			

Fig. 28.16: Cashier's report envelope

Some POS systems allow the swipe of a guest room key as sufficient verification for posting a charge.

No-Post Status

A term used to indicate a guest who is not given charge privilege facility which means he has to settle all his bills at various POS in cash.

Chance Sales

It is a term applied to sales of any point of sale to non-resident guests. It is commonly applied to the local guests who use the hotel facility without taking up accommodation in the hotel. These sales may be cash or credit sales.

Discount

These are concessions in original charges, offered either to stimulate sales and service or to ensure payment within stipulated time. The rate or amount of discount is normally fixed or determined in advance, i.e. pre-planned. Discounts are pre-requisite for sales promotion.

Concessionaires

In hotels, some of the services are not under the direct control of the management. Some hotels, in order to avoid the hassles, may entrust the operations of florist, travel desk, book shop, etc. to outside agencies. This is done by the hotel by renting the space in the lobby to experienced agencies to operate such departments/outlets for the hotel, entirely independent of the hotel management, supervision and control, so long as they conform to the hotel policies. These outlets which are leased by the hotel are referred to as concessions and the persons who enjoy such privileges of operating their business in the hotel premises are called concessionaires. The hotels are compensated by the concessionaires by way of rent or commission on sale or by any other manner, depending upon the contract between the management and the concessionaires.

Service Charges

Some hotels may follow the practice of charging on the bills of the guest in order to save the trouble of having to give tips to the individual staff for services rendered. In some hotels, all charges incurred by the guest are subject to service charges, whereas in others, they are levied on some specific items depending upon the policy of the management. Service charges must be brought to the notice of the guests. Service charge is not revenue of the hotel but mere recovery on behalf of employees, which is distributed later among the employees of the hotel or restaurant as the case may be.

Bucket Check

Internal control requires cross reference and cross tallying of the front office folio charges with the records of checks maintained at point of sales in respect of the charges originating there at and entering the same in the front office guest folio. In the process, discrepancies or error may creep in. The cross tallying of charges, referred to as bucket check, helps in detecting the errors and rectifying the same promptly, thereby setting right, the folios and records. Bucket is a container or provision for storing the front office guest folios, and therefore, bucket check is checking of entries in the folios stored in the bucket with that of the sources of entries appearing in the folios.

Split Billing

An arrangement whereby a guest's charges are separated into two or more folios. *Split billing can be of two types:*

- Vertical split: When the room and charges are shared by two guests, who would like to have their bill split into two identical halves.

- Horizontal split: When certain charges are paid by the company or agency and the rest are to be borne by the guest himself, the split is made on the basis of charges.

Tips to Employees

It is the amount that is paid by the cashier to an employee of the hotel at the request of the guest. When the guest is given the service check at any food and beverage outlet and he wants to give some tip to the waiter, he adds the amount of tips to the check when signing it. The cashier separates the food charges and tips amount and posts the tips amount under cash advance category and food charges under food and beverage department account. The cashier pays the tips amount to the waiter after he puts his initials on the cash advance voucher. Hotels find it difficult in case of credit card settlement by the guest because the tips amount which is paid to the waiter is reimbursed may be after a few weeks, and these ties up hotel funds. The tips amount is not recorded in the front office cash sheet and must be separated from items on the folio that are recorded in and eventually reconciled to front office cash sheet.

REVIEW QUESTIONS

1. Describe the importance of front office accounting system.
2. What is VTL? Draw a VTL and discuss its advantages.
3. What are different types of Accounts?
4. Explain the different types of vouchers prepared by the front office.
5. What are the four types of folios commonly used in front office accounting?
6. What policy is adopted by the front office for monitoring credit?
7. What is credit control and what are the measures to monitor credit control system in hotels?
8. What is the basic front office accounting formula?
9. List at least five types of accounts that come under city ledger other than the non-guest accounts. What common difficulties a hotel faces in collecting city ledger accounts?
10. What items are recorded in a front office cash sheet? How does a cash sheet help ensure internal control in the front office?
11. Explain the reasons and circumstances that result in:
    - Hotel asking the guest for partial pre-payment
    - Hotel classifying a guest as 'no post status'
    - A corporate customer asking for a split folio
    - A hotel categorizing an account as delinquent
12. Write a note on the following with illustrations of vouchers.
    - Account correction
    - Account allowance
    - Charge purchase
    - Cash payment
13. Differentiate between:
    - Guest folio and Master folio
    - Net outstanding balance and Net cash receipt
    - Shortage and Due back
    - Cash bank and Due back
    - Shortage and Overage
    - Master folio and Split folio
    - Visitors paid out and VTL
    - Floor limit and House limit
    - Guest account and Non-guest account
    - Folio and Voucher
    - Guest ledger and City ledger
    - Account correction and Account allowance
14. Draw the formats of:
    - Correction voucher
    - Front office cash sheet
    - Visitor's paid out
    - High balance report
    - Miscellaneous charge voucher
    - Transfer voucher
    - VTL
15. Write short notes on:
    - Guest account
    - No post
    - Guest ledger
    - City ledger
    - Charge privileges
    - High balance account
    - POS
    - Split folio
    - Accounts receivable
    - Voucher
    - Folio
    - Cash bank
    - Credit monitoring
    - Internal controls at front office

Night Auditing

Learning Objectives

After reading this chapter, you will be able to understand the following:

- Night audit—its purpose and usefulness
- Night auditor—duties and responsibilities
- Night audit process—establishing the end of the day, completing outstanding postings and verifying transactions, reconciling transactions, verifying no-shows, preparing reports, deposit cash, updating the system, distribute reports
- Operating modes for night audit—manual, semi-automated and fully-automated
- Common errors during the night audit
- Terms and Terminology related to night auditing

NIGHT AUDIT

According to Oxford Advanced Learner's Dictionary, audit is 'an official examination of business and financial records to see that they are true and correct'. An audit is generally carried out at the end of every financial year in most businesses. Since hotels operate on 24 × 7 × 365 pattern, continuous auditing is a pre-requisite to safeguard the loss of revenue that may occur due to any errors in the posting of charges in the folios.

Since hotel business is a round-the-clock business, the auditing is carried out during slack time; normally this time is between midnight and early morning, hence the audit is known as night audit.

The process of verifying and providing the accuracy and completeness of guest and non-guest account against departmental transactions reports is called night audit.

Night audit is actually the audit process of taking inventory of the day's work. In other words, it is the activity of checking and confirming that whatever transactions have been during the day are correct and complete. It helps in calculating the total revenue generated for that day. The night auditor also monitors guest credit limits, balances all accounts, sorts out any discrepancies in room status and prepares reports for the management.

The night audit requires attention to accounting detail, procedural controls, and guest credit restrictions.

NIGHT AUDITOR—DUTIES AND RESPONSIBILITIES

A night auditor is the person who audits the hotel accounts daily at night. The audit team generally comprises members of the accounts department. The number of people in the audit team depends upon the size of the establishment, its accounting practices and the use of property management system.

A night auditor should be a skilled book-keeper as he is required to track all the financial transactions between the hotel and its guests and to calculate the total revenue generated during the day. A night auditor should also possess the skills of a receptionist, as in many small and medium hotels, he may be required to carry out the check-in/ checkout function at night.

Night auditors monitor the current status of guest accounts vis-à-vis the credit limits, and verify discounts, allowances, and promotional programs that are offered to guests. They prepare reports about the front office operation for the management. Fully automated hotels may not require a team of night auditors as most of the functions that a night auditor performs are carried out automatically by the computerized system, but a person is still required to physically verify the accounts and vouchers.

Does the night auditor belong to the accounts department or is he an integral part of front office?

This will depend on the organizational setup of the hotel, but there are some instances where the night auditor is responsible to the accountant or financial controller—especially where the auditor is simply employed to run the night audit and perform audit checks.

In cases where the auditor also performs front office tasks—including check-in, guest cashiering and checkouts he will almost certainly report to the front office manager.

A night auditor performs the following duties and responsibilities:

- Establishes the end of the day.
- Ensures the accuracy of front office accounting records and balances them.

- Reconciles all the financial transactions between a hotel and its guests.
- Calculates the total revenue generated during the day.
- Verifies and validates the cashier's posting of charges in the guest and non-guest accounts.
- Posts the room charges and taxes in the guest folio.
- Transfers the unpaid guest accounts, i.e. accounts of guests who have left the hotel without settling their bills—to city ledger.
- Monitors the credit limit of guests.
- Prepares a high balance report of guest accounts exceeding house limit.
- Monitors the current status of discounts, meal coupons, and other promotional activities that are carried out by the front desk employees.
- Prepares important reports which detail the results of operations for the management of the department and the hotel at large.
- Tracks room occupancy percentage and other front office statistics.
- Prepares a daily summary of all the cash and credit card settlements that took place at the front desk.

NIGHT AUDIT PROCESS

The night audit is conducted in every hotel to maintain an accurate and efficient accounting system that keeps proper records of all the transactions. This prevents the loss of revenue as well as leads to higher levels of guest satisfaction (as guests are presented error-free and up-to-date bills). The night audit is conducted on a daily basis as the creditors (hotel guests to whom the hotel extends credit facility) are mostly unknown and if their accounts remain unsettled due to some error in bookkeeping, the hotel will lose that revenue. The daily authentication of accounts, by verifying support documents and the posting of charges in the guest accounts, is necessary to prevent any possible loss of revenue.

The night audit process is complete when the totals for guests, non-guests and departmental accounts are 'in balance' or proven correct and is not showing any 'out of balance' position. As long as the audit process presents an out of balance position, the audit is considered incomplete. An out-of-balance position occurs when the charges and credits posted to guest and non-guest accounts throughout the day do not match the charges and the credits posted to the departmental revenue sources. An out-of-balance condition may require a thorough review of all account statements, vouchers, support documents and departmental source documentation.

The steps that are commonly involved in night audit process are as follows.

Establishing the End of the Day

The end of the day is simply an arbitrary stopping point for business day. Typically the business day ends when the night audit begins. Usually the period from 11 p.m. (when the audit work starts) to the time when audit work is completed is called 'audit work time' and any transactions during this period are posted in the next business day. For example, if a guest checks-in at 11.15 p.m. at a hotel where the night audit begins at 11 p.m., the guest's account is included in the next day's business.

Completing Outstanding Postings and Verifying Transactions

One of the primary functions of the night audit is to ensure that all transactions affecting guest and non-guest accounts are posted to appropriate folios before the end of the day. It is important to accurately post and account for all transactions on the day these occurred. This actually means waiting until all food and beverages outlets, including the banquets facilities are closed. Incomplete posting will cause error in account balancing and complicate summary reporting.

The night auditor physically verifies all the financial transactions by cross examining the account with supporting vouchers like credit vouchers, debit vouchers, visitors paid-out vouchers, allowance vouchers, etc. There might be errors in the posted charges, which should be caught by the night auditor. The night auditor should catch this mistake and correct it, saving the hotel from a lot of embarrassment and from the trouble of amending many records. The night auditors verify the postings in all accounts by going through all the documents carefully. They also validate the discounts given to guests. They examine and authenticate all the transactions that involve monetary elements.

Reconciling Transactions in Guest Accounts, City Accounts, and Points of Sale

The next step is the reconciliation of each financial transaction with the original source documents (generally vouchers). The reconciliation of accounts may be carried out in the following sections:

- **Guest accounts**

  All the financial transactions that occur between a hotel and its guests are examined by the night audit team. The folio of each guest is matched with the original documents of transactions for the verification of the account. Before the end of the day the night auditor also checks for any discrepancy in room status reports of the front office and housekeeping departments. Errors in room status can lead to loss of room revenue.

  The reconciliation of charges in guest accounts is done in the following order:

  – Room charges: The night auditor posts the room charges and room taxes in the guest folio at the beginning of the night audit process. The basis of posting of the room charges depends upon the house customs—check-in/checkout basis or twenty-four hour basis. If rates charged are less than rack rates, check if the rates are discounted? Is the discount correct?

– Food and beverage charges: The food and beverage charges posted in the guest folio by the front desk cashier are verified by the night auditor against the restaurant checks, which bear the guest's signature as a token of the guest's acceptance that the services were availed by him and that the amount is correct. Every such entry in the guest folio is reconciled by the night auditor.

– Other charges: The night auditor also reconciles other financial transactions that take place between the hotel and its guests such as visitors paid-out, miscellaneous expenses, discounts, etc.

• **City accounts**

Night auditors also examine the entries in non-resident guest accounts or city accounts. They check these accounts for completeness and accuracy. If any instance of high outstanding balance is found during the night audit process, the night auditors prepare the report of high balance, which is presented to the management, and the responsibility of collecting payment (part/full) shifts to the accounts department.

• **Points of sale**

The cashiers of all the points of sale (POS) prepare a summary sheet of financial transactions that occur at their POS and keep the same in an envelope, along with the second copy of bills. The same envelope, along with the cash and credit transactions, is generally deposited with the cashier at the front desk cash and bills section. The night auditor pulls out all the cheque/vouchers (including void cheque) of individual POS and reconciles them with the POS cashiers' sales summary sheets.

Verifying No-shows

A guest who does not arrive on the scheduled date of arrival after making a room reservation is called a no-show. The night auditor verifies

all such no-shows, and in the case of guaranteed reservations, posts the retention charges in the guests' folios. Usually, hotels charge one night's retention for no-shows.

In posting retention charges the night auditor must be careful to verify that the reservation was guaranteed and the guest never registered with the hotel. Sometimes duplicate reservations may be made for a guest. If the front office does not identify this, the guest may actually arrive but appear to be a no-show under the second reservation.

Preparing Reports for the Management

The next step in the night audit process is to prepare daily reports for managerial use, which help the management to review the profitability of the hotel operations and plan future goals. In manual system, this is one of the typical tasks carried out by the night auditor. In fully-automated hotels, these reports are automatically prepared by the system. *The night auditor generally prepares the following reports for the management.*

• **High balance report**

This is a report which summarizes all the high balance guests and the amount due on them for settlement. It is prepared by night auditor after screening all guest folios. The front office cashier can ask the guest for partial or full settlement of the outstanding balance in order to re-establish charge privileges.

• **Occupancy reports**

The night auditor also generates the following occupancy related reports:

– **Occupancy percentage**

The occupancy percentage is the ratio of the number of rooms sold to the total number of saleable rooms. It determines the level of revenue that will be generated by the hotel and is an indicator of the performance of the hotel. Higher the rate of occupancy percentage, better the room sales revenue of the hotel.

The formula to calculate occupancy percentage is as under:

Occupancy percentage =

$$\frac{\text{Number of rooms sold}}{\text{Total number of saleable rooms}} \times 100$$

In case some rooms are under renovation, repair, or are out of order, the same are not included in the total number of saleable rooms.

– **House count**

The house count is the total number of resident guests present in the hotel. It is used to determine the average room rate per person.

The formula to calculate the house count is as under:

House count = House count of previous day brought forward + Today's arrival – Today's departure

House count can also be calculated as follows:

House count = Single rooms + 2 × (double rooms) + Extra beds

– **Bed occupancy percentage**

It is the ratio of the number of beds occupied to the total number of available beds in the property. In ascertaining the total number of beds available, all the beds in different types of rooms must be counted to arrive at the bed occupancy percentage.

The formula to calculate bed occupancy percentage is as under:

Bed occupancy percentage =

$$\frac{\text{Number of beds occupied}}{\text{Total no. of beds available for guests}} \times 100$$

– **Domestic occupancy percentage**

It is the ratio of the total number of domestic guests to the house count. This is the indicator of the type of clients visiting the hotel.

The formula to calculate domestic occupancy percentage is as under:

Domestic occupancy percentage =

$$\frac{\text{Total number of domestic guests}}{\text{House count}} \times 100$$

Alternatively, it may be calculated by subtracting the foreigner's occupancy percentage from 100.

Domestic occupancy percentage = 100 – Foreigner's occupancy percentage

– **Foreigner's occupancy percentage**

It is the ratio of the total number of foreign nationals to the house count. This is an indicator of the type of clients visiting the property.

The formula to calculate foreigner's occupancy percentage is as under:

Foreigner's occupancy percentage =

$$\frac{\text{Total number of foreigner guests}}{\text{House count}} \times 100$$

Alternatively, it may be calculated by subtracting the domestic occupancy percentage from 100.

Foreigner's occupancy percentage = 100 – Domestic occupancy percentage

• **Manager's report/Daily operations report**

This is a very useful report (Fig. 29.1) for the general manager as it provides an overview of the hotel's financial efforts of the previous day. It summarizes the day's business and provides insight into revenue, receivables, operating statistics and cash transactions related to the front office. These data are necessary for monitoring the operation of a financially viable business.

• **Flash sales summary report**

It gives information about the total income generated from the hotel from its food and beverage division as well as from the other revenue centers and also gives information

HOTEL XYZ MANAGER'S REPORT					
As on 24 Aug, 2015					
Particulars	This Year			Last Year		
	Today	MTD	YTD	Same Day	MTD	YTD
Guest In-House						
Rooms Sold						
Complimentary Rooms						
House Use						
Rooms Available						
OOO Rooms						
Arrival Rooms						
Arrival Persons						
Departure Rooms						
Departure Persons						
Occupancy %						
Double Occupancy %						
Hotel Revenue						
Revenue per Guest						
Room Revenue						
Average Room Rate						
ARR without Comps						
Average Person Rate						
RevPAR						
No of VIP Guests						
Reservations made today						
Cancellations for today						
No-show Rooms						
Day-Use						
Walk-In Rooms						

Fig. 29.1: Manager's report

about the occupancy of the rooms of the hotel for the particular day (Fig. 29.2).

- **Final departmental detail and summary report**

  These are produced and filed with the source documents for the accounting division. These reports help prove that all transactions were properly posted and accounted for.

Deposit Cash

The night auditor counts all the cash collected in the cashier's drawer and separates the cash bank/float, which had been allotted to him at the beginning of his shift for carrying out the departmental transactions. After removing the cash bank from the total amount collected, the night auditor keeps the money in a front office cash deposit envelope and mentions the details of all the money kept in the envelope by stating their denominations and numbers and also mentions if there is any overage, shortage or due back balances.

Updating the System

Once the outstanding charges have been posted and reconciled, no-shows have been verified, and reports and summaries for the management have been prepared, the night auditor updates the system and ends the night audit process. In fully automated hotels, the computerized system automatically carries out system updates.

Distribute Reports

Due to the sensitive and confidential nature of front office information, the night auditor must promptly deliver appropriate reports to the authorized and designated individuals.

OPERATING MODES FOR NIGHT AUDIT

Night audit process may be performed manually (non-automated), mechanically (semi-automated) or electronically (auto-mated).

The manual method makes no use of electronic systems; the semi-automated

	HOTEL XYZ					
	FLASH BUSINESS					
	As on 24 Aug, 2015					
Particulars	This Year			Last Year		
	Today	MTD	YTD	Same Day	MTD	YTD
Total Room Revenue						
Total F&B Revenue						
Total Telephones						
Total Laundry						
Total Health Club						
Total Business Center						
Total Shop Rental						
Total Travel Desk						
Total Tax Collection						

Fig. 29.2: Flash sales summary report

process refers to the use of the account posting machines whilst the automated method refers to the use of Management Information Systems (Property Management System).

The following sections briefly discuss each of these three operating modes in relation to the night audit routine.

Manual

In a non-automated (manual) system four audit forms are typically used to complete the audit. These are:

- Daily and supplemental transcript
- Guest and non-guest folio
- Front office cash sheet
- Audit recapitulation sheet

The night auditor prepares daily and supplemental transcripts by copying the day's activity from each guest and non-guest account folio to the appropriate line on one of the transcripts. The transcript columns are then summarized to determine the total charge transaction for each day. Information from these two transcripts along with data from the cost sheet may be transferred to a recapitulation sheet to provide a summary of daily front office activity. Manual auditing is cumbersome and error prone and should be used for small hotels only.

Semi-Automated

One of the most important developments in the history of front office accounting has been account posting machine. Posting machines record guest charges on folios and simultaneously perform a number of other activities which simplify the work of the front office desk agents and the night auditor.

Posting machines may be electro-mechanical or electronic. Electro-mechanical posting machines are only capable of producing limited number of departmental total; they do not retain folio balances and do not interface with other POS. On the other hand, electronic posting machines are capable of all these. These machines may retain folio balances to eliminate the need to enter the previous balance (pick-up balance) and the errors associated with those entries. Many electronic posting machines can store previous account balances, which help eliminate 'pick-up errors'.

Steps in machine posting

- Locating the folio.
- Take folio out of the bucket.
- Place folio along with the check in the posting machine.
- Enter previous closing balance.
- Post the check amount with department code.
- Machine balances the folio.
- Folio is refiled in bucket.

Simultaneously, the following occurs:

- The voucher used to indicate the posting is imprinted with the same information as posted to the accounts folios (this acts as a verification that the voucher has been posted).
- The same information is printed on an internal machine paper tape to save as a permanent journal record and as part of the hotel's internal audit trial (known as *Audit Trial*).
- Amount of each charge is added to or subtracted from the running departmental total.

Forms/Formats used

- Front office cash report
- Night auditor's report/D-card: It mentions the total opening balance, debits, credits and net outstanding balances. Used for reconciliation with departmental summaries.

Fully-Automated

It is a simple, quick and very accurate system in which automatic posting to guest and non-guest folios can be done. It is very fast and various audit functions can be performed. A fully-automated system can be interfaced with point of sale equipment, call accounting systems and other revenue center devices for quick, accurate and automatic postings to guest and non-guest account. Monitoring account balances and verifying posting is done simply by comparing guest ledger and non-guest ledger audit data to front office daily report for balancing. When these documents are out-of-balance, it usually indicates an internal computer problem or an unusual data handling error. Many computer-based front office systems can perform nearly continuous system audit routine and provide summary reports at a predetermined time providing greater flexibility in front office operation.

COMMON ERRORS DURING THE NIGHT AUDIT

Many types of posting, mathematical, and clerical errors occurred in the various transactions of the front office can be easily identified during the night audit process. The following are some of the common errors made in the front office.

Pick-up Errors

This type of error is common in the manual and semi-automated systems. Mistakes can be committed in bringing forward the balance of the previous day as the opening balance of the current day. Such previous day balances are referred to as 'pick-up balances' and mistakes or errors committed while bringing forward the balances are known as 'pick-up errors'. Such mistakes cause incorrect closing balance and this trend will continue till the error is detected and set right. Precautions must therefore be taken not only by the front office staff but also by the night auditor to check pick-up balances to ensure correctness. Such errors can also be caused due to missing folios. Errors may also be caused due to carrying forward the balance to a wrong guest folio or if the folio is improperly filled. Missing

folios must be traced on the basis of previous day's balance to ensure accuracy in bringing forward the balances.

Transposition Errors

This type of error is also common in the manual and semi-automated systems. An error that occurs when numbers in a transaction are reversed. For example, writing 189 instead of 198.

Missing Folios

This type of error is also common in the manual and semi-automated systems. This type of error occurs when a folio has been filed incorrectly or removed from the folio bucket. This normally happens when the front office cashier forgetfully allows the guest folio to remain behind the room tab instead of bringing the same to the front of the room tab after the departure of the guest. Folios can also seem to misplace if the front office cashier of a particular shift forgets to inform the front office cashier of the next shift that the same has been sent to the credit manager for his review. When a folio is missing, the night audit will not balance.

TERMS AND TERMINOLOGY

Cross Referencing

For internal control purpose, an accounting system should provide independent supporting documentation to verify each transaction in non-automated or semi-automated operations. Transactional document identifies the nature and the monetary value of transaction and is the basis for data input in the front office accounting system. This documentation consists of guest checks and charge vouchers. Front office cashier posts an entry to the appropriate guest or non-guest folio based on the documentation he receives.

The night auditor relies on these transaction documents to prove that proper accounting procedures were followed.

Account Integrity

Sound internal control techniques help ensure the accuracy, completeness and integrity of front office accounting process. Proper internal control techniques suggests that different front office staff members should be involved in posting charges, verifying charges and collection for sales transactions at front desk.

The night auditor establishes guest and non-guest account integrity by cross-referencing account posting with departmental source documentation.

Daily and Supplemental Transcript

A daily transcript is a detailed report of all transactions for guest account for that particular day. A supplemental transcript is often used to record the day's transactional activities for non-guest account. Together the daily transcript and the supplemental transcript represent all transactions for a particular day and forms the basis of a consolidated report of front office accounting data against which department totals can be matched.

The total of charged purchases reported by various revenue generating centers should equal to total amount of charged purchases posted to guest and non-guest accounts.

Daily and supplemental transcripts are simply worksheets designed to detect various types of posting errors. They can facilitate the night audit routine by identifying out-of-balance figures.

Front Office Accounting Formula

Previous Balance + Debits (purchases)– Credits (payments) = Net Outstanding Balance

$$PB + DR - CR = NOB$$

Date Roll

A certain point in the night to establish a change in date.

Clear or Back up the System

After the night audit is complete the totals must be cleared from the system. Manual systems are cleared by simply moving the closing balance from the night audit report to the opening balance of the next day's report.

In semi-automated system, the totals in the posting machine must be brought to a zero balance after the night audit is over. As each account is reduced to zero, a separate card (sometimes called Z-card) is used to verify the zero balance. Z-card is usually submitted with the night audit report to show that all accounts have been properly reset.

In fully-automated systems, a system back up should be conducted after each night audit and stored in a safe place. Usually there are two kinds of system back ups. The daily back up creates copies of the electronic files on magnetic tape or disc in case of computer failure while the second type of systems back up is performed once or twice a week. This back up not only backs up the daily information but also eliminates account and transactions information no longer needed. The system back up is also called 'fail-safe'.

REVIEW QUESTIONS

1. What do you understand by the term night audit? Why is it generally performed at night?
2. All the efforts of hotel operations can become meaningless if guest faces problems at departure and settlement process. Justify this statement on the basis of the concept called 'account integrity'.
3. 'Night audit can prevent revenue loss due to frauds and corrupt practices by employees.' Explain.
4. List at least five common accounting and billing frauds and explain how night audit can stop them from occurring.
5. Explain the various functions involved in night auditing.
6. Explain the night audit process in detail.
7. What is the role of night auditor in hotel?
8. Compare and contrast the night audit process in manual, semi-automated and fully-automated hotels.

9. What are the reports and statistics prepared by night auditor during night audit process?

10. Why does the night auditor verify room status and no-show before posting room rate and taxes? Why is it important that these postings occur as late as possible?

11. Differentiate between:
    - Room occupancy and Bed occupancy
    - Daily transcript and Supplemental transcript

12. Write short notes on:
    - Trial balance
    - End of day
    - Audit trial
    - Pick-up error
    - Night audit
    - Cross-referencing
    - Front office accounting formula

Part IV

Front Office Management

Introduction to Hotel Revenue Management

Learning Objectives

After reading this chapter, you will be able to understand the following:

- The historic development of yield management
- Yield management in the hotel industry
- Yield management users: The current situation
- Tools of revenue maximization
- Elements of yield management
- Benefits of yield management
- Yield management strategies
- Challenges or problems in yield management
- Yield management team
- Basic yield management system requirements
- Measuring yield

THE HISTORIC DEVELOPMENT OF YIELD MANAGEMENT

Yield is the revenue generated per statistical unit. It is the airlines industry that spearheaded the development of yield management system in the early 80s. To maximize the revenue generated from selling the seats in a flight, the airlines adopted a technique based on demand and supply. When the demand for seats in a particular flight exceeded the supply of seats, the airlines charged higher rates. But when the supply exceeded demand, the airlines offered various types of discounts and packages, resulting in the lowering of prices, which would lead to the selling of more seats on that particular flight. This way of maximizing revenue generation is termed yield management.

What exactly is Yield Management

Yield management or *Revenue management* is the process of understanding, anticipating, and influencing consumer behavior in order to maximize revenue or profits from a perishable resource such as airline seats or hotel rooms. Revenue management can also be defined as 'the science and art of enhancing firm revenues while selling essentially the same amount of product'. Revenue management can also be defined as the set of demand forecasting techniques, optimization models, and implementation procedures which collectively determine which reservation requests to accept and which to reject in order to maximize revenue.

The primary aim of revenue management is selling the right product to the right customer at the right time for the right price and with the right pack (Fig. 30.1). This process can result in price discrimination, where a firm charges different prices from customers consuming otherwise identical goods or services. For example, airlines charge different airfares from travelers who are traveling in the same class of the same flight, depending upon the number of days in advance the tickets have been booked. In general, the tickets purchased much earlier than the date of travel are less expensive than bookings made a little in advance. Of late it is being followed in railways as well. This way of maximizing revenue generation is termed yield management.

Selling the right room
at the right price

at the right time

Fig. 30.1: Yield management

YIELD MANAGEMENT IN THE HOTEL INDUSTRY

Hotels fulfill the essential conditions for yield management to be applicable. They are:

- **Relatively fixed capacity**: Hotels have a limited inventory of rooms available for sale to guests.
- **Perishable inventory**: In case the rooms are not sold on a particular day, it cannot be reused or stored for future use. This means that there is a time limit to selling the rooms, after which the revenue for the day is lost.
- **Market can be segmented**: Demand for the service can be divided into clear market

segments and sensitivity to prices varies among the market segments.

- **Rate differentiation:** Different customers are willing to pay a different price for using the same amount of resources. Business customers would not mind paying higher room rates and may book rooms closer to the stay dates, whereas families and guests on vacation are price sensitive and may book much ahead of the stay dates and would want to pay lower room rates.
- **Uncertainty of demand:** There are definite peaks and valleys in demand, which can be predicted, but not with a high degree of certainty.
- **High fixed costs, low variable costs:** The cost of selling an additional unit of the existing capacity is low relative to the price of the service.
- **Advance sales:** Reservations for rooms are accepted days, weeks and months in advance (even years for major conventions).

A guest room is one of the highly perishable products of the hospitality sector. Rooms if not sold on a particular day, cannot be reused or stored for future use, and hence the entire potential revenue that could be generated from it is lost forever. Hotels have come to realize that mere *volume sales* do not generate the desired revenue, and that they have to think of *quality deals* in terms of revenue generated per sale. Their focus is shifting from *high volume reservations* to *high profit reservations*. To maximize the revenue generated from rooms, hotels now sell their rooms at varying prices. Hence the process of knowing how many bookings to take at what rate, for how many days so that the maximum numbers of rooms are sold and maximum possible revenue is generated for that period is called yield management. Since yield management is the management in terms of maximizing revenue generation, it can also be called revenue management or revenue enhancing technique.

Thus we can define yield management in the hotel industry as:

- 'A technique based on the principle of demand and supply, used to maximize the revenue generation of any hotel by lowering prices to increase sales during off season (low demand period) and raising the prices during peak season (high demand periods)'.

- 'A technique to determine the right number of units of a product type (single, double, suite) to be allocated for sale to the right customer type (business or leisure travelers) on the right day (weekday or weekend) at the right price (rack rate, corporate rate, group rate) in order to maximize revenue'.

- 'It is a set of demand forecasting techniques, optimization models, and implementation procedures which collectively determine which reservation request to accept and which to reject in order to maximize revenue'.

Why Revenue Management is a Better Performance Measure

Most of us consider the Average Daily Rate (ADR) and Occupancy Percentage as indicators of how successful a hotel is. Whilst they certainly provide us with the basic tools, occupancy percentage and average daily rate are both one-dimensional analysis. As a result, neither of these measuring sticks captures the relationship between these two factors and the room revenue they produce. Revenue management presents a more precise measure of performance than either occupancy percentage or average daily rate because it combines occupancy percentage and ADR into a single statistic: Yield (Fig. 30.2).

YIELD MANAGEMENT USERS: THE CURRENT SITUATION

The principles of yield management had their origins in the airline industry, but have also taken hold widely throughout the passenger

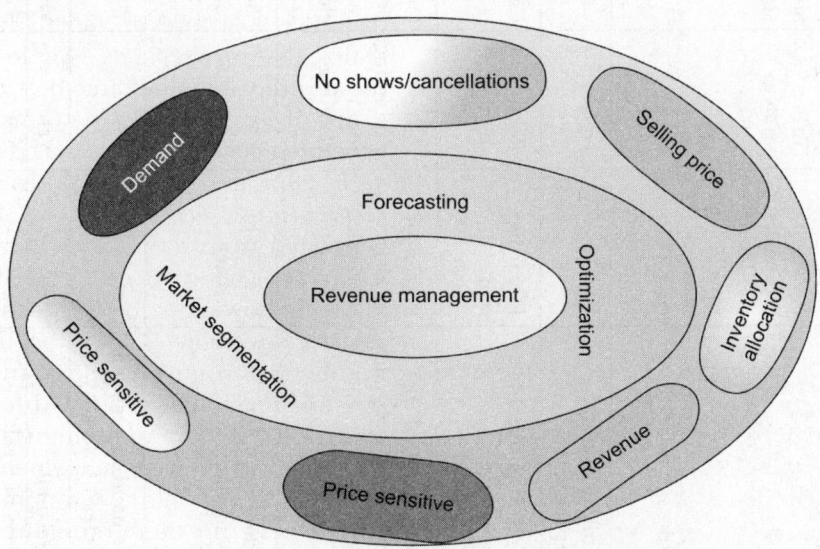

Fig. 30.2: Concepts of revenue management

Revenue management criterion/industry	Airlines	Hotels	Car rental	Freight transportation	Healthcare	Broadcasting	Telephone	Golf
Market segmentation	Market is segmented between business and leisure traveler using discount fare restrictions	Can pursue airline strategy	Can pursue airline strategy	Market is segmented by type of commodity to be transported	Time sensitive care vs Postponable care. Clinical care vs Surgical care	Guaranteed spots vs Preemptable spots vs Rotatable spots	Businesses vs Residences Wire or Landline vs Wireless or Satellite	Member vs Guest or Walk-in Senior vs Adult vs Child
Unit of fixed capacity	Flight	Hotel	Car fleet	Truck, train	Hospital	Television show	Phone network	Golf course
Unit of perishable inventory	Departing seat	Room night	Car Day	Trailer/boxcar departure labor	Room hour/ bed night	Advertising second or minute	Line minute vs airtime	Tee time
Low marginal cost for incremental sales	Passenger meals, processing	Order processing, room cleaning	Order processing, car cleaning	Order processing, freight handling, gas	Order processing meals, supplies	Order processing distribution channel	None	Order processing
Bookings taken in advance	Yes (up to 1 year)	Yes	Yes	Yes, but often close to departure	Yes, for elective procedures	Yes	Not usually	Yes
Demand forecasting cycles	Yes	Yes	Yes	Yes		Yes	Yes	Yes
Seasonal	Yes	Yes	Yes	Yes		Yes	Yes	Yes
Days of week	Yes	Not usually	Sometimes	Sometimes		Yes	Yes	Yes

Fig. 30.3: Revenue management criterion across industries

transport, travel and lodging industries and other sectors of the service industries (Fig. 30.3):

- Healthcare
- Blood bank
- Broadcasting
- Financial institutions
- Telecommunications
- Advertising
- Freight transportation
- Car rentals
- Cruise lines
- Tour operators
- Railroads
- Hotels

TOOLS OF REVENUE MAXIMIZATION

Yield management seeks to maximize revenue by using forecast information with the help of the three tools.

Capacity Management

Capacity management involves various methods of controlling and limiting room supply. In some cases a hotel may well choose to accept more bookings than it can actually accommodate. For example, hotels will typically accept a statistically supported number of reservations (based on the front office manager's experience and historical data available) in excess of actual room availability in an attempt to offset the effects of early departures/understays, cancellations, and no-shows. Capacity management (*also called selective overbooking*) balances the risk of overselling guest rooms against the potential loss of revenue arising from room spoilage (rooms going unoccupied after the hotel stopped taking reservations for a given date).

Overbooking is not done by mere guesswork. Selective overbooking is done by considering the following factors:

- **Past history of data related to**
  - Cancellation statistics: The number of cancellations received during that period in previous years.
  - Under-stay statistics: The number of guests who stayed for less than their reserved days during that period in previous years.
  - No-show statistics: The number of no-shows (guests with confirmed bookings who do not turn up at the hotel on the expected date of arrival, without any prior information) during that period in previous years.
  - Turn away statistics: The number of guests who were turned away or denied reservations due to non-availability of rooms during that period in previous years.
- **Activities in town**
  - Sporting events: Events like international cricket, tennis, football matches, etc. scheduled to be held during that period in and around the city.
  - Cultural events: Events like cultural fairs, festivals, etc. scheduled to be held during that period in and around the city.
  - Business events: Events like trade fair, business conferences, etc. scheduled to be held during that period in and around the city.
  - Protest/unrest/emergency: Events like curfews, bandhs, etc. scheduled to be held during that period in and around the city.
- **The experience of the reservation manager**
  The reservation manager can tell from his experience how many of the reserved guests will actually turn up.

Capacity management usually varies with room type, i.e. it might be economically advantageous to overbook more in lower-priced rooms, because upgrading to higher-priced rooms is an acceptable solution to an oversell problem. The amount of such overbooking depends, of course, on the demand for the higher-priced rooms in that duration. Overbooking should be avoided in cases where the hotel has only one or two rooms of the requested category, like presidential suites or penthouse. In sophisticated computerized yield management systems, capacity management may also be influenced by the availability of rooms at neighboring hotels or competing properties.

Differential Pricing/Discount Allocation

Price of goods or services may be defined as 'the value of the goods or services expressed in terms of money'. Price is a major criterion for a guest while choosing a hotel for stay. The pricing of a hotel's accommodation products is based on its demand in the market. Yield management attempts to get the right sales mix. It is next to impossible for a hotel to sell its rooms at rack rate at all times. A hotel must therefore have a sales strategy that will allow it to sell the maximum number of rooms at the best rates (to satisfy the projected demand for rooms at that rate), while at the same time filling the rooms that would have otherwise remained unsold at a discounted rate.

The basic idea is that during periods of high demand for hotel rooms, the price is set at the highest rate so as to maximize room revenue and at times of lower demand, the lower rates are set so as to encourage higher occupancy. Hotels may offer off-season rates, package plans, special offers, etc. to attract more business during lean season, whereas during peak season, hotels may not offer any discount on room tariff and will prefer to charge rack rate as they are confident that the demand for rooms would be more than the supply of available rooms.

Duration Restriction/Duration Control

Duration control places time constraints on accepting reservations, in order to protect sufficient space for multi-day requests. For example, a hotel may refuse a reservation request for one-night stay, even though rooms are available for that night. This is done with an objective that an expected request for more nights' reservation by another guest, shall be accepted. A hotel may exercise the length of stay restriction to control the imbalance of occupancy during the week. If the rooms of a hotel are sold out on one day of the week, the hotel will not be able to take bookings for two or more days including the sold-out day. As this will lead to a loss of potential revenue, hotels may turn down requests for single night booking on days of higher occupancy.

Normally, a hotel practices minimum length stay restrictions in case of a special event happening in the city. For example, if Wednesday is close to selling out but other nights are not, a hotel may want to protect the last few rooms on Wednesday for requests for Tuesday through Thursday or Wednesday through the weekend, even at a discounted rate, rather than accept reservations for Wednesday only, because the multi-day reservations represent more total revenue to the hotel.

ELEMENTS OF YIELD MANAGEMENT

While developing a successful yield strategy, the following elements are very important.

Group Room Sales

It is common to have reservations for group sales three months to two years in advance. Therefore, understanding group booking trends and requirements can be critical to the success of yield management.

To clearly understand how group sales affect overall room revenue, the hotel should collect the following information:

- **Groups already booked:** Information regarding how many groups is already booked and how many rooms for each one of them are blocked is very important. On the basis of the same, management can determine whether the group blocks should be reduced because of anticipated cancellations or overestimation of the group's size. If the group has booked at the hotel before, management can often determine this information by looking at the group's booking history. Groups often block 5% to 10% more rooms than they need, to ensure that they will have sufficient space for their members. The deletion of unnecessary group rooms from the group block is called the 'wash out factor'.

- **Group booking pace:** The group's booking pace indicates the rate at which group business is being booked as per the historical records. The rate at which initial agreement between the group coordinators and hotel takes place over a period of time, which may be a month, quarter, half yearly, etc.

- **Anticipated group business:** Sometimes the reservation manager has to keep in consideration that business of group which has not even contacted the hotel yet, but the manager is almost sure of this business. This assumption is normally made on the basis that the group comes to the hotel regularly after a particular period.

- **Booking lead time:** Lead time is the time gap between the date of booking and expected date of arrival. Group booking lead time measures how far in advance of a stay the group bookings are made. Some hotels have average lead times of two months. This is very important in determining whether to accept an additional group and at what room rate to book the new group.

- **Displacement of transient business:** Displacement occurs when a hotel accepts group business at the expense of turning away transient guests. Since transient guest often pay higher room rates than group business, this situation should be looked at very carefully. This might cause profitability problems and bad reputation. The hotel must take good care of turned away guests and make sure to accommodate

them in some other hotel of the same standard and keep in constant touch with them so as to give them a feeling that the hotel really cares for them.

Transient or Individual Room Sales

Transient rooms are those rooms sold to guests who are not affiliated with the group staying in the hotel (i.e. free independent travelers). Transient business is usually booked closer to the date of arrival than group business. A commercial hotel may book a majority of its group business three to six months before arrival, but book transient business only one to three weeks before arrival. As with group business, management must monitor the booking pace and lead time of transient business in order to understand how current reservations compare with historical and anticipated trends. This leads to the more complex subject of transient room pricing.

Food and Beverage Activities

While catering functions are really considered food and beverage revenue generators, they can have an effect on room revenue as well. For example, if a banquet with no guest room requirements is occupying the hotel's ballroom, a group needing 50 guest rooms and a ballroom may have to be turned away. In most cases, the group needing both catering and guest room space will be more profitable. Therefore, local food and beverage functions should be viewed in light of the potential for booking groups that need meeting space, food and beverage service, and guest rooms. Cooperation and communication between departments is necessary to maximize revenue from all revenue centers in the hotel.

Local and Area-wide Activities

Local and area-wide conventions can have dramatic effects on the yield management strategies of a hotel. Even when a hotel is not in the immediate vicinity of a convention, individual guests and small groups, who have been displaced by the convention, may be referred to your hotel (as an overflow facility) and this may have a tremendous impact on the hotel's revenue. The front office manager should be aware of the convention and the demand for guest rooms it is creating in the area to take full advantage of the opportunity.

Special Events

During special events (like concerts, festivals, and sporting events), hotels might decide to benefit from high demand by restricting room rate discounts or requiring a minimum length of stay.

BENEFITS OF YIELD MANAGEMENT

There are a lot of benefits associated with the use of yield management in the hotels. These benefits include the following:

- Improved forecasting
- Improved seasonal pricing and inventory decisions: It helps in deciding the peak season and off season pricing for accommodation products and also in making important inventory decision like renovation.
- Identification of new market segments
- Identification of market segment demands
- Enhanced coordination between the front office and sales departments: As the two departments work together to forecast and manage revenue and yield, it helps enhance coordination between them.
- Determination of discounting activity: Yield management helps determine the amount of discounts to be offered, depending on the dates and periods.
- Establishment of a value-based rate structure
- Improved development of short-term and long-term business plans: Revenue management helps develop business plans as the management can forecast the revenue that can be generated and take measures to generate those figures.

- Increased business and profits
- Savings in labor costs and other operating expenses: As most of the revenue management tools are computerized, it helps in saving labor costs and other operating expenses.
- Initiation of consistent guest contact scripting: Revenue management helps initiate consistent contact with guests.

YIELD MANAGEMENT STRATEGIES

Yield management requires each hotel to determine its strategies for both high demand periods and low demand periods.

Tactics—How to Act During High Demand

In case of high demand periods, the theory is to increase room revenue by maximizing average room rate.

- Close or restrict discounts.
- Apply a minimum of nights, but with caution.
- Reduce group room allocations as groups get very low room rates
- Reduce or eliminate 6 p.m. holds to avoid last moment no-shows or cancellations.
- Reduce 'late checkout'.
- Tighten guarantee and cancellation policies to avoid last moment no-shows or cancellations.
- Raise rates as consistent with competitors to generate optimum revenue.
- Increase the cost of packages and hotel services.
- Apply rack rates to higher category of rooms like suites and executive rooms.
- Select dates that are to be closed-to-arrivals.
- Invoice for all of the nights in case the guest shortens their stay

Tactics—How to Act During Low Demand

The underlying strategy during low demand periods is to increase revenue by encouraging higher occupancy.

- Sell added value and special benefits.

- Offer packages and special promotional rates.
- Keep discount categories like advance purchase rates, corporate rates, etc. open.
- Encourage upgrades.
- Offer weekend getaway or low cost packages to the local transient market.
- Offer stay-sensitive price incentives.
- Do not use 'stay restrictions'.
- Establish relationships with competitors.
- Lower room rates to attract more guests and to generate more revenue for the hotel.
- Solicit group business from companies or market segments known to be rate sensitive.

CHALLENGES OR PROBLEMS IN YIELD MANAGEMENT

The yield management techniques and the models of overbooking, if applied aptly, would definitely maximize the revenue of the hospitality industry. But there are some challenges or problems in this, which include:

- **Measuring performance of a yield management system**: Occupancy rates and yield are measures that are affected by external competition. Therefore, an ideal measurement can be done using the opportunity model, i.e. if the hotel segments the market and fixes different rates for different guests, then it has to see that the revenue is generated from those rooms and it has to be utilized ideally.

- **Guest satisfaction**: Some guests do not like the practice of differential pricing. In evaluating the efficiency of yield management system, the trade off between generating short-term profits and creating long-term guest loyalty needs to be studied carefully.

- **Employee malpractice**: Revenue management may influence the employees to follow wrong practices. For example, hotels might offer incentives to the staff for selling higher category rooms and this might

motivate the reservation agents to upsell while making reservations. So the agents might not sell the basic category rooms and offend certain guests.

YIELD MANAGEMENT TEAM

Before an effective yield management system can be put into place and used, it is necessary to have an operations team. They will be the driving force behind the successful implementation of yield management and will be the group of people who regularly meet to forecast the forthcoming business of the hotel.

The yield management team will typically consist of the front office manager, the sales manager and the reservations manager. This does not mean that anyone else excluded, but for a speedier and more effective decision making progress it is wise not to make the management team too big. The front office manager has the overall control of the department with targets for maximizing both occupancy and revenue. The sales department must be a part for it is they who go out on a daily basis to sell the guest rooms, and whilst they are fully aware of the need to maximize revenue their primary thrust is to get guests into the hotel. Thus by working with the forecast team it ensures that the sales and marketing staff are fully aware of the peaks and troughs of the hotel's business. The last but probably the most important member of the team is the reservations department. The reservations manager is the person who has a complete understanding of all of the hotel's bookings, the future booking patterns and the past histories of the hotel's arrivals and occupancies, and who is most up to date with bookings.

The role of the yield management team is fourfold:

- To predict the demand of rooms
- To allocate the right number of rooms to various market segments on the basis of reasoning and possible revenue generation calculation. To assess whether to take

bookings of transient or groups, or displace transient in preference to the group.

- To open or close rates as seen fit
- To conduct feedback sessions

BASIC YIELD MANAGEMENT SYSTEM REQUIREMENTS

The majority of YMS need the same kind of data and information and shares certain requirements for successful operation:

- An appropriate Data Processing System: Software and hardware
- A Reservation System (CRS/PMS)
- A Decision Support System (DSS)
- A database of past reservation and occupancy information, including the pattern of reservation build-up prior to arrival date
- A Forecasting Model/System: A system to forecast the demand for reservations between now and arrival date/departure time by rate class/ room type, based on historical demand data
- An Optimization Model: An optimization model to determine which combination of expected reservation requests and authorization limits results in the greatest expected revenue
- A Monitor and Performance Control System
- A clearly defined Marketing Policy
- A demand oriented Rate Structure
- A rational set of effective Fences
- A clearly defined policy concerning Overbooking and Service Quality
- A medium and long-term Yield Management Strategy

MEASURING YIELD

The revenue generated by a hotel can be measured by yield management. Yield is the ratio of actual revenue generated to the potential revenue. Actual revenue is the revenue generated by the number of rooms sold. Potential revenue is the revenue that could be earned by the hotel if all rooms were sold at rack rates.

The hotel may determine its potential revenue in more than one way:

- Some properties calculate their potential revenue as the amount that would be earned if all rooms were sold at the rack rate on double occupancy. However, it is hypothetical case that all rooms are sold at rack rate on double occupancy.
- Other properties calculate their potential revenue by taking into account the percentage of rooms normally sold at both single and double occupancy.

The second method results in a lower figure. In fact, while it is unlikely that a hotel will attain a potential that is based on 100% double occupancy (first method), a hotel using the second method may actually be able to exceed its 'potential' if demand for double rooms exceeds sales mix projections. Since the yields vary with the method used, once a preferred method has been chosen, it should be used consistently.

Potential Average Single Rate

It is the ratio of the single occupancy room revenue to the total number of rooms. The potential average single rate can be calculated as under:

Potential average single rate

$$= \frac{\text{Single occupancy room revenue}}{\text{Number of rooms}}$$

Potential Average Double Rate

It is the ratio of the double occupancy room revenue to the total number of rooms. The potential average double rate can be calculated as under:

Potential average double rate

$$= \frac{\text{Double occupancy room revenue}}{\text{Number of rooms}}$$

Multiple Occupancy Percentage

It is calculated as the ratio of the number of rooms occupied by more than one guest to the number of occupied rooms. It is important because it indicates sales mix and helps balance rates with future occupancy demand. The multiple occupancy percentage can be calculated as under:

Multiple occupancy percentage

$$= \frac{\text{No. of rooms occupied by more than one guest}}{\text{Number of occupied rooms}}$$

Or

Multiple occupancy percentage

$$= \frac{\text{No. of rooms occupied by more than one guest}}{\text{Total no. of rooms} \times \text{Occupancy ratio}}$$

Rate Spread

The determination of a room rate spread among various room types is essential to the use of decision making in targeting a hotel's specific market.

Rate spread = Potential average double rate potential average single rate

Potential Average Rate

It is a collective statistics that effectively combines the potential average single rate, multiple occupancy percentage, and rate spread. The potential average rate is calculated as under:

Potential average rate = (Multiple occupancy% × Rate spread) + Potential average single rate

Room Rate Achievement Factor

The percentage of the rack rate that the hotel actually receives is contained in the hotel's **achievement factor (AF)**, also called the **rate potential** percentage. It can be calculated as under:

Room rate achievement factor

$$= \frac{\text{Actual average rate}}{\text{Potential average rate}}$$

Yield

Yield is the ratio of actual revenue generated to the potential revenue. It can be calculated as under:

$$\text{Yield} = \frac{\text{Actual revenue generated}}{\text{Potential revenue}}$$

$$\text{Yield} = \frac{\text{Total rooms sold}}{\text{Total available rooms}} \times \frac{\text{Actual average rate}}{\text{Potential average rate}}$$

Yield = Occupancy% × Achievement factor

Identical Yields

It involves calculations of different combinations of occupancy and actual average room rate which may result in identical room revenue and yields. In other words, it means equivalent gross revenue.

Suppose a hotel is considering increasing its average room rate from ₹ 2000 to ₹ 3500. What occupancy percentage it must achieve to match the current yield?

Identical yield

$$= \text{Current occupancy\%} \times \frac{\text{Current rate}}{\text{Proposed rate}}$$

$$= \frac{70}{100} \times \frac{2000}{3500} = 40\%$$

Equivalent Occupancy

A more effective way of evaluating whether a change in room rates is justifiable involves determining an equivalent occupancy. The equivalent occupancy formula can be used when management wants to know; what other combinations of room rate and occupancy percentage provide equivalent net revenue.

The marginal cost of providing a room is the cost the hotel incurs by selling that room (for example, cleaning and supplies); this cost would not be incurred if the room were not sold.

Equivalent occupancy

$$= \text{Current occupancy} \times \frac{\text{Current contribution margin}}{\text{New contribution margin}}$$

Required Non-Room Revenue per Guest

Non-room revenue means revenue generated from centers other than room such as food and beverage. If a manager, in order to increase occupancy percentage, decides to give discount or reduces room rate, then this would render an offsetting change in non-room revenue and involves calculating or estimating a number of elements.

- The net loss in room revenue due to room rate discounting
- The amount of non-room revenue needed to offset this loss.
- The average amount each guest spends in non-room revenue centers.
- The increase in occupancy likely to result from rate discounting

The breakeven calculation is based on the **weighted average contribution margin ratio (CMR$_w$)** for all non-room revenue centers. While a detailed discussion of this topic is beyond the scope of this text, a simple formula for determining the CMR$_w$ for all non-room revenue centers is as follows:

$$\text{CMR}_w = \frac{\begin{array}{l}\text{Total non-room revenue} - \text{Total}\\ \text{non-room revenue center}\\ \text{variable costs}\end{array}}{\text{Total non-room revenue}}$$

Knowing the CMR$_w$ and the average amount that guest spends in non-room revenue and having estimated the probable change in occupancy, the front office manager can then determine whether the net loss caused by discounting room rates is likely to be more than offset by the net gain in non-revenue. The formula is as follows:

Required non-room revenue per guest

$$= \frac{\begin{array}{l}\text{Required increase in net}\\ \text{non-room revenue}\end{array}}{\text{Number of additional guests}} \div \text{CMR}_w$$

The front office manager can compare the result of this equation with the actual average non-room spending per guest. If this number is higher than the actual average non-room spending per guest, the hotel is likely to lose net revenue by discounting; that is, the additional guests brought in by discounting will not spend enough to offset the net loss in room revenue. If the amount needed per additional guest is lower than the actual average amount spent, the hotel is likely to increase its net revenue by discounting.

REVIEW QUESTIONS

1. What is yield management? Why do hotels use this system?
2. Differentiate between actual revenue and potential revenue.
3. What elements may be included while planning successful yield strategy?
4. What are the tools and strategies used for yield management?
5. Mention the objectives of yield management, justifying its applicability to rooms division.
6. Explain what are the various techniques used by front office to maximize yield.
7. Enlist all the formulae required to measure yield.
8. Discuss the role of yield management team in enhancing room revenue.
9. Enlist the potential high and low demand tactics, hotels use to increase their yield.
10. What is yield management software?
11. Explain how yield management enhances forecasting and seasonal pricing of inventory in hotel industry.

12. 'Applying yield management improvises the co-ordination between front office and sales department'. Justify.

13. What effect can group reservations have on the hotel operations?

14. What is group booking pace? What role does it play in yield management?

15. What is wash factor and how does it affect yield management?

16. How can duration control be used to increase room revenue?

17. Explain how selective overbooking improves revenue generation.

18. Explain in which situations hotels should practice duration restriction for room reservation.

19. Write short notes on the following:
    • Capacity management
    • Discount allocation
    • Duration control
    • Displacement
    • Rate spread
    • Wash factor

• Rate potential
• Contribution margin
• Equivalent occupancy

20. What steps can front office employees take to control understays and unwanted overstays?

21. What is break even analysis? Explain how it can be used in rooms division to maximize room revenue, suggesting the role of non-room revenue.

22. State two types of formula to calculate equivalent occupancy.

23. Give formula for the following:
    • Yield %
    • Rate spread
    • CMR_w
    • Potential average rate
    • Multiple occupancy %
    • Potential average single rate
    • Potential average double rate
    • Room rate achievement factor
    • RevPAR
    • Potential average rate

Forecasting Room Availability

FORECASTING ROOM AVAILABILITY

According to New Oxford Intermediate Learner's Dictionary, forecast means 'to say (with the help of information) what will probably happen in future'. Thus, forecasting is the prediction of future happenings, based on the precise analysis of the data available rather than guesswork. Forecasting is a time-sensitive process. The farther out the forecast, the less accurate it becomes.

The most important short-term planning that front office managers engage in is forecasting the number of rooms available for sale on any future date. A room forecast is a prediction of the hotel's occupancy over a period of time, which may be weekly, monthly or yearly, to enable the management to take effective decisions regarding reservations, room rates, marketing, etc. A room availability forecast can also be used as occupancy forecast. Since there is a fixed number of rooms available in any hotel, forecasting the number of rooms available for sale and the number of rooms expected to be occupied can be useful in computing an expected occupancy percentage for a given date.

BENEFITS OF FORECASTING

In the hotel industry, reservation forecasting is very useful in the following ways:

• Helps the reservation or revenue manager to project future volume of business and the revenue that would be generated by the hotel.

• The volume of reservations will help the front office manager and the management to plan the following:

  – Scheduling employees in each department for the smooth functioning of the hotel.

  – Minimum inventory of items required by each department to carry out their tasks efficiently.

  – Allocation of resources to serve the guests in the best possible way.

  – Maintenance and replacement requirements of the furniture, fixtures and ultimately the property, as the wear and tear of these depends on the number of people using it.

– Special arrangements to be made for the arrival of groups, commercially important persons (CIPs) and VIPs.

• Provides the necessary data to the reservation manager to practise yield management.

• Helps to take selective overbooking, based on the reservation forecast.

• Provides information about the lean days so that the sales department may take necessary actions to attract the business for those durations.

• Reveals the sold out dates, which will ensure that the reservation agent does not accept reservations for those days.

DATA REQUIRED FOR FORECASTING

Forecasting, which is a difficult skill to develop, can be acquired through the effective and efficient tracking of records, by using accurate mathematical calculations, and through experience. The front office managers, by virtue of their experience, have found that the following information is necessary for making an accurate forecasting:

• Thorough product knowledge.

• A good judgment about what could happen in the future.

• Thorough knowledge about their area of operation in the hotel.

• The profile of the target market to which the hotel is catering.

• The events that are scheduled in the area during the forecasted period.

• Percentage of no-shows.

• Overstay percentage.

• Understay percentage.

• Turn-down statistics.

• Future plans for renovation or addition of more rooms in the property.

• Future plans regarding the opening of any new property in the vicinity of the hotel.

• A precise knowledge about the room status in competitor's property.

• Knowledge about competitor's plans with respect to activities (like renovations), which will reduce the supply of rooms in their property.

• Cancellation statistics.

• Wash out percentage

RECORDS REQUIRED FOR FORECASTING ROOM AVAILABILITY

The forecasting of room availability is not done on mere guesswork. The process of forecasting room availability generally relies on historical occupancy data. To facilitate forecasting, the following occupancy data during the same period should be recorded:

• The number of arrivals on each day during the same period

• Number of checkouts

• Number of stay-overs (rooms which will continue being occupied)

• Number of no-shows

• Number of walk-ins

• Number of understays (guests who check-out prior to original checkout date)

• Number of overstays (guests staying beyond original checkout date)

Percentage of No-shows

It is the percentage of those guests who did not turn up in spite of confirmed reservation. This ratio helps the front office manager to decide the early release of rooms to chance guests or walk-ins. Non-guaranteed reservations typically have a higher no-show percentage than guaranteed reservations since the potential guest has no obligation to pay if he does not check-in.

The percentage of no-shows is calculated by dividing the number of no-shows for a specific period of time, by the total number of reservations for the same period.

Percentage of no-shows

$$= \frac{\textbf{Total number of no-shows}}{\textbf{Total no. of confirmed reservations}} \times 100$$

Percentage of Walk-ins

The percentage of walk-ins is calculated by dividing the total number of walk-ins for a specific period by the total number of arrivals for the same period. A higher percentage of walk-ins make it difficult for the hotel to forecast accurate room availability as the number of walk-ins cannot be predetermined.

Percentage of walk-ins

$$= \frac{\textbf{Total number of walk-ins}}{\textbf{Total number of arrivals}} \times 100$$

Percentage of Overstays

Overstays represent rooms occupied by guests who continue their stay beyond their originally scheduled departure dates.

The percentage of overstays is calculated by dividing the number of overstays for a specific period by the total number of check-outs for the same period.

Overstays result in lowering the room availability. This condition is favorable in lean season as hotels generally run on low occupancy levels and can accommodate the request for overstay. It alerts front office managers to potential problems when the hotel is near full occupancy and rooms have been reserved for arriving guests.

Percentage of overstays

$$= \frac{\textbf{Total number of overstays}}{\textbf{Total number of checkouts}} \times 100$$

Percentage of Understays

Understays represent rooms occupied by guests who checkout before their scheduled departure dates. These guests are also called *early departure*. It alerts front office manager to probable additional room availability when the hotel is near full occupancy. The percentage of understays is calculated by dividing the number of understays for a specific period by the total number of checkouts for the same period.

Percentage of understays

$$= \frac{\textbf{Total number of understays}}{\textbf{Total number of checkouts}} \times 100$$

Guests leaving before their stated departure date create empty rooms that typically are difficult to fill. Thus, understay rooms may represent permanently lost room revenue. Overstays, on the other hand, are guests staying beyond their stated departure date and may boost room revenues. When the hotel is not operating at full capacity, overstays result in additional, unexpected room revenues. In an attempt to regulate understay and overstay rooms, front office staff should:

- Confirm or reconfirm each guest's departure date at registration. Some guests may already know of a change in plans, or a mistake may have been made in the original processing of the reservation. The sooner erroneous data are corrected, the greater the chance for improved planning.

- Present an alternate guest room reservation form to a registered guest, explaining that an arriving guest holds a reservation for his or her assigned room. A note card may be placed in the guest's room the day before or the morning of the scheduled day of the registered guest's departure.

- Review group history. Many groups, especially associations, hold large closing events for the entire group on the last day of the meeting. Guests may make reservations to include attending the final event. However, changes in plans or other priorities may require guests to leave early. While it is difficult for the hotel to hold guests to the number of nights they reserved, managers may be better able to plan for early departures, based on the group's departure history. Some hotels that have a lot of association business or a history of transient guests departing before their scheduled date may apply the reservation deposit to the last night of the stay, not the first night.

- Contact potential overstays guests about their scheduled departure date to confirm their intention to checkout. Room occupancy data should be examined each day; rooms with guests expected to check-out should be flagged. Guests who have not left by checkout time should be contacted and asked about their departure intentions. This procedure permits an early revised count of overstays and allows sufficient time to modify previous front office planning, if necessary.

Percentage of Cancellations

It is the percentage of total number of cancellations as against total number of reservations.

Percentage of cancellations

$$= \frac{\text{Total number of cancellations}}{\text{Total no. of confirmed reservations}} \times 100$$

Room Availability Formula

Once the above statistics are gathered, the number of rooms available for sale on any given date can be determined by the following formula:

Total no. of guest rooms

Minus Out of Order rooms

Minus Number of Stay-overs

Minus Number of Reservations

Plus (Number of Reservations × Percentage of No-shows)

Plus Number of Understays

Minus Number of Overstays

Equals Number of Rooms Available for Sale

SAMPLE FORECAST FORMS

The front office may prepare several different forecasts, depending on its needs. Occupancy forecasts are typically developed on a monthly basis and reviewed by food and beverage and rooms division management to forecast revenues, project expenses, and develop labor schedules. A ten-day forecast, for example, may be used to update labor scheduling and cost projections and may later be supplemented by a more current three-day forecast. Together, these forecasts help many hotel departments maintain appropriate staff levels for expected business volumes and thereby help contain costs.

Ten-Day Forecast: At most lodging properties, the ten-day forecast is developed jointly by the front office manager and the reservations manager, possibly in conjunction with a forecast committee. Many properties develop their ten-day forecast from their yearly forecast. A ten-day forecast usually consists of:

- Daily forecasted occupancy figures, including room arrivals, room departures, rooms sold, and number of guests
- The number of group commitments, with a listing of each group's name, arrival and departure dates, number of rooms reserved, number of guests, and perhaps quoted room rates
- A comparison of the previous period's forecasted and actual room counts and occupancy percentages

A special ten-day forecast may also be prepared for food and beverage, banquet, and catering operations. This forecast usually includes the expected number of guests, which is often referred to as the *house count*. Sometimes the house count is divided into group and non-group categories so that the hotel's dining room managers can better understand the nature of their business and their staffing needs.

Ten-Day Occupancy Forecast

Location _____ # _____ Week Ending _____

Date _____ Prepared by _____

To be submitted to all deparment heads at least one week before the first day listed on forecast.

	Fri.	Sat.	Sun.	Mon.	Tues.	Wed.	Thur.	Fri.	Sat.	Sun.
1. Date and day (start week and end week the same as the payroll schedule)										
2. Estimated Departures										
3. Reservation Arrivals—Group (taken from log book)										
4. Reservation Arrivals—Individual (taken from log book)										
5. Future Reservations (estimated reservations received after forecast is completed)										
6. Expected Walk-ins (% of walk-ins based on reservations received and actual occupancy for past two weeks)										
7. Total Arrivals										
8. Stay-overs										
9. Total Forecasted Rooms										
10. Occupancy Multiplier (based on number of guests per occupied room for average of the same day for last three weeks)										
11. Forecasted Number of Guests										
12. Actual Rooms Occupied (taken from daily report for actual date to be completed by front office supervisor)										
13. Forecasted Variance (difference between forecast and rooms occupied on daily report)										
14. Explanation (to be completed by front office supervisor and submitted to general manager; attach additional memo, if necessary)										

Approved _____ Date _____
(General Manager's Signature)

Fig. 31.1: Ten-day forecast form

To help various hotel departments plan their staffing and payroll levels for the upcoming period, the ten-day forecast should be completed and distributed to all department offices by mid-week for the coming period. This forecast can be especially helpful to the housekeeping department. A ten-day forecast form (Fig. 31.1) is typically developed from data collected through several front office sources.

First, the current number of occupied rooms is reviewed. The estimated numbers of overstays and expected departures are noted. Next, relevant reservation information is evaluated for each room (and guest) by date of arrival, length of stay, and date of departure. These counts are then reconciled with reservation control data. Then, the actual counts are adjusted to reflect the projected percentage of no-shows, anticipated understays, and expected walk-ins. These projections are based on the hotel's recent history, the seasonality of its business, and the known history of specific groups scheduled to arrive. Finally, conventions and other groups are listed on the forecast to alert various department managers to possible periods of heavy, or light, check-ins and check-outs. The number of rooms assigned each day to each group may also be noted on the sheet.

Most automated systems provide a summary of recorded data in a report format for the front office manager to use. However, only revenue management systems are programed to 'forecast' business. Programming hotel property management systems to successfully analyze historical trends and market conditions has been attempted in the past with a little success. Revenue management systems are much more sophisticated, with special trend analysis and regression analysis programming built in. Even with revenue management system forecasting, it is the front office manager's knowledge and skill that ultimately determines the accuracy of the forecast. Table 31.1 presents a checklist that some revenue managers use when revising forecasts.

Table 31.1: Refining a forecast

A yearly forecast provides an excellent starting point for developing shorter-term, more accurate forecasts. Managers can better access business by reviewing current reservation and booking pace. The closer the forecast is, the more accurate it will be. Here is a checklist for reviewing forecasts:

- List all group bookings and transient reservations on the books.
- Examine arrivals, departures and group information for the given period.
- Determine if demand for this particular off time is high or low.
- Chart the peaks and valleys on a graph to better identify high/low demand.
- Having sales agents call competing hotels for rates and consider adjusting your rates.
- Make decisions to maximize revenue during each time period.

Three-Day Forecast: A three-day forecast is an updated report that reflects a more current estimate of room availability. It details any significant changes from the ten-day forecast. The three-day forecast is intended to guide management in fine-tuning labor schedules and adjusting room availability information. Figure 31.2 shows a sample three-day forecast form. In some hotels, a brief daily revenue meeting is held to focus on occupancy and rate changes for the next few days. The results of this meeting are often included in the three-day forecast.

ROOM COUNT CONSIDERATIONS

Control books, charts, software applications, projections, ratios, and formulas can be essential in short- and long-range room count planning. Each day, the front office performs several physical counts of rooms occupied, vacant, reserved, and due to checkout, to complete the occupancy statistics for that day.

Three-Day Forecast

Date _____   Prepared by _____
Total Rooms in Hotel _____

	Tonight	Tomorrow	3rd Night
Day			
Date			
Previous night occupied rooms[1]			
- Expected departures			
- Early departures			
+ Unexpected stayovers			
+ Unoccupied rooms[2]			
= Rooms available for sale			
+ Expected arrivals			
+ Walk-ins and same day reservations			
- No-shows			
= Occupied rooms			
= Occupancy %			
= Expected house count[3]			

Distribution: General Manager, Front Office Manager, Executive Housekeeper, Food and Beverage Manager, Accounting, Sales and Marketing, Security, Maintenance

1. Previous night occupied rooms are determined from either the actual number of rooms occupied last night or the forecasted number of rooms from the previous night.

2. Unoccupied rooms equal the total number of rooms in the hotel less the number of rooms occupied.

3. Expected house count equals the forecasted occupied rooms times the multiply occupancy percentage for the day (found on the computer report).

Fig. 31.2: Three-day forecast form

An automated system may reduce the need for most final counts, since the system can be programmed to continually update room availability information.

It is important for front desk agents to know *exactly* how many rooms are available, especially if the hotel expects to operate at nearly 100 percent occupancy. Once procedures for gathering room count information are established, planning procedures can be extended to longer periods of time to form a more reliable basis for revenue, expense, and labor forecasting. Table 31.2 illustrates a daily checklist for accurate room counts.

Table 31.2: Daily checklist for accurate room counts

- Make counts of the rack and reservations. On tight days; a count should be made at 7:00 am, 3:00 pm, and 6:00 pm. On normal days, a 7:00 am and 6:00 pm count will suffice.
- Check room rack against the folio bucket to catch sleepers and skippers.
- Check housekeeping reports against the room rack to catch sleepers and skippers.
- Check for rooms that are due out, but still have balances on their folios, especially where credit cards are the indicated source of payment.
- Check reservations for any duplication.
- Call the reservations system to make sure all cancellations were transmitted.
- Check the switchboard, telephone rack, and/or alphabetical room rack to make sure that the guest is not already registered.
- Call the local airport for a report on cancelled flights.
- Check the weather reports for cities from which a number of guests are expected.
- Check reservations against convention blocks to catch duplications.
- Check with other hotels for duplicate reservations if a housing or convention bureau indicated the reservation was second choice.
- Check arrival dates on all reservation to be sure none was misfiled.
- Check the rooms' cancellation list.
- If a reservation was made through the reservations manager, sales manager, or someone in the reservations office and property is close to full, call that staff. Often, such guests are personal friends and are willing to help out by staying somewhere else.
- Close to the property's cut-off time, consider placing a person-to-person phone call to any guest with a non-guaranteed reservation who hasn't arrived. If the person accepts the call, confirm whether or not he will arrive yet that night.
- After the property's cut-off time, if it becomes necessary, pull any reservations that were not guaranteed or pre-paid.
- If any rooms are out-of-order or not presently in use, check to see if they can be made up; let housekeeping know when a tight day is expected, so that all possible rooms are made up.
- Before leaving work, convey in writing all pertinent information to the oncoming staff. Two-way communications is essential.

REVIEW QUESTIONS

1. What steps can front office employees take to control understays and unwanted overstays?

2. How do ten-day and three-day forecasts help ensure efficiency in front office operations? What is the relationship between these forecasts? What departments in the hotel rely on these forecasts, other than the front office?

3. What information do front office managers require to develop room availability forecasts? Why are these forecasts important? How reliable are such forecasts?

4. State the different data which are used in the different ratios that help in determining the rooms' availability position.

5. Explain in brief the 10-day forecast in front office. Draw its format.

6. What do you mean by forecasting room availability? Explain the procedure of forecasting in hotel. Discuss room availability forecast formula.

7. What are the benefits of forecasting?

8. What skills and information are required to make an accurate forecast?

9. Give formula for the following:
   - Overstay %
   - Understay %
   - No-show %
   - Walk-in %

10. **Based on the following data, calculate the hotel position:**

Total rooms	–	200
House use	–	3
Occupied	–	170
Reservations	–	80
Expected departures	–	97
% of No-shows	–	4%
% of Overstays	–	2%
% of Understays	–	3%

11. Hotel XYZ has 300 rooms. On 1st December 2015, it has two 'OOO' rooms. The hotel has 80 confirmed reservations with a no-show percentage of 5%. It also has 100 guaranteed reservations with a no-show percentage of 3%. There are 60 stayovers that day and the predicted number of overstay is 12, while the expected

number of walk-ins are 22. About 15 guests are expected to curtail their stay in the hotel that day. Fifty-four guests are expected to checkout. Calculate the number of rooms available for sale on 1st December 2015. (Show all steps of calculation clearly)

12. Hotel Luxury has 400 rooms. On 1st May 2016, there are 10 out of order rooms and 160 stayovers. There are 120 guests with reservations expected to arrive and the no-show percentage has been calculated at 10%. It is forecasted that 10 understays and 18 overstays are to be expected that day.

- Give the room availability forecast formula.
- Using that formula, calculate the number of rooms still to be sold for 100% occupancy on 01.05.2016 using the above given information showing each step.

Planning and Evaluating Operations

Learning Objectives

After reading this chapter, you will be able to understand the following:

▫ The management process in terms of the functions performed by front office managers to achieve organizational objectives

▫ The process of budgeting for operations by front office managers

▫ Evaluating front office operations

MANAGEMENT FUNCTIONS

Most front office managers will readily admit that they rarely have all the resources they would like. Resources available to managers include people, money, time, work methods, materials, energy and equipment. All these resources are in limited supply. An important part of front office manager's job involves planning how to use these limited resources to attain the organization's objectives. An equally important part of a front office manager's job is evaluating the success of front office activities in meeting the organization's objectives.

Planning

Planning is probably the most important management function in any business (Fig. 32.1). Planning gives us direction and focus. Without competent planning, front office work would be chaotic. A front office manager's first step in planning should be to

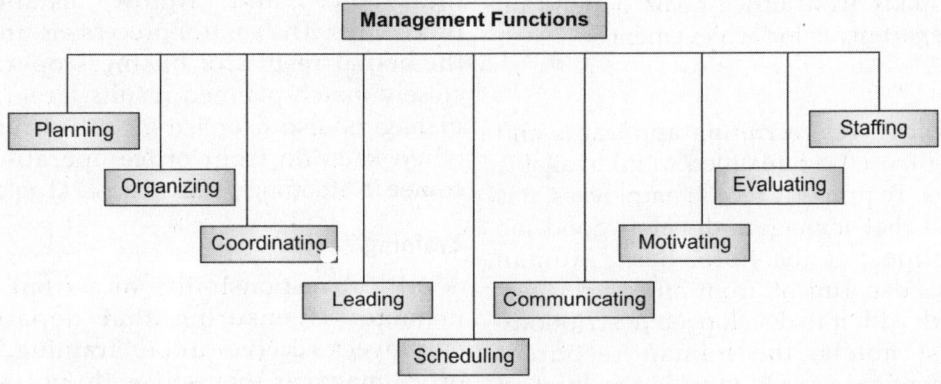

Fig. 32.1: Management functions

establish department goals (both short-term and long-term) which should then be used as a guide to planning more specific, measurable objectives. An important component of planning is communicating preliminary plans with all those involved. When staff members have a chance to contribute to the plan, they gain ownership of the plan, give managers a chance to address their concerns, and have a more comprehensive understanding of the plan and how it came about.

Organizing

Using planned goals as a guide, front office managers can organize the department by dividing the work among front office staff. Managers should distribute work so that everyone participates and the work can be completed in a timely manner. Organizing includes determining the order in which tasks should be performed and establishing completion deadlines for each group and subgroup of tasks.

Coordinating

Coordinating involves bringing together and using available resources to attain planned goals. Front office managers must be able to coordinate the efforts of many individuals to ensure that work is performed efficiently. Coordinating front office procedures may involve engaging with other hotel departments; many front office goals depend on other departments for achievement.

Staffing

Staffing involves recruiting applicants and selecting those best qualified to fill available positions. To properly recruit employees, it is essential that managers develop good job descriptions. If the hotel has a human resources department, front office managers will work with it to develop job descriptions. At most hotels, the human resources department are usually involved in the first level of qualifying and interviewing job applicants. The staffing process also involves scheduling employees. Most front office managers develop staffing guidelines based on formulas for calculating the number of employees required to meet guest and operational needs under specific conditions.

Leading

Leading is a complicated management skill that is exercised in a wide variety of situations. Leading is closely related to other management skills such as organizing, coordinating, and staffing. For front office managers, leadership involves overseeing, motivating, training, disciplining, and setting an example for the front office staff. A front office manager's leadership impact often extends beyond the front office, because so much of the hotel's business activity flows through the front desk. One of the best ways to lead a department is by example. When a manager leads effectively, this behavior demonstrates what is expected of all department employees. A good front office manager will be able to step into situations where his staff cannot deal with the workload.

Controlling

Every front office department has a system of internal controls to protect the hotel's assets. However, internal control systems only work when managers believe in the systems' importance and follow established procedures. The control process ensures that the actual results of business operations closely match planned results. Front office managers also exercise a control function when keeping front office operations on course in attaining planned goals (Table 32.1).

Training

A critical responsibility of a front office manager is ensuring that department employees receive proper training. Front office manager themselves do not always assume the duties and responsibilities of a

Table 32.1: Management as a process	
Planning	It means determining the objectives of the unit or activity. It also involves deciding in advance as to what is to be done, how and when.
Organizing	It refers to identification of activities to be carried out, grouping of similar activities and creation of departments. Organization also leads to creation of authority and responsibility relationships throughout the enterprise.
Staffing	It involves manpower planning, employment of personnel and their training, appraisal, remuneration, etc.
Directing or Leading	It is a very broad function concerned with the interpersonal relations. It includes communication with subordinates, providing them leadership and also motivating them.
Controlling	It refers to comparing the actual performance with the plans or standards. Corrective steps are taken when the actual performance is not up to the mark.

trainer. The actual training function may be delegated to supervisors or even to talented employees. Regardless of who does training, the front office manager should be the person responsible for the department's ongoing training programs. Job breakdowns can be used to train new employees and retrain experienced employees to perform tasks according to established standards.

Scheduling

Scheduling employees is one of the most difficult tasks a front office manager faces. Scheduling can be extremely complex—especially when employees are trained to do only specific tasks. For example, a telephone operator cannot be scheduled as a front office cashier when he has not been properly trained to checkout the guests. The more cross-trained employees a front office manager has, the

fewer employees are needed to perform tasks. Staffing flexibility can be achieved by having enough employees properly trained in several areas. Cross-training lowers labor costs and provides employees with expanded job knowledge and a broader range of skills.

Motivating

Motivation is the end result of satisfying human needs associated with personal worth, value and belonging. In an organization, the outcome of motivation should be that an employee's sense of worth, value and belonging has improved from taking part in a particular activity. Employees who receive recognition for the contributions they make to a hotel's success are typically highly motivated, top performers.

Communicating

Keeping employees informed about what goes on in the department and hotel cannot help but reap positive results. Employees who are aware of events taking place and learn about such events through channels other than the grapevine, feel a greater sense of belonging and value. Department newsletters are an excellent way to open and maintain clear lines of communication. Some hotels allow front office employees to develop newsletters and provide their own articles. A bulletin board can provide a place to post schedules, memorandum and other pertinent information. Bulletin boards are most effective when they are in an area accessible to all employees and when employees review the information daily.

Evaluating

Evaluating determines the extent to which planned goals are attained. Unfortunately this function is often overlooked or performed without much thought in many front office operations. Evaluating also involves reviewing and, when necessary, revising front office goals.

BUDGETING FOR OPERATIONS

The most important long-term planning function that front office managers perform is budgeting for front office operations. The hotel's annual operations budget is a profit plan that addresses all revenue sources and expense items. Annual budgets are commonly divided into monthly plans that, in turn, are divided into weekly (and sometimes daily) plans. These budget plans are standards against which management can evaluate the actual results of operations. In most hotels, room revenues and profits are usually greater than for other hotel revenue areas, so an accurate rooms' division budget is vital to creating a hotel's overall budget.

The budget planning process requires the closely coordinated efforts of all management personnel. While the front office manager is responsible for rooms' revenue forecasts, the accounting division staff will be counted on to supply department managers with statistical information essential to the budget preparation process. The accounting division staff is also responsible for coordinating the budget plans of individual department managers into a comprehensive property-wide operations budget for top management's review. The general manager and controller typically review departmental budget plans and prepare an overall hotel budget report for approval by the hotel's owners. If the budget is not satisfactory, elements requiring change may be returned to the appropriate division managers for review and revision.

The front office manager's primary responsibilities in budget planning are forecasting rooms' revenue and estimating related expenses. Rooms' revenue is fore-casted with input from the reservations manager, while expenses are estimated with input from all department managers in the rooms division.

Forecasting Rooms' Revenue

Historical financial information often serves as the foundation on which front office managers build rooms' revenue forecasts. One method of rooms' revenue forecasting involves an analysis of rooms' revenue from past periods. For example, if for the past four years rooms' revenue increased an average of ten percent, for the next year rooms' revenue might be budgeted at a ten percent increase over the previous year's revenue.

Another approach to forecasting rooms' revenue bases the revenue projection on the trends of past room sales and average daily room rates. Detailed approaches to forecasting rooms' revenue consider the variety of different rates corresponding to room types, guest profiles, days of the week, seasonality of business, and other factors.

Forecasted annual rooms revenue = Rooms available × Occupancy percentage × Average daily rate

Rooms available = Total rooms × 365 days

Estimating Expenses

Most expenses for front office operations are variable expenses (they vary in direct proportion to rooms' revenue). Historical data can be used to calculate an approximate percentage of rooms' revenue that each expense item may represent; these percentage figures can then be applied to the total amount of forecasted rooms' revenue, resulting in rupees estimates for each expense category for the budget year.

Typical rooms' division expenses are payroll and related expenses, guest room laundry, guest supplies, hotel merchandising (in-room guest directory, hotel brochure), travel agent commissions and direct reservation expenses, and other expenses. When rooms' division expenses are totaled and divided by the number of occupied rooms, the cost per occupied room is determined. The cost per occupied room is often expressed in rupees and as a percentage. Another method of estimating expenses is to estimate variable costs per room sold and then multiply these costs by the number of rooms expected to be sold.

Refining Budget Plans

Departmental budget plans are commonly supported by detailed information gathered in the budget preparation process and recorded on worksheets and summary files. These support documents should be saved to provide an explanation of the reasoning behind the budget. They may help resolve issues that arise during the budget review and may assist in the preparation of future budgets. Many hotels refine their budgets as they progress through the budget year. Re-forecasting is normally suggested when actual operating results start to vary significantly from the budget. Significant variations may indicate that conditions have changed since the budget was first prepared.

EVALUATING FRONT OFFICE OPERATIONS

Evaluating the results of front office operations is an important management function; without evaluation, managers will not know whether the front office is attaining planned goals. Front office managers should evaluate the results of department activities on a daily, monthly, quarterly, and yearly basis.

Important tools that front office managers can use to evaluate the success of their operations include the following:

Daily operations report: The daily operations report is also known as the *manager's report*, the *daily report*, and the *daily revenue report*. The daily operations report summarizes the hotel's financial activities during a twenty-four-hour period. The daily operations report provides a means of reconciling cash, bank accounts, revenue, and accounts receivable.

This report also serves as a posting reference for various accounting journals and provides important data that must be input to link front and back office automated functions. Daily operations reports are often uniquely structured to meet the specific needs of individual hotels. Copies of the daily report are usually distributed to the hotel's general manager and all department heads. Basically the daily operations report gives an insight into the performance of the hotel on a daily basis with comparisons to the previous month and year.

Occupancy ratios: Occupancy ratios measure the effectiveness of the front office and reservations staffs in selling guest rooms. The following rooms' statistics must be gathered to calculate basic occupancy ratios: Number of rooms available for sale, number of rooms sold, number of guests, number of guests per room, and net rooms' revenue. (This information is usually available in the daily operations report).

Occupancy ratios are typically calculated on a daily, weekly, monthly, and yearly basis.

The front office system typically generates occupied rooms data and calculates occupancy ratios for the front office manager, who analyzes the information to identify trends, patterns, or problems. When analyzing the information, the front office manager must consider how a particular condition may produce different effects on occupancy. For example, as multiple occupancy increases, the average daily room rate may also increase. This is because, when a room is sold to more than one person, the room rate may be greater than when the room is sold as a single. However, since the room rate for two people in a room is usually not twice the rate for one person, the average room rate per guest decreases.

- **Occupancy percentage**

  The occupancy percentage is the ratio of the number of rooms sold to the total number of saleable rooms. It determines the level of revenue that will be generated by the hotel and is an indicator of the performance of the hotel. Higher the rate of occupancy percentage, better the room sales revenue of the hotel.

  The formula to calculate occupancy percentage is as under:

  Occupancy percentage

  $$= \frac{\text{Number of rooms sold}}{\text{Total no. of saleable rooms}} \times 100$$

  In case some rooms are under renovation, repair, or are out of order, the same are not included in the total number of saleable rooms. Saleable rooms are also called *lettable rooms* or *available rooms*.

- **House count**

  The house count is the total number of resident guests present in the hotel. It is used to determine the average room rate per person.

  The formula to calculate the house count is as under:

  House count = House count of previous day brought forward + Today's arrival – Today's departure

  House count can also be calculated as follows:

  House count = single rooms + 2 x (double rooms) + extra beds

- **Bed occupancy percentage**

  It is the ratio of the number of beds occupied to the total number of available beds in the property. In ascertaining the total number of beds available, all the beds in different types of rooms must be counted to arrive at the bed occupancy percentage.

  The formula to calculate bed occupancy percentage is as under:

  Bed occupancy percentage

  $$= \frac{\text{Number of beds occupied}}{\substack{\text{Total no. of beds available} \\ \text{for guests}}} \times 100$$

- **Domestic occupancy percentage/ Local occupancy percentage**

  It is the ratio of the total number of domestic guests to the house count. This is the indicator of the type of clients visiting the hotel.

  The formula to calculate domestic occupancy percentage is as under:

  Domestic occupancy percentage

  $$= \frac{\text{Total number of domestic guests}}{\text{House count}} \times 100$$

  Alternatively, it may be calculated by subtracting the foreigner's occupancy percentage from 100.

  Domestic occupancy percentage = 100 – Foreigner's occupancy percentage

- **Foreigner's occupancy percentage**

  It is the ratio of the total number of foreign nationals to the house count. This is an indicator of the type of clients visiting the property.

  The formula to calculate foreigner's occupancy percentage is as under:

  Foreigner's occupancy percentage

  $$= \frac{\text{Total number of foreigner guests}}{\text{House count}} \times 100$$

  Alternatively, it may be calculated by subtracting the domestic occupancy percentage from 100.

  Foreigner's occupancy percentage = 100 – Domestic occupancy percentage

- **Double occupancy percentage**

  This ratio is worked out to understand the trend in the sale of double rooms. The percentage reveals how the hotel is utilizing the rooms' capacity to increase its room revenue by ensuring sale of double rooms. More the sale of double rooms, more is the revenue from rooms.

The formula to calculate double occupancy percentage is as under:

Double occupancy percentage

$$= \frac{\text{Number of double rooms occupied}}{\text{Number of rooms sold}} \times 100$$

Double occupancy percentage can also be calculated as follows:

Double occupancy percentage

$$= \frac{\text{House count}}{\text{Number of rooms sold}} - 1 \times 100$$

- **Single occupancy percentage**

  The formula to calculate single occupancy percentage is as under:

  Single occupancy percentage

  $$= 2 - \frac{\text{House count}}{\text{Number of rooms sold}} \times 100$$

  Alternatively, it may be calculated by subtracting the double occupancy percentage from 100.

  Single occupancy percentage = 100 − Double occupancy percentage

- **Complimentary occupancy percentage**

  This ratio is worked out to find out the proportion of rooms given as complimentary in relation to the number of rooms available during the year.

  The formula to calculate complimentary occupancy percentage is as under:

  Complimentary occupancy percentage

  $$= \frac{\text{Number of complimentary rooms}}{\text{No. of rooms available during the year}} \times 100$$

- **Average Daily Rate (ADR)/Average Room Rate (ARR)**

  Average daily rate is a statistical unit that is often used in the hospitality industry. It is the average rental income per occupied room for a given time period. It is calculated by dividing the total room revenue generated in a specific duration of time by the total number of rooms sold in that duration.

The formula to calculate ADR is as under:

Average Daily Rate

$$= \frac{\text{Total room revenue generated in a specific period}}{\text{Total number of rooms sold in that period}} \times 100$$

Some hotels include complimentary rooms in the denominator to show the true effect of complimentary rooms on the average daily rate. This may also be called the *average house rate*.

- **Average Rate per Guest (ARG)/Average Person Rate**

  Average room rate per guest is calculated by dividing the total room revenue by the total number of guests in the hotel, including children above five years.

  The formula to calculate the average room rate per guest is as under:

  Average room rate per guest

  $$= \frac{\text{Total room revenue generated in a specific period}}{\text{Total no. of guests in the hotel}} \times 100$$

- **Average Guest per Room (AGR)**

  It is also known as *Guest per occupied room* or *Average occupancy per room* or *Average number of guests per room sold* (also called the *Occupancy multiplier*).

  The formula to calculate the average number of guest per room is as under:

  $$\text{AGR} = \frac{\text{Number of guests}}{\text{Number of rooms sold}}$$

- **Revenue per Available Room (RevPAR)**

  RevPAR is the revenue per available room. It is one of the most important hotel statistics, because it provides a statistical benchmark for comparison with similar hotels.

RevPAR is calculated by multiplying the average daily rate with the occupancy percentage.

RevPAR = ADR × Occupancy percentage

RevPAR can also be calculated as under:

RevPAR

$$= \frac{\text{Total room revenue generated in a specific period}}{\text{Total number of rooms available in the hotel}} \times 100$$

RevPAR can analyze the performance of a hotel over any timeframe—daily, weekly, monthly, quarterly, or yearly. The RevPar analysis can also be used to compare hotels with different tariffs and number of rooms.

- **Revenue per Available Customer (Rev PAC)**

RevPAC divides the total revenue generation of the hotel by the number of guests staying overnight, thereby showing the average revenue generated by each guest. For hotels with high multiple occupancy, RevPAC is especially important, since it provides an average spending figure per guest.

RevPAC can be calculated as under:

RevPAC

$$= \frac{\text{Total revenue generated in a specific period}}{\text{Total number of guests in the hotel}}$$

- **Gross Operating Profit per Available Room (GopPar)**

Gross operating profit is defined as the profit of the hotel before allocating central charges such as bank interest, depreciation and property taxes.

The formula to calculate the GopPar is as follows:

$$\text{GopPar} = \frac{\text{Total revenue} - \text{Management controllable operating expenses}}{\text{Number of rooms available}}$$

- **Average stay per visitor**

At the end of the day, the total number of nights divided by the number of departures for that day shall give us the average stay per visitor. For example, there are 5 departures (A, B, C, D and E) in a day in a particular hotel. The duration of stay of each was 7, 10, 15, 2 and 6 days respectively. Average stay per visitor can be calculated as under:

Average stay per visitor

$$= \frac{7 + 10 + 15 + 2 + 6}{5} = 8 \text{ days}$$

- **Market Share Index/RevPAR Index**

Market share is defined as a hotel's occupancy performance in relation to other hotels within a predetermined competitive set.

A major task in calculating the market share is the determination of the competitive set. The answer to the question—if a guest is not staying at our hotel, where can he possibly stay?— constitutes the competitive set. The total market potential is the sum total of the number of rooms that are available in the total number of participating hotels.

Market share index enables the managers to assess their hotel's performance with respect to the competitors. It assists the managers to develop plans to combat the loss of fair market share and also to gain market share from the competitors.

RevPAR Index can be calculated as under:

RevPAR Index

$$= \frac{\text{Hotel RevPAR}}{\text{Competitive Set RevPAR}}$$

Competitive set RevPAR figures can be obtained from the STAR Report or the HRM (HotelRevMax) Report.

If the market share index is below 100%, it implies that it is an under performing hotel. If it is 100%, it implies that the hotel has a fair share and if it is above 100%, it implies that the hotel is over performing.

Rooms Revenue Analysis: Front office staff members are expected to sell rooms at the rack rate unless a guest qualifies for an authorized discounted room rate. A *room rate variance report* lists those rooms that have been sold at other than their rack rates. Managers use the room variance report to review the use of various special rates to determine whether staff has followed front office policies and procedures. One way for front office managers to evaluate the sales effectiveness of the front office staff is to generate a *yield statistic*, which is actual rooms' revenue as a percentage of potential rooms' revenue.

Potential rooms' revenue is the amount of rooms' revenue that can be generated if all the rooms in the hotel are sold at rack rate on a given day, week, month, or year. The yield statistic can be calculated as follows:

$$\text{Yield statistic} = \frac{\text{Actual rooms' revenue}}{\text{Potential rooms' revenue}}$$

The Hotel Income Statement: A hotel's income statement provides important financial information about the results of hotel operations for a given period of time. The period may be one month or longer, but should not exceed one business year. The income statement reveals the amount of net income for a given period, so it is one of the most important financial statements management uses to evaluate the hotel's overall success. Although front office managers may not directly rely on the hotel's statement of income, it is an important financial indicator of operational success and profitability. The hotel income statement relies in part on detailed front office information that is supplied through the rooms' schedule. (*The rooms' schedule is discussed in the next section.*)

The hotel's statement of income is often called a consolidated income statement because it presents a composite picture of all the hotel's financial operations. Rooms' division information appears on the first line, under the category of operated departments.

The amount of income generated by the rooms division is determined by subtracting 'payroll and related expenses' and other expenses from the amount of net revenue produced by the rooms division. Payroll expenses charged to the rooms division may include those associated with the front office manager, front desk agents, reservations agents, housekeepers, and uniformed service staff. Since the rooms division is not a merchandising facility, there is no 'cost of sales' to subtract from the net revenue amount. Revenue generated by the rooms division is usually the largest single amount produced by revenue centers within a hotel.

The Rooms Schedule: The hotel's income statement primarily contains summary information; the separate departmental income statements prepared by each revenue center, called 'schedules', provide more detail. The hotel accounting division, not the front office accounting staff, usually prepares the rooms' schedule.

By carefully reviewing the rooms' schedule, the front office manager may be able to develop action plans to improve the division's financial condition and services. For example, the income statement may indicate that the hotel's telecommunications revenue is down, due to the application of a long-distance surcharge. This analysis reveals that guests are choosing to make fewer telephone calls using the hotel's telecommunications system, because the cost per call was increased by the surcharge. Therefore, even though the revenue per call may have increased, overall telecommunications revenues have decreased.

Rooms Division Budget Reports: The hotel's accounting division prepares monthly budget reports that compare actual revenue and expense figures with budgeted amounts. Budget reports can provide timely information for evaluating front office operations. Front office performance is often judged according to how favorably the rooms division's monthly income and expense figures compare with budgeted amounts.

A typical budget report format should include both monthly variances and year-to-date variances for all budget items. Front office managers are more likely to focus on the monthly variances, since year-to-date variances merely represent the accumulation of monthly variances. Percentage variances are determined by dividing the rupees variance by the budgeted amount. Both rupees and percentage variances are shown because either type of variance alone may not indicate the significance of the variances reported. It should be expected that actual results will differ somewhat from budgeted amounts, since budgeting is not a perfect science. Therefore, only significant variances from budget should be investigated by management.

Operating ratios: Operating ratios assist managers in evaluating the success of front office operations.

Payroll and related expenses tends to be the largest single expense item for the rooms division as well as the entire hotel. For control purposes, labor costs are analyzed on a departmental basis. Dividing the payroll and related expenses of the rooms division by the division's net room revenue yields one of the most frequently analyzed areas of front office operations, i.e. labor cost.

Operating ratios should be compared against standards such as budgeted percentages. Any significant differences between actual and budgeted labor cost percentages must be carefully investigated, since payroll and related expenses represent the largest single expense category.

Figure 32.2 shows a number of operating ratios that may be used.

Ratio standards: Operating ratios are meaningful only when compared against useful criteria such as planned ratio goals, corresponding historical ratios, and industry averages.

Ratios are best compared against planned ratio goals. For example, a front office manager may more effectively control labor and related expenses by projecting a goal for the current month's labor cost percentage that is slightly lower than the previous month's. The expectation of a lower labor cost percentage may reflect the front office

	Net revenue	Payroll and related expenses	Other expenses	Departmental income
% of total hotel revenue	×			
% of departmental revenue		×	×	×
% of departmental total expenses		×	×	
% of total hotel payroll and related expenses		×		
% of change from prior period	×	×	×	×
% of change from budget	×	×	×	×
Per available room	×	×	×	×
Per occupied room	×	×	×	×

Fig. 32.2: Useful rooms division operating ratios

manager's efforts to improve scheduling procedures and other factors related to the cost of labor. By comparing the actual labor cost percentage with the planned goal, the manager can measure the success of his efforts to control labor costs.

Industry averages may also provide a useful standard against which to compare operating ratios. These industry averages can be found in publications prepared by the national accounting firms and trade associations serving the hospitality industry.

Experienced front office managers realize that operating ratios are only indicators; they do not solve problems or necessarily reveal the source of a problem. At best, when ratios vary significantly from planned goals, previous results, or industry averages, they indicate that problems may exist. Considerably more analysis and investigation are usually necessary to determine appropriate corrective actions.

REVIEW QUESTIONS

1. How do the seven functions of management fit into the overall management process? How do these functions apply to the front office manager's position?

2. List and explain any five tools that front office managers can use to evaluate the success of front office operations.

3. Which are the historical data used to forecast rooms' revenue?

4. Which are the different methods by which rooms' revenue is forecasted?

5. Which are different expenses of the rooms' division department and to what are they directly related to?

6. Give the formula to estimate the various rooms' division expenses.

7. Explain the daily operations report. What information it provides to the hotel management?

8. How do you go about analyzing rooms' revenue for the hotel?

9. Describe the importance of:
   • Rooms division statement
   • Rooms division budget report

10. What are some useful standards against which front office managers should compare operating ratios? What is the significance of a variance from standards?

11. Discuss operating ratios.

12. Differentiate between:
    • ARR and RevPAR
    • RevPAR and RevPAC

13. Give formula for the following:
    • Occupancy %
    • House count
    • Local occupancy %
    • Bed occupancy %
    • RevPAR
    • ARR
    • Average rate per guest
    • Occupancy multiplier

14. Solve the following:
    • If a room revenue for March, 2016 is ₹ 10,45,350 and the no. of rooms sold is 950, then calculate the ARR.
    • Blue star hotel has 350 lettable rooms, out of which 250 are single and 100 are twin. On 15th Oct 2015, 180 single rooms and 60 twin rooms are occupied by guest. Calculate bed occupancy percentage for the day.

15. A hotel with 100 saleable rooms just finished a 30 days accounting period and had 1500 rooms sold on double occupancy and 600 rooms sold on single occupancy. The room sales totaling to ₹ 30,00,000/-. Calculate the following:
    • Room occupancy %
    • Bed occupancy %
    • Average room rate
    • Average number of guests per room

The Electronic Front Office

Learning Objectives

After reading this chapter, you will be able to understand the following:

- The electronic front office
- Property management systems
- PMS applications in front office
- Back office interfaces
- System interfaces
- Different property management systems like Micros, Amadeus, IDS Fortune, and ShawMan

ELECTRONIC FRONT OFFICE

As advancements in the technology of electronics and computerization gained popularity so did the developments in the use of computers in the hotel industry. Hotel industry is one place that has undergone a tremendous change in the past few years with the advent of computers. The customers now can explore various options, compare prices, and make reservations from the comfort and privacy of their home rather than having to walk to the office of a tour operator.

In the modern age of automation, front office is becoming very important to provide efficient, effective and reliable services to the guest with the optimum use of technology. However, this does not mean that the services of front desk staff is no longer required, but rather they are relieved of some of the routine and repetitive clerical task which is a part of the day-to-day operation of every front office.

The electronic front office (EFO) is quite expensive, and to justify its cost it must offer substantial benefits not only to the hotel but also to its guests. Many tasks in the front office are best suited for the use of a computer because of their speed and accuracy. These include the processing of reservations and registering of guests; the updating of room status; posting guests' charges; settling guest bills; updating of guest history; and gene-rating typical reports for the use of deparment heads and top management. A computer system also increases and enhances the availabilty of statistical data, thus enabling better management decisions through faster and more accurate information.

An electronic front office does not mean that it operates in isolation. On the contrary, the front office now has the ability to work directly with many or all of the other departments within the hotel. Departments such as housekeeping, restaurants, bars, coffee shops, night club or a health club all have access to the information normally held at the front desk. This ability to call up front desk information is known as being *interfaced* with the front desk.

The computerization of the front office is a part of the Property Management System, wherein all the departments of a hotel are linked to the mainframe.

PROPERTY MANAGEMENT SYSTEMS

A PMS is a computer-based management system. It is a specific program meant for carrying on specific tasks depending on the need of the end-user in a particular environmental set up. In the hospitality industry, a PMS manages all aspects of hotel operations like guest bookings, online reservations, check-ins, guest accounting, point of sale, telephone and checkouts. There are many types of PMSs—it is up to a hotel to choose the one that best fits its needs.

There are different modules of PMS to manage individual departments of a hotel, such as front office module, housekeeping module, restaurant management system, back office module, etc. These modules are supplied by a large number of vendors, who

modify PMS solutions to meet the requirements of the hotel. Each individual user is given access to different modules, depending upon its area of work and its level in the hierarchy of the hotel. The general manager and the system administrator have full access to all PMS modules.

Hotel PMS software offers both front and back office modules. Front desk modules are designed for guest and room management while the back office modules are designed to manage hotel resources, assets and inventory.

All property management systems do not operate identically. However, some generalizations about property management systems may illustrate the nature of front office applications. A property management system comprises sets of software packages that are capable of supporting a variety of tasks in front and back office areas. The four common front office software packages are designed to help front desk employees perform various functions related to (Fig. 33.1):

• Reservations management
• Rooms' management
• Guest account management
• General management

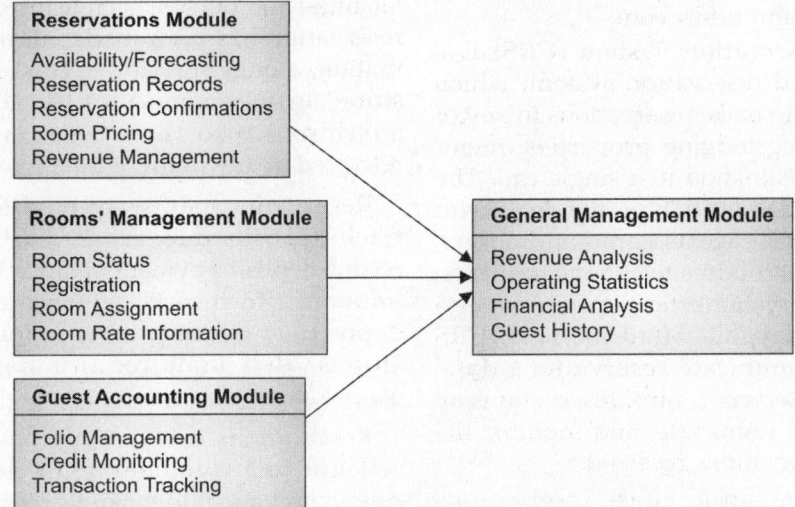

Fig. 33.1: Front office computer applications

PMS APPLICATIONS IN FRONT OFFICE

Reservations Management Software

An in-house reservations software module enables a hotel to rapidly process room requests and generate timely and accurate rooms, revenue, and forecasting reports. Most hotel chains participate in remote reservation networks known as Global Distribution Systems (GDS), Internet Distribution Systems (IDS), and Central Reservation Systems (CRS).

Global Distribution System (GDS) is a worldwide computerized reservation network, which is used as a single point of access for reserving hotel rooms, airline seats, rental cars, and other travel-related items by travel agents, online reservation sites, and large corporations. The largest and best-known worldwide GDSs are SABRE, Galileo International, Amadeus IT, and Worldspan. Internet Distribution System (IDS) are intermediary websites that represent hotel companies and sell overnight accommodations on a commission or mark-up fee basis. IDS capture guest reservation data and may be automatically connected to the hotel's reservation management system through a GDS or CRS. Examples of internet distribution sites include travelocity.com, travelweb.com, priceline.com, and orbitz.com.

Central Reservation System (CRS) is a computer-based reservation system, which enables guests to make reservations in any of the participating lodging properties (members) at any destination in a single call. The central reservation office typically deals with direct guests, travel agents, corporate bookers, etc. by means of toll-free telephone numbers. The central reservation offices operate twenty-four hours a day, all round the year. CRS typically communicate reservations data, track rooms reserved, control reservations by room type and room rate, and monitor the number of reservations received.

A hotel using an in-house reservations software module can receive data sent directly from any remote reservations networks. Automated in-house reservations records, files, and revenue forecasts are instantaneously updated as reservation data are received at the hotel. It is through electronic file updating that the in-house system remains current and in control of reservations activities. Many systems allow real time, two-way communication between remote reservations networks and hotel computers, enabling instantaneous updates of inventory and guest information. This way, accurate guest room inventories and pricing are updated between both systems.

In addition, previously received reservation data can be automatically reformatted into pre-registration applications, and a series of appropriate reports can be generated. Various reservation management reports, containing a summary of reservation data, occupancy forecasts, and guest account status information, can also be generated. Reservations management software also includes upgraded rate control features, guest history modules, and more detailed hotel information such as bed types, guest room views, guest room location, in-room facilities, meeting space, public areas, entertainment and recreational facilities, and other special features. Once the reservation has been made, all of the information, along with any special requests, is stored in the memory of the software for anything up to three years and can be retrieved at any point of time.

Reservations software module can also track deposits due, request deposits, and record deposit payments made. This is very important to resorts, which often require deposits to confirm reservation requests. Hotels also may require deposits for confirming group reservations.

Reservations software module facilitates revenue management as it enables the reservation agent to make the best choices as to what rooms to book and at what rate. This gives the ability to instantly analyse the profit potential of each booking. This is done by adjusting the room rate to suit the demand for rooms at any particular time.

To stay ahead of the competition, more and more hotels are concentrating on data mining, also known as 'warehousing'. This is where hotel chains centralize their guest history records. Previously, individual hotels used to keep a record of their customers. Now, this information is being centralized. This means that any hotel within that chain can have access to the guest details. Therefore, the front desk staff can recognize the guest's needs prior to their arrival and keep track of their preferences.

Rooms' Management Software

Rooms' management software module facilitates pre-registration activity. Because all of the guest's information has been taken by the reservations agent at the time of reservation, all that the receptionist has to do is call up that information. Upon the arrival of the guest the receptionist calls on the system to print the guest registration card, confirms all the details with the guest, confirms the mode of payment, and then simply gets the guest to sign the registration card.

Rooms' management software module maintains updated information on the status of rooms and their rates and assists front desk staff in room assignment at the time of registration, and helps him in coordinating guest services. A rooms' management software module can also provide rapid access to room availability data during the reservation process. This information can be especially useful in short-term reservation confirmation and rooms revenue forecasting. Nowadays the rooms' management software module has replaced most traditional front office equipment.

Rooms' management software module can provide front desk employees with a summary of each room's status. For example,

a front desk employee may enter a room number at a front office computer terminal, and the current status of the room will be displayed on the terminal's screen. Once the room has been cleaned and readied for occupancy, the housekeeping staff can communicate the room's updated status by means of computer terminal located in the housekeeping department. With a property management system, changes in room status can be instantaneously communicated between both the departments (front office and housekeeping). In addition, front desk staffs can enter a guest's specific requests into the system to find a room that suits his requirements. For example, a front desk staff can request information on all vacant and clean rooms facing the Taj Mahal that have king-size beds. Finding a room with specific information is very difficult in a manual system.

Of late, some rooms' management software module also features maintenance and special-request dispatch capabilities. For example, a room with a television set problem or needing extra pillows can be recorded through this module and a hotel engineer or room boy can be dispatched immediately to fulfill the request.

Rooms' management software also assists the reservations function. When rooms are temporarily taken out of inventory for repair, maintenance or deep cleaning, the number of available rooms in the reservation office is automatically reduced. This helps control future room inventory and ensure that all guests have rooms ready when they arrive.

Rooms' management software facilitates self-check. A guest can check-in to a hotel without necessarily contacting the front desk. They initiate the self-registration process by inserting their credit card into a terminal located at the airport or in the hotel lobby, and then enter their details. As the self-registration terminal is electronically linked to the reservation management software and rooms' management software, the terminal is easily able to track the reservation information of the guest in the hotel. They are assigned a room by the terminal and the room keys are also dispensed by the machine. Room status gets automatically updated at the front desk.

Guest Account Management Software

Guest account management software greatly helps to reduce the drudgery of the guest checkout procedures. It prints, in a matter of seconds, an accurate bill for the final settlement by the guest.

Guest account management software also increases the hotel's control over guest accounts and significantly simplifies the traditional night audit process. Guest accounts are maintained electronically, thereby eliminating the need for folio cards, folio trays, and account posting machines. The guest accounting module monitors predetermined guest credit limits and provides flexibility through multiple folio formats. At checkout, previously approved outstanding account balances may be automatically transferred to an appropriate back office accounts receivable file for subsequent billing and collection. Account management capabilities represent some of the major benefits of a property management system. For example, a credit manager in a large business hotel can monitor the credit limits of all guests and report all accounts that are approaching or have exceeded their credit limits.

When the hotel's point-of-sale outlets are connected to the front office system, guest charges can be communicated instantly to the property management system. These charges will be automatically posted to appropriate electronic guest folios. Automatic posting procedures are intended to improve efficiency while reducing the number of late charges.

General Management Software

General management software cannot operate independently of other front office software packages. General management applications tend to be report-generating packages that depend on data collected through reservations management, rooms' management, and guest account management programs. For example, general management software may be able to generate a report showing the day's expected arrivals and the number of rooms available for occupancy—a combination of reservations and rooms' management data. With this module, the management can quickly access information such as room occupancy, average room rate, bed occupancy, no-show, outstanding balance reports, etc. In addition to generating reports, the general management module serves as the natural link between front and back office system interface applications.

BACK OFFICE INTERFACES

A comprehensive property management system involves integrating the hotel's front and back office areas. Although front and back office software packages can be independent of one another, integrated systems offer the hotel a full range of control over a variety of operational areas. Such areas include room sales, telephone call accounting, payroll, and account analysis. An integrated system cannot produce complete financial statements unless all the required data are stored in an accessible database. Many reports generated by the back office system depend on the front office system's collection of data.

Most property management system vendors offer several back office application modules as well. The four most popular back office applications are:

• *General ledger accounting software,* consisting of accounts receivable and accounts payable application packages. Accounts receivable software monitors guest accounts and account billing and collection when integrated with the front office guest accounting module. Accounts payable software tracks hotel purchases and helps the hotel maintain sufficient cash flow to satisfy its debts.

- *Human resources software* may include payroll accounting, personnel record-keeping, and labor scheduling. Payroll accounting includes time and attendance records, pay distribution, and tax withholdings. Personnel records include withholding and deductions, labor history, and job performance evaluations. Labor scheduling involves tracking employee skills and availability in relation to the hotel's staffing requirements.

- *Financial reporting software* helps the hotel to develop a chart of accounts in order to produce balance sheets, income statements, and transaction analysis reports.

- *Inventory control software* monitors stocking levels, purchase ordering, and stock rotation. Additional computations include inventory usage, variance, valuation, and extensions.

SYSTEM INTERFACES

A variety of property management system interface applications (guest- and non-guest-operated) are available in an automated environment for enhanced efficiency.

Non-guest-operated Interfaces: Common interfaces that are not initiated by guest activity include the following:

- An *electronic point-of-sale (EPOS) system* allows guest account transactions to be quickly transmitted from different revenue centers to the property management system for automatic posting to electronic folios. For example, a guest can have breakfast and the charges for the same will have been posted to his room account before he actually leaves the coffee shop. Similarly, the EPOS can work in the lounge area, bar, night club, restaurants, barber shop, etc.

- A *call accounting system (CAS)* tracks outgoing calls made by the guest from his room for pricing and automatic posting to electronic folios. The main advantage of CAS is that the guest can make local as well as long distance calls from the comfort of their rooms without having to go through the telephone operator. It enhances the guest satisfaction and also results in lesser number of guest complaints.

- An *electronic locking system (ELS)* typically interfaces with the rooms' management module to provide enhanced guest security and service. The ELS is used widely nowadays in hotels. This system helps the hotels to control access to guest rooms. Only a person with the proper card key coded for the specific room can enter that room. When the electronic locking system is networked with the PMS, the front desk person is able to code the room keys for the guests. The coding is such that the key will become non-functional after the checkout time on the date of departure.

- An *energy management system (EMS)* automatically controls the temperature, humidity, and air movement in public areas and guest rooms through a rooms management interface.

An EMS is a computer-based control system designed to automatically control the operation of mechanical equipment in a hotel, so as to achieve the optimum savings on utilities such as gas and electricity. An EMS interfaced to a property management system offers a number of opportunities for energy control. The system determines when equipment such as heating, air conditioning and ventilation can be turned on or off, or regulated up or down. An example would be the control of lighting and central heating or air conditioning of a guest room. When the guest enters the room they insert their room key into a slot located next to the door. This notifies the energy system to activate all power to that room, which in turn means that the room lights can be turned on and the room ventilation system works. When the guest leaves the room, the system will automatically shut down after a short delay period, thus saving on electricity that would be wasted if room lights were left on, and saving gas or electricity from heating a room which is not in use. By interfacing an EMS with a front office rooms' management system, it is possible to automatically control room assignments and achieve desired energy cost savings.

Guest-operated Interfaces: Modern hotels have gone beyond basic property management systems by providing automated conveniences and services to its guests by installing a variety of guest-operated devices. In some hotels, guests may inquire about in-house events and local activities through automated information kiosks located in public areas, or through the television or a portable communication device in their rooms. Connecting a printer to an information terminal enables guests to print out customized information.

Recent technological developments enable guests to review their folios and complete the checkout formalities from the comfort and privacy of their guest rooms. In-room televisions or other communication devices interfaced with a guest accounting module enable guests to simultaneously access folio data and to approve and settle their accounts by selecting a pre-approved method of settlement. Guest room telephones interfaced with the property management system may also be used for this purpose. In-room computers linked to external computer information services permit guests to access e-mail, Internet websites, transportation schedules, local restaurant and entertainment guides, stock market reports, news and sports updates, shopping catalogs, and video games.

An *in-room entertainment system* can be interfaced with a front office accounting module or can function as an independent system. In-room entertainment systems allow guests to access various forms of entertainment through their guest room televisions. If there is a charge for the service, such as a pay-per-view movie, video game, or Internet access, the charge can be automatically calculated and posted to the guest's electronic folio. To keep guests from inadvertently tuning to a pay channel, the television is usually preset to a non-pay channel or preview channel. Incorporating a preview channel can significantly reduce the number of guest disputes about the validity of pay TV or movie charges. In-room entertainment systems may even require the guest to dial an in-house department to request that a pay channel be activated. In addition, a preview channel provides the hotel with advertising opportunities. The preview channel can display information about the services and facilities offered by the hotel. The hotel can also sell advertising for local attractions, thereby creating a new revenue source.

There are two types of *in-room vending systems. Non-automated honor bars* consist of beverage and snack items in both dry and cold storage areas within a guest room. The bar's beginning inventory level is recorded, and changes in the inventory are noted by staff members on a daily basis. Appropriate charges for missing or consumed items are posted to the guest's folio. Since honor bars are available at all times, this system often results in an unusually high volume of late charge postings. *Automated honor bars* or *in-room fully automated vending machines* may contain fiber-optic sensors that record the removal of stored products from designated compartments. When a sensor is triggered, these machines assume a sale and transmit point-of-purchase information to a POS microprocessor that, in turn, communicates to the front office accounting module for electronic folio posting.

Other technology-based guest amenities include the *in-room fax machine* and *centralized printer/server*. Faxes can be sent and received directly from the guest room without taking the help from any staff member. This facility is popular in hotels serving meeting, convention, and business travelers, and may link with the hotel's telephone system, which automatically calculates the cost of the fax and forwards the charge to the front desk for folio posting. With a centralized printer, guests upload their documents through an Internet connection, then leave their guest rooms and go to the hotel's business centre to download and print their documents.

Guest Services and Technology: Outstanding guest services can provide a major competitive advantage in attracting guests. For example, in-room entertainment companies are developing systems to provide local information through guest room software browsers and guest room television. Guests can locate restaurants, museums, shopping malls, and other places of interest through the guest room television. In a similar way, hotels can promote their own services or other hotels in the chain. If a guest is interested in a particular hotel, the system will automatically connect to the chain's reservation system and provide access to a reservation-booking engine.

High-speed Internet access (HSIA) service has evolved from an upgrade amenity to a 'must-have' for most hotels. HSIA service, whether wired or wireless, is usually provided in guest rooms as well as in public areas and meeting facilities. At resort locations, wireless HSIA may also be available at pools and recreational facilities, as well as on the beach. Hotels may also offer Internet access through guest room televisions, the hotel business center, and informational kiosks located throughout the property. In some properties, guests may opt to use the guest room telephone for alternative Internet access.

DIFFERENT PROPERTY MANAGEMENT SYSTEMS

There are several companies that provide the PMS software to hotels, with their own unique features to suit the different requirements of various hotels. Some of the companies providing PMS software also provide other tailor-made products to their clients. Let us discuss briefly about some of these softwares.

Micros

Micros Systems, Inc. (NASDAQ: MCRS), headquartered in Columbia, Maryland, is the world's leading developer of enterprise applications serving the hospitality and specialty retail industries. Micros Systems provides PMS software solutions to the hospitality industry through: Opera Enterprise Solutions and Micros Fidelio.

- **Opera Enterprise Solutions**: The Opera Enterprise Solution (OES) is a fully integrated suite of products that can be easily combined for deployment at any size of organization—from a single-property hotel to a global, multi-branded hotel chain. Opera modules include:
  - **Opera Reservation System:** The Opera Reservation System (ORS) is a centrally managed computer reservation system that handles all types of reservations—individual, group and party, company, travel agent, multi-legged, multi-rate, and waitlisted.
  - **Opera Customer Information System:** This gathers and manages profile data of guests, travel agents, sources, groups, and companies in a central database that can be shared by multiple properties.
  - **Opera GDS Interface:** The Opera GDS Interface links the hotel's database and the global distribution systems (GDS) and web booking engines (WBEs) through a third-party switching company.
  - **Opera Sales Force Automation:** Opera Sales Force Automation (SFA) is a central sales support tool for a hotel chain's regional or national sales team. SFA is fully integrated with Opera Reservation System (ORS).

- **Micros Fidelio**: Micros Fidelio range of software products includes property management systems (PMS), reservation systems, and points of sale (POS) systems.
  - **Property Management System:** It takes care of tasks like guest information, night audits, inventory control, profit management, and report generation.
  - **Reservation systems:** Configured to a hotel's specification, it gives the hotel staff instant room status and availability report, leading to instant reservations.
  - **POS systems:** Micros Fidelio points of sale (POS) software systems facilitate faster information retrieval, which aids faster transactions.

Amadeus PMS

Amadeus property management system can be integrated with front office, sales/marketing, and financial management functions. It allows the user to move faster in all core aspects of guest experience management.

- **Front Office Module:** This module offers full availability, reservation, yielding and billing functionalities, which in turn generate useful performance statistics. It provides data on performance indicators such as sales, accounts, source, and segment activity, which is monitored to analyze business efficiency and used to generate management reports.
- **Sales and Marketing Module:** This module aids the sales and marketing professionals to target potential guests and effectively manage customer relationships.
- **Conference and Banqueting Module:** Event planning is a feature that helps generate revenue for hotels. The key features of the conference management system include real-time conference/meeting room availability and equipment management. An interface with front office

helps the sales agent to book rooms according to conference dates and guest preferences.

- **Financial Management:** It is designed for liquidity planning and control along with comprehensive accounting, financial reporting, and analysis.

IDS Fortune

IDS Software Pvt. Ltd. is Asia's largest dedicated provider of integrated, full-service enterprise property management software for the hospitality and leisure industry. IDS have established its presence in 40 countries worldwide. IDS have Property Management System with Back Office, one of which is Fortune Next (Fig. 33.2).

IDS offer the following three PMS for all categories of hotels:

- **Fortune Enterprise:** It provides centralized data integration along with colorful displays to aid the absorption of the important facts at a glance.
- **Fortune Express:** This has the flexibility needed to manage mid-segment and budget hotels. A single database keeps all aspects of management, speeding up information sharing for greater efficiency.
- **Fortune Genie:** It has been designed to cater to the needs of limited service hotels, motels, and serviced apartments. A single database gives a clear overview of the entire enterprise, ensuring transparency.

IDS Fortune Modules

- Front office management
- Point-of-sale
- Accounts receivable
- Sales and marketing
- Banquets
- Telephone management
- Material management
- Food and beverage costing

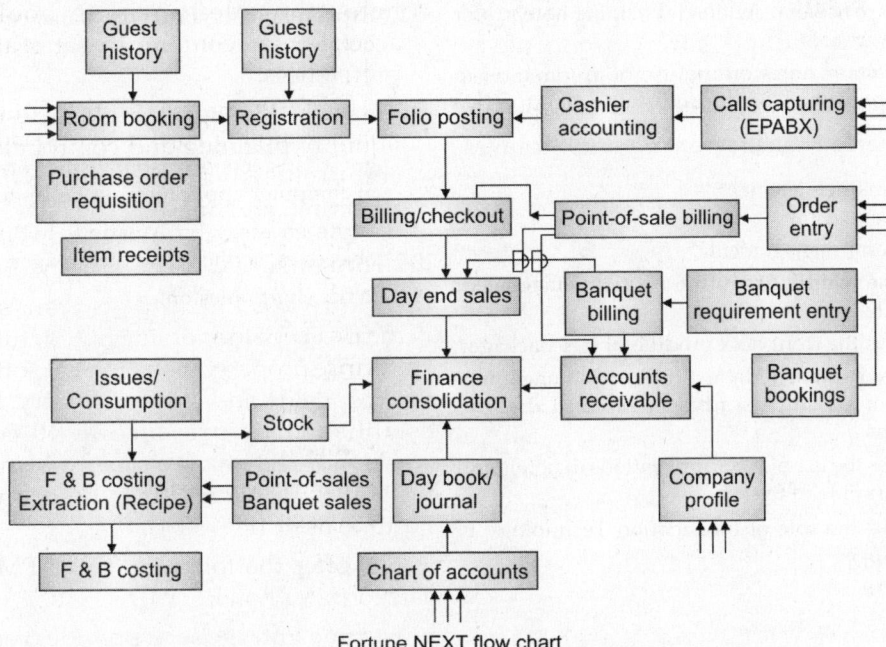

Fortune NEXT flow chart

Fig. 33.2: Fortune Next flow chart (*Source*: http://idsnext.com/)

- Financial management
- HR and payroll
- Maintenance management
- Quality management
- SMS alert

ShawMan

It started with vision of a man, the Late Pesi M. Shaw, for meeting the needs of end-to-end software to manage hotel business. It obtained large-scale recognition in late nineties for its cutting edge development efforts and is now a leader in Hospitality solutions across India and overseas.

The ShawMan PMS is an integrated front office management software that can host multiple properties and handle guest reservations simultaneously across many units, including an integrated web-based reservation agent and an auto confirmation manager.

The ShawMan Hotel Centre is a unique and powerful interface design that handles registration, cashiering, and night audit functions, treating guests and rooms as two manageable objects and supports right click quick task launch of forms and reports, undo and redo, eraser and task history to effortlessly customize the desktop for each user and department to manage a slew of jobs undertaken daily by the attendants, supervisors and managers both at the front desk and back office.

REVIEW QUESTIONS

1. Define PMS. What is the importance of PMS?
2. What is a Property Management System (PMS)? Explain various modules of any PMS used in hotel.
3. List and describe the four most common front office software packages of a PMS in a hotel. How does a general management module depend on the other three?

4. How are newer technologies helping hotels offer more services?

5. Write short notes on the use of automation in the operations of the following front office functions:
   - Reservations
   - Rooms management
   - Guest accounting
   - General management

6. List the main functions of a room management module.

7. What is the front desk module of IDS package?

8. Briefly explain Fidelio Property Management System stating its advantages and disadvantages.

9. Discuss the use of PMS applications in hotels and explain Fidelio System.

10. Discuss the role of Information Technology in hotels.

11. Explain how the Property Management System helps in smooth functioning of front office department?

12. 'Hotel automation leads to the efficient management of hotel resources and to greater guest satisfaction'. Elaborate the various aspects of computer applications in hotels today.

13. Write an essay on Amadeus highlighting its objectives, services and advantages.

14. Write short notes on:
   - Fidelio system
   - IDS
   - Amadeus
   - Galileo
   - Point-of-sales module
   - ShawMan
   - GDS

Front Office Budgeting

Learning Objectives

After reading this chapter, you will be able to understand the following:

- Budget
- Front office budgeting
- Types of budget
- Budgetary control

BUDGET

Budgeting is one of the main planning activities of the front office manager. It is the process by which, based on the actual performance of establishments in the past, estimates of expenditure and receipts are made and adjusted for forecasting future outcomes. Budgets can be defined in different ways:

A budget is a financial and/or quantitative statement, prepared prior to a defined period of time, a policy to be pursued during that period for the purpose of attaining a given objective.

It can also be defined as a plan by which resources required to generate revenues are allocated.

Budget is a detailed estimated plan of operations for a specific future period. It acts as a yardstick for the evaluation of the performance of the organization as it is a complete program of activities of the business for the period covered.

A budget is a financial plan to accomplish the business objectives, covering all phases of operations for a definite period in future and is expressed in terms of money. It is also referred to as a *financial plan*.

A budget is a formal expression of the objectives in terms of finance, specifying the manner in which the course of action is to be followed to achieve it.

Why is Budget Required?

The budget provides an opportunity for taking a critical look on the cost incurred by the department, reviewing past planning and present accomplishments, and then taking appropriate steps to accomplish better outcomes in the next *financial year*.

The budget thus acts as a guide that provides the managers with the standards by which they can measure the success of operations. A budget provides a financial framework within which a particular department operates. The budget also acts as a guide as to which things need repair or replacement. It also helps to determine what valuable pieces of equipment may be purchased and to pinpoint the areas which need to be emphasized for the coming year. It can be said that a budget is an instrument used by the management for controlling and directing activities, especially purchasing activities.

The budget is just like an architect's plan for a construction of a building, which guides the architect as well as the building contractor to know the manner in which they are expected to proceed with their work of constructing the building, so that the building when ready, is exactly of the same dimensions as shown in the architect's plan.

Approaches to Budgeting

There are basically two approaches to budgeting:

- **Top-down budgeting:** As per top-down budgeting approach, the organizational budget is prepared first. This budget is then passed down to the lower levels with each department or section preparing its own budget in order to be in line with the organizational budget. Thus, the departments or sections prepare their own budgets to achieve the targets set out or defined in the organizational budget.

- **Bottom-up budgeting**: This is the most common method used for budgeting. Under this approach, the departmental or sectional budgets are prepared first and later integrated to form the organizational budget. This kind of budgeting approach has a better acceptance due to the participation of the lower level management in the budgetary process.

FRONT OFFICE BUDGETING

Front office budgeting refers to the budget of various sections and sub-sections of front office department. It means the 'departmental master budget', which includes the budget of each section of the front office, i.e. reservation, reception, communications, bell desk, information, cash and bills. Budget for each section is prepared by each section heads, i.e.

reservations manager, lobby manager, bell captain. The section budget includes the cost incurred of various pieces of equipment and stationery used by each section such as reservation racks, room racks, information racks, front desk, filing cabinets, telephone lines, computers and accessories, PMS, cost of stationery in terms of various forms and formats used in each section and also the cost incurred on front desk staff.

Then, these section budgets are submitted to the front office manager. Finally, the front office manager will prepare the departmental budget. The departmental budget is prepared on the basis of section budgets submitted by the various section heads of the front office department. Past figures and statistics should also be considered while planning front office budget. Even the sales and marketing department plays an active role in planning the front office budget as it can also forecast the expected future business. Once the final budget is prepared, it is to be approved by the general manager and then by the board of directors. Usually, front office budget is prepared on quarterly basis or as per the policy of the hotel.

TYPES OF BUDGETS

Budget may be of different kinds, based on type of expenses involved, the departments, and the flexibility of expenses (Fig. 34.1).

Categorization by Types of Expenditure

Based on the types of expenses involved, budget may be categorized into:

- Capital Expenditure Budget: These budgets take into account the amount of capital needed to acquire capital assets that have a life span considerably in excess of one year—these are assets that are not normally used up in day-to-day operations. This budget may also be required for replacement purpose or for the purpose of expansion of the organization. Some of the

Types of Budget		
Types of Expenses	Department involved	Flexibility of expenditure
• Capital Expenditure budget • Operating budget • Pre-opening budget	• Master budget • Departmental budget	• Fixed budget • Flexible budget

Fig. 34.1: Types of budget

common items of the front office included in the capital expenditure budget are pieces of equipment such as EPABX, telephones, fax, computers and its accessories, software (PMS), self check-in/checkout terminals, etc.

- Operating Budget: These forecast expenses and revenues associated with the routine operations of the front office department during a certain period. Operating expenditures are those costs that are incurred in order to generate revenue in normal course of doing business. It includes salaries and wages of front office staff, telephone bills, cost of stationery, and internet bills.

- Pre-opening Budget: These force the planning necessary for the smooth opening of a new hotel. These budgets allocate resources for opening parties, advertising, generating of initial goodwill, liasions, and public relations. Pre-opening budget also include the initial cost of employee salaries and wages, crockery, cutlery and other items.

Categorization by Department Involved

Based on departments involved, budgets may be categorized into:

- Master Budget: These represent the forecasted targets set for the whole organization and incorporate all incomes and expenditures estimated for the organization. Also known as *final budget*, it is the summary of all the budgets that are prepared in the organization.

- Departmental Budget: Departmental budget is the budget that is prepared by individual departments of a hotel, containing the proposed estimates of revenue and expenditure for a specific period of time. Examples of departmental budget are front office budget, housekeeping budget, food and beverage budget, maintenance budget, etc.

Categorization by Flexibility of Expenditure

Budget may be classified on the basis of flexibility of expenditure:

- Fixed Budget: These budgets remain unchanged over a period of time and are not related to the level of revenues. Such budgets include budgets for advertising and administration.

- Flexible Budget: A flexible budget is a budget that adjusts or flexes for changes in the volume of activity. The flexible budget is more sophisticated and useful than a static budget, which remains at one amount regardless of the volume of activity. Flexible budget is suitable when level of activity during the year varies from time to time either due to seasonal nature of business or due to variation in demand. This budget is also helpful in case of a new business, where future demand cannot be predicted accurately.

BUDGETARY CONTROL

Budgetary control is the management and control exercised over the activities of the organization with the purpose of achieving the targets or estimated results as laid down in the budget. The main objective of budgetary control is to help achieve the objectives of the business in accordance with the formulated policies. Most of the hotels are highly departmentalized and all revenue as well as non-revenue earning departments work in coordination with one another towards achieving their common objectives. The functions of budgetary control, is therefore, to break down the organizational objectives into departmental objectives and set norms for each department to achieve their respective objectives, and in the process, the organization achieves its overall objectives. The objectives can only be achieved if the available resources are used effectively and in a planned manner.

Advantages of Budgetary Control in Front Office Department

- It helps periodic evaluation of performance, comparison of actual performance with planned performance to know the deviation and to decide corrective measures to improve the same. It thus helps to exercise control over operations of the front office department.

- All employees of the front office department especially the managerial and supervisory staff get an opportunity to learn and get used to the planning and control process and work according to plans.

- It is a powerful tool available to the front office manager for reducing wastage, fraud and cost, and thereby maximizing the profits.

- It enables the front office department to plan their activities in advance, thereby enhancing efficiency in their work.

- It motivates and activates front office employees and habituates them to work efficiently in accordance with the plans and policies towards achieving hotel's objectives.

- It enables maintaining proper records and statistics of front office operations for more realistic forecasting required for formulation of budgets.

- By employee participation in budget preparation and communication of objectives to them, proper direction is established and conveyed to all those concerned in the front office department, to focus not only on their personal goals but also on the department's objectives.

- It is a powerful tool available to the fornt office manager for performance appraisal. The staff responsible for those functions where there is a favorable variation may be rewarded; whereas the staffs responsible for those functions where there is adverse variation may be corrected.

REVIEW QUESTIONS

1. Define budget. What are the different types of budget? Explain with the help of examples.
2. What do you understand by the term 'budgetary control'?
3. Explain the advantages of budgetary control in front office department.
4. Explain front office budgeting. What is the need of front office budget?
5. What are the different approaches to budgeting?

Managing Human Resources in Front Office

Learning Objectives

After reading this chapter, you will be able to understand the following:

❑ Importance of human resources in the hospitality industry
❑ HRM vs. HRD
❑ Basic human resources activities
❑ Organization of human resources department
❑ Importance of human resource planning
❑ Job analysis – job description and job specification
❑ The process of recruitment, selection, orientation and training
❑ HR challenges in the hospitality industry

IMPORTANCE OF HUMAN RESOURCES IN THE HOSPITALITY INDUSTRY

First, let us understand what we mean by the term 'human resources'. The term 'human resources' may be defined as the total knowledge, skills, creative abilities, talents and aptitudes of an organization's workforce, as well as the values, attitudes, approaches and beliefs of the individuals involved in the affairs of the organization. It is the sum total or aggregate of inherent abilities, acquired knowledge and skills represented by the talents and aptitudes of the persons employed in an organization.

Several terms have been used by various management thinkers to represent human resources. These include 'personnel', 'people at work', 'manpower', 'staff', 'employees', 'human assets' and 'human capital'. Whatever may be the term used, the human resources of an organization include all individuals (both men and women) engaged in various organizational activities at different levels. They include the permanent employees, part-time employees, and trainees. These human resources work in the organization for the fulfillment of their needs. These needs may be physiological, social and psychological.

The hospitality industry is one part of the larger travel and tourism industry that, in addition to hospitality, consists of transportation services organizations and retail businesses. While the hospitality industry is broad and diverse, organizations within it share some things in common. One is the need for staff members with a variety of knowledge, skills, and experience to produce the products and services that are needed or desired by consumers.

Organizations in the hospitality industry tend to be labor-intensive. Technology cannot provide the level of service that is integral to the expectations of many consumers. It is universally acknowledged that the employees of a service-oriented industry make the difference between a good and bad experience. Even the phrase, *hospitality*, refers to the friendly treatment of one's guests, and this human touch must be provided by the organization's staff members.

Hospitality and tourism organizations require employees; the greater their level of revenue and the more consumers they serve, the more staff members these operations require. Success requires a full complement of staff members from owners/managers to entry-level employees who consistently attain required quality and quantity standards.

The goal of the hospitality managers in all types and sizes of organizations and in locations around the world is to employ persons with the attitudes and abilities required to best meet the needs of those being served. Organizations and managers in the hospitality industry face real challenges in recruiting, developing and maintaining a committed, competent, well managed and well-motivated workforce which is focused on offering a high-quality 'product' to the increasingly demanding and discerning customer.

The hospitality industry has often been described as a 'people business'. In this context, the people typically referred to are both the employees who produce the products and services, and those who purchase and consume them. In this chapter, we will be focusing on one of the two groups of people just discussed, i.e. employees.

In this industry, the emphasis must be on human resources, and leaders must practice human resources management principles and practices. It requires professional human resource practices to get the right people for the right job and thereafter, maintain them at a high level of motivation to serve the guest well and re-train them.

HRM vs. HRD

During and after 1970s, several changes took place in many countries which led to the use of the term 'Human Resource Management' in place of the traditional term 'Personnel Management'. These changes include technological advances, declining importance of trade unions, shift from industrial employment to service sector employment, growing competition, de-regulation and globalization of economies, etc.

HRM means many things to many people, depending on whether you are a manager, an employee or an academician and there is no one definition that adequately captures the potential complexity of the topic.

HRM is being broadly recognized about how organizations seek to manage their employees in the pursuit of organizational success.

HRM is a distinctive approach to employment management which seeks to achieve competitive advantage through the strategic deployment of a highly committed and capable workforce, using an integrated array of cultural, structural and personnel techniques.

These are the processes used by an organization to enhance its performance by effectively using all of its staff members.

It is that part of management process that helps managers plan, recruit, select, train, develop, motivate, remunerate, and maintain human resources for an organization.

Human Resource Development (HRD) in the organization context refers to the process whereby the employees are continuously helped in a planned way to:

• Acquire competencies and skills so as to ensure their usefulness to the organization in terms of both present and future organizational conditions.

• Develop their general enabling capabilities as individuals so that they are able to discover and exploit their own and/or organizational development purposes.

- Develop an organizational culture where superior–subordinate relationships, teamwork and collaboration among different sub-units are strong and contribute to the professional well-being, motivation and pride of employees.

HRD looks at employees as an asset with potential for development for the individual and organization good. HRD, therefore, believes that:

- Every person has a potential arising from his strengths.
- Potential are of different types in different people suiting different roles and responsibilities.
- In the same role people may have more or less potential.
- Potential changes in complexion with better utilization.
- Potential can temporarily erode due to disuse or misuse.

To raise potential, HRD has to create an organizational environment where it can be fostered and developed.

Now we understand the difference between HRD and HRM. Simply put, HRM focuses on *productivity*, while HRD focuses on *potential*.

BASIC HUMAN RESOURCES ACTIVITIES

Figure 35.1 illustrates different activities performed by the HR department.

Human Resource Planning: Determining the size of the workforce and quality of staff required to fill various positions in the organization.

Recruitment and Selection: These tasks include tactics and procedures to attract applicants to the organization (recruiting) and choosing the very best persons among them (selecting).

Orientation and Socialization: Orienting and guiding new recruits to fit the social milieu in order to make them productive in the shortest period of time.

Training and Development: Preparing new staff members to do required work, updating their experienced peers, and providing opportunities for all interested staff members to assume more responsible positions are integral to the efforts of most organizations to attain goals and address competitive pressures, if applicable. It is the duty of the HR Manager to train each employee properly to develop technical skills for the job for which he has been employed and also to develop him for the higher jobs in the organization. A good training program should include a mix of both types of methods— on-the-job and off-the-job.

Compensation and Appraisal: Personnel should receive pay and benefits commensurate with their contributions to the organization. Performance appraisal provides input to help employees attain the on-job success that can yield promotions and transfers with higher compensation levels. Performance appraisal system should ensure that staffs are evaluated fairly and rewarded justly.

Safety and Protection: Safety and security concerns are of obvious importance to all employees. Merely appointment and training of people is not sufficient; they must be provided with good working conditions so that they may like their work and work place and maintain their efficiency. Working conditions certainly influence the motivation and morale of the employees. The HR department also provides for welfare services which may include provision of cafeteria, restrooms, group insurance, education for children of employees, etc.

Motivation: Employees work in the organization for the satisfaction of their needs. In most of the cases, it is found that they do not contribute towards the organizational goals as expected from them. This happens because employees are not adequately motivated. The

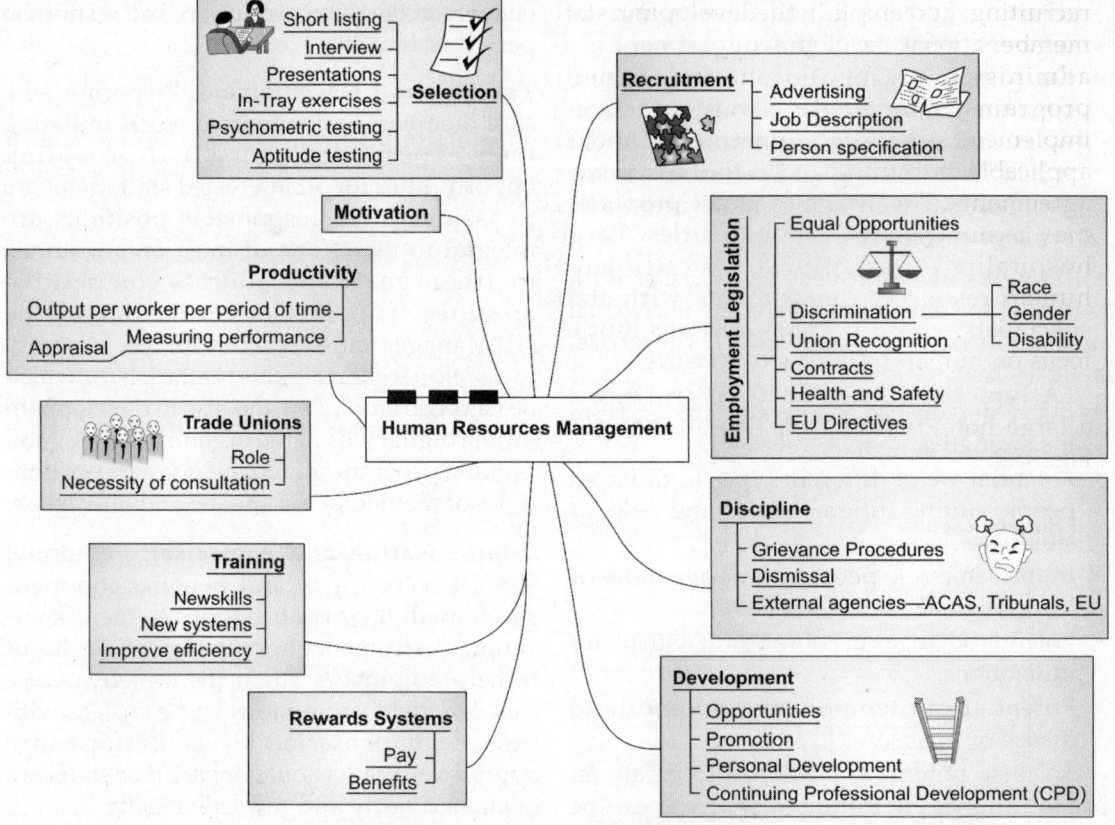

Fig. 35.1: Functions of HR department

human resource manager helps the departmental heads to develop and implement a reward (monetary as well as non-monetary) and recognition schemes to motivate the employees.

Employee Relations: The responsibility of maintaining good employee relations is mainly discharged by the human resource department. The human resource manager can help in collective bargaining, joint consultation and settlement of disputes, if the need arises. This is because of the fact that he is in possession of full information relating to staff and has the working knowledge of various labor enactments. The human resource manager can do a great deal in maintaining industrial peace in the organization as he is deeply associated with various committee on discipline, labor welfare, safety, grievance, etc. He helps in laying down the grievance procedure to redress the grievances of the employees. He also gives authentic information to the trade union leaders and conveys their views on various labor problems to the top management.

ORGANIZATION OF HUMAN RESOURCES DEPARTMENT

It is the department within a large organization, under the leadership of human resources manager, with the responsibility for recruiting, screening, and developing staff members. Persons in this department also administer compensation and benefit programs, coordinate safety practices, implement labor law requirements, and, if applicable, administer collective bargaining agreements. An owner of small properties may assume these responsibilities. Large hospitality organizations typically have human resources departments with staff specialists whose primary responsibilities focus on human resources concerns.

A typical HR organization structure of a large hotel would look like as shown in Fig. 35.2.

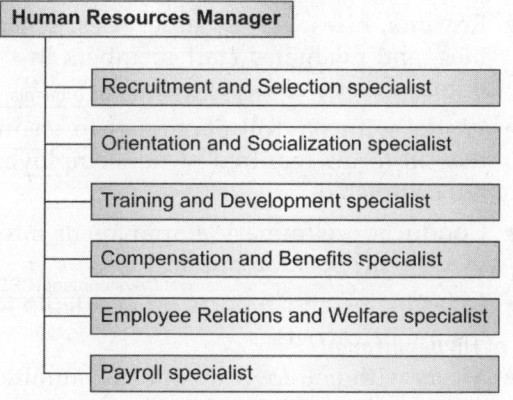

Fig. 35.2: HRD organization structure of a large hotel

Job Description of Director of Human Resources

Job Title: Director of Human Resources
Department: HR Department
Reports to: General Manager
Supervises: Human Resources Associates
Coordinates with: All Head of the Departments
Job Summary: Assists department management staff with recruitment, selection, and orientation of new staff members. Administers payroll records, directs the processing of wage and salary payments, and ensures that all applicable federal, state, and local wage and

hour, worker's compensation, and other labor laws are consistently complied with. Implements data collection systems and manages the organization's health, employee protection, retirement, and other benefits programs. Conducts labor analyses, staff planning, and other studies as requested. Serves on the organization's executive committee.

Scope of Job

- Administers employee compensation, benefits, performance management systems, and safety and recreation programs.
- Advises managers about organizational policies and recommends needed changes.
- Develops and places recruitment ads, plans recruitment strategies, screens applicants, and makes hiring recommendations.
- Conducts and reviews wage and benefit surveys, and proposes employee benefit modifications to the general manager.
- Analyzes data and reports to identify and determine causes of personnel-related problems, and develops recommendations for improvement.
- Analyzes training needs and designs applicable to employee development, language training, and health and safety programs.
- Conducts exit interviews to identify reasons for employee termination.
- Maintains organization's policy manual, and communicates policy changes to applicable staff members.
- Develops, administers, and evaluates applicant tests.
- Coordinates all employee (personnel) recordkeeping functions.
- Continually reviews and assists in updating the organization chart and employee handbook.
- Manages the organization's group insurance, unemployment, and related benefits programs; communicates benefits information

to staff, and ensures compliance with legal requirements.

- Maintains records and compiles statistical reports concerning personnel-related data such as hires, transfers, performance appraisals, and absenteeism rates.
- Negotiates collective bargaining agreements, helps interpret labor contracts, and administers the formal labor relations program with unionized staff.
- Oversees the evaluation, classification, and rating of occupations and job positions.
- Undertakes special projects relating to job description and specification updates, performance appraisal improvements, wage and salary comparison surveys, long-range staff planning, and other personnel issues.
- Keeps abreast of laws and regulations relating to employees, ensures compliance with these laws and regulations, and advises managers as necessary.
- Advises line managers about discipline, discharge, and related employment matters.
- Manages educational and referral programs for alcohol and substance abuse.
- Assists department heads in planning professional development and training programs for employees.
- Develops forecasts of short- and long-term staffing needs.
- Coordinates transfer, promotion, and layoff strategies.
- Benchmarks employee recruitment and selection processes with others in the industry, and explores new strategies as appropriate.
- Develops and maintains a library of training resources specifically designed for each position.
- Plans and implements employee motivation, recognition, and retention programs.

- Organizes employee activities such as the holiday party and other activities as appropriate.
- Provides current and prospective en.ployees with information about policies, job duties, working conditions, wages, opportunities for promotion, and employee benefits.
- Provides terminated employees with outplacement or relocation assistance.
- Represents the organization at personnel-related hearings and investigations.
- Oversees all work-related injury claims to ensure integrity, ongoing case management, and reporting compliance.
- Recruits, hires, trains, supervises, schedules, and evaluates staff members in the human resources department.
- Works with payroll personnel to ensure that all forms required of new employees are completed.
- Conducts preliminary employment interviews with position applicants.
- Investigates and reports on accidents for insurance carriers.
- Meets with employee relations committee on a regular, scheduled basis.
- Coordinates, monitors, and suggests improvements for employee performance appraisal system.
- Schedules and conducts employee safety meetings.
- Recommends drug-testing procedures for employee applicants.
- Interacts with the general manager and department heads to investigate employee violations of policies and to recommend correction actions, if necessary.
- Attends staff meetings as scheduled.
- Serves as a member of the organization's executive committee.

IMPORTANCE OF HUMAN RESOURCE PLANNING

According to Decenzo and Robbins, 'Human resource planning is the process by which an organization ensures that it has the right number of people (quantitative) and right kind of people (qualitative), at the right place, at the right time, capable of effectively and efficiently completing those tasks that will help the organization achieve its overall objectives.' Human resource planning aims at ascertaining the manpower needs of the organization both in right number and of right kind. It further aims at the continuous supply of right kind of personnel to fill various positions in the organization.

According to Dale S. Beach, 'Human resource planning is the process of determining and ensuring that the organization will have an adequate number of qualified persons, available at the proper times, performing jobs which meet the needs of the enterprise and which provide satisfaction for the individuals involved'.

Human resource planning is a continuous process. It cannot be rigid or static; it is amenable to modifications, review and adjustments in accordance with the needs of the organization or the changing circumstances.

Human resource planning is a two-phased process by which management can project future manpower requirements and develop manpower plans. It helps in proper recruitment and selection so that the right kind of people is available to occupy various positions in the organization. It also facilitates designing of training programs for the employees to develop the required skills in them. Thus, human resource planning plays an important role in the effective management of staff.

Systematic human resource planning provides lead time for the acquisition and training of employees to meet future requirements. It is all the more crucial because the lead time for procuring staff is a time consuming process and in certain cases, one may not always get the requisite kind of staff needed for the jobs. Non-availability of suitable manpower may result in postponement or delays in executing new projects and expansion programs, which ultimately lead to lower efficiency and productivity. To overcome this, an organization must plan out its manpower requirements well in advance so that it could compete effectively with its competitors in the market.

JOB ANALYSIS

It is the process of collecting, analyzing and setting out information about the contents of jobs in order to provide the basis for a job description and data for recruitment, training, job evaluation and performance management. Job analysis is a detailed and systematic study of jobs to know the nature and characteristics of people to be employed for each job. Job analysis is a systematic process of gathering information about work, jobs and the relationships between jobs as also the knowledge, skills and competencies required to perform the job successfully.

Undertaking a job analysis may not be necessary for every time a vacancy arises, especially in organizations that have high levels of labor turnover. Organizations may use one or more of the following techniques for deriving information about job analysis:

- Observation of the job
- Work diaries
- Interviews with job holders
- Questionnaires and checklists
- Brainstorming
- Review of existing job descriptions
- Guest comments
- Analysis of records
- Role clarification
- Hiring outside consultants
- Critical incident
- Video-recording

The output from such job analysis is the job description and job specification (Fig. 35.3).

Job Analysis
A process of obtaining information about job

Job Description	Job Specification
It is a statement that outlines the purpose of job. Usually includes details like	It is a statement of human qualifications necessary to do the job. Usually includes details like
• Job title	• Job title
• Location	• Category
• Job summary	• Education
• Duties	• Experience
• Machines, tools and equipment	• Age limit
• Materials and form used	• Gender
• Supervision given or received	• Physical ability
• Working conditions	• Personality
• Hazards	• Communication skills
	• Special requirements

Fig. 35.3: Job description and job specification in job analysis

Job Description

It is an organized factual statement that outlines the purpose of the job, the duties and responsibilities involved, additional responsibilities and the reporting relationships. It will give details of the working conditions, necessary equipment and materials, and other important information specific to the place of employment including the hours of work. Job description discloses what is to be done, how it is to be done and why it is to be done. In many respects the job description can be thought of as a functional document which outlines the 'what' elements of a job. It should aim to provide clear information to candidates about the organization and the job itself, such that it acts as a realistic preview of the job. Importantly, as well as offering a realistic description of the nature of the job, the job description should also act as a marketing document that seeks to make the job look attractive to potential applicants.

To be most effective, job descriptions should be customized to the operational procedures of a specific lodging property. Job descriptions should be task-oriented; they should be written for a position, not for a particular employee. Job descriptions will become dated and inappropriate as work assignments change, so they should be reviewed at least once a year for possible revision. Typically, front office managers write job descriptions. Employees should also be involved in writing and revising their job descriptions. Properly written job descriptions can minimize employee anxiety by specifying the chain of command and the responsibilities of the job. The job description of a bell captain is shown in Fig. 35.4.

Job Title: Bell Captain
Department: Front Office
Reports to: Lobby Manager
Supervises: Bell Boys
Coordinates with: Front Office and Housekeeping staff
Job Summary: The bell captain is in charge of bell desk and is responsible for organizing, supervising and controlling all lobby services with the help of bell boys.
Scope of Job:
• He controls the movement and activities of bell boys on the lobby control sheet which is a summary of the activities of all bell boys.
• He prepares duty rota and schedules leave for bell boys.
• Conducts daily briefing of bell boys.
• Handles left luggage room formalities and supervises left luggage register, baggage check and luggage inventory sheet.
• Trains bell boys to maximise their efficiency. Responsible for implementation of SOP as laid down by the management.
• Controls the sale of postage stamps to guests.
• He maintains record of all guests with scanty baggage and informs front desk staff.
• He reports irregularities of suspicious persons to the lobby manager.
• He prepares errand cards for the bell boys.
• Coordinates and control the distribution of morning newspaper.
• Organizes and supervises check-in/out baggage, formalities of groups, crews, etc.

Fig. 35.4: Job description of a bell captain

Well-written job descriptions can also be used:

- In evaluating job performance.
- As an aid in training or retraining employees.
- To prevent unnecessary duplication of duties.
- To help ensure that each job task is performed.
- To help determine appropriate staffing levels.

Job Specification

Job specification is also called *person specification*. It is a document which states the minimum acceptable qualifications an incumbent must possess to perform a given job successfully. It identifies the knowledge, skills, and abilities needed to do the job effectively.

Whilst the job description considers the 'what' aspects of the job, the job specification is concerned with the 'who'. In this way the job specification should aim to provide a profile of the 'ideal' person for the job. In reality, the ideal person may not exist, but the job specification provides a framework to assess how close candidates come to being the ideal. Conventionally the job specification is a document which describes the personal skills and characteristics required to fill the position, usually listed under 'essential' and 'desirable' headings. In that sense essential criteria form the minimum standard expected for any given job and will form the basis for potentially rejecting applicants. For example, if an advert for a tour company manager stipulates a degree in a travel and tourism-related area, then non-degree holders would be automatically excluded. On the other hand, the desirable criteria are those things which are considered over and above the minimum and should provide the basis for selection. For example, an organization may stipulate that for the same managerial job we have just outlined that a foreign language is desirable. If a candidate has knowledge of a foreign language he may be at an advantage to other candidates who do not, though ultimately the company may appoint somebody who does not have a proficiency in language.

Basically, front office job specifications spell out front office management's expectations for current and prospective employees; they are typically prepared by the front office manager with input from front office employees. Job specifications are usually developed after job descriptions, since a particular job may require special skills and traits. Factors considered for a job specification are: Formal education, work experience, general knowledge, previous training, physical requirements, communication ability, and equipment skills. Job specifications often form the basis for advertising job opportunities and identifying eligible applicants; they may also help to identify current employees for promotion. Job specification of a bell boy is shown in Fig. 35.5.

Job Title: Bell Boy

Category: Non-supervisory

Educational Qualification: High school

Age Limit: 18–30 years

Equipment Skills: Should know how to use luggage trolley, public address system

Physical Ability: Sound health and ability to carry luggage

Personality Consideration: Pleasant personality, well groomed

Language Skills: Ability to communicate in English and local language

Experience: Not necessary

Special Requirements: Eye for detail, honesty, calm demeanour, good memory, courtesy, tactful and diplomacy

---

Fig. 35.5: Job specification of a bell boy

RECRUITMENT

According to Edwin B. Flippo, 'Recruitment is the process of searching the candidates for employment and stimulating them to apply for jobs in the organization.' Recruitment includes all the activities an organization may use to attract a pool of viable candidates. It is a linking activity that brings together those offering jobs and those seeking jobs. Recruitment refers to the attempt of getting interested applicants and creating a pool of prospective employees so that the management can select the right person for the right job from this pool. Recruitment precedes the selection process.

The process of recruitment begins when new recruits are sought and ends when their applications are submitted.

Recruitment serves the following purposes:
- It attracts highly qualified and competent people.
- It provides a pool of potentially qualified candidates for various job positions at minimum cost.
- It meets the organization's legal and social obligations regarding the composition of its workforce.
- It identifies suitable candidates for various job positions.

Of course, recruitment is a dynamic process as within organizations people are constantly retiring, resigning, being promoted or, at times, being dismissed. Equally, changes in technology, procedures or markets may all mean that jobs are re-configured and become available to the external labor and thereby trigger the recruitment and selection process.

Sources of Human Resources Supply in the Hospitality Industry

Hospitality managers at all levels and in all sizes of organizations will continually find that they must actively recruit employees. From company presidents to the lowest-skilled entry-level employee, candidate recruitment will usually be an ongoing activity.

Although a variety of methods could be used to examine the employee search process, one way to categorize it is based on the approach utilized by the organization conducting the search. Using this method, an employee search may be categorized as being one of the following (Fig. 35.6):
- Internal search
- External search
- Outsourced search

Internal Search

The internal sources of recruitment include present employees (who can be promoted or transferred) and employee referrals. An

Fig. 35.6: Sources of HR for the purpose of recruitment

internal search is undertaken when a manager or organization believes that the best candidates for upper-level positions will be found among those employees who are currently employed by the organization.

Applied properly, a *promote-from-within* approach can be very effective. If, for example, when seeking a bell captain, a front office manager conducting the search felt that the best job candidates would be found among the hotel's current bell boys, an internal search could prove to be very effective.

Current employees may be informed about pending job openings in conversations with their supervisors or through the public posting of the information on employee bulletin boards, websites, or newsletters. The advantages associated with utilizing internal searches when seeking to fill positions are many, and include the facts that internal searches:

- Build employee morale.
- Can be initiated very quickly.
- Improve the probability of making a good selection because much is already known about the individual who will be selected.
- Are less costly than initiating external or outsourced searches.
- Result in reduced training time and less training costs because the individual selected need not be trained in organizational culture with which he is already familiar.
- It also generates higher employee commitment, development, and satisfaction because it offers opportunities for career advancement to employees rather than outsiders. Encourage talented individuals to stay with the organization.

Despite many advantages of internal searches, managers utilizing them to fill their position vacancies have also reported distinct disadvantages. These include:

- Inbreeding and a lack of new ideas can occur when an organization relies only on its own current workers to fill advanced positions.
- Resentment among employees can occur when one worker is chosen for advancement while others are not.
- Increased recruitment and training efforts result when a position is filled internally because the position vacated by the promoted employee must also be filled with a new staff member (who must also be trained).

Despite the added training effort required, most experienced hospitality managers firmly believe that the advantages of a promote-from-within policy far outweigh this and the other disadvantages that can sometimes be associated with it. While it is not, strictly speaking, an internal search system, the use of *employee referral systems* extends the network of potential applicants from an organization's own employees to those potential workers whom the current employees recommend.

Employee referral systems tend to work well because employees rarely recommend someone unless they feel that person can do a good job and will fit well into the organization. Another advantage to having employees directed to the organization via an employee referral system is that these employees tend to have a more accurate view of the job to be done and the organization's unique culture. This information reduces unrealistic expectations and can help lead to reduced new employee turnover.

In some operations, employee referral systems provide a financial bonus to a staff member who recommends an applicant who is hired by the organization and remains with it for a specified time. Such a system financially rewards employees for their suggestions, and saves the operation time and money that would otherwise be invested in the search process.

Just as there can be disadvantages associated with internal searches, potential problems may exist with employee referral systems. In some cases, recommenders may suggest their own friends or relatives for positions regardless of their qualifications. It is simply human nature to want to work with those people whose companionship is also enjoyed outside of work. Therefore, hiring managers must remember to apply the same standards of employment consideration to those candidates referred by employees as they would to any other individual being considered for a vacant position.

Interdepartmental *transfers* are an option that makes employees more flexible. Employees can be then accommodated in any department when need arises. Transfer is a good source of filling vacancies with employees from overstaffed departments.

External Search

Despite the advantages of internal searches, many managers find that an external search is a good way to help identify a pool of applicants who are qualified for their organizations' vacancies. They are more expensive as the cost of advertisement is higher. However, the external sources generate more applicants with the desired qualifications and make the selection process more effective.

When organizations seek candidates externally, they rely on a variety of sources:

- Professional or Trade Associations
- Advertisements
- Employment exchange/employment bureaus
- Educational institutions or campus recruitments
- Unsolicited applications
- Walk-ins
- Job fairs
- Headhunting from competition
- Labor contractors
- Acquisitions and mergers

- Poaching
- Websites or e-recruiting
- Unions
- Industrial trainees and apprentices

Outsourced Search

In some cases, HR managers decide that an outsourced search is the best method they can use to find the candidates they are looking for. An outsourced search is one in which an organization chooses an executive search firm (management consultants/head hunters) to find potential candidates for a job.

Executive search firms impose significant charges for their services, with typical fees ranging from one-third to approximately half of the annual wages that will be paid to the employee who will be hired. These fees are typically paid by the employer, not the employee.

Executive search firms make it their business to monitor executive-level talent so they can advise their clients about the best candidates available. In most cases, the executive search firm will identify potential candidates from their lists of contacts and do preliminary screening. These firms are adept at seeking out executives with proper skills and who fit well with the hiring organization. Even though outsourced searches rely heavily on the expertise of the private employment agency chosen, the final hiring decisions still remain with the employer, not the executive search firm.

It is important to understand that, in most cases, HR managers do not choose from among internal, external, and outsourcing as their sole method of recruiting. Rather, the best managers select the approach appropriate for the vacancy they seek to fill. In some cases, this will result in using more than one of the strategies, or even all three of them when seeking to fill a specific position.

SELECTION

Selection involves a series of steps (Fig. 35.7) by which the candidates are screened for choosing the most suitable persons for vacant positions. The process of choosing the right candidate from the pool of applications received in the recruitment process is called selection. It is the process of assessing job applicants using one of a variety of methods with the purpose of finding the most suitable person for the organization.

Factors affecting Selection efforts

After HR managers have assembled a pool of qualified candidates, they must select the applicant they wish to hire. When choosing potential applicants for employment, hospitality managers will generally utilize some or all of the major selection activities. These are:

- Screening applications and resumes
- Conducting preliminary round of interviews
- Pre-employment testing: Integrity tests, personality tests, performance simulation

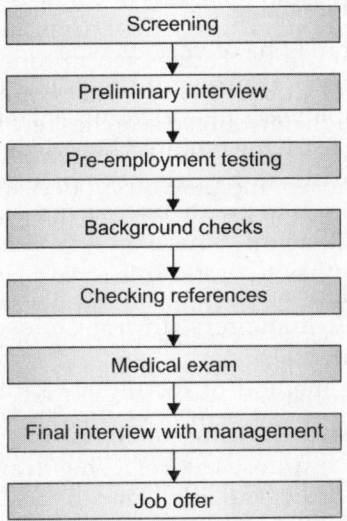

Fig. 35.7: Selection process

tests, knowledge tests, assessment centers, work sampling, trade tests, Aptitude tests, psychological tests

- Background checks: Criminal history, credit reports, driving records, academic credentials
- Checking references: Many organizations seek information from past employers about an employee's previous work performance to verify information supplied by the candidate.
- Medical examination identifies health problems that increase absenteeism and accidents, as well as detecting diseases that may be unknown to the applicant.
- Final interview with management
- Job offer

ORIENTATION AND SOCIALIZATION

Orientation may be defined as 'a planned introduction of new employees to their jobs, their co-workers, the management, and the organization'. Orientation is also known as *induction*.

According to Edwin B. Flippo, orientation is the welcoming process to make the new employee feel at home and generate in him a feeling of belongingness to the organization. It covers the activities involved in introducing a new employee to the organization and his department. It is the beginning of the fusion process which helps integration between the organizational goals and the personal goals of the new employee. Planned orientation welcomes the new employee, creates a favorable attitude, reduces labor turnover and increases commitment and productivity. Further, the employee feels at home right from the beginning.

Socialization is the teaching of the organization's culture, values, norms, and behaviors that are consistent to the success of the company.

The range of information that may be covered under orientation program is as follows:

- The history of the organization
- Products of the organization
- Organizational structure
- Different departments and their locations
- Policies and practices of the organization
- Rules and regulations regarding working hours, attendance, leave, code of conduct, etc.

The orientation may be formal and structures through presentations, demonstrations, and lectures or informal where employees get introduced to co-workers while performing their job. Most hotels have a training department, which takes care of the orientation program in general.

Orientation serves the following purposes:

- It makes new employees comfortable in their new roles.
- It familiarizes new employees with their co-workers. It gives the employees an overview of the organization—work culture, functions of all departments, and organizational structure.
- It reduces the anxiety of the new employees and makes them feel at home in the new workplace.
- It facilitates informal relations and team work among the employees.

TRAINING

Training is the overall enhancement of human ability by developing knowledge, skills, attitude, and behavior in order to achieve individual goals. Training relates not only to new employees, but is an ongoing process for the entire team.

Simply hiring and placing employees in jobs does not ensure their success. In fact, even tenured employees may need training, because of changes in the business environment. Here are some changes that may signal that current employees need training:

- Introduction of new equipment or processes
- A change in the employee's job responsibilities
- A drop in an employee's productivity or in the quality of output
- An increase in safety violations or accidents
- An increased number of questions
- Complaints by customers or co-workers

Benefits of Training

The benefits of training are as follows:

- Quick learning
- Higher productivity
- Standardization of procedures
- Less supervision
- Fewer accidents
- Reduces labor turnover
- Economical operations
- Higher morale
- Preparation of future managers
- Better management

Types of Training

Training may be of various types:

- **Induction Training**: It is carried out when an employee is new to the organization and has to learn the required knowledge, skills, and attitude for his new position. The purpose is to give a 'bird's eye view' of the organization where newly employed person has to work. It is a very short and informative training given immediately after hiring. It creates a feeling of involvement in the minds of newly appointed employees.
- **Refresher Training or Retraining**: As the name implies, the refresher training is meant for the old employees of the organization. This is carried out when an old employee has to be re-trained to refresh his memory. The basic purpose of refresher

training is to acquaint the existing workforce with the latest methods of performing their jobs and improve their efficiency further. These are designed to avoid personnel obsolescence.

- **Remedial Training**: This is carried out for old employees when there is a change in the present working style, which may be related to a competitive environment, technological changes or guest expectations.

- **Cross Training:** This training enables employees to work in departments other than their specialization in periods of staff shortage.

Methods of Training

Different methods of training are illustrated in Fig. 35.8.

On-the-job training: On-the-job training is considered to be the most effective method of training the staff. Under this method, the worker is given training at the workplace by his immediate supervisor. In other words, the worker learns in the actual work environment. It is based on the principle of 'learning by doing'.

On-the-job training is suitable for imparting skills that can be learnt in a relatively short period of time. This method of training is relatively cheaper and less time consuming.

Two popular types of on-the-job training include the following:

- **Job rotation:** By assigning people to different jobs or tasks to different people on a temporary basis, employers can add variety and expose people to the dependence that one job has on others. Job rotation can help stimulate people to higher levels of contributions, renew people's interest and enthusiasm, and encourage them to work more as a team.

- **Mentoring programs:** A new employee frequently learns his job under the guidance of a seasoned veteran. In the trades, this type of training is usually called an *apprenticeship*. In white-collar jobs, it is called a *coaching* or mentoring relationship. In each, the new employee works under the observation of an experienced worker.

Off-the-job training: It requires the worker to undergo training for a specific period away from the workplace. This takes place in a classroom, by means of workshops, demonstrations, lectures, discussions, seminars, audio-visual presentations, case studies and role-playing.

HR CHALLENGES IN THE HOSPITALITY INDUSTRY

There are many human resource challenges in the hotel and tourism industry.

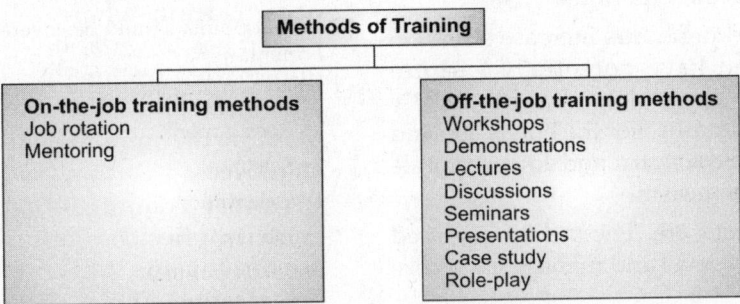

Fig. 35.8: Methods of training

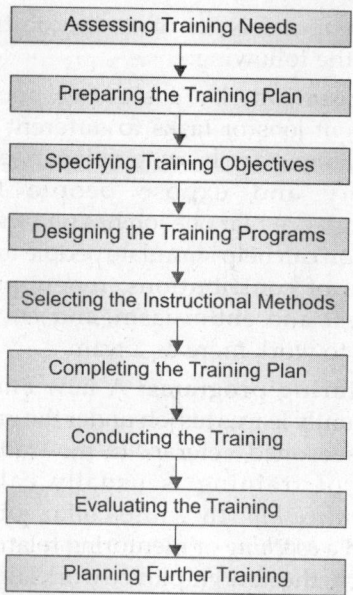

Fig. 35.9: Steps in planning a training program

The characteristic features of work in hotel industry—long, anti-social hours, low pay, instability, orthodox opinion and low status—make it unattractive as a career choice, and as a result the sector continues to suffer from high labor turnover and difficulties in recruiting qualified professionals. The hotel staff works especially hard on those days when the rest of the world is enjoying vacation and holidays, such as Diwali, Holi, New Year, and long weekends. The hotels are generally booked to capacity during such festivals, occasions and seasons and the situation gets worse when the hotel staff is required to work overtime during these periods.

The long working hours increase stress on individuals, and have potentially harmful effects on their psychology, health, family affairs and social status. For the hotels, it could lead to poor productivity and lower morale resulting in absenteeism.

Other key issues are: The lack of qualified staff at both entry level and managerial levels,

the unwillingness of university graduates and postgraduates to enter the industry, and the gap between what is taught in institutes and the realities of the industry itself. Together, these negative factors damage the reputation and perception of the entire tourism and hospitality sector. The primary challenge is to change this negativity, first by improving working conditions to attract suitable staff and retain them, and second by increasing the pay scale and investing in training and development of the manpower.

REVIEW QUESTIONS

1. Define human resources. Explain the importance of HR in the hospitality industry.
2. Differentiate between HRM and HRD.
3. Explain the importance of human resource planning.
4. Differentiate between job description and job specification.
5. What is training? Discuss different types of training.
6. What are the benefits of training?
7. Differentiate between on-the-job training and off-the-job training.
8. Discuss the sources of labor that can be tapped for filling front office positions.
9. Define recruitment. Discuss the recruitment process in detail.
10. What do you understand by the term 'selection'? Explain the selection process in detail.
11. What are the challenges faced by the HR department in the hospitality industry?
12. What points should be covered in an employee orientation?
13. Define the following:
    • Recruitment
    • Selection
    • Orientation
    • Job analysis
    • Cross training

Appendices

General Knowledge about the Hospitality Industry and Front Office

1. Appendix I: Pioneers of the Hospitality Industry

2. Appendix II: Major Players in Hotel Industry in India

3. Appendix III: Domestic Airlines in India

4. Appendix IV: Tourist Attractions in India

5. Appendix V: Guidelines for Approval of Hotels at Project Stage and Classification and Re-classification of Hotels

6. Appendix VI: Front Office Standard Operating Procedures

7. Appendix VII: Glossary of Terms Used in Hotel Front Office

8. Appendix VIII: Abbreviations

Pioneers of the Hospitality Industry

We all love to be catered to at hotels, but to truly appreciate the tireless individuals that make our stay perfect, we must pay homage to the pioneers that started it all.

Kemmons Wilson

Family road trips can be quite the adventure, but when there is not one quality motel to unwind in after being trapped in a car for hours, you'll be better off inventing your own top-notch hotel.

That is exactly what drove Kemmons Wilson to design the reputable Holiday Inn chain. For him it was a trip to Washington DC that led to the grand opening of the first Holiday Inn in Memphis, Tennessee in 1952. He started off with popcorn business. In 1959, the hotel's locations grew to be about 100, and just a year later it became an international sensation. As one of the top hoteliers during this time, Kemmons attracted the likes of *Time Magazine* who featured the entrepreneur on its cover. He also spread his Christian beliefs by leaving a Bible in every room of his motels. Kemmons Wilson is widely considered to be the first hotelier to put two beds in one hotel room. He incorporated the theory of *brand loyalty* to his chain hotels. Wilson started Holidex central reservation system in order to streamline the entire reservation process and to handle the immense volume of reservation pouring in the various properties of the chain. Holiday Inn Worldwide is now one of the largest hotel chains in the world. Wilson died in February 2003 at the age of 90.

J. Willard Marriott (1900–1985)

Anyone that can transform a root beer stand into a leading hotel company deserves a place in this list. J. Willard Marriott started as the owner of A and W fast food restaurant in Washington DC in 1927. After obtaining the franchise rights to A and W Root Beer in a couple of states, Marriott decided to transform it into a booming family restaurant

called The Hot Shoppe. Then he started the hamburger operations, Bob's Big Boy. In 1937, Marriott, exhibiting his trademark innovation, offered the first ever in-flight food service to airlines servicing the old Hoover Airfield in Washington. In 1957, founded his hotel empire with the Twin

Bridges Marriott Motor Hotel in Virginia, near Washington DC. Marriott Hotel chain includes Marriott Hotels and Resorts, Courtyard by Marriott, Renaissance, JW Marriott, Residence Inn, Fairfield Inn, SpringHill Suites, etc.

Ellsworth Milton Statler (1863–1928)

The next time you are taking a soothing hot shower or bath in your gorgeous suite, you may want to thank Ellsworth Statler. In 1878, at the age of 13, he started as a bell boy in Wheeling, West Virginia. In 1901, he opened a 2100 room temporary hotel, the 'Outside Inn' to house visitors to the Pan American Exposition in Buffalo. He had his second opportunity to operate a 2257 room temporary hotel, the 'Inside Inn', at St. Louis World's Fair.

In 1908, Ellsworth Milton Statler opened what many believed to be the first 'modern' hotel, the Buffalo Statler. This hotel featured telephone in every room, modern plumbing, ice water, full size closets with lights, private bathrooms, full length mirrors, circulating hot and cold water in each room, etc. He also provided free morning newspaper, provided radios free of charge.

However, the hotel businessman did not stop there. He continued to build new properties in various American cities including St. Louis, Detroit and Cleveland. Currently, his name echoes proudly throughout the hospitality industry Hall of Honor as one of the greats to change the world of travel.

He also developed 'The Statler Service Code' which all employees had to memorize and carry with them. He is known and credited as the person with the saying 'the guest is always right'. The Hotels Statler Company, Inc., was sold to Conrad Hilton's (Hilton Hotels) in 1954 for $111,000,000 in what was then the world's largest real estate transaction.

Conrad Hilton (1887–1979)

He was born in San Antonia, Mexico, known as 'King of Innkeepers' and Master of Hotel Finance.

In 1919, Conrad Hilton purchased his first hotel, The Mobley, in Cisco, Texas. In 1925, he built his first hotel to carry the Hilton name in Dallas, Texas. In 1938, Conrad Hilton opened his first hotel outside Texas, the Sir Francis Drake in San Francisco. In 1942, Town House in LA became Hilton's headquarters. In 1943, took over Roosevelt and Plaza Hotels in New York. In 1945, as World War II ended, Hilton purchased what was the largest hotel of its time, The Stevens Hotel, and renamed it the Chicago Hilton and Towers. In 1949, Hilton acquired one of the most famous hotels of all time founded by the legendary William Waldorf Astor, The Waldorf-Astoria. In 1950, Hilton Hotel Corporation (organized in 1946) bought over the 2 largest hotel chain at that time; Sheraton and Statler. In 1954, Hilton purchased the Statler Hotel Company in what was then the largest real estate transaction to date with the amount of 111 million dollars. Throughout the 1950s and 1960s, Hilton expanded domestically and internationally. Hilton believed in absolute cost control. He is credited with introducing new methods of forecasting and control techniques into the hotel business. In 1979, the founder, Conrad Hilton, died and his son Baron Hilton took over. Hilton Hotels Corporation and Hilton International Company now owns more than 500 properties in more than 50 countries and includes Hilton Garden Inns, Doubletree, Embassy Suites, Hampton Inns, Harrison Conference Centers, Homewood Suites by Hilton, Red Lion Hotels and Inns, Conrad International.

Caésar Ritz (1850–1918)

Caésar Ritz was a famous Swiss hotelier and founder of several hotels, most famously The Ritz Hotel. His nickname was 'king of hoteliers, and hotelier to kings', and it is from his name and that of his hotels that the term *ritzy* derives.

By the age of 17 he was already a manager of the Grand National Hotel, Lucerne Switzerland (the most luxurious hotel in the world when it was built in 1870). Ritz worked as the first general manager of the Savoy Hotel before he opened the Hôtel Ritz in Paris, France in 1898. He went on to open The Ritz Hotel in London, United Kingdom and the Hotel Ritz Madrid in Madrid, Spain. Ritz enjoyed a long partnership with Escoffier, the famous French chef and father of modern French cooking. The partnership lasted until Ritz's breakdown. He made evening dress compulsory and introduced orchestras in the restaurant.

Ray Kroc (1902–1984)

Raymond Albert Ray Kroc was an American fast food businessman who joined McDonald's in 1954 and built it into the most successful fast food operation in the world. Kroc was included in *Time 100: The Most Important People of the Century*, and amassed a fortune during his lifetime.

In 1954, a fifty-two-year-old milk-shake machine salesman saw a hamburger stand in San Bernardino, California, and envisioned a massive new industry: fast food. In what should have been his golden years, Raymond Kroc, the founder and builder of McDonald's Corporation, proved himself an industrial pioneer no less capable than Henry Ford. He revolutionized the American restaurant industry by imposing discipline on the production of hamburgers, french fries, and milk shakes.

By developing a sophisticated operating and delivery system, he ensured that the french fries customers bought in Topeka would be the same as the ones purchased in New York city. Such consistency made McDonald's the brand name that defined American fast food.

Ray Kroc was intelligent enough to understand the value of public relations and thus included them in the marketing and business of McDonald's.

Ray Kroc is famously called the 'Father of the Fast Food Industry'.

Ernest Henderson

Ernest Henderson and Robert Moore started the Sheraton chain of hotels in 1937, when they acquired their first hotel, The Stonehaven, in Springfield, Massachusetts. Sheraton hotel chain was the first hospitality company to be listed in the New York Stock Exchange. In 1968, Sheraton hotels were acquired by ITT Corporation as a wholly owned subsidiary to create a global network of properties. In 1980s, under the leadership of John Kapi-oltas, the company received international recognition as an industry innovator in modern hotel accommodations. Sheraton chain is currently owned by the Starwood Hotels and Resorts Worldwide and is perhaps one of the largest hotel chains of the world.

Howard Fearing Johnson (1898–1972)

In 1925, he bought a drugstore and soda fountain with the money $500 he borrowed from his friend. He converted his drugstore to first restaurant by adding hamburgers, hotdog and sandwiches to the menu. In 1940, he opened a motor lodge in Savannah, Georgia. In 1972, he had already opened 500 Howard Johnson's motor lodges and 850 restaurants. The Johnson family's company-owned restaurant were sold to Marriott Corporation, and turned to Big Boy restaurant.

Ralph Hitz (1891–1940)

Running away from home, you might think Ralph Hitz would be a flight risk but his innovative ideas and moneymaking strategies have earned him a spot in this list. Working his way up into management from spending years as a busboy, Ralph Hitz proved himself to be a worthy employee when he tripled the net worth of the Hotel Gibson in Cincinnati just after two years of working there. This caught the eye of the newly built New Yorker Hotel that opened its door a few weeks after the stock market crash. Nevertheless, Hitz worked his magic, which made the hotel stay afloat during the economy's collapse.

Although he died at the young age of 50, he contributed much more than turning profits. Hitz also introduced using a circuit radio system that allowed guests to listen to broadcasts about programmed entertainment and food selections that were provided, which have greatly influenced the way TV channels provide guests with useful information in modern hotels across the globe.

Hitz had an excellent marketing brain as he was the first intellectual to develop a customer database for providing the guests of the hotel with personalized service leading to guest satisfaction and overall profitability of the hotels.

Richard C. Kessler

As Chairman and CEO of The Kessler Enterprise, Richard C. Kessler has dabbled in everything under the sun when it comes to hotels, even creating various themes for his properties with the Kessler Collection project.

Under the Kessler Collection, various hotels were created and operated, as well as given specific concepts to complement their guests' needs precisely. His other developments include Days Inn Resorts and Suites and Sheraton Studio City Hotel in Florida, in addition to Kehoe House Bed and Breakfast in Savannah. Kessler was also the President and CEO of Days Inn of America, Inc, which unquestionably makes him an honorable leader.

Gerard Blitz and Gilbert Trigano

'Great minds think alike' is the notion that brought the original all-inclusive resort, Club Med, to life. The idea was first introduced by Gerard Blitz, a water polo champion whose experience in the French resistance in World War II inspired him to create a utopic destination for veterans. With the help of Gilbert Trigano, a French businessman whose father supplied tents and canopies used for camping, they set out to change the face of the 'all-inclusive' holiday vacation.

That's where Club Med comes in. Their luxurious villas create a 'happy place' amid exotic locations that create the perfect backdrop for travelers wanting to de-stress. Since 1950, the resort has done just that and much more while continuing to stay true to Blitz's and Trigano's utopic vision.

Jack DeBoer

There's a reason Jack DeBoer is dubbed the 'originator of extended stay'. It is simply because he noticed that there was something missing when it came to accommodations for small business owners who were fronting their own bills during their hotel stay. So DeBoer set out to design the perfect lodging for these individuals, which is why the Value Place came into being in Kansas in October 2003. Not only were its prices unbeatable, but it also created a clean and safe environment that would entice its guests to stay longer.

And with the Residence Inn, Candlewood Sui's and Summerfield Suites already under DeBoer's belt, he was on a roll. Soon after the establishment of Value Place and thanks to DeBoer, a franchising project was launched less than a year later.

Barry Sternlicht

Of course we cannot leave out 'The king of hotels' himself, Barry Sternlicht. This Harvard graduate has a knack for putting a million dollar spin on real estate, especially when he started the Starwood Capital Group, which held his long portfolio of family-friendly apartments that he purchased during the government bailout.

Sternlicht's success in real estate business would prove beneficial when he began to focus his attention on the hotel industry. And by utilizing his expertise he was able to acquire the Westin Hotels and Resorts. To add to his long list of accomplishments, he also established the St. Regis brand as well as launched the Starwood Preferred Guest program, which is arguably the first of its kind to hit the hospitality industry.

Major Players in Hotel Industry in India

1589 HOTELS

1589 Hotels is owned and run by the **Clarks Family**. Keeping in mind the legacy of the parent group and their contribution to the Indian hospitality, 1589 caters to a generation of modern Indians, who are global in their outlook yet traditional in essence.

Brands of 1589 Hotels

Brij

Arte

GenX

RnB Select

RnB

List of 1589 Hotels in India

RnB Yaduraj Barmer, Rajasthan

Fort Bijainagar, Ajmer

GenX Bhavnagar, Bhavnagar, Gujarat

RnB, Chittorgarh

RnB Ramcons, Goa

GenX Sundaram, Haldwani

GenX Banjara Hills, Hyderabad

RnB, Jaipur

RnB Select, Jaipur

GenX, Jodhpur

1589 Royal Heritage, Kishangarh, Rajasthan

GenX Casaya Hotel, Lucknow

GenX Plaza, Mughalsarai

Sultan Bagh Forest Camp Resort, Rajasthan

GenX, Vadodara

GenX Valley View, Udaipur

ACCOR HOTELS

Accor, the world's leading hotel operator and market leader in Europe, is present in 92 countries with more than 3,500 hotels and 450,000 rooms. With more than 160,000 employees in Accor brand hotels worldwide, the group offers to its clients and partners nearly 45 years of know-how and expertise. Accor provides an extensive offer from luxury to economy that are recognized and appreciated around the world for their service quality.

Accor entered the Indian market in 2006 with the commencement of operations at the Novotel Hyderabad Convention Centre and the Hyderabad International Convention Centre. Committed to increasing its presence in the region, today the group operates 31 hotels in India. The aim is to have a network of 90 hotels in India, Bangladesh, Nepal, Sri Lanka and Bhutan under operation or under active development by 2015, with at least 50 hotels under operation.

Brands of Accor Group

Sofitel

Pullman

MGallery

Novotel

Suite Novotel

Mercure

Grand Mercure

Ibis

Ibis Styles

Ibis budget

HotelFormule1

Adagio

Adagio Access

Thalassa Sea and Spa

LUXURY AND UPSCALE	SOFITEL LUXURY HOTELS pullman HOTELS AND RESORTS		REGIONAL BRANDS GRAND MERCURE THE SEBEL
MIDSCALE	NOVOTEL NOVOTEL	Mercure HOTELS	adagio
ECONOMY	ibis ibis budget	ibis STYLES	adagio FORMULE1 En Asie hotelF1 En France

List of Accor Group of Hotels in India

Ibis Bangalore, Bangalore

Hotel Ibis, Gurgaon

Hotel Ibis Delhi Airport

Hotel Ibis Jaipur

Hotel Ibis Mumbai Airport, Mumbai

Ibis Navi Mumbai, Navi Mumbai

Hotel Ibis Nashik Ibis, Pune

Ibis Bengaluru City Centre Hotel, Bengaluru

Ibis Bengaluru Techpark, Bengaluru

Hotel Formule 1, Gurgaon

Hotel Formule 1, Ahmedabad

Hotel Formule 1, Greater Noida

Hotel Formule 1, Pune Pimpri

Hotel Formule 1, Pune Hinjewadi

Hotel Formule 1, Bengaluru Whitefield

Hotel Pullman New Delhi Aerocity

Hotel Novotel New Delhi Aerocity

Hotel Novotel Goa Shrem Resort

Hotel Novotel Ahmedabad

Novotel Hyderabad Convention Centre, Hyderabad

Novotel Hyderabad Airport, Hyderabad

Novotel Mumbai Juhi Beach, Mumbai

Novotel, Visakhapatnam

Hotel Novotel, Pune

Hotel Novotel Visakhapatnam Varun Beach

Novotel Bengaluru Techpark, Bengaluru

Novotel Kolkata Hotel and Residences

Mercure Lavasa, Lavasa

Grand Mercure Bangalore, Bangalore

Hotel Grand Mercure Goa Shrem Resort

Royalton Hyderabad Abids, Hyderabad

Sofitel Mumbai BKC, Mumbai

AMBASSADOR GROUP OF HOTELS

Efficiency, comfort, luxury, and excellence is synonymous with the Ambassador Hotels. This group of hotels is a perfect example of hospitality and service. The Ambassador Group of Hotels features world class facilities and renowned cuisines. The lip smacking delicacies prepared in the in-house restaurant of this group definitely leaves a tangling taste in the minds of the travelers. Hotels are just a part of the Ambassador Group. They have also triumphed in the flight catering business. They boast of the largest flight catering unit in south-east Asia. Growing pretty fast and dynamically the

Ambassador Hotels have added their name in the list of the best hotels around the world.

List of Ambassador Group of Hotels

The Ambassador, Mumbai
The Ambassador Pallava, Chennai
The Ambassador Ajanta, Aurangabad

ABAD HOTELS

ABAD group is a leading business concern headquartered at Cochin, the commercial capital of the state of Kerala. From the humble beginning as an exporter of sun dried Shrimps to Burma in 1894, the group has steadily grown into a towering presence in the business and industrial landscape of Kerala. ABAD group runs a chain of 12 Resorts and Hotels in Kerala asserting its presence in Kerala's tourism and hospitality scenario.

List of ABAD Hotels

Hotel ABAD Fort Cochin
Hotel ABAD Plaza Cochin
ABAD Atrium Hotel Cochin
ABAD Metro Cochin
ABAD Airport Hotel Cochin
Pepper Route by ABAD, Cochin.
ABAD Harmonia Ayurveda Beach Resort, Kovalam
ABAD Copper Castle, Munnar
ABAD Turtle Beach, Marari
ABAD Whispering Palms, Kumarakoms
ABAD Green Forest, Thekkady
Hotel Asopalav, Ambaji

AMAN RESORTS

Aman Resorts have 18 luxury resorts across the world under its kitty. All the resorts of this Aman Resorts group have some common characters, including a beautiful location, exceptional facilities, excellent services along with warm hospitality, and fewer rooms to ensure exclusivity and privacy. Each Aman resort has been built with locally sourced materials that showcase elements of the natural surroundings and the traditions of local cultures.

List of Aman Resorts in India

Aman-i-Khas
Amanbagh

BEST WESTERN GROUP OF HOTELS

The Best Western Group of Hotels are located at the prime tourist locations of India like Mumbai, Chennai, Kolkata, Hyderabad, Khajuraho, Dehradun, Varanasi and Kovalam. The Best Western Group is a brand name ensuring reliable service, affordable rates, quality accommodation and professionalism. Other than in India, the Best Western International runs more than 4100 independently owned and operated hotels across 84 countries including North America, South America, Europe, Asia, Africa, the Middle East and South Pacific.

List of Best Western Hotels in India

The Emerald Best Western, Mumbai
The Kenilworth Best Western, Kolkata
Best Western Swagath Holiday Resort, Kovalam
The Amrutha Castle, Hyderabad

THE CARLSON REZIDOR HOTEL GROUP

The Carlson Rezidor Hotel Group is one of the world's largest and most dynamic hotel companies. It has a fantastic portfolio of more than 1350 hotels in operation and under development, a global footprint covering over 105 countries and territories.

Brands of Carslon Rezidor Group

Radisson: Radisson's properties are conveniently situated in major urban and suburban settings, in resorts, business districts and airport gateways. The brand takes great pride in its deep understanding of how it feels to travel in this increasingly fast forward world, and its a brand that is focused on the art of truly human hospitality.

As of September 2014, Radisson has 154 hotels operating throughout the world with 31,448 rooms, and 29 hotels under development with an additional 4,927 rooms.

Radisson Blu: Radisson Blu is an iconic, stylish and sophisticated brand with leading-edge style where the delight is in the detail. It offers guests a holistic hospitality experience that is totally relevant to now.

One of the world's leading hotel brands, Radisson Blu creates iconic buildings with individual interiors invoking an inviting, exciting ambiance. Pioneering bold and innovative lobbies, guest rooms and public spaces with the latest

AVERAGE DAILY RATE — chart showing Radisson brand positioning across service levels (LUXURY, UPPER UPSCALE, UPSCALE, UPPER MIDSCALE, MIDSCALE, ECONOMY) and service types (LIMITED SERVICE, SELECT SERVICE, FULL SERVICE): Quorvus Collection (Luxury, Full Service); Radisson Blu (Upper Upscale, Full Service); Radisson RED (Upscale, Select Service); Park Plaza and Radisson (Upscale, Full Service); Country Inns & Suites (Upper Midscale, Limited Service); park inn by radisson (Upper Midscale, Select Service).

technology and a range of highly individual solutions, Radisson Blu offers a guest experience that is truly unique in the world of hotels today.

As of September 2014, Radisson Blu has 286 hotels operating throughout the world with another 101 projects under development. The brand's flagship properties can be found in prime locations, including major cities, airport gateways, and leisure destinations around the world.

Park Plaza: An upscale hotel brand for business and leisure travelers, Park Plaza offers stylish guest rooms, outstanding meeting spaces and dedicated staff who live the brand's value proposition by going out of their way to show their appreciation to guests. Park Plaza hotels are located in key cities, regional and commercial areas, and resort destinations.

As of September 2014, Park Plaza has 48 hotels operating throughout the world and 13 under development.

Country Inn and Suites By Carlson: Country Inns and Suites By Carlson[SM] is a leading mid-market brand committed to providing a caring, consistent and comfortable hospitality experience delivered with a touch of home. Welcoming and genuine, with a Be Our Guest service philosophy, Country Inns and Suites delivers high levels of service and guest satisfaction.

As of September 2014, Country Inns and Suites has 470+ hotels operating throughout the world,

primarily in the United States, Canada, India and Latin America.

Park Inn by Radisson: Park Inn® by Radisson is a colorful and dynamic, midscale hotel brand aimed at tech savvy, young at heart travelers who know what they want from a hotel stay and are willing to pay for it, but also seek value for their money. Accessible and inclusive, Park Inn by Radisson is friendly, fresh, vibrant and uncomplicated. Park Inn by Radisson can now be found all around the world—in capital cities and economic hubs—often close to city centers, airports and railway stations.

As of September 2014, Park Inn by Radisson has 207 hotels operating throughout the world and 52 hotels under development.

Radisson Red

Quorvus Collection

List of Radisson Hotels in India

Radisson Hotel Jalandhar

Radisson Hotel Kandla

Radisson Hotel Khajuraho

Radisson Jass Hotel Shimla

Radisson Hotel Varanasi

Radisson Hyderabad Hitec City

List of Radisson Blu Hotels in India

Radisson Blu Agra Taj East Gate

Radisson Blu Hotel Ahmedabad

Radisson Blu Hotel Amritsar
Radisson Blu Hotel Chennai
Radisson Blu Hotel Chennai City Centre
Radisson Blu Hotel Ghaziabad
Radisson Blu Hotel Guwahati
Radisson Blu Hotel Haridwar
Radisson Blu Hotel Hyderabad Banjara Hills
Radisson Blu Hotel Indore
Radisson Blu Hotel Jaipur
Radisson Blu Hotel Ludhiana
Radisson Blu Hotel Greater Noida
Radisson Blu Hotel Noida
Radisson Blu Hotel Nagpur
Radisson Blu Hotel Ranchi
Radisson Blu Hotel New Delhi Dwarka
Radisson Blu Hotel New Delhi Paschim Vihar
Radisson Blu Marina Hotel Connaught Place
Radisson Blu Plaza Delhi
Radisson Blu Resort Goa Cavelossim Beach
Radisson Blu Hotel Pune Kharadi
Radisson Blu Resort Temple Bay Mamallapuram
Radisson Blu Plaza Hotel Mysore
Radisson Blu Udaipur Palace Resort and Spa

List of Park Plaza Hotels

Park Plaza Ahmedabad, Ellis Bridge
Park Plaza Bengaluru
Park Plaza Chandigarh
Park Plaza Delhi, CBD Shahdara
Park Plaza Faridabad
Park Plaza Gurgaon
Park Plaza Jodhpur
Park Plaza Kolkata Ballygunge
Park Plaza Ludhiana
Park Plaza Noida
Park Plaza Salem
Park Plaza Zirakpur

List of Country Inn and Suites by Carlson

Country Inn and Suites by Carlson, Gurgaon Udyog Vihar
Country Inn and Suites by Carlson, Gurgaon, Sector 29
Country Inn and Suites by Carlson, Gurgaon, Sector 12
Country Inn and Suites by Carlson, Gurgaon, Sohna Road

Country Inn and Suites by Carlson, Ajmer
Country Inn and Suites by Carlson, Goa, Candolim
Country Inn and Suites by Carlson, Goa, Panjim
Country Inn and Suites by Carlson, Navi Mumbai
Country Inn and Suites by Carlson, Mysore
Country Inn and Suites by Carlson, Amritsar, Queens Road
Country Inn and Suites by Carlson, Sahibabad, Ghaziabad
Country Inn and Suites by Carlson, Indore
Country Inn and Suites by Carlson, Jaipur
Country Inn and Suites by Carlson, Ahmedabad
Country Inn and Suites by Carlson, Jalandhar
Country Inn and Suites by Carlson, Katra at Vaishno Devi
Country Inn and Suites by Carlson, Mussoorie
Country Inn and Suites by Carlson, Delhi, Satbari
Country Inn and Suites by Carlson, Delhi, Saket
Country Inn and Suites by Carlson, Bhiwadi
Country Inn and Suites by Carlson, Meerut
Country Inn and Suites by Carlson, Bhatinda
Country Inn and Suites by Carlson, Haridwar

List of Park Inn By Radisson

Park Inn by Radisson, Gurgaon
Park Inn by Radisson, Jaipur Jai Singh Highway
Phoenix Park Inn Resort Candolim, Goa
Park Inn by Radisson, New Delhi, IP Extension

CITRUS HOTELS

The essence of Citrus, India's newest chain of hotels is one-of-a-kind experience, specially created for the world traveller. Smart and sophisticated with a distinct global life style, every Citrus Hotel is unique in location, look, mood and service.

Citrus Hotels is a part of the **Mirah Group**, a well diversified group engaged in Real Estate development, Hospitality, Travels, Wind Energy Generation, Computer, Education, Textiles, Corporate Gifts and International Trading.

List of Citrus Hotels

Citrus Retreat, Alleppey
Citrus Bengaluru, Bangalore
Citrus Sriperumbudur, Chennai
Citrus Hotel, Goa

Citrus Hotel Gurgaon Central, Gurgaon

Citrus Hotel, Lonavala

Citrus Chambers, Mahabaleshwar

Citrus Hotel, Pune

Citrus Hotel, Bhiwadi

Citrus East Coast Road, Chennai

CLARKS GROUP OF HOTELS

An internationally renowned hotel group, Clarks Group of Hotels owned by **UP Hotels Limited** has made a name for itself in the Hospitality Industry by virtue of its blue ribbon service and warm Indian hospitality. Clarks Group of Hotels has defined a benchmark of fine living and expressed high standards in the art of hospitality in northern India. Each hotel offers an unforgettable Indian experience. The hotels that the Clarks Group of Hotels run provide its guests with the best of the luxury, comfort, world-class amenities and services to make their stay a perfect one.

List of Clarks Group of Hotels in India

Hotel Clarks Ajmer, Jaipur

Hotel Clarks Shiraz, Agra

Hotel Clarks Avadh, Lucknow

Hotel Clarks, Khajuraho

Hotel Clarks, Varanasi

Hotel Clarks Tower, Varanasi

CLARKS INN GROUP OF HOTELS

Clarks Inn Group of Hotels is one of the fastest growing Hotel Management Company in India with 57 operational properties in 16 states. Within a short span of 9 years, Clarks Inn has been recognized as a well-known brand in India and abroad.

Brands of Clarks Inn Hotels

Clarks Inn: Clarks Inn provides full-service 3 and 4 star hotels, which are located in some of the most visited leisure and business destinations. These hotels enable convenient access to the tourist and commercial hubs of the city, and offer prompt and courteous service to their guests. They ensure guests get unbeatable value for money, complemented by welcoming and comfortable rooms, modern amenities as well as reliable security standards.

Clarks Exotica: Clarks Exotica offers luxury resorts at famed leisure destinations. The backdrop, resplendent with natural beauty and local culture, perfectly complement the exceptional service and amenities provided by these hotels.

Clarks Inn Suites: Clarks Inn Suites, leading business and leisure hotels, are primarily located in premium leisure spots and business centers. At par with the tastes of global travelers, simply walking into these hotels help in uplifting the spirits of the guests; at the end of a long tiring day.

Clarks Residences: Clarks Residences offer a chain of fully serviced apartments in key business districts and metropolises. It is designed keeping in mind the need for providing an extended stay facility that is ideal for business travelers, as well as for vacations and relocation needs. These residential-style hotels offer a sense of sophistication and timeless class, with a distinctive décor and rich textures that spell luxury. Inspired

by the essence of the city, the serviced apartments are full of amenities that will relax even the weariest of travelers.

List of Clarks Inn Group of Hotels

Clarks Inn Suites, East Delhi
Clarks Inn Apple Tree, Ghaziabad
Clarks Inn Pacific Mall, East Delhi
Clarks Inn Nehru Place, New Delhi
Clarks Inn Kailash Colony, New Delhi
Clarks Inn, Gurgaon
HK Clarks Inn, Amritsar
Amrapali Clarks Inn, Bareilly
Clarks Inn Airport Hotel, Hubli, Karnataka
VijayaTej Clarks Inn, Patna
Blue Saphire Clarks Inn, Haldwani
Keshav Clarks Inn Gadag, Karnataka
Amrapali Clarks Inn, Deoghar, Jharkhand
MB Greens Clarks Inn, Moradabad
Bhawna Clarks Inn, Agra
Clarks Inn, Mathura
Sapna Clarks Inn, Lucknow
Velvet Clarks Exotica, Jirakpur, Punjab
Clarks Inn, Panchkula
Tarawade Clarks Inn, Pune
Clarks Inn Ambala, Haryana
Clarks Inn Muzaffarpur, Bihar
Clarks Inn Phagwara, Punjab
Clarks Inn Alwar
Clarks Inn Binsar, Uttarakhand
Clarks Inn Corbett Resort and Spa
Clarks Inn Cytrus, Noida
La Sunila Clarks Inn Suites, Goa
Clarks Inn Brinjal, Haridwar
Vision Clarks Inn, Samode, Jaipur

CAMBAY HOTELS

Cambay Hotels and Resorts began its ambitious endeavor into the world of hospitality with the launch of Cambay Spa Resort in 2004 at Gandhinagar, the capital of Gujarat. Cambay Hotels is a part of the **Neesa Group**, a diversified industrial group with interests in hospitality, education, agritech, information technology, infrastructure and food processing.

Recognizing the importance of quality training in hospitality management, The Neesa Group established **Cambay Institute of Hospitality Management (CIHM) in Gandhinagar, Jaipur, Udaipur and Neemrana** that offers courses in hospitality management.

List of Cambay Hotels

Cambay Sapphire, Ahmedabad
Cambay Grand, Ahmedabad
Cambay Sapphire, Gandhinagar
Cambay Resort, Jaipur
Cambay Resort, Udaipur
Cambay Palm Lagoon, Kollam
Cambay Grand, Jaipur

CLUB MAHINDRA HOLIDAYS

Mahindra Holidays and Resorts India Ltd., (MHRIL) is a part of the Leisure and Hospitality sector of the **Mahindra Group** and brings to the industry values such as Reliability, Trust and Customer Satisfaction. Started in 1996, the company's flagship brand 'Club Mahindra Holidays', today has a fast growing customer base of over 170,000 members and 40 beautiful resorts at some of the most exotic locations in India and abroad. Over the last decade, MHRIL has established itself as a market leader in the family holiday business.

List of Club Mahindra Resorts

Club Mahindra Ashtamudi, Kerala
Club Mahindra Baiguney, Sikkim
Club Mahindra Binsar Valley, Uttarakhand
Club Mahindra Binsar Villa, Uttarakhand
Club Mahindra Cherai Beach, Kerala
Club Mahindra Corbett, Uttarakhand
Club Mahindra Derby Green, Ooty
Club Mahindra Dharamshala, Himachal Pradesh
Club Mahindra Emerald Palms, Goa
Club Mahindra Gangtok, Sikkim
Club Mahindra Gir, Gujarat
Club Mahindra Jaisalmer, Rajasthan
Club Mahindra Kanatal, Uttarakhand
Club Mahindra Kandaghat, Himachal Pradesh
Club Mahindra Kumarakom, Kerala
Club Mahindra Kumbhalgarh, Rajasthan
Club Mahindra Madikeri, Coorg – Karnataka
Club Mahindra Mahabaleshwar Sherwood Resort, Maharashtra

Club Mahindra Manali, Himachal Pradesh
Club Mahindra Mashobra, Himachal Pradesh
Club Mahindra Masinagudi, Tamil Nadu
Club Mahindra Munnar, Kerala
Club Mahindra Mussoorie, Uttarakhand
Club Mahindra Naukuchiatal, Uttarakhand
Club Mahindra Nawalgarh, Rajasthan
Club Mahindra Poovar, Kerala
Club Mahindra Puducherry, Pondicherry
Club Mahindra Thekkady, Kerala
Club Mahindra Udaipur, Rajasthan
Club Mahindra Varca Beach, Goa
Club Mahindra Virajpet, Coorg – Karnataka
Lake Forest Hotel, Yercaud, Tamil Nadu
Whote Meadows, Manali

CONCEPT HOSPITALITY PVT. LTD. (CHPL)

Concept Hospitality Pvt. Ltd. was formed on 10th July, 1996 and made a deemed public limited company on 16th February, 1998, and was recently deemed as Concept Hospitality Pvt. Ltd. CHPL sets up and operates Restaurants, Hotels, Clubs and Resorts for different owners. Today the company is managing 13 hotels across India and has signed up 15 hotels which are scheduled to open in the coming two years. The company is actively looking out for new management opportunities and is focusing on becoming the leading management oriented hotel chain in India. Concept hospitality has just finished the launch of its new brand **The Fern**, which will soon become a trademark for environmentally sensitive hotels across India.

Brands of Concept Hotels

The Fern: It is the luxury badge for hotels and resorts with a 5-star and above classification. It is a full-service product with premium accommodation and services.

The Fern Residency: It is a mid-tiered badge for hotels and resorts in the 3 and 4-star categories. It is a full-service product with upmarket accommodation and services.

Beacon Hotels and Resorts: This brand provides a budget hotel experience in the 2-star category. It offers limited services with the essential amenities for a comfortable stay.

List of the Fern Hotel and Resorts

Meluha The Fern, Mumbai
The Fern, Jaipur
The Fern, Ahmedabad

Mansarovar The Fern, Hyderabad
Beaumonde The Fern, Kochi
The Fern Gir Forest Resort, Sasan Gir
The Fern Gardenia Resort, Goa
The Fern Beira Mar Resort, Goa
The Fern Courtyard Resort, Ganpatipule
The Fern Surya Resort, Mahabaleshwar
The Fern Samali Resort, Dapoli
The Fern Citadel, Bengaluru
The Fern Residency, Rajkot
The Fern Residency, Mumbai
The Fern Residency, Chandigarh
The Fern Residency, Vadodara
The Fern Residency, Jodhpur
The Fern Residency, Tezpur

List of Independently Branded Hotels

Royal Melange Hotel, Ajmer
The Uppal, New Delhi
Desert Scape, Jodhpur
Floatel, Kolkata
Rodas, Mumbai
The Wallstreet, Jaipur
Quality Hotel Regency, Pune
Maia Beacon Residences, Bengaluru

CHOICE HOTELS INDIA

Choice Hotels India was established in 1987 with the objective of setting up first class, mid-market franchised hotels in metropolitan and secondary cities. Choice Hotels India is a wholly-owned subsidiary of Choice Hotels International, one of the largest and most widespread hotel franchisor in the world with over 6300 hotels across the globe. Ranging from limited service to full service hotels in the economy, mid-scale and upscale segments, Choice-branded properties provide business and leisure travelers with a range of high-quality, high-value lodging options throughout the world.

Brands of Choice Hotels in India

Quality: At Quality hotels, you know you will get your money's worth. Our signature "Value Qs" are your "cues" that the most important things are done well at every Quality brand hotel —Q Breakfast, Q Bed, Q Shower, Q Service and Q Value. Relax in our signature Q Bed featuring a comfortable mattress, 200 thread-count linens and plenty of fluffy pillows. Start your morning with a

hot, fresh and healthy Q Corner Café breakfast. Enjoy all the free Q Value amenities like high-speed Internet, newspaper, local calls and 24-hour coffee and tea in the lobby. All that, plus our helpful and friendly Q Service too — that's how Quality gives you value for your hard-earned money.

Comfort Inn: At Comfort Inn hotels, you will find everything you need to create your perfect stay. From welcoming rooms and cozy beds to a new, free hot breakfast with lots of choices—all at a great rate. Our friendly staff is also available to help you with whatever you need to make your trip more enjoyable. You will wake up refreshed and ready for a great day.

Sleep Inn: You will have sweet dreams every time you choose Sleep Inn. Modern rooms, cozy beds and large signature showers leave you energized for the day ahead. Our friendly staff will take good care of you with a bountiful array of free amenities like the Morning Medley hot breakfast buffet, high-speed Internet, pool or exercise room and morning newspaper.

Clarion: Clarion helps people come together for occasions ranging from business meetings and special events to vacations and weekend getaways. Clarion provides the full-service facilities, attentive services and social atmosphere that make it easy to connect, achieve and enjoy. Clarion facilities include meeting space and catering, a casual restaurant or a social bar and lounge. Plus, we offer a wide range of supportive amenities like free high-speed Internet, in-room dining, pool, fitness center, 24-hour business center and, of course, comfortable rooms with high-quality bedding, large desk and ergonomic chair—everything you need in a social atmosphere with affordable rates.

Cambria hotels and suites: A thoughtfully designed all-suites hotel, Cambria Suites offers spacious rooms with separate workspace. Enjoy luxurious bath amenities and bedding, flat-panel LCD TVs, swimming pool and state-of-the-art fitness center — plus, contemporary dining, hot breakfast, dinner menu and a 24/7 convenience store. And all Cambria hotels are 100% smoke-free.

List of Choice Hotels in India

Comfort Inn Sunset, Ahmedabad
Comfort Inn President, Ahmedabad
Comfort Inn Alstonia in Amritsar
Comfort Inn Vijay Residency, Bengaluru
Quality Inn Shravanthi, Bengaluru
Comfort Inn Tulip Heights, Bathinda
Quality Inn Sabari, Chennai
Clarion Hotel Chennai
Clarion Hotel Coimbatore
Quality Inn Bliss, Gurgaon
Quality Inn Sabari Resorts—Kodaikanal
Quality Inn Viha, Kumabakonam
Comfort Inn, Lucknow
Quality Inn River Country Resort, Manali
Comfort Inn Heritage, Mumbai
Quality Inn Regency, Nashik
Clarion Collection, New Delhi
Quality Inn Himdev, Shimla
Quality Inn DV Manor, Vijayawada
Quality Inn Bez Krishnaa, Vishakhapatnam

CGH EARTH

Formerly the **Casino Group of Hotels**, CGH Earth was founded by Dominic Joseph Kuruvinakunnel on the harbour island of Willingdon, in Cochin, Kerala, in 1957. Today, the Group is a leading hotel and resort operator in southern India. The CGH Earth Group covers nine destinations that capture many moods of Kerala and the Lakshadweep Islands. The beautiful location of these hotels satiates the nature lovers. All the resorts are located at beautiful places and each has its own theme and atmosphere. Under the personal management of the founder's six sons, the group shares a deep commitment towards responsible and ecological tourism and tours in India.

List of CGH Earth Hotels in India

Spice Village, Thekkady
Coconut Lagoon, Kumarakom
The Marari Beach Resort, Mararikulam
Spice Coast Cruises, Vembanad
The Brunton Boatyard, Fort Kochi
Casino Hotel, Willingdon Island
Bangaram Island Resort, Lakshadweep

GOLDEN TULIP HOTELS, SUITES AND RESORTS

Golden Tulip Hotels, Suites and Resorts is a hotel chain, part of **Louvre Hotels Group**, which was founded in 1976.

Brands of Golden Tulip Hotels

Royal Tulip
Golden Tulip
Tulip Inn

List of Golden Tulip Hotels in India

Tulip Inn, Mussoorie
Golden Tulip, Manali
Golden Tulip, Gurgaon
Golden Tulip, Bhiwadi
Golden Tulip, Amritsar
Golden Tulip, Haridwar

Golden Tulip, Ghaziabad
Golden Tulip, Chandigarh
Golden Tulip, Udaipur
Golden Tulip, New Delhi
Golden Tulip, Ahmedabad
Golden Tulip, Bangalore
Golden Tulip, Jaipur
Golden Tulip, Lucknow
Golden Tulip, Neemrana
Golden Tulip, Goa
Golden Tulip, Navi Mumbai
Golden Tulip, Shimla
Golden Tulip, Vasai

HYATT HOTELS

Hyatt group of hotels is a global hospitality company with widely recognized, industry leading brands and a tradition of innovation developed over their more than 50-year history. Their mission is to provide authentic hospitality by making a difference in the lives of the people they touch every day. They focus on this mission in pursuit of the goal of becoming the most preferred brand in each segment that Hyatt serves for their associates, guests, and owners. They support the mission and goal by adhering to a set of core values that characterizes their culture. Hyatt manages, franchise, own and develop Hyatt branded hotels, resorts and residential and vacation ownership properties around the world. As of December 31, 2013, the company's worldwide portfolio consisted of 549 properties.

Brands of Hyatt Group

Park Hyatt
Grand Hyatt

PARK HYATT	ANDAZ	GRAND HYATT	HYATT	HYATT REGENCY	HYATT PLACE	HYATT HOUSE	HYATT RESIDENTIAL
PARK HYATT	ANdAZ	GRAND HYATT	HYATT	HYATT REGENCY	HYATT PLACE	HYATT house	HYATT RESIDENCE CLUB
29 HOTELS	8 HOTELS	38 HOTELS	29 HOTELS	146 HOTELS	169 HOTELS	54 HOTELS	23 PROPERTIES
5,815 ROOMS	1,701 ROOMS	21,505 ROOMS	7,478 ROOMS	67,740 ROOMS	21,957 ROOMS	7,603 ROOMS	2,193 ROOMS
19 COUNTRIES	3 COUNTRIES	19 COUNTRIES	2 COUNTRIES	31 COUNTRIES	1 COUNTRY*	1 COUNTRY*	8 COUNTRIES
LUXURY	BOUTIQUE-INSPIRED	FULL SERVICE	FULL SERVICE	FULL SERVICE	SELECT SERVICE	EXTENDED-STAY	RESIDENCE TIMESHARE

Hyatt Hotels
Hyatt Regency
Hyatt Place
Hyatt House
Hyatt Zilara
Hyatt Ziva
Hyatt Residence Club
Andaz

List of Hyatt Hotels in India

Hyatt Ahmedabad
Hyatt Amritsar
Hyatt Bangalore
Hyatt Hyderabad
Hyatt Pune
Hyatt Raipur
Hyatt Regency, Chennai
Hyatt Regency, New Delhi
Hyatt Regency, Pune
Hyatt Regency, Ahmedabad
Hyatt Regency, Chandigarh
Hyatt Regency, Gurgaon
Hyatt Regency, Kolkata
Hyatt Regency, Mumbai
Hyatt Regency Ludhiana

Grant Hyatt, Goa
Grant Hyatt, Mumbai
Park Hyatt, Hyderabad
Park Hyatt, Chennai
Park Hyatt Goa Resort and Spa
Hyatt Place, Hampi
Hyatt Place, Gurgaon
Hyatt Place, Pune

HILTON WORLDWIDE

Hilton Worldwide is one of the largest and fastest growing hospitality companies in the world, spanning the lodging sector from luxury and full-service hotels and resorts to extended-stay suites and focused-service hotels, with more than 4,300 hotels, resorts and timeshare properties comprising more than 715,000 rooms in 94 countries and territories.

In the nearly 100 years since their founding, Hilton Worldwide has defined the hospitality industry and established a portfolio of 12 world-class brands, including their flagship full-service Hilton Hotels and Resorts brand, which is the most recognized hotel brand in the world.

List of Hilton Hotels in India

Hilton Chennai Guindy, Chennai
Hilton Bangalore Embassy Golflinks
Hilton Chennai

Brands of Hilton Worldwide		Hotels	Countries and territories
Hilton Hotels and Resorts	The stylish, forward-thinking global leader in hospitality.	557	82
Waldorf Astoria Hotels and Resorts	Offers unforgettable experiences at iconic destinations around the world.	26	12
Conrad Hotels and Resorts	Offers smart luxury travelers inspiring connections and intuitive service in a world of style.	24	18
Canopy by Hilton	Canopy by Hilton is designed as a natural extension of the neighborhood—with local design, food and drink, culture, guest-directed service, and surprisingly comfortable spaces.		
Curio: A Collection by Hilton	The global collection of authentic, independent and remarkable properties that are woven into the fabric of their destinations. While each one is unique, they are united in their individuality.	5	1
DoubleTree by Hilton	Fast-growing, global collection of upscale hotels in gateway cities, metropolitan areas and vacation destinations.	415	35
Embassy Suites by Hilton™	Full service, upscale hotels offering two-room suites, free, cooked—to order breakfast and complimentary evening receptions with snacks and drinks.	222	6
Hilton Garden Inn	Offers the amenities and services that allow guests to discover and connect while on the road.	626	22
Hampton by Hilton®	Quality experience, great value and friendly service in its signature Hamptonality style.	2025	16
Homewood Suites by Hilton	For guests seeking home-like accommodations when traveling for an extended stay.	364	3
Home2 Suites by Hilton	Offering flexible guest room configurations and inspired amenities for the cost-conscious guest.	48	3
Hilton Grand Vacations (HGV)	High-quality vacation ownership resorts in celebrated destinations	44	4

Hilton Mumbai International Airport

Hilton Jaipur

Hilton Shillim Estate Retreat and Spa, Pune

Hilton Garden Inn, Gurgaon

Hilton Garden Inn, New Delhi

Hilton Garden Inn, Trivandrum

DoubleTree by Hilton, Agra

DoubleTree Suites by Hilton, Bangalore

DoubleTree by Hilton, Goa

DoubleTree by Hilton, Gurgaon

DoubleTree by Hilton, Pune

Hampton by Hilton, Vadodara

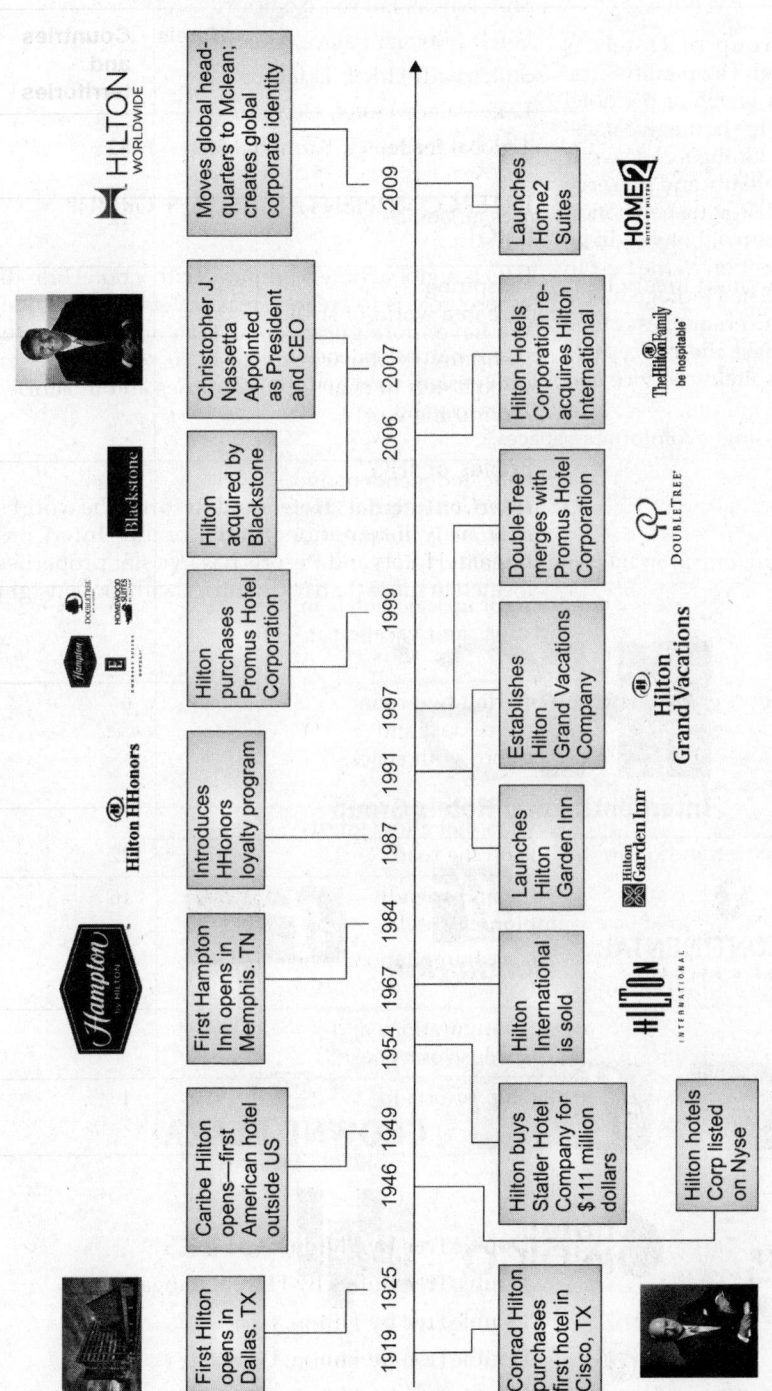

History of Hilton Hotels

| 1919 | 1925 | 1946 | 1949 | 1954 | 1967 | 1984 | 1987 | 1991 | 1997 | 1999 | 2006 | 2007 | 2009 |

First Hilton opens in Dallas, TX

Conrad Hilton purchases first hotel in Cisco, TX

Caribe Hilton opens—first American hotel outside US

Hilton buys Statler Hotel Company for $111 million dollars

Hilton hotels Corp listed on Nyse

First Hampton Inn opens in Memphis, TN

Hilton International is sold

Introduces HHonors loyalty program

Launches Hilton Garden Inn

Establishes Hilton Grand Vacations Company

Hilton purchases Promus Hotel Corporation

DoubleTree merges with Promus Hotel Corporation

Hilton acquired by Blackstone

Christopher J. Nassetta Appointed as President and CEO

Hilton Hotels Corporation re-acquires Hilton International

Moves global head-quarters to Mclean; creates global corporate identity

Launches Home2 Suites

HRH GROUP OF HOTELS

The mission of the HRH Group of Hotels is 'Preservation of Heritage through Hospitality'. The HRH Group of Hotels in India is one of the hotel groups offering hotel packages for heritage palace-hotels and resorts in India. HRH Group's endeavour to maintain the tradition of hospitality and preserve the royal heritage is a matter of distinguished honor. The recognition of the HRH group of hotels will go a long way in reaffirming its position as the best in the Heritage category in India. The Heritage hotels of the HRH Group symbolize the uniqueness of the group by presenting to its guest the concept of experiencing the original in hospitality service and grandeur of the past.

List of HRH Group of Hotels

Gajner Palace, Bikaner

Karni Bhawan Palace, Bikaner

Maan Bilas Hotel, Bikaner

Gorbandh Palace, Jaisalmer

Shiv Niwas Palace, Udaipur

Fateh Prakash Palace, Udaipur

Shikarbadi Hotel, Udaipur

Lake Palace Hotel, Udaipur

The Aodhi Hotel, Kumbhalgarh

INTERCONTINENTAL HOTELS GROUP (IHG)

IHG is a global hotel company with 9 hotel brands whose goal is to create Great Hotels Guests Love. We have more guest rooms than any other hotel company in the world—674,000 rooms in over 4,700 hotels in nearly 100 countries and territories around the world.

Brands of IHG

InterContinental® Hotels and Resorts: The world's first truly international hotel brand, InterContinental Hotels and Resorts has five star properties located in more than 60 countries with local insight

that comes from over 60 years experience. Each property offers its own distinctive style, from heritage elegance to urban chic and resorts on tropical shores, making it an ideal brand for travelers with discerning tastes.

HUALUXE™ Hotels and Resorts: HUALUXE™ is dedicated to providing luxurious, close-to-nature surroundings and attentive, considerate service which help our guests achieve their accomplishments their way. We celebrate the essence of world-class Chinese hospitality underpinned by internationally renowned consistency.

Crowne Plaza® Hotels and Resorts: At Crowne Plaza® Hotels and Resorts, we make sure that everything you need to succeed is right at your fingertips when traveling for business. Each property offers comprehensive meetings facilities, the Crowne Plaza Sleep Advantage®, 24-hour business services and fully-equipped fitness centers ensuring you can get your work done and be at the top of your game.

Hotel Indigo®: Now in major cities worldwide, Hotel Indigo® hotels are known for fresh design, personalized service and an endless ability to inspire through our distinctively local personality. Offering a genuine boutique experience with the reliability and benefits of a brand, each hotel is designed to reflect the sights, sounds and character of the neighborhoods where we are located. From the art on the walls to the locally influenced food and drink.

EVEN™ Hotels: The EVEN™ Hotels brand offers a fresh perspective on travel to wellness-minded guests. Throughout their travel journey, we help guests Keep Active, Rest Easy, Eat Well, and Accomplish More so they can find the balance they seek while away from home. Wellness-savvy staff and modern, natural spaces enable guests to maintain their routines by providing uplifting choices, calming influences, and healthier options —on their terms.

Holiday Inn® Hotels: Fresh, modern and better than ever. Today's Holiday Inn® hotels offer a comfortable, familiar atmosphere where guests can relax and enjoy amenities such as internet, restaurants, fitness centers and comfortable lounges—plus kids eat and stay free. The perfect mix of business and pleasure for today's comfort-seeking traveler.

Holiday Inn Express® Hotels: The smart choice for travelers who need a simple, engaging hotel to rest, recharge and get a little work done. Whether you are on the way to somewhere or with us for a week, you will like that Holiday Inn Express® hotels provide just the things that you need most and nothing more.

Holiday Inn Resort® Family fun and relaxation from the name you know and trust. Located in key family vacation destinations around the world, the Holiday Inn Resort® brand puts you in the middle of the fun. Every resort features a pool, activities and live entertainment, so everyone in the family can make the most of their vacation with us.

Holiday Inn Club Vacations® Making a vacation memorable is easier than ever in these family-style villas located in top US destinations. Room for everyone and home-away-from-home amenities create a great jumping off point close to area attractions. It is the opportunity for an experience your family and friends will return to again and again.

Staybridge Suites® Spacious suites with fully equipped kitchens and roomy workstations help you settle in and get things done. Plus, the complimentary hot breakfast buffet, free wireless anywhere and the social evening receptions get you going and keep you connected in every way.

Candlewood Suites: With a focus on comfort, space and value, Candlewood Suites® hotels offer studio and one-bedroom suites with more space to stretch out and relax. Fully equipped kitchens, large work areas and free laundry mean you will feel completely at home, no matter how long your stay.

List of Hotels of IHG Group

Crowne Plaza Ahmedabad City Centre

Crowne Plaza Bengaluru

Crowne Plaza Kochi

Crowne Plaza Mayur Vihar, Noida

Crowne Plaza Okhla, New Delhi

Crowne Plaza Rohini, New Delhi

Crowne Plaza Gurgaon

Crowne Plaza Greater Noida

Crowne Plaza Jaipur

Holiday Inn Express Ahmedabad

Holiday Inn Express and Suites Hyderabad

Holiday Inn Amritsar

Holiday Inn Cochin

Hoiday Inn Mayur Vihar, Noida

Holiday Inn New Delhi International Airport

Holiday Inn Resort Goa

Holiday Inn Jaipur

Holiday Inn Jaipur City Centre

Holiday Inn Mumbai International Airport

Holiday Inn Panchkula Chandigarh

Holiday Inn Pune

InterContinental Marine Drive, Mumbai

ITC WELCOMGROUP

Symbolised by its distinctive 'Namaste' logo, ITC Hotels integrated India's fine tradition of hospitality with globally benchmarked services. Launched on October 18, 1975 ITC WelcomGroup started its hotel division with the 'Chola Sheraton in Chennai'.

ITC WelcomGroup has lots of awards and achievements added to its cap. These hotels have played host to many world leaders and business tycoons like Tony Blair-the British Prime Minister, Bill Clinton—the US President, Vladimir Putin—the Russian President, Barack Obama—the US President, Bill Gates. Well appreciated and approved by these world leaders and tycoons ITC WelcomGroup grew one step up in its hospitality business.

Each and every hotel has a design which gets etched in the minds of the guests and the onlookers. The architectural structure and designs of the chain of hotels of ITC WelcomGroup rekindles the bygone dynasties that ruled India. Today, ITC Hotels is one of largest hotel chains in the Country with over 100 hotels across 70 destinations.

ITC Hotels opened the **Welcomgroup Graduate School of Hotel Administration in 1994 at Manipal** in association with the TMA Pai Foundation. The Institute enshrines the vision of the ITC Hotels by developing Management Trainees and retraining Practicing Managers.

Brands of The ITC Hotels

ITC Hotels—The Luxury Collection: ITC Hotels—The Luxury Collection are super deluxe and premium hotels located at strategic business and leisure locations. The association of ITC Hotels and 'The Luxury Collection' of Starwood presents a unique set of hotels in a bouquet of enriching experiences that celebrate the spirit and distinctive character of each destination. Eleven exceptional properties of ITC Hotels bring you the architectural grandeur of ancient dynasties and the cultural ethos of different regions of the Indian peninsula. Distinguished by their magnificent décor and

impeccable service, the ITC Luxury Collection Hotels and Resorts, all with timeless grace and style are rich with unique qualities that defy easy description.

WelcomHotels | Sheraton: An enduring expression of efficiency and luxury, WelcomHotels provide the comfort of home in a delightfully distinctive ambience. Synonymous with customer centricity and efficiency, these hotels have been aligned under the renowned Sheraton brand, offering warm, comforting services to the global traveler. With the advantage of peerless location, they offer easy access to business hubs and historical landmarks and cater efficiently to the needs and aspirations of today's traveler. Located at Kerala, Delhi, Chandigarh, Jodhpur, Aurangabad, Vadodara, Visakhapatnam and Chennai, WelcomHotels are the gateway to these special destinations and are rich with the warmth of personalized service, making every visit memorable.

Welcom Heritage Hotels: Welcom Heritage, a joint venture between ITC Ltd. and Jodhana Heritage, represents some of the best traditions of heritage hospitality and tourism in India. It offers more than 50 exclusive heritage destinations in India where you could savor the best of Indian traditions. Welcom Heritage Hotels contains, within its fold, palaces, havelis and forts, ancestral mansions and resorts, each a residence as well as a living monument to India's regal past.

Fortune Hotels: Fortune Hotels was established in 1995 to cater to the mid-market and upscale segment in corporate as well as leisure destinations. It has emerged as one of the fastest growing 'first-class, full-service business hotel' chains in India. A wholly-owned subsidiary of ITC Hotels Ltd., Fortune Hotels is one of the renowned hotel groups in the hospitality industry.

In keeping with the demand for hotels providing quality accommodation at affordable rates, these hotels offer multiple brand extensions to cater to various segments.

Brands of Fortune Hotels

Fortune Select: Upscale hotels offering contemporary product and services, located at business locations in metros, mini metros and key leisure locations. These hotels offer best in class services in 4 to 5 star category.

Fortune Park: Business hotels located in metro/non-metro cities. These are mid-scale hotels between 3 to 4 star category.

Fortune Resorts: Hotels located at popular holiday and leisure destinations with products catering to mid-market and upscale segment. These resorts offer a variety of exciting family friendly holiday packages.

Fortune Inns: Full service business hotels with 30 to 50 rooms in each property and limited F and B outlets.

My Fortune: Stylish lifestyle hotels, where the 'comfort of home' and 'efficient service' come seamlessly together, thereby creating a 'sense of belongingness'. These hotels offer a perfect blend of traditional Indian hospitality with new age technology, while catering to the upscale business traveler.

List of ITC - The Luxury Collection Hotels

ITC Grand Bharat, A Luxury Collection Resort, Gurgaon

ITC Gardenia, A Luxury Collection Hotel, Bengaluru

ITC Windsor, A Luxury Collection Hotel, Bengaluru

ITC Grand Central, A Luxury Collection Hotel, Mumbai

ITC Maratha, A Luxury Collection Hotel, Mumbai

ITC Maurya, A Luxury Collection Hotel, New Delhi

ITC Rajputana, A Luxury Collection Hotel, Jaipur

ITC Grand Chola, A Luxury Collection Hotel, Chennai

ITC Kakatiya, A Luxury Collection Hotel, Hyderabad

ITC Mughal, A Luxury Collection Hotel, Agra

ITC Sonar, A Luxury Collection Hotel, Kolkata

List of WelcomHotels | Sheraton Hotels

Welcom Hotel Rama International, Aurangabad

Welcom Hotel, Vadodara

Sheraton, New Delhi

WelcomHotel Dwarka, New Delhi

WelcomHotel, Kollam

WelcomHotel, Kozhikode

WelcomHotel Panchkula, Chandigarh

WelcomHotel, Jodhpur

WelcomHotel Grand Bay, Visakhapatnam

List of Welcom Heritage Hotels

Corbett Ramganga Resort, Corbett

Solang Valley Resort, Manali

Woodville Palace, Shimla

Bal Samand Palace, Mandore Road, Jodhpur

Maharani Bagh Orchard Retreat, Ranakpur

Sardar Samand Palace, Jodhpur

The Judge's Court, Pragpur, Kangra

Taragarh Palace, Palampur, Kangra

Umaid Bhawan Palace, Kota

Connaught House, Mount Abu

Royal Camp-Nagaur Fort Jodhpur

The Fort-Nalagarh, Solan

Karni Fort, Udaipur

Panjim Inn, Goa

Grace Hotel, Dharamsala, Kangra

Golf View, Pachmarhi

Ranjit's Svaasa, Amritsar

Bassi Fort, Chittorgarh

Bijay Niwas Palace, Bijaynagar

Lal Niwas, Phalodi

Mandir Palace, Jaisalmer

Sirsi Haveli, Jaipur

Regency Villas palace, Ooty

Jukaso Ganges, Varanasi

Rao Bagh Palace, Charkhari

Kasmanda Palace, Mussoorie

Windsor Lodge, Ranikhet

Palace Belvedere, Nainital

List of Fortune Hotels

Fortune Select Global, Gurgaon

Fortune Select Exotica, Navi Mumbai

Fortune Inn Jukaso, Pune

Fortune Hotel Landmark, Ahmedabad

Fortune Park Center Point, Jamshedpur

Fortune Landmark, Indore

Fortune Park Galaxy, Vapi

Sullivan Court, Ooty

Fortune Pandian Hotel, Madurai

Fortune Chariot Beach Resort, Mahabalipuram

Fortune Resort Bay Island, Port Blair

Fortune Hotel The South Park, Thiruvananthapuram

Fortune Hotel Calicut, Kozhikode

Fortune Select JP Cosmos, Bengaluru

Fortune Kences Hotel, Tirupati

Fortune Murli Park, Vijayawada

Fortune Inn Sree Kanya, Visakhapatnam

Fortune Resorts, Darjeeling

Fortune Inn Riviera, Jammu

Fortune Park Klassik, Ludhiana

Fortune Ummed, Jodhpur

Fortune Select Trinity, Bengaluru

Fortune Park JP Celestial, Bengaluru

Fortune Select Palms, Chennai

Fortune Inn Haveli, Gandhinagar

Fortune Inn Grazia, Ghaziabad

Fortune Park Orange, Sidhrawali, Gurgaon

Fortune Select Excalibur, Gurgaon

Fortune Park Vallabha, Hyderabad

Fortune Park Panchwati, Kolkata

Fortune Inn Valley View, Manipal

Fortune Park Boulevard, New Delhi

Fortune Inn Exotica, Hinjewadi, Pune

Fortune Park Lake City, Thane

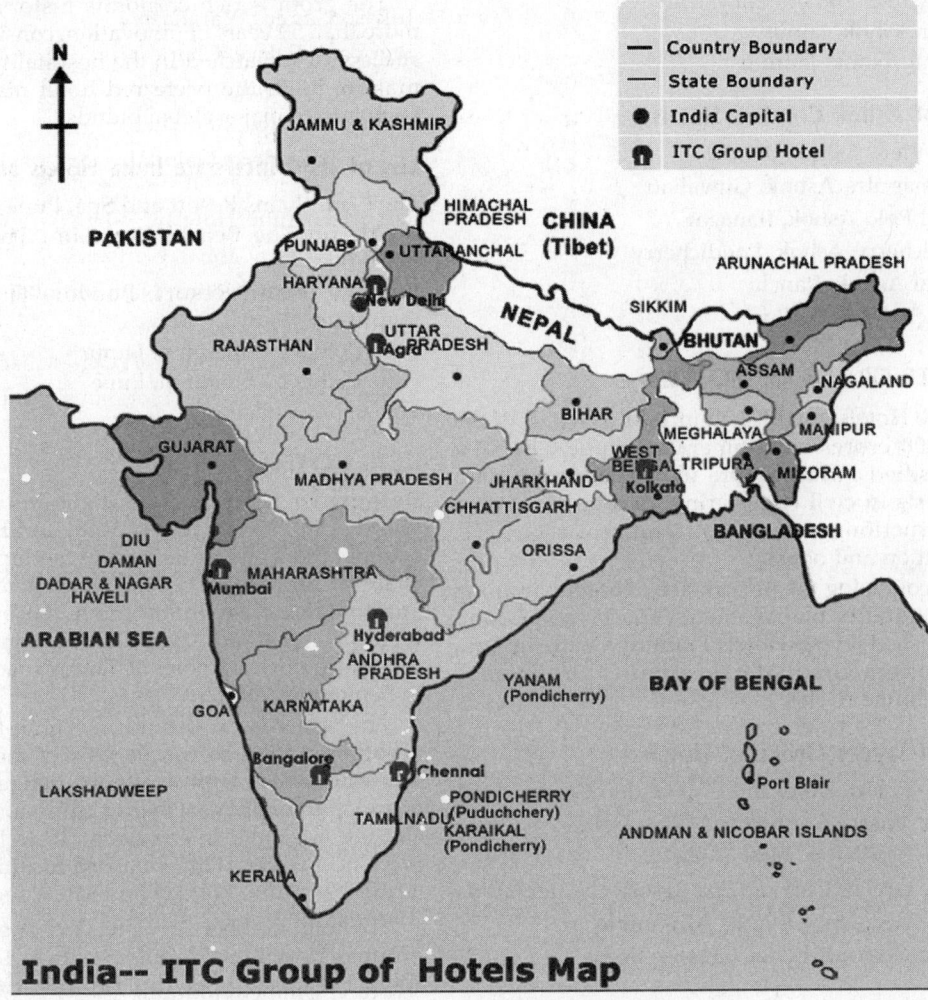

India-- ITC Group of Hotels Map

ITDC—ASHOKA GROUP OF HOTELS

Owned by the India Tourism Development Corporation (ITDC), the 'Ashoka Group of Hotels' has been India's gracious and regal host to leading national and international visitors. The Group is known for its unique mix of traditional Indian hospitality and modern-day systems and facilities that are beyond compare. The group has 16 hotels in 12 major destinations.

Recognizing the importance of quality training in hospitality management, The ITDC Group established **Ashok Institute of Hospitality and Tourism Management in New Delhi** in 1971.

Brands of The ITDC Hotels

Ashok Elite Hotels
Ashok Classic Hotels
Ashok Comfort Hotels

List of Ashok Elite Hotels

Ashok Hotel, New Delhi
Samrat Hotel, New Delhi
Lalitha Mahal Palace, Mysore

List of Ashok Classic Hotels

Hotel Janpath, New Delhi
Bharatpur Ashok, Bharatpur

Kalinga Ashok, Bhubaneswar

Jaipur Ashok, Jaipur

Jammu Ashok, Jammu

List of Ashok Comfort Hotels

Lake View Ashok, Bhopal

Brahmaputra Ashok, Guwahati

Donyi Polo Ashok, Itanagar

Pondicherry Ashok, Pondicherry

Ranchi Ashok, Ranchi

Nilanchal Ashok, Puri

JAYPEE GROUP OF HOTELS

Jaypee Hotels, established in 1981, is a part of the ₹ 20000 crores conglomerate Jaypee Group, a diversified infrastructure industrial group with interests in civil engineering, hydropower, dam construction, cement, golf and real estate, IT, education and hotels.

Recognizing the importance of quality training in hospitality management, The Jaypee Group established **Jaypee Hotels Training Centre in Agra** that offers Graduate Hospitality Proficiency Programme (GHPP).

List of Jaypee Group of Hotels

Jaypee Palace Hotel and Convention Centre, Agra

Jaypee Vasant Continental, New Delhi

Jaypee Siddharth, New Delhi

Jaypee Greens Golf and Spa Resort, Greater Noida

Jaypee Residency Manor, Mussoorie

Jaypee Delcourt-Hotel, Greater Noida

JHM INTERSTATE INDIA HOTELS AND RESORTS

JHM Interstate India Hotels and Resorts is a joint venture between two of the world's most experienced hotel management companies. Combining the global expertise of Interstate Hotels and Resorts (the largest and leading US based global hotel management company), with the local knowledge and resources of JHM Hotels and its founders the Rama family of Gujarat, JHM Interstate India is uniquely qualified to bring its skill and background to hotel owners and developers across India. JHM Interstate India has successfully brought this expertise to India in 2008, and is now one of the country's most rapidly growing hospitality management companies.

The group's rich corporate history spanning more than 50 years of innovation, consistency and success is unmatched in the hospitality industry, making them the preferred hotel management company of major global brands.

List of JHM Interstate India Hotels and Resorts

The Corinthians Resort and Spa, Pune

Turtle on The Beach, Kovalam, Thiruvananthapuram

Estuary Island Resort, Poovar, Thiruvananthapuram

Four Points by Sheraton, Jaipur

Four Points by Sheraton, Pune

The Astor, Kolkata

KEYS HOTELS

Berggruen Hotels, a young hotel company founded in 2006 by Sanjay Sethi and his team of experienced hotel professionals, set up a leading chain of mid-market hotels, resorts and service apartments across India under the brand Keys Hotels, Resorts and Apartments. The group rolled out its first owned hotel at Thiruvananthapuram in September 2009.

The focus is on providing value-for-money accommodation to the upwardly mobile mid-segment traveler with an offering that reflects their brand personality of being Stylish, Cheerful, Cutting Edge and International. Keys Resorts are more up-market, with enhanced facilities to cater to the discerning holiday and MICE segments.

The total numbers of hotels operated by Keys group is 15, with an inventory of over 1200 rooms. Keys Hotels is growing rapidly through its successful hotel management vertical, with over 100 hotels targeted to join the Keys portfolio by 2019.

List of Keys Hotels

Keys Hotel and Services Apartments, Whitefield, Bengaluru

Keys Hotel Katti-Ma, Chennai

Keys Hotel Nestor, Mumbai

Keys Hotel The Aures, Aurangabad

Keys Hotel, Hosur Road, Bengaluru

Keys Hotel, Ludhiana

Keys Hotel, Pimpri, Pune

Keys Hotel Templetree, Shirdi

Keys Hotel, Thiruvananthapuram

Keys Hotel Marigold, Jaipur

Keys Hotel Vihas, Tirupati

List of Keys Resorts

Pratap Palace A Keys Resort, Ajmer

Evershine A Keys Resort, Mahabaleshwar

Keys Hotel – Ronil, Goa

Keys Ras Resorts, Silvassa

LORDS HOTELS AND RESORTS

Lords intends to operate on a wide spectrum which includes full service hotels and inns, in the 3 stars plus category across the length and breadth of the country covering popular business and leisure destinations, with a clear focus on the budget market segment.

Brands of Lords Hotels

Aleenta

Lime Tree

L Café

Blue Coriander

Distil

Indulge

List of Lord Hotels in India

Parker Lords Eco Inn, Ahmedabad

Lords Eco Inn, Porbandar

P L Palace Lords Inn, Agra

Mirage Lords Inn, Kathmandu

Lords Plaza, Ankleshwar

Lion Lords Inn, Rajula

Lords Plaza, Bengaluru

Aakar Lords Inn, Saputara

Top3 Lords Resort, Bhavnagar

Patang Lords Eco Inn, Saputara

Lords Eco Inn, Dahej

Lords Eco Inn, Dwarkadish

Lords Resort, Silvassa

Lords Eco Inn, Gandhidham

Lords Inn, Somnath

Lords Plaza, Jaipur

Lords Plaza, Surat

Jagrati Lords Inn, Jaipur

Goradias Lords Inn, Shirdi

Lords Inn, Jammu

Revival Lords Inn, Vadodara

Lords Inn, Jodhpur

THE LALIT SURI HOSPITALITY GROUP

The Lalit Suri Hospitality Group, an enterprise of **Bharat Hotels Limited** is India's largest and the fastest growing privately owned Hotel Company.

Headquartered in New Delhi, the company opened its first hotel here in 1988 under the dynamic leadership of Founder Chairman Mr Lalit Suri, who spearheaded the Group's unprecedented expansion plans.

Rapid expansion and consolidation of its leadership position continues under the enterprising stewardship of Dr Jyotsna Suri, who took over as Chairperson and Managing Director in 2006.

All hotels within the group operated under the brand The Grand – Hotels, Palaces and Resorts. It was re-branded as 'The Lalit' on November 19, 2008 as a tribute to the company's Founder Chairman Mr Lalit Suri.

The company offers Seventeen luxury hotels, with 3600 rooms in the five-star deluxe segment with eleven operational hotels and six under development/restoration (including three overseas).

The Group has also forayed into mid-segment hotels under the brand—'The Lalit Traveller'. The first two hotels under this brand opened in Jaipur and Khajuraho with 25 more hotels planned in the next five years in Amritsar, Pune, Dehradun, Ludhiana, to name a few.

Dr Jyotsna Suri recently launched **The Lalit Suri Hospitality School in Faridabad** offering customized programmes in Hotel Management.

List of Lalit Group of Hotels

The Lalit New Delhi

The Lalit Mumbai

The Lalit Ashok Bangalore

The Lalit Jaipur

The Lalit Laxmi Vilas Palace Udaipur

The Lalit Temple View Khajuraho

The Lalit Golf and Spa resort Goa

The Lalit Resort and Spa Bekal (Kerala)

The Lalit Grand Palace Srinagar

The Lalit Chandigarh

The Lalit Great Eastern Kolkata

The Lalit Traveller Jaipur

The Lalit Traveller Khajuraho

THE LEELA PALACES, HOTELS AND RESORTS

The Leela Palaces, Hotels and Resorts is owned and managed by **Hotel Leelaventure Limited** which was established in 1983 in Mumbai. The company is a part of The Leela Group whose portfolio includes hotel and resort properties; IT and business parks; as well as, real estate development.

The Group was founded by Late Capt. C. P. Krishnan Nair, Chairman Emeritus and Founder Chairman of Hotel Leelaventure Limited, who envisioned The Leela as one of the finest Indian luxury hospitality brands. Under his leadership, within a short span of 25 years, the Group has evolved from one hotel, on the outskirts of Mumbai, into a range of eight award-winning properties across the country that celebrate India's diverse geography and architectural history. In the process, The Leela redefined standards of luxury hospitality.

Currently, The Leela owns and operates eight properties in prime urban cities and magical holiday destinations—Mumbai, Goa, Bangalore, Kovalam, Udaipur, Gurgaon, New Delhi and Chennai. Other Leela properties under development include Jaipur; Bangalore at Bhartiya city near the international airport; Agra, where every room will face The Taj Mahal and Lake Ashtamudi in Kerala near The Leela Kovalam resort.

List of Leela Group of Hotels

The Leela Palace, New Delhi
The Leela, Mumbai
The Leela Palace, Bengaluru
The Leela Ambience, Gurgaon
The Leela Kovalam A Raviz Hotel, Kerala
The Leela Palace, Chennai
The Leela Palace, Udaipur
The Leela, Goa

LEMON TREE HOTELS

Fresh, spirited and youthful, the Lemon Tree Hotel Company is India's fastest growing chain of upscale, midscale and economy hotels. This award winning Indian hotel chain was founded in September 2002 and currently owns and operates 26 hotels in 15 cities with 3000 rooms and over 3000 employees. This speedy growth has made the group the 3rd largest hotel chain in India by owned rooms, currently. By 2017-18, Lemon Tree will own and operate over 8000 rooms in 60 hotels across 30 major cities of India including Ahmedabad, Aurangabad, Bengaluru, Chandigarh, Chennai, Coimbatore, Dehradun, Ghaziabad, Gurgaon, Goa, Hyderabad, Indore, Jaipur, Kolkata, Muhamma (Kerala), Mumbai, New Delhi, Pune, Shimla and Udaipur to name a few.

Brands of Lemon Tree Hotels

The group offers three brands to meet hotel needs of guests across all levels:

Lemon Tree Premier *(Upscale segment)*: The plush and spacious interiors at Lemon Tree Premier take the zing up a notch. This chain of upscale business and leisure hotels elevates the Lemon Tree experience while retaining the same freshness, quirkiness and energy that Lemon Tree is well known for. Lemon Tree Premier pampers the style conscious and upbeat traveler with its personalized services, premium in-room amenities, award winning restaurants and fun experiences.

Lemon Tree Hotels *(Midscale segment)*: Lemon Tree Hotels are the only midscale business and leisure hotels that uplift your spirits at the end of a long day. Like the fruit they are named after, Lemon Tree Hotels are fresh, cool and sparkling with zest. Cheery greetings, a friendly smile and a whiff of the signature lemon fragrance welcome you at Lemon Tree. This stylish business hotel with fresh and bright interiors refreshes you with its witty humor and spirited environment. Lemon Tree's 'close to home' comfort helps you unwind with its smart in-room amenities, vibrant café, recreation bar, pool and fitness center.

UPSCALE

MIDSCALE

ECONOMY

Red Fox Hotels (*Economy segment*): Red Fox Hotels welcome you with its fresh bold interiors as well as crisp and clean rooms. These economy hotels delight you with its unbeatable value and reliable safety standards. Here friendly smiles and a lively environment go hand in hand with professional service. The business facilities at Red Fox include hi-speed WiFi, Cyber Kiosk, Clever Fox Café, an efficient meeting room, a well-equipped gym and laundry service.

List of Lemon Tree Premier Hotels

Lemon Tree Premier, Gurgaon

Lemon Tree Premier, Bengaluru

Lemon Tree Premier, Hyderabad

Lemon Tree Premier, Delhi Airport

Lemon Tree Premier, Jaipur

List of Lemon Tree Hotels

Lemon Tree Hotel, Bengaluru

Lemon Tree Hotel, Ahmedabad

Lemon Tree Hotel, Chennai

Lemon Tree Hotel, Kaushambi

Lemon Tree Hotel, Aurangabad

Lemon Tree Hotel, Indore

Lemon Tree Hotel, Pune

Lemon Tree Hotel, Chandigarh

Lemon Tree Hotel, Gurgaon

Lemon Tree Hotel, Chennai

Lemon Tree Hotel, Bengaluru

Lemon Tree Hotel, Hyderabad

Lemon Tree Hotel, Dehradun

Lemon Tree Amarante Beach Resort, Goa

Lemon Tree Vembanad Lake Resort, Kerala

List of Red Fox Hotel

Red Fox Hotel, Hyderabad

Red Fox Hotel, Delhi Airport

Red Fox Hotel, Jaipur

Red Fox Hotel, East Delhi

MARRIOTT INTERNATIONAL

Marriott International, Inc. (NASDAQ: MAR) is a global leading lodging company based in Bethesda, Maryland, USA, with more than 4,300 properties in 85 countries and territories. The company operates and franchises hotels and licenses vacation ownership resorts under 19 **brands**, including:

The Ritz-Carlton®

Bvlgari®

EDITION®

JW Marriott®

Autograph Collection® Hotels

Renaissance® Hotels

Marriott Hotels®

Delta Hotels and Resorts®

Marriott Executive Apartments®,

Marriott Vacation Club®

Gaylord Hotels®

AC Hotels by Marriott®

Courtyard®

Residence Inn®

SpringHill Suites®

Fairfield Inn and Suites®

Towne Place Suites®

Protea Hotels®

Moxy Hotels®

List of Marriott Hotels in India

Courtyard by Marriott, Ahmedabad

Courtyard by Marriott, Agra

Courtyard Bengaluru Outer Ring Road

Courtyard by Marriott, Bhopal

Courtyard by Marriott, Bilaspur

Courtyard by Marriott, Kochi

Courtyard by Marriott, Chennai

Courtyard by Marriott, Gurgaon

Courtyard by Marriott, Hyderabad

Courtyard by Marriott, Raipur

Courtyard by Marriott, Pune

Courtyard by Marriott Pune City Centre, Pune
Courtyard by Marriott Hinjewadi Pune, Pune
Courtyard by Marriott Mumbai International
Airport, Mumbai

JW Marriott Hotel, Bengaluru
JW Marriott Hotel, Chandigarh
JW Marriott Hotel, Pune
JW Marriott New Delhi Aerocity, New Delhi
JW Marriott Hotel, Mumbai
JW Marriott Hotel, Mumbai Sahar
JW Marriott Mussoorie Walnut Grove Resort and
Spa, Mussoorie

Jaipur Marriott Hotel, Jaipur
Kochi Marriott Hotel, Cochin
Hyderabad Marriott Hotel & Convention Centre,
Hyderabad
Goa Marriott Resort & Spa, Panaji (Goa)
Bengaluru Marriott Hotel Whitefield, Bangalore

Renaissance Mumbai Convention Centre Hotel,
Mumbai
Renaissance Lucknow Hotel
The Ritz Carlton, Bangalore

Fairfield by Marriott Bengaluru Outer Ring Road
Fairfield by Marriott Bengaluru Rajajinagar
Fairfield by Marriott Lucknow
Lakeside Chalet, Mumbai—Marriott Executive
Apartment

MANSINGH GROUP OF HOTELS

Mansingh Hotels and Resorts (MHRL) is part of
an Industrial Group founded by Late Seth Mukand
Lal in the mid 1920s. The group's foray into Hotels
business stated in early 1980s when Mansingh
Hotels and Resorts Limited, owners of Mansingh
Hotel at Jaipur was taken over by the family. The
first hotel of the Group, 'Mansingh' at Jaipur gave
the Group its present name, MANSINGH GROUP.
The Mansinghgroup's valued inheritance and
contemporary luxury attracts many tourists again
and again to its various hotels.

List of Hotels of Mansingh Group

Hotel Mansingh, Jaipur
Hotel Mansingh Towers, Jaipur
Hotel Mansingh Palace, Agra
Hotel Mansingh Palace, Ajmer
Jodhpur Hotel

MANGO HOTELS

IntelliStay Hotels owns and operates Mango
Hotels—a chain of independent small hotels that
are big on service.

List of Mango Hotels

Mango Hotels, Agra
Mango Hotels, Bengaluru
Mango Hotels, Goa
Mango Hotels, Hyderabad
Mango Hotels, Jodhpur
Mango Hotels, Mysore
Mango Hotels, Nagpur
Mango Hotels, Navi Mumbai

NEEMRANA GROUP OF HOTELS

The Neemrana Group of Hotels manages some of
the best hotels in India that not only ensure a
luxurious stay for its guests but also gives them
the real taste of India and its diverse culture. The
group has twenty-nine elegant hotels scattered
across various locations in India and each of these
hotels is a perfect blend of modernism and old
heritage.

Most of the hotels of the Neemrana Group are
heritage properties that have been converted into
luxury hotels. In fact, the Neemrana Group of
Hotels is the first to take on the massive task of
rebuilding old forts and palaces and converting
them into beautiful hotels. One such example is
the Neemrana Fort-Palace in Rajasthan, whose
beauty and elegance sets it a class apart.

Neemrana Hotels call themselves as **non-hotel
hotels** as none of their properties were originally
meant to be hotels. They are all monuments of
historical significance with different stories to tell
about their glorious history and culture. And, they
have been restored by Neemrana group to their
past magnificence and glory.

List of Neemrana Hotels

Neemrana Fort-Palace, Rajasthan
Le Colonial, Cochin, Kerala
The Verandah in the Forest, Maharashtra
The Piramal Haveli, Shekhawati, Rajasthan
Wallwood Garden, Coonoor, Tamil Nadu
The Tower House, Cochin, Kerala
The Baradari Palace, Patiala, Punjab
Le Maison Tamoule, Pondicherry

Divan's Bungalow, Ahmedabad
Deo Bagh, Gwalior
The GlassHouse on the Ganges, Rishikesh
Cabana Dempo, Goa
Arci Iris, Goa
Deco House, Dehradun
Hill Fort-Kesroli, Alwar, Rajasthan
The Bungalow on the Beach, Tamil Nadu
Gate House, Tamil Nadu
Hotel De L'Orient, Pondicherry
The Ramgarh Bungalows, Kumaon Hills, Uttara-khand
Villa PotiPatti, Bengaluru
IshaVilas, Goa
The Pataudi Place

OBEROI GROUP OF HOTELS

The Oberoi Group of Hotels, founded in 1934, by Rai Bahadur Mohan Singh Oberoi, operates 35 hotels and three cruisers in seven countries under two principal brands—*The Oberoi* brand in the 5-star deluxe category and the *Trident* brand in the 5-star category. The Group is also engaged in flight catering, airport restaurants, travel and tour services, car rentals, project management and corporate air charters. **East India Hotel Limited** is the parent company of the Oberoi Group of Hotels.

Oberoi Group of Hotels is synonymous across the globe for providing the right blend of service, luxury and efficiency. A distinctive feature of the Oberoi Group of Hotels is their highly motivated and well-trained staff that provides the kind of

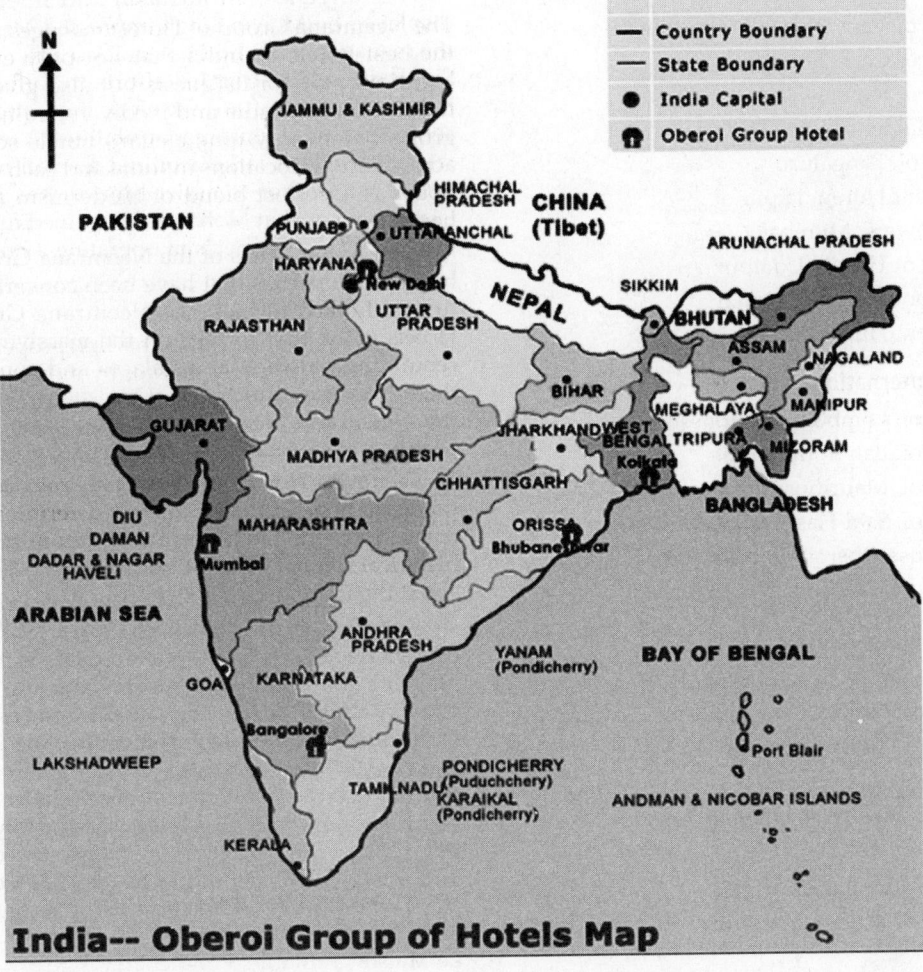

India-- Oberoi Group of Hotels Map

attentive, personalized and warm service that is rare today. Internationally recognized for all-around excellence and unparalleled levels of service, the Oberoi Hotels have been the recipients of innumerable awards and accolades.

Recognising the importance of quality training in hospitality management, The Oberoi Group established **The Oberoi Centre of Learning and Development in New Delhi** in 1966.

List of Oberoi Group of Hotels in India

The Oberoi Amar Vilas, Agra

The Oberoi Udaivilas, Udaipur

The Oberoi, Mumbai

The Wildflower Hall, Shimla

The Oberoi Grand, Kolkata

The Oberoi, New Delhi

The Oberoi Vanyavilas, Ranthambore

The Oberoi Cecil, Shimla

The Oberoi Rajvilas, Jaipur

Oberoi Maidens Hotel, Delhi

The Oberoi, Bengaluru

The Trident Hilton, Jaipur

Hilton Towers, Mumbai

The Trident Hilton, Udaipur

The Trident Hilton, Chennai

The Trident Hilton, Kochi

Oberoi International Hotels

The Oberoi, Lombok in Indonesia

The Oberoi, Bali in Indonesia

The Oberoi, Mauritius

The Oberoi, Sahl Hasheesh in Egypt

Mena House Oberoi, Cairo in Egypt

The Oberoi, Madina in Saudi Arabia

List of Oberoi Cruises

The Oberoi Motor Vessel, Vrinda, Kerala

The Oberoi Zahra, Luxury Nile Cruiser, Egypt

The Oberoi Philae, Nile Cruiser, Egypt

Rai Bahadur Mohan Singh Oberoi, Founder Chairman, EIH Limited (1898–2002)

Early Life

Rai Bahadur Mohan Singh Oberoi was born on 15th August, 1898 in erstwhile undivided Punjab, which is now in Pakistan. He was only six months old when his father died. Success and fortune did not, therefore, come easily to him. Initiative, resourcefulness and hard work, combined with the capability to face and overcome the most overwhelming odds can best characterise this phenomenal entrepreneur.

Mohan Singh completed his primary education in Rawalpindi and moved to Lahore for his Bachelor's degree. Shortly thereafter, to flee the ravages of a virulent plague, he went to seek his fortune in Shimla, the summer capital of British India. Arriving penniless, he found a job at a monthly salary of INR 50, as the front desk clerk at the Cecil Hotel. Today, The Oberoi Group owns the hotel The Oberoi Cecil where the young Mohan Singh found his metier.

The diligence, enthusiasm and intelligence displayed by Mohan Singh impressed Mr. Grove, the manager of the hotel. A quick learner, Mohan Singh did not restrict his efforts to fulfilling the job description of a desk clerk but sought and shouldered additional responsibilities. A few years later, when Mr Clarke acquired a small hotel, he asked Mr Oberoi to assist him. It was here, at Clarkes Hotel, that Mohan Singh gained first hand experience in all aspects of operating a hotel.

Budding Entrepreneur

In 1934, Mr. Oberoi acquired his first property, The Clarkes Hotel, from his mentor by mortgaging his wife's jewellery and all his assets. Four years later, he signed a lease to take over operations of the five hundred rooms Grand Hotel in Calcutta that was on sale following a cholera epidemic. With his customary confidence and sheer determination to succeed, he was able to convert this hotel into a highly profitable business venture.

Over several years, Mr. Oberoi had purchased shares in Associated Hotels of India (AHI), which owned Cecil and Corstophans hotels in Shimla, Maidens and Imperial hotels in Delhi and a hotel each in Lahore, Murree, Rawalpindi and Peshawar. In 1943, Mr. Oberoi acquired controlling interest in AHI. He thus became the first Indian to run the country's largest and finest hotel chain. In the tumultuous years just prior to Indian independence, Mr. Oberoi met and intimately interacted with the would-be leaders of Free India, all of whom were, at one time or other, guests at his hotels.

International Pioneer

Having consolidated his early ventures, Mr. Oberoi became the first Indian hotelier to enter into an agreement with an internationally renowned hotel chain, to open the first modem, five-star hotel in the country. The Oberoi Inter Continental, in New Delhi opened in 1965.The I-Con, as it became popularly known, offered facilities that no other hotel in the country matched and was India's first luxury hotel.

This achievement was enhanced with the opening of the 35-storey Oberoi Sheraton in Bombay, in 1973. Mr. Oberoi was the first Indian to work in association with international chains to woo international travelers to India. This led to a heavy influx of international travelers and foreign occupancy soared to an average of 85%. This enabled The Oberoi Hotels to significantly contribute to India's foreign exchange earnings.

OYO ROOMS

OYO Rooms is India's largest branded network of hotels spread across 65 cities with 1000+ hotels offering standardized rooms in different locations across the country.

The hotels are located at Agra, Ahmedabad, Ajmer, Allahabad, Ambala, Amritsar, Aurangabad, Bangalore, Bhopal, Chandigarh, Chennai, Dehradun, Delhi, Faridabad, Ghaziabad, Goa, Gurgaon, Guruvayur, Gwalior, Haridwar, Hyderabad, Indore, Jaipur, Jaisalmer, Jalandhar, Jodhpur, Kanpur, Karnal, Kasauli, Kochi, Kolkata, Kota, Kovalam, Lonavala, Lucknow, Ludhiana, Manali, Manesar, Mathura, Meerut, Mount-Abu, Mumbai, Mussoorie, Mysore, Nagpur, Nashik, Neemrana, Noida, Pondicherry, Pune, Pushkar, Rishikesh, Shimla, Shirdi, Surat, Tirupati, Trivandrum, Udaipur, Ujjain, Vadodara, Varanasi, Vijaywada, Vrindavan and Warangal.

PRIDE GROUP OF HOTELS

The Pride Group of Hotels owes its genesis and successful growth to Mr. S.P. Jain—a young Chartered Accountant from a remote town in Madhya Pradesh who arrived in Mumbai in the year 1975 with a vision, ambition and determination to make it big. He ventured by starting a Management Consultancy Firm, S. P. Capital Financing and later diversified into construction. Mr. Jain steered his conglomerate into the Hospitality Sector, after carrying out a

thorough study in the year 1988 by inaugurating the group's first hotel in Pune. Today, the Pride Group operates 15 hotels across different locations aggregating over 2000 rooms.

Brands of Pride Hotels

Pride Hotels
Pride Biznotels
Pride Resorts

List of Pride Hotels

The Pride Hotel Ahmedabad
The Pride Hotel Bengaluru
The Pride Hotel Chennai
The Pride Hotel Nagpur
The Pride Hotel Pune
The Pride Hotel, Kolkata
The Pride Hotel, Delhi Aerocity
The Pride Hotel, Goa

List of Pride Biznotels

Pride Biznotel, Salem
Pride Biznotel Ranipet
Pride Biznotel Aurangabad

List of Pride Resorts

Pride Amber Villas Resort, Jaipur
Pride Tiger Woods Resort and Spa, Kanha
Pride Sun Village Resort and Spa, Goa
Pride Sherwood Resort, Mahabaleshwar

PARK GROUP OF HOTELS

The Park Hotels are part of the **Apeejay Surrendra Group**, a business conglomerate with diverse interest. The Apeejay Surrendra Group was established in 1910, with its foundation in steel. Its current activities include shipping, tea, real estate and construction, hospitality and financial services. The hotels have achieved global standards of product quality and service excellence over 35 years of industry experience. A destination of choice for corporate and leisure travelers from India and abroad, these hotels, due to their downtown location, give easy access to key commercial and entertainment districts. Luxury and elegance with friendly services offered at Park Hotels give you a never-to-be forgotten stay.

The Park Hotels located in India is a collection of premium boutique hotels. The Park group of hotels epitomize the highest standards of elegance and hospitality, juxtaposing luxury with modernity.

List of The Park Group of Hotels

The Park, Bengaluru

The Park, Chennai

The Park, Kolkata

The Park, New Delhi

The Park, Visakhapatnam

ROYAL ORCHID HOTELS

Incorporated in 1986 by Mr. Chender K. Baljee, Royal Orchid Hotels Limited (Royal Orchid) is the flagship company of the Royal Orchid Group of Hotels. The group, comprises 14 subsidiaries, five joint ventures and one associate company, with an inventory of 1,724 rooms and operate **28** business and leisure hotels in **20** popular destination. The flagship five star property of the company, Hotel Royal Orchid is located in Bangalore. Historically skewed towards the Bangalore market, the company in the last few years has been setting its foothold in other cities, viz. Mysore, Pune, Jaipur, Goa, Ahmedabad, Mumbai, Mussorie, Gurgaon, Shimoga, Vadodara and Hospet. In 2013, the group made its first international foray with the Royal Orchid Malaika Beach resort at Mwanza. The group started **Presidency College of Hotel Management in Bengaluru in 1994.**

Brands of Royal Orchid Hotels

Hotel Royal Orchid (5-star)

Royal Orchid Central (4-star)

Royal Orchid Suites (4-star long-stay)

Regenta (4-star)

List of Royal Orchid Hotels

Royal Orchid, Bangalore

Royal Orchid Central, Bangalore

Ramada, Bangalore

Royal Orchid Resort and Convention Centre, Bangalore

Royal Orchid Suites, Bangalore

Royal Orchid Metropole, Mysore

Royal Orchid Brindavan Garden, Mysore

Royal Orchid Central, Jaipur

Royal Orchid Central, Pune

Royal Orchid Golden Suites, Pune

Royal Orchid Beach Resort and Spa, Goa

Royal Orchid Central Grazia, Navi Mumbai

Royal Orchid Central Kireeti, Hospet

Royal Orchid Central, Shimoga

Royal Orchid, Jaipur

Royal Orchid Central, Vadodara

Central Bluestone, Gurgaon

Regenta Resort MPG Club, Mahabaleshwar

Regenta Central Ashok, Chandigarh

Regenta Central Hari Mangala, Bharuch

Regenta Resort B-Cube, Bhuj

Regenta Central, Haridwar

Regenta Central, Rajkot

Regenta, Ahmedabad

Regenta Central, Jaipur

Royal Orchid Azure, Nairobi

Malaika Beach Resort by Royal Orchid Hotels, Tanzania

RAMEE GROUP OF HOTELS, RESORTS AND APARTMENTS

Ramee Group of hotels, resorts and apartments consist of 35 leading business and leisure hotels located in the cities Dubai, Abu Dhabi, Mumbai, Bangalore, Tirupati, Bahrain and Oman. The group is known as a strongest upcoming hotel chain in Middle East and India.

List of Ramee Group of Hotels

Ramee Guestline Hotel, Juhu Mumbai

Ramee Guestline Hotel, Dadar Mumbai

Ramee Guestline Hotel, Khar Mumbai

Ramee Guestline Hotel, Bangalore

Ramee Guestline Hotel, Tirupati

Ramee Grand Hotel, Pune

SAROVAR HOTELS PVT LTD

Sarovar Hotels and Resorts, pioneers in the mid-market hotel segment, ventured into the Indian hospitality landscape in 1994. Over a very short time period, the Company has not only been successful in meeting the demand in this segment, but is today the fastest growing hotel management company in India, with 70+ hotels across 48 destinations in India. Sarovar Hotels operate Park Plaza and Park Inn in India under a franchise agreement with Carlson Hotels Worldwide.

Sarovar Hotels has a diverse portfolio encompassing hotels, resorts, restaurants and corporate hospitality services. Apart from over 70+ operational hotels, additional 20 hotels are at various stages of development. By 2020, Sarovar aims to be a 100-hotel chain.

Brands of Sarovar Hotels

Sarovar Premiere: It offers upscale 5 star properties which resonate with a cosmopolitan vibe. Modern aesthetics make their presence felt in every detail from the architecture and interiors to innovative food and beverage concepts. The hotels, located primarily in cities and select premium leisure spots, are designed to the refined tastes of the informed local and global traveler.

Sarovar Portico: It offers full-service 3 and 4 star hotels located in the city's most coveted spots that offer convenient access to commercial and tourist centers. A range of business and leisure facilities and innovative food and beverage concepts make the hotel versatile. Sophistication and comfort are the highlights which give you an experience that transcends to higher levels of satisfaction.

Hometel: It is Sarovar's vibrant value brand that redefines the concept of an economy hotel. The hotels are defined by intelligent amenities and essential services which speak to the needs of the budget-conscious traveler who seeks quality and affordability. The compact rooms blend comfort with functionality to provide comfortable, restful stays.

The Oriental Blossom, specialty Chinese restaurant chain that serves a rich repertoire of authentic Szechwan and Cantonese cuisine.

Geoffrey's The Pub, a metro centric pub chain which is a contemporary interpretation of a typical old world English Pub.

Brands Managed by Sarovar Hotels

Radisson
Park Plaza
Park Inn

List of Sarovar Hotels

Sarovar Portico, Ahmedabad

Park Plaza, Ahmedabad

Legend Sarovar Portico, Baddi

Sarovar Portico, Badrinath

Davanam Sarovar Portico Suites, Bengaluru

Radha Hometel, Bengaluru

Park Plaza, Bengaluru

Optus Hometel Bhiwadi

Hometel, Chandigarh

Park Plaza, Chandigarh

Park Plaza Zirakpur, Chandigarh

Radha Regent, Chennai

Abu Sarovar Portico - Kilpauk, Chennai

Peerless Sarovar Portico, Durgapur

Park Plaza, Faridabad

Express Sarovar Portico Surajkund

Ambar Sarovar Portico, Gandhidham

The Royal Plaza, Gangtok

Mahagun Sarovar Portico Suites, Ghaziabad-Vaishali

Lazylagoon Sarovar Portico Suites, Goa

Phoenix Park Inn, Goa

Optus Sarovar Premiere, Gurgaon

Park Inn, Gurgaon

Park Plaza, Gurgaon

Ambrosia Sarovar Portico, Haridwar

Claresta Sarovar Portico, Hosur

Radisson Hyderabad Hitec City, Hyderabad

Aditya Park, Hyderabad

Aditya Hometel, Hyderabad

Sarovar Portico, Indore

Sarovar Portico, Jaipur

Park Inn, Jaipur

Nirwana Hometel, Jaipur

Sarovar Portico, Jalandhar

Park Plaza, Jodhpur

Paradigm Sarovar Portico, Kakinada

Vasundhara Sarovar Premiere, Kerala-Vayalar

The Gokulam Park, Kochi

The Peerless Inn, Kolkata

Sahil Sarovar Portico, Lonavala

City Heart Sarovar Portico, Ludhiana

Park Plaza, Ludhiana

La Place Sarovar Portico, Lucknow

Marine Plaza, Mumbai

Grand Sarovar Premiere, Mumbai

Residency Sarovar Portico, Mumbai

Majestic Court Sarovar Portico, Navi Mumbai

Grand Hometel, Mumbai

Lily Sarovar Portico, Nashik

The Muse Sarovar Portico, New Delhi

The Ashtan Sarovar Portico, New Delhi

Sarovar Portico Naraina, New Delhi

Park Plaza East Delhi

Park Plaza, Noida

Le Dupleix, Pondicherry

The Promenade, Pondicherry

Peerless Sarovar Portico, Port Blair

Noorya Hometel, Pune

Marasa Sarovar Portico, Rajkot

Hometel, Roorkee

Shraddha Sarovar Portico, Shirdi

Royal Sarovar Portico, Siliguri

Balaji Sarovar Premiere, Solapur

R K Sarovar Portico, Srinagar

Poetree Sarovar Portico, Thekkady

Marasa Sarovar Premiere, Tirupati

STARWOOD HOTELS AND RESORTS WORLDWIDE

Starwood Hotels and Resorts Worldwide, Inc. is one of the leading hotel and leisure companies in the world with more than 1,200 properties in 100 countries and 181,400 employees at its owned and managed properties. Starwood is a fully integrated owner, operator and franchisor of hotels, resorts and residences.

NINE DISTINCT, GLOBAL LIFESTYLE BRANDS [a]					
ST REGIS	• 33 hotels • 7,300 rooms	THE LUXURY COLLECTION	• 92 hotels • 18,100 rooms	W HOTELS WORLDWIDE	• 46 hotels • 13,100 rooms
WESTIN HOTELS & RESORTS	• 203 hotels • 76,600 rooms	Sheraton	• 436 hotels • 153,400 rooms	LE MERIDIEN	• 98 hotels • 26,100 rooms
aloft HOTELS	• 91 hotels • 15,400 rooms	FOUR POINTS BY SHERATON	• 193 hotels • 34,300 rooms	element	• 14 hotels • 2,200 rooms

Brands of Starwood Hotels

St. Regis® *(luxury full-service hotels, resorts and residences)* is for connoisseurs who desire the finest expressions of luxury. They provide flawless and bespoke service to high-end leisure and business travelers. St. Regis hotels are located in the ultimate locations within the world's most desired destinations, important emerging markets and yet to be discovered paradises, and they typically have individual design characteristics to capture the distinctive personality of each location.

The Luxury Collection® *(luxury full-service hotels and resorts)* is a group of unique hotels and resorts offering exceptional service to an elite clientele. From legendary palaces and remote retreats to timeless modern classics, these remarkable hotels and resorts enable the most discerning traveler to collect a world of unique, authentic and enriching experiences indigenous to each destination that capture the sense of both luxury and place. They are distinguished by magnificent decor, spectacular settings and impeccable service.

W® *(luxury and upscale full-service hotels, retreats and residences)* is where iconic design and cutting-edge lifestyle set the stage for exclusive and extraordinary experiences. Each hotel and retreat is uniquely inspired by its destination, where innovative design converges with local influences to create energizing spaces for guests to play or work by day or mix and mingle by night.

Westin® *(luxury and upscale full-service hotels, resorts and residences)* provides innovative programs and instinctive services which transform every aspect of a guest's stay into a revitalizing experience. Whether an epic city center location or refreshing resort destination, Westin ensures guests leave feeling better than when they arrived.

Le Méridien® *(luxury and upscale full-service hotels, resorts and residences)* is a Paris-born hotel brand, currently represented by 96 properties in 37 countries worldwide.

Sheraton® *(luxury and upscale full-service hotels, resorts and residences)* is the largest brand serving the needs of business and leisure travelers worldwide. For over 75 years this iconic brand has welcomed guests, becoming a trusted friend to travelers and one of the world's most recognized hotel brands. From being the first hotel brand to step into major international markets like China, to completely captivating entire destinations like Waikiki, Sheraton understands that travel is about bringing people together. Sheraton transcends lifestyles, generations and geographies and will continue to welcome generation after generation of world travelers as The World's Gathering Place.

Four Points® by Sheraton *(select-service hotels)* delights the self-sufficient traveler with what is needed for greater comfort and productivity. Four Points by Sheraton was started in the year 1995 by Sheraton group. The Four Points hotels are present in across 24 countries. The hotel group is running about 142 hotels presently at different levels with the further plan of expansion. Distinct from its sister concern, the Four Points Chain of Hotels offers unmatched services and latest comfort solutions to its esteemed guests.

Aloft® *(select-service hotels)* first opened in 2008 and now has more than 85 properties. Aloft provides new heights: an oasis where you least expect it, a spirited neighborhood outpost, a haven at the side of the road. Bringing a cozy harmony of modern elements to the classic on-the-road tradition, Aloft offers a sassy, refreshing, ultra effortless alternative for both the business and leisure traveler.

Element® *(extended stay hotels)* first opened in 2008, provides a modern, upscale and intuitively designed hotel experience that allows guests to live well and feel in control. Inspired by Westin, Element hotels promote balance through a thoughtful, upscale environment. Decidedly modern with an emphasis on nature, Element is intuitively constructed with an efficient use of space that encourages guests to stay connected, feel alive, and thrive while they are away.

List of The Luxury Collection Hotels in India

ITC Grand Bharat, A Luxury Collection Resort, Gurgaon

ITC Gardenia, A Luxury Collection Hotel, Bengaluru

ITC Windsor, A Luxury Collection Hotel, Bengaluru

ITC Grand Central, A Luxury Collection Hotel, Mumbai

ITC Maratha, A Luxury Collection Hotel, Mumbai

ITC Maurya, A Luxury Collection Hotel, New Delhi

ITC Rajputana, A Luxury Collection Hotel, Jaipur

ITC Grand Chola, A Luxury Collection Hotel, Chennai

ITC Kakatiya, A Luxury Collection Hotel, Hyderabad

ITC Mughal, A Luxury Collection Hotel, Agra

ITC Sonar, A Luxury Collection Hotel, Kolkata

List of W Hotels in India

W Retreat and Spa Goa

W Gurgaon Wtc

List of Westin Hotels in India

The Westin Gurgaon

The Westin Sohna Resort and Spa, Gurgaon

The Westin Bekal Resort and Spa, Kerala

The Westin Mumbai Garden City, Mumbai

The Westin Koregaon Park, Pune

The Westin Velachery, Chennai

The Westin Mindspace, Hyderabad

The Westin Rishikesh Resort and Spa, Uttaranchal

The Westin Rajarhat, Kolkata

List of Le Meridien Hotels in India

Le Méridien Ahmedabad

Le Méridien, Gurgaon - NH8

Le Méridien Bangalore

Le Méridien Kochi

Le Méridien Mahabaleshwar Resort and Spa

Le Méridien Pimpri, Pune

Le Méridien Pune

Le Méridien New Delhi

Le Méridien Jaipur

Le Royal Méridien Chennai

Le Méridien Coimbatore

Le Méridien, Noida

List of Sheraton Hotels in India

Sheraton Hyderabad Hotel

Sheraton Bangalore Hotel at Brigade Gateway, Bengaluru Sheraton Bengaluru Whitefield Hotel and Convention Center, Bengaluru

Sheraton New Delhi Hotel

Sheraton Jalandhar Resort and Spa

Sheraton Park Hotel and Towers, Chennai

Sheraton Greater Noida Hotel

List of Four Points by Sheraton Hotels in India

Four Points by Sheraton Tirupati

Four Points by Sheraton Visakhapatnam

Four Points by Sheraton Ahmedabad

Four Points by Sheraton Dahej

Four Points by Sheraton Vadodara

Four Points by Sheraton Gurgaon, Faridabad Road

Four Points by Sheraton Bengaluru, Whitefield

Four Points by Sheraton Aurangabad

Four Points by Sheraton Navi Mumbai, Vashi

Four Points by Sheraton Hotel and Serviced Apartments, Pune

Four Points by Sheraton New Delhi, Airport Highway

Four Points by Sheraton Jaipur, City Square

Four Points by Sheraton Tiruchirappalli

Four Points by Sheraton Hyderabad

Four Points by Sheraton Agra

Four Points by Sheraton Dehradun

List of Aloft Hotels in India

Aloft Goa Calangute

Aloft Ahmedabad SG Road

Aloft Bengaluru Cessna Business Park

Aloft Bengaluru International Airport

Aloft Bengaluru Whitefield

Aloft New Delhi Aerocity

Aloft Chandigarh Zirakpur

Aloft Chennai Omr – IT Expressway

Aloft Coimbatore Singanallur

SINCLAIRS HOTELS

The Group owns and operates hotels and resorts at Siliguri, Darjeeling, Kalimpong Dooars, Ooty and Andamans.

List of Sinclairs Hotels

Sinclairs Siliguri

Sinclairs Darjeeling

Sinclairs Bayview Port Blair

Sinclairs Retreat Ooty

Sinclairs Retreat Dooars Chalsa

Projects under implementation

Sinclairs Retreat Kalimpong

Sinclairs Tourist Resort, Burdwan

Sinclairs Kolkata

TAJ HOTELS RESORTS AND PALACES

The Indian Hotels Company Limited (IHCL) and its subsidiaries are collectively known as Taj Hotels, Resorts and Palaces and is recognized as one of Asia's largest and finest hotel company. Incorporated by the founder of the Tata Group, Mr. Jamsetji Nusserwanji Tata, the company opened its first property, The Taj Mahal Palace Hotel, Bombay in 1903. The Taj, a symbol of Indian hospitality, completed its centenary year in 2003.

Taj Hotels, Resorts and Palaces comprises 93 hotels in 55 locations across India with an additional 18 international hotels in the Maldives, Malaysia, Australia, UK, USA, Bhutan, Sri Lanka, Africa and the Middle East.

Spanning the length and breadth of the country, gracing important industrial towns and cities, beaches, hill stations, historical and pilgrim centers and wildlife destinations, each Taj hotel offers the luxury of service, the apogee of Indian hospitality, vantage locations, modern amenities and business facilities.

IHCL operates **Taj Air**, a luxury private jet operation with state-of-the-art Falcon 2000 aircrafts designed by Dassault Aviation, France; and Taj Yachts, two 3-bedroom luxury yachts which can be used by guests in Mumbai and Kochi, in Kerala.

IHCL also operates **Taj SATS Air Catering Ltd.**, the largest airline catering service in South Asia, as a joint venture with SATS (formerly known as Singapore Airport Terminal Services).

The Taj group in collaboration with the Maulana Azad Educational Trust established the **Indian Institute of Hotel Management at Aurangabad** in 1989.

Brands of The Taj Group of Hotels

IHCL operate in the luxury, premium, mid-market and value segments of the market through their following brands:

Taj (*luxury full-service hotels, resorts and palaces*) is the flagship brand for the world's most discerning travelers seeking authentic experiences given that luxury is a way of life to which they are accustomed. Spanning world-renowned landmarks, modern business hotels, idyllic beach resorts, authentic Rajput palaces and rustic safari lodges, each Taj hotel reinterprets the tradition of hospitality in a refreshingly modern way to create unique experiences and lifelong memories.

Taj also encompasses a unique set of iconic properties rooted in history and tradition that deliver truly unforgettable experiences. A collection of outstanding properties with strong heritage as hotels or palaces which offer something more than great physical product and exceptional service. This group is defined by the emotional and unique equity of its iconic properties that are authentic, non-replicable with great potential to create memories and stories.

- **Taj Luxury Hotels:** These are located in the main political and metropolitan cities of India. These hotels offer world class services to the guests with exquisitely decorated rooms and modern supply and amenities.

- **Taj Business Hotels:** Located in the heart of India's key commercial cities and towns, the Taj Business Hotels provide every modern facility at attractive room rates. These international style hotels meet the growing needs of business travelers visiting cities, which are rapidly industrializing and expanding. Vibrant and progressive, they retain the warmth and spirit of India.

- **Taj Leisure Hotels:** At the Taj Leisure Hotels, you get everything to fulfill your content. These properties include idyllic beach resorts, genuine palaces, turn-of-the-century garden retreats, hotels located close to historic monuments, pilgrim centers and some of India's best life sanctuaries. While offering you with experiences entirely unique in themselves, the Taj Leisure Hotels leave you with the unmistakable feeling of warmth.

Taj Exotica: It is resort and spa brand found in the most exotic and relaxing locales of the world. The properties are defined by the privacy and intimacy they provide. The hotels are clearly differentiated by their product philosophy and service design. They are centered around high end accommodation, intimacy and an environment that allows its guest unrivalled comfort and privacy. They are defined by a sensibility of intimate design and by their varied and eclectic culinary experiences, impeccable service and authentic Indian Spa sanctuaries.

Taj Safaris: These are wildlife lodges that allow travelers to experience the unparalleled beauty of the Indian jungle amidst luxurious surroundings. They provide guests with the ultimate, interpretive, wild life experience based on a proven sustainable eco-tourism model.

Vivanta by Taj Hotels and Resorts (*upper upscale hotels*): Stylish and sophisticated, Vivanta by Taj delivers premium hotel experiences with imagination, energy and efficiency. It is the flavor of contemporary luxury, laced with cool informality and the charming Taj hospitality. With innovative cuisine, energetic spaces, unique motifs,

distinct avatars, the smart use of technology and experiences that seek to constantly engage, invigorate and relax, it appeals to the cosmopolitan world-traveler immersed in a sensory lifestyle.

The Gateway Hotels and Resorts (*upscale/mid-market full service hotels and resorts*): It is a pan-India network of hotels and resorts that offers business and leisure travelers a hotel designed, keeping the modern nomad in mind. The Gateway Hotel believes in keeping things simple. This is why these hotels are divided into 8 simple zones: Enter, Stay, Hangout, Meet, Work, Workout, Unwind and Explore. 24/7 services such as 24/7 breakfast, 24/7 'active studio' and 24/7 laundry are all designed to cater to guests round-the-clock. Flexible, dynamic and warm service, 'in-room yoga' amenities and 'explore' packages all make The Gateway Hotels and Resorts sanctuaries that refresh, refuel and renew.

Ginger Hotels (*economy hotels*): It is the revolutionary concept in hospitality for the value segment. Intelligently designed facilities, consistency and affordability are hallmarks of this brand targeted at travelers who value simplicity and self-service. These hotels cater to the economy or 'value for money' segment and being predominantly domestic clientele based, are far less prone to fluctuations than the luxury and upscale segments.

Roots Corporation (RCL) the wholly-owned subsidiary of Indian Hotels Company Limited (IHCL) operates the Ginger chain of hotels. Established in 2003, RCL runs the first-of-its-kind Smart Basics™ chain of hotels across India. Launched in June 2004, the Smart Basics™ concept kicked off a revolution in the Indian hospitality industry. A GenNext category of hotels, Ginger signifies warmth and friendliness, simplicity with style, convenience and affordability—calling all "New-Age" executives and travelers to experience luxury on a budget. Ginger aims to provide a superior product with no compromise on quality, and at no additional cost!

Jiva Spas: Taj Hotels also promise a whole new experience of tranquillity and total 'wellness', through Jiva Spas a unique concept, which brings together the wisdom and heritage of the Asian and Indian philosophy of wellness and well-being. Rooted in ancient Indian healing knowledge, Jiva

Spas derive inspiration and spirit from the holistic concept of living. There is a rich basket of fresh and unique experiences under the Jiva Spa umbrella of offering, Yoga and Meditation, mastered and disseminated by accomplished practitioners, authentic Ayurveda, and unique Taj signature treatments. Royal traditions of wellness in service experiences, holistic treatments involving body therapies, enlivening and meaningful rituals and ceremonies and unique natural products blended by hand, come together to offer a truly calming experience.

Taj Luxury Residences: For over a hundred years, Taj Hotels, Resorts and Palaces have played host to discerning travelers from around the world. Renowned for its warm hospitality, the spirit of the Taj is now found embodied in the Taj Luxury Residences.

Elegant, comfortable and exquisitely finished, Taj Luxury Residences combine warmth with a flawless service that can only be the Taj. Taj Wellington Mews, Mumbai, offer luxuriously furnished apartments for short visits or extended stays.

List of Taj Hotels Resorts and Palaces

Taj Bengal, Kolkata

Taj Lands End, Mumbai

The Taj Mahal Palace, Mumbai

Taj Palace, New Delhi

The Taj Mahal Hotel, New Delhi

The Taj West End, Bangalore

Taj Coromandel, Chennai

Taj Falaknuma Palace, Hyderabad

Taj Krishna, Hyderabad

Rambagh Palace, Jaipur

Umaid Bhawan Palace, Jodhpur

Mahua Kothi, Bandhavgarh National Park

Banjaar Tola, Kanha National Park

Pashan Garh, Panna National Park

Baghvan, Pench National Park

Taj Lake Palace, Udaipur

Taj Wellington Mews, Mumbai

Taj Exotica Resort and Spa, Maldives

Taj Cape Town, South Africa

Taj Samudra, Colombo, Sri Lanka

Taj Dubai, UAE

Taj Palace Hotel, UAE

St. James' Court, A Taj Hotel, London, UK

Taj 51 Buckingham Suites and Residences, UK

Taj Boston, USA

The Pierre, New York

Taj Campton Place, San Francisco

Taj Pamodzi, Lusaka

List of Vivanta by Taj Hotels

Vivanta by Taj, Aurangabad

Vivanta by Taj, M G Road, Bangalore

Vivanta by Taj, Whitefield, Bangalore

Vivanta by Taj, Yeshwantpur, Bangalore

Vivanta by Taj, Bekal, Kerala

Vivanta by Taj, Connemara, Chennai

Vivanta by Taj, Fisherman's Cove, Chennai

Vivanta by Taj, Malabar, Cochin

Vivanta by Taj, Surya, Coimbatore

Vivanta by Taj, Madikeri, Coorg

Vivanta by Taj, Fort Aguada, Goa

Vivanta by Taj, Holiday Village, Goa

Vivanta by Taj, Panaji, Goa

Vivanta by Taj, Gurgaon

Vivanta by Taj, Begumpet, Hyderabad

Vivanta by Taj, Hari Mahal, Jodhpur

Vivanta by Taj, Green Cove, Kovalam, Kerala

Vivanta by Taj, Kumarakom, Kerala

Vivanta by Taj, Gomti Nagar, Lucknow

Vivanta by Taj, President, Mumbai

Vivanta by Taj, Ambassador, New Delhi

Vivanta by Taj, Surajkund, Faridabad

Vivanta by Taj, Blue Diamond, Pune

Vivanta by Taj, Sawai Madhopur Lodge, Ranthambore

Vivanta by Taj, Dal View, Srinagar

Vivanta by Taj, Trivandrum, Kerala

Vivanta by Taj, Coral Reef, Maldives

Vivanta by Taj, Bentota, Sri Lanka

Vivanta by Taj, Rebak Island, Malaysia

List of The Gateway Hotels

The Gateway Hotel Fatehabad Road, Agra

The Gateway Hotel Ummed, Ahmedabad

The Gateway Hotel Residency Road, Bangalore

The Gateway Hotel Beach Road, Calicut

The Gateway Hotel IT Expressway, Chennai

The Gateway Hotel KM Road, Chikmagalur

The Gateway Hotel Church Road, Coonoor

The Gateway Hotel Marine Drive, Ernakulam

The Gateway Hotel Gir Forest, Gir Forest

The Gateway Hotel Lakeside, Hubli

The Gateway Hotel Ramgarh Lodge, Jaipur

The Gateway Hotel Rawalkot, Jaisalmer

The Gateway Hotel, Jodhpur

The Gateway Hotel EM Bypass, Kolkata

The Gateway Hotel Pasumalai, Madurai

The Gateway Hotel Old Port Road, Mangalore

The Gateway Hotel Ambad, Nashik

The Gateway Hotel, GE Road, Raipur

The Gateway Hotel, Damdama Lake, Gurgaon

The Gateway Hotel, Balaghat Road, Gondia

The Gateway Hotel Athwalines, Surat

The Gateway Hotel Akota Gardens, Vadodara

The Gateway Hotel Ganges, Varanasi

The Gateway Hotel Janardhanapuram, Varkala

The Gateway Hotel M G Road, Vijayawada

The Gateway Hotel Beach Road, Visakhapatnam

The Gateway Hotel Airport Garden, Colombo

The Gateway Hotel Kuteeram, Bangalore

List of Ginger Hotels

Ginger Agartala

Ginger Ahmedabad

Ginger Bangalore

Ginger Bangalore (Inner Ring Road)

Ginger Bhubaneshwar

Ginger Chennai

Ginger Chennai (Vadapalani)

Ginger Faridabad

Ginger Jamshedpur

Ginger Goa

Ginger Guwahati

Ginger Mangalore

Ginger Manesar

Ginger Mysore

Ginger Mumbai

Ginger Nashik

Ginger New Delhi

Ginger Pantnagar

Ginger Pune (Pimpri)

Ginger Pune (Wakad)

Ginger Pondicherry

Ginger Surat

Ginger Trivandrum

Ginger Vadodara

Ginger East Delhi

Ginger Indore

Ginger Tirupur

Ginger Noida

Ginger Jaipur

Ginger Chandigarh

List of Taj – Other Hotels

Hotel Chandela, Khajuraho

Jai Mahal Palace, Jaipur

Nadesar Palace, Varanasi

Savoy Hotel, Ooty

SMS Hotel, Jaipur

Taj Chandigarh, Chandigarh

Taj Club House, Chennai

Taj Banjara, Hyderabad and Taj Deccan, Hyderabad

Taj Exotica, Goa and Usha Kiran Palace, Gwalior

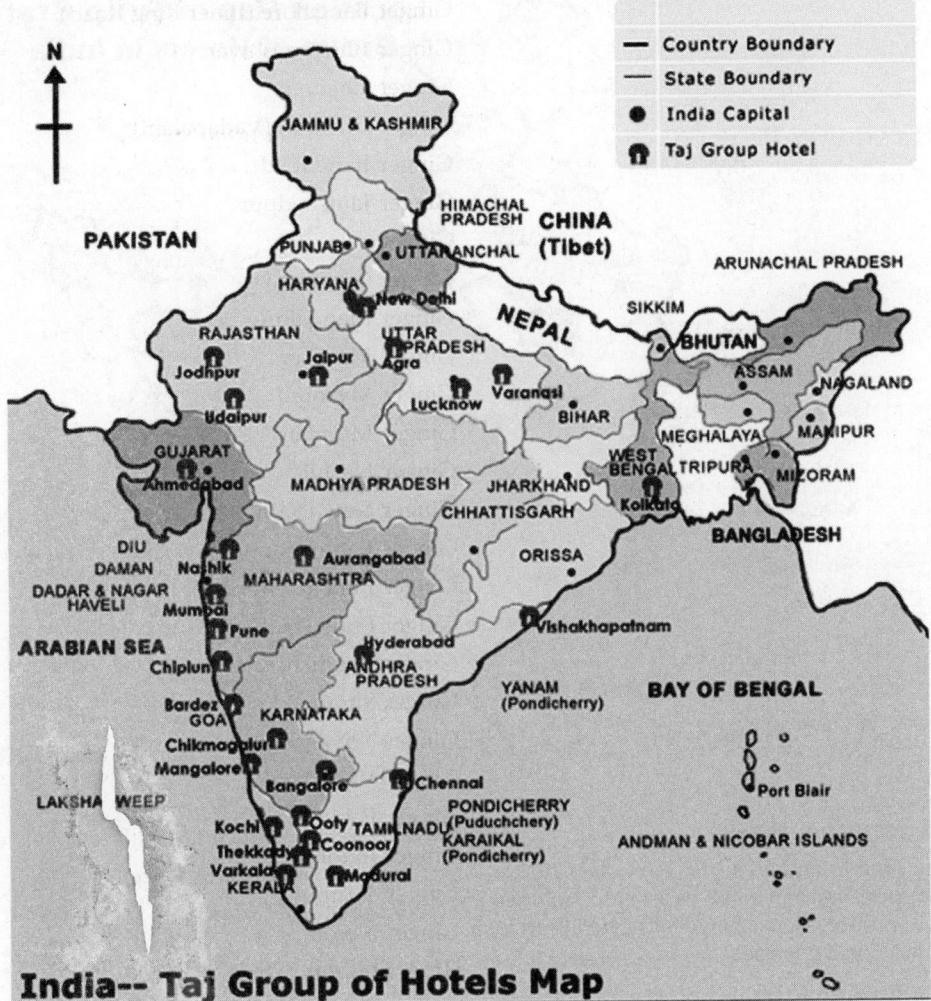

India-- Taj Group of Hotels Map

TULIP STAR GROUP OF HOTELS

The Tulip Star Group of Hotels in India boasts of some of the best hotels in India. The hotels are located at the prime tourist locations that dot the various states of India like the gorgeous Goa, Kerala, Maharashtra, Rajasthan, Delhi, Uttar Pradesh, Uttaranchal, Orissa and Gujarat.

List of Tulip Star Group of Hotels in India

Bogomallo Resort, Goa
Coconut Grove, Goa
Kohinoor Continental, Mumbai
Hans Plaza, New Delhi

TREEHOUSE HOTELS, RESORTS, SERVICED APARTMENTS

Karma Hospitality is a young hotel company with a rich and diverse experience of operating and managing full scale hotels, resorts and serviced apartments under the brand name of Treehouse. The company operates with team of experienced professionals drawn from reputed Indian and International brands. Karma forayed into management of hotels in 2007 and currently manages 275 keys in 6 hotels and an employee strength of close to 400 staff across all locations of NCR, Bhiwadi, Goa, Neemrana and Ranthambore.

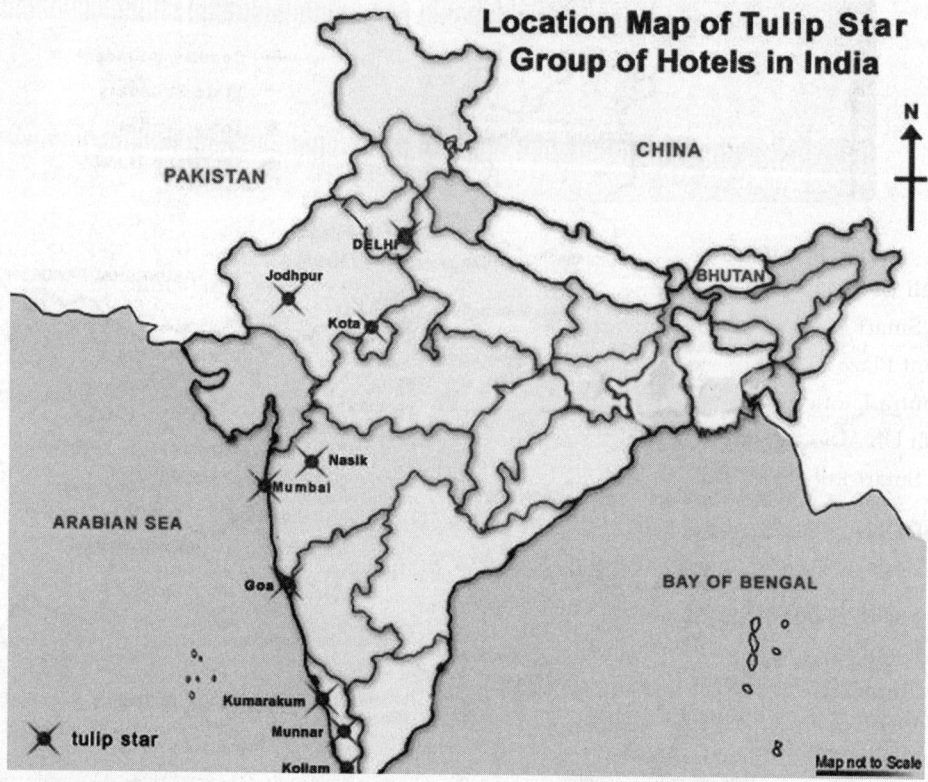

Brands of Treehouse Hotels

Treehouse Hotels and Resorts: First class hotels with large spaces and in some cases club facilities for the local non-residents as well. Located both in business and leisure locales.

Treehouse Business Hotel: Located in business hubs, the business hotels has functionality to be able to cater to the corporate traveler ensuring a perfect stay.

Serviced Apartments: Large and spacious apartments with a full kitchenette in locations with availability of entertainment and restaurants in the vicinity.

List of Treehouse Hotels

Treehouse Hotel, Club and Spa—Bhiwadi

Anugraha Palace—A Treehouse Palace Hotel—Ranathambore

Treehouse Dwaraka, Hotel and Club—Neemrana

Treehouse Blue, Serviced Apartment—Goa

Bijolai Palace—A Treehouse Palace Hotel, Jodhpur

Treehouse Queen Pearls, Gurgaon

Treehouse, Chail Himachal

UNA GROUP OF HOTELS

UNA Group of Hotels is one of the leading hospitality partner and hotel chain with 31 hotels, resorts, homestays in 22 destinations covering 9 states across India.

Brands of UNA Hotels

UNA Smart: Mid Market Business Hotels

UNA Comfort: Extended Stay Resorts

UNA Xpress: Limited Service Hotels

UNA Homestay: Boutique Bed and Breakfast Homestay

List of UNA Smart Hotels

Humble UNA Smart, Amritsar

UNA Smart Mandarin, Zirakpur

Aravali UNA Hotel, Alwar

UNA Smart Matsya Aravali, Alwar

Radiant Plaza UNA Smart, Barnala

Patliputra Exotica UNA SMART Patna

Kapish UNA Smart, Jaipur

UNA Smart Ludhiana

List of UNA Comfort Hotels

The Earls Court UNA Hotels, Nainital

UNA Comfort Kasauli Exotica, Kasauli

UNA Comfort Nandini, Dharamshala

UNA Comfort Great Himalayan Adventure Resort, Mcleodganj

UNA Comfort Dee Jay, Chintpurni (Pragpur)

UNA Comfort Kasauli Hills Resorts, Kasauli

Corbett Wild IRIS Spa And Resorts, Jim Corbett

UNA Comfort Vanrai Palace, Ranthambore

ACE UNA Comfort, Goa

UNA Comfort Swiss Meadows, Khajjar

UNA Comfort The Bheelgarh, Udaipur

UNA Comfort Bhowali, Nainital

UNA Xpress

Buddha Residency UNA Xpress, Lucknow

Hotel Paul UNA Xpress, Ludhiana

Aurum—A Boutique Hotel, Jaipur

Swathi UNA Xpress, New Delhi

Abode UNA Xpress, Amritsar

List of UNA Homestay Hotels

UNA Homestay UNA House, Lucknow

UNA Homestay Mahal Frams, Kasauli

UNA Homestay Villa Arcangela, Goa

UNA Homestay Sheenwynds, Srinagar

UNA Homestay Sheenwynds, Pahalgam

WYNDHAM HOTEL GROUP

Wyndham Hotel Group is the world's largest and most diverse hotel company, encompassing approximately 7,590 hotels and more than 655,300 rooms in 71 countries under fifteen hotel brands. Wyndham Hotel Group is one of three business units of **Wyndham Worldwide Corporation** (NYSE: WYN), one of the world's largest hospitality companies providing hotels, timeshare resorts, vacation rentals and timeshare exchange.

Brands of Wyndham Hotels

Wyndham Hotels and Resorts®: Wyndham Hotels and Resorts is an upscale, full-service brand located in key business and vacation destinations across the world. It offers the comfort and amenities you would expect in a world-class hotel, including beautifully appointed public areas, thoughtfully detailed guest rooms and distinct dining options. Business locations feature well-designed meeting

space flexible enough to accommodate anything from an executive board meeting to a major sales conference, as well as business and fitness centers.

Wyndham Garden® Hotels: Located primarily in business, airport and suburban locations, Wyndham Garden hotels offer the comfort and amenities expected from the Wyndham brand—thoughtful services, flexible meeting space that can accommodate a wide range of functions, inviting lobbies and cozy lounges.

Wyndham Grand® Hotels and Resorts: The crown jewel of the Wyndham family, the Wyndham Grand ensemble of hotels and resorts offers guests one-of-a-kind experiences in spectacular resort and urban destinations. Each hotel offers refined guest accommodations, attentive service, relaxing surroundings and thoughtful touches designed to satisfy and delight business and leisure travelers alike.

TRYP by Wyndham®: TRYP by Wyndham is a select-service, mid-priced hotel brand located in key urban markets throughout North America, Europe, Central America and South America, including Madrid, Barcelona, Paris, Frankfurt, New York and Buenos Aires. The brand's "Own the City" culture, which openly displays passion for guests and for the cities in which TRYP by Wyndham properties are located, aims to help guests make the most of their travels.

Wingate by Wyndham: Hotels feature all the extras at no extra cost, including complimentary hot breakfast and free high-speed wired and wireless Internet access in all guest rooms and public areas. Every Wingate by Wyndham® property has exercise facilities, whirlpools, meeting space and a 24-hour business center with free copying, printing and faxing.

Hawthorn Suites by Wyndham®: Hawthorn Suites by Wyndham extended stay hotels offer guests studio, one-and-two-bedroom suite accommodations.

Microtel Inn and Suites by Wyndham®: Located throughout North America and the Philippines, Microtel Inn and Suites by Wyndham hotels are designed to provide guests with a better stay.

Dream® Hotels: The Dream® Hotels brand is an upper-upscale, full-service, boutique brand targeted at prime, city center or destination resort locations. Hotels offer a progressive and unexpected list of services and amenities with a sense of humor and whimsy. Design is at the forefront of these concept lifestyle hotels, yet comfort and convenience is never compromised.

Planet Hollywood® Hotels: Planet Hollywood Hotels are upscale properties with a stunning collection of rooms, each with its own one-of-a-kind touch of Hollywood memorabilia and richly appointed amenities.

Ramada Worldwide

- **Ramada Plaza:** Ramada Plaza hotels are the premiere offering of the Ramada brand and are located near city centers and major airports throughout the world. Designed for discerning travelers, these full-service hotels offer the very best the brand has to offer by way of both service and style.

- **Ramada:** Ramada hotels are located around the globe and offer high-quality, attractive accommodations with a greater number of amenities than their Ramada Limited counterparts.

- **Ramada Limited:** Ramada Limited hotels are located throughout North America and offer high-quality accommodations at value prices. (The brand is currently in the process of gradually phasing out this tier.)

- **Ramada Hotel and Suites:** Ramada Hotel and Suites properties are located in key destinations throughout North America and Europe and offer contemporary accommodations.

- **Ramada Resort/Ramada Hotel and Resort (Outside the US):** Ramada Resort and Ramada Hotel and Resort properties cater to leisure travelers in key destinations throughout the world.

- **Ramada Encore (Outside the US):** Ramada Encore hotels are stylish, midscale hotels that cater to value-conscious guests throughout Europe and Asia.

Baymont Inn and Suites: Baymont Inn and Suites hotels are midscale hotels that offer an array of complimentary amenities including free breakfast with waffles, free Wi-Fi access and fitness centers.

Night Hotels: The Night® Hotels brand is positioned as a limited-service, "affordably chic" hotel serving urban, and collegiate and key resort locations. Guests can expect engaging and intelligent design in both guest room and public spaces coupled with innovative service offerings.

Days Inn and Suites: Days Inn and Suites properties are economy hotels located throughout the world and offer guests single, double and suite accommodations.

Super 8®: Super 8 hotels offer an array of complimentary amenities including free Wi-Fi access and free continental breakfast.

Howard Johnson Plaza/Howard Johnson Hotel: Howard Johnson Plaza and Howard Johnson Hotels are located in key destinations throughout world and provide guests with single, double and suite accommodations.

- **Howard Johnson Inn:** Located primarily in the United States, Canada and some parts of Mexico, Howard Johnson Inn hotels are midscale properties that offer guest single and double room accommodations.

- **Howard Johnson Express Inn:** Located primarily in the United States and some parts of Canada, Howard Johnson Express Inn hotels are economy properties that offer guest single and double room accommodations. (The brand is currently in the process of gradually phasing out this tier.)

Travelodge®/Travelodge Hotels/Travelodge Suites/Thriftlodge®: Operating primarily in the economy segment in the US and in the midscale segment in Canada, hotels carrying either the Travelodge or Thriftlodge (Canada only) name offer a variety of standard amenities designed to keep guests close to adventure.

Knights Inn: Knights Inn hotels are economy hotels located throughout the United States and Canada that offer guests single, double and suite accommodations.

List of Hotels of Wyndham Group in India

Ramada Ahmedabad

Ramada Alleppey

Ramada Amritsar

Ramada Bangalore

Ramada Jalandhar City Centre

Ramada Jamshedpur

Ramada Jaipur

Ramada Khajuraho

Ramada Encore Bangalore Domlur

Ramada Resort Cochin

Ramada Gurgaon Central

Ramada Plaza Palm Grove, Juhu Mumbai

Ramada Powai Hotel and Convention Centre, Mumbai

Ramada, Navi Mumbai

Ramada Caravela Beach Resort, Goa

Ramada Neemrana Jaipur Highway

Ramada Udaipur Resort and Spa

Ramada Plaza JHV Varanasi

Ramada Chennai Egmore

Wyndham Grand Agra

Howard Johnson Bengaluru Hebbal

Days Hotel Neemrana Jaipur Highway

Days Hotel Panipat

**Most Preferred INTERNATIONAL
BRAND OVERALL**

**Most Preferred DOMESTIC
BRAND OVERALL**

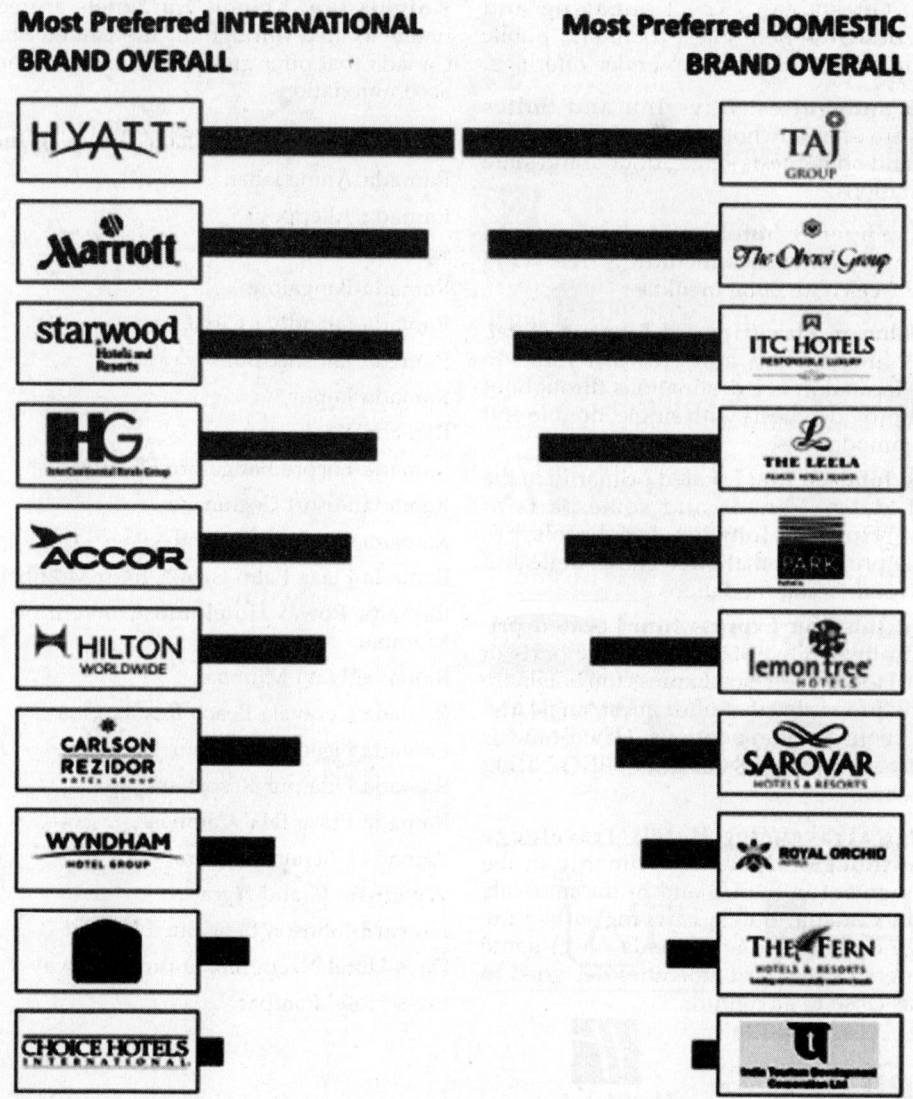

Domestic Airlines in India

The boom in aviation industry has led to rapid growth in domestic flights in India. Cheap airfares, low cost airline tickets, good connectivity, online bookings and excellent in-flight service had altogether make domestic airline in India very popular in the recent time. Air Deccan, Air India, Alliance Air, GoAir, Indian Airlines, IndiGo, Jagson Airlines, JetLite, Jet Airways, Kingfisher Airlines, Paramount Airways, SpiceJet, MDLR Airlines, Kingfisher Red and Jet Konnect are the domestic airlines flights operating in India. The domestic airlines operate as Full Service Airlines and Low-Cost Carriers (LCC) as well. While Full Service Airlines operate flight to major cities in India, Low-Cost Carriers mainly focus on second tier cities that are not usually covered by Full Service Airlines operating as domestic airlines. Introduced a couple of years back as a new concept by domestic airlines, low cost airlines offer attractive cheap airlines fares as low as the First Class and Second Class fares of Rajdhani Express. Few of the low cost airlines like Kingfisher Red even serves complimentary meals to passengers on board.

SPICEJET

It is a low fare airline based in New Delhi and is one of India's newest start-up private airlines promoted by Ajay Singh, Sanjay Malhotra and the Kansangra family.

Recently it was acquired by the media king Kalanithi Maran. SpiceJet was earlier known as Royal Airways, now sets its standards high as it competes with the Indian Railway passengers traveling in the AC coaches; this speaks tremendously for the cheap and discounted SpiceJet airfares. SpiceJet Airlines marked their entry in the service with INR. 99 fares for the first 99 days and then followed with an INR. 999 promotional fare for select sectors. With all these marketing strategies and booming business, SpiceJet leaves no loose ends. SpiceJet offers 'every day spicy fares' to budget conscious travelers. This business also thrives on online air ticket bookings with a detailed list of SpiceJet flight status and flight schedules. Low cost airlines that offers budget travel at discounted rates with the best of services; SpiceJet all fits in perfectly. SpiceJet is the only profit making low cost airlines listed on the NSE/BSE stock exchanges.

SpiceJet Domestic Sectors

There are numerous routes served by SpiceJet flights, which are preferred by budget conscious travelers. It is counted amongst the few airlines that operate flights to every nook and cranny of the country. From Srinagar in north, Tuticorin in south, Guwahati in east to Ahmedabad in west, all prominent Indian cities are served by SpiceJet flights. Amritsar, Chandigarh, Goa, Hyderabad, Indore, Jaipur, Kolkata, Lucknow, Nanded, Port Blair, Rajahmundry, Surat, Thiruvananthapuram, Udaipur and Visakhapatnam are names of some of these cities. Moreover, it connects to all the metropolitan cities of India, such as Bangalore, Chennai, Mumbai, Delhi and Kolkata. Travelers can easily avail the online booking facility to purchase tickets for flights operating to any of these domestic destinations.

SpiceJet International Sectors

SpiceJet is one of the low cost airlines that offer flights to a couple of international destinations. Sharjah, Dubai, Kabul, Male, Colombo, and Kathmandu are served by SpiceJet flights. Booking tickets for flights operating on these international routes is quite easy with the availability of online reservation facility.

VISTARA

It is a Delhi based Indian Airline Company which commenced operations on 9 January 2015 and operates 14 daily flights with 3 Airbus A320 aircraft. It is established as a joint venture between  India's Tata Sons Ltd. and Singapore Airlines. The new carrier was announced in September 2013. The two companies will initially invest a combined US$100 million to start the airline, with Tata Sons owning 51 percent and Singapore Airlines (SIA) owing the remaining 49 percent.

It is the first private Indian carrier to offer a three-class configuration on domestic routes with the introduction of 36 Premium Economy seats on its aircraft apart from Business and Economy class.

AIR INDIA

Many consider October 15, 1932 as the birth of civil aviation sector in India, when the founder of Air India —JRD. Tata flew  from Karachi to Mumbai. Since then, Air India has served domestic as well as international travelers through regular flights. It is an enterprise of the government-owned National Aviation Company of India Limited. The Air India airlines not only connect to metropolitans and state capitals but also to smaller cities. The national carrier is a preferred traveling option of many as it offers a range of benefits—be it cheap flights for budget conscious travelers, state-of-the-art first class or services like online booking, mobile booking, web check-in and round-the-clock support. There are also special offers provided by Air India, such as Companion Schemes, Airport Upgrades and Corporate House Benefits, among others. Besides the crew is courteous and takes care of every need of travelers. All this and much more is offered by the flag bearer Air India to ensure that travelers have a smooth and comfortable flight to their destination.

Air India Domestic Sectors

Air India is one of the few airlines that connect different parts of the country through regular flights. These short-haul routes are connected by Airbus A320 family and recently, by Boeing 787. There are around 50 cities and towns connected by Air India flights, including Delhi, Mumbai, Chennai, Kolkata, Bangalore, Hyderabad and Pune. It also connects capital cities, such as Patna, Jaipur, Imphal, Agartala, Bhopal, Chandigarh, Lucknow and Thiruvananthapuram. Moreover, all the popular tourist destinations are connected by Air India flights, such as Agra, Goa and Leh. For traveling to these domestic destinations, travelers can easily book tickets online.

Air India International Sectors

Air India connects to major international destinations across continents. The airlines international flights are available for American cities like Los Angeles, New York and Washington, as well as for European destinations like London, Paris and Amsterdam. Air India international routes extend to as far as Sydney as well as to nearby Colombo and Kathmandu. For booking, passengers can always take assistance of Air India call center.

GOAIR

GoAir is a low-fare airline that entered the aviation sector in the year 2005. A part of the Wadia Group, it was established with a mis-  sion to offer safe, secure and efficient air travel option. The fleet size of the GoAir is 19 aircraft and serves in more than 20 destinations. It offers flights to prominent business metropolis (like Delhi and Mumbai) and holiday destinations (like Goa and Jaipur) across India. Throughout its history, the airline has provided facilities and services to make travel simple and fun for passengers. Online booking facility, on-board and airport amenities, friendly staff and lucrative offers are some of the benefits that travelers gain while flying with GoAir. Moreover, understanding the special requirements of certain passengers, the airline offers unaccompanied minor assistance, carriages to travelers with limited mobility and meet-and-assist service to visually and hearing impaired passengers. Besides, passengers have the option to contact the airline anytime for getting any information or assistance regarding their travel. Hence, owing to cheap fare and optimum services, GoAir has been able to compete with well-established airlines and make a name for itself in the aviation sector.

GoAir Domestic Sectors

GoAir is one of the popular airlines that offer regular flights on domestic routes. The airline serves airports of metropolitan cities of Delhi, Mumbai, Chennai, Kolkata, Bangalore, Pune, Jaipur and Ahmedabad. Srinagar, Leh, Jammu and Chandigarh are the northern cities served by GoAir flights. Guwahati and Bagdogra airports are also connected by the airline's flights regularly. The airline also provides flights to various Indian capital cities like Patna, Ranchi and Lucknow. Many other cities are also a part of airline's well-connected network, like Goa, Kochi, Nagpur and Port Blair. Individuals considering traveling to any of these destinations with GoAir can book their tickets easily through online booking.

GoAir International Sectors

At present, the airline caters to domestic sector only and thus, has no flights for international destinations.

JET AIRWAYS

JET AIRWAYS 🦅

Jet Airways has become a renowned name since its first flight in 1993. Over its two decades of existence, the airlines has taken immense measures to make traveling convenient to domestic as well as international destinations. It offers direct or/and indirect flights to over 70 cities (domestic and international).

It is one of the largest employers in the airline sector in India with more than 13900 employees.

The fleet size of Jet Airways is 100 aircraft and ordered another 86 new fleets to expand its network.

Besides its extensive fleet, the prominent reason for its popularity in the airline sector is the range of services it offers. From online booking of cheap tickets and web check in to JetPrivilege offers, these services only enhance the overall travel experience of passengers. Furthermore, it offers an array of on-ground services like ticket reservation and check in. The in-flight services provided by Jet Airways are also commendable, wherein convenience, safety, entertainment and various other aspects are taken care of. Understanding special needs of its passengers like infants, travelers with disability and seniors, Jet Airways takes adequate steps to cater to their requirements. All these passenger centric facilities have made Jet Airways popular with passengers in India as well as abroad.

Jet Airways Domestic Sectors

The routes covered by the airlines in India extend from northernmost state of J&K (Leh) to southernmost Indian capital Thiruvananthapuram in Kerala, as well as from north-eastern state of Manipur (Imphal) to Bhuj in Gujarat. Furthermore, passengers can book their tickets for metropolitan cities like Delhi, Mumbai, Chennai and Kolkata, along with a number of other prominent cities like Bangalore, Hyderabad and Chandigarh. Those traveling to popular Indian tourist destinations, such as Dehradun, Jaipur, Udaipur, Goa, Khajuraho and Leh, will find a number of direct and indirect flight options, for which they can purchase tickets online. In all, there are over 51 domestic cities connected by direct and indirect flights of Jet Airways.

Jet Airways International Sectors

Jet Airways is one of the prominent airlines that connect to a number of international cities, 20 to be precise. It connects to near and far international destinations, including Toronto, London, Singapore, Dubai, Colombo and Dhaka. Moreover, booking tickets for flights on these routes is quite easy if one is traveling with Jet Airways.

AIR ASIA

Air Asia India is an Indo-Malayan low cost carrier. Announced on 19 February 2013, the airline is a joint venture with Air Asia Berhad holding 49% of the airline, Tata Sons holding 30% and Telstra taking up the remaining 21% in the airline. The joint venture would also mark

Tata's return to aviation industry after 60 years.

Air Asia Berhad was established with a single objective—making flying possible for everyone. Since its inception Air Asia has become not only Asia's largest low-cost airlines, it has also been consistently ranked as one of the leading airlines in the world. In its quest for making flying possible for everyone, Air Asia has broken travel norms all across the globe and is currently one of the world's most reliable low cost airlines. Air Asia Berhad is headquartered in Kuala Lumpur, Malaysia. Through the mantra of innovative solutions, efficient processes and a passionate approach to business, Air Asia has established route networks over 20 countries. Air Asia's associate companies include reliable names such as Air Asia X, Thai Air Asia, Indonesia Air Asia, Philippines' Air Asia Inc and Air Asia Japan. Air Asia is set to take low-cost flying to an all new high with its belief in the motto 'Now Everyone Can Fly'.

AIR INDIA EXPRESS

Air India Express, a premier low-cost airline is a subsidiary of Air India. With its headquarters in

Kochi and engineering center in Thiru-vananthapuram, the airline was started to cater to the Malayalee expatriate communities in the Middle East. Though the airline was established in the beginning of May 2004, it began its operations only from April 2005. Its first international flight was from Thiruvananthapuram to Abu Dhabi. The Air India Express fleet size is 20 aircraft and serves in more than 29 destinations. This budgeted airline has been welcomed by many as it offers efficient service, world clas˙ hospitality, comfortable seats and standardized meal boxes. All passengers are entitled to complementary refreshments and mineral water. Other snacks and reading material can be purchased on-board. Air India Express also offers entertainment facility. Air India Express operates in all international

destinations that can be reached within a short duration of four hours from the Indian cities. As of June 2014, it has daily flights to 13 international destinations, mostly Middle East and South Asia. Singapore, Colombo, Bahrain, Kuwait, Abu Dhabi, Sharjah, Dhaka, Dubai, Salalah, Al Ain, Muscat and Kuala Lumpur are the few destinations. This airline operates close to 100 flights per week mainly from the South Indian states such as Tamil Nadu, Karnataka and Kerala.

Air India Express Domestic Sectors

Air India Express also flies the domestic sectors to ensure connectivity between the different commercial and non-commercial cities. Maximum numbers of flights fly out Mumbai, connecting this city to Kochi, Kozhikode, Delhi, Chennai, Pune and Tiruchirapalli. The other cities it caters to include Mangalore, Jaipur, Amritsar, Lucknow and Thiruvanthapuram.

Air India Express International Sectors

There is a range of international destinations served by Air India Express. The airlines fleet connects to UAE destinations like Dubai, Abu Dhabi, Al Ain and Sharjah, besides Muscat, Salalah, Kuwait and Doha. The flights also operate on other international routes leading to Singapore, Dhaka, Colombo and Bahrain.

JET KONNECT

Jet Konnect is owned and operated by Jet Airways. Jet Konnect is engaged in offering cheap flights to numerous cities across India. Not only does the airline boasts of a comfortable fleet but also a courteous staff, which are a prerequisite for hassle-free travel. In addition, it offers numerous facilities that benefit passengers who choose to travel with Jet Konnect. From the point of purchasing the ticket to on-board experience, the airline takes measures to ensure that the overall experience is safe and satisfying. Online booking, flight status, schedule and web check in facilities prove extremely beneficial for travelers. The airline also provides numerous special offers like SkyMart: Buy On-The-Go, SkyMart: Scratch & Win, Jet2Kerala and

Concessional Fares. Besides, JetPrivilege is probably the best offer by the airline that entitles guests to enjoy several benefits. The airline also takes care of special needs of its passengers, including infants, unaccompanied minors and travelers with disability. All these facilities and services have helped Jet Konnect secure a place in the list of popular airlines in India.

Jet Konnect Domestic Sectors

Jet Konnect, one of the low-fare airlines, provides individuals the option to travel on Indian routes at an affordable price. The airline boasts of an extensive network that spans from one corner of the country to another. Whether individuals are looking for flights to business metropolis or popular holiday destinations, Jet Konnect has flights for almost every prominent Indian city and for which, travelers can easily book tickets online. Airports in Delhi, Mumbai, Chennai, Kolkata, Ahmedabad, Bangalore, Hyderabad and other metro cities in India are served by Jet Konnect flights. Besides, it offers flights to Goa, Dehradun, Jaipur, Leh and Tirupati, among other famous tourist destinations. Thus, Jet Konnect is the perfect travel partner for both business and leisure travelers.

Jet Konnect International Sectors

Jet Konnect is a domestic airline that offers flights to cities across the country and yet not forayed into the international sector.

AIR COSTA

Air Costa is an Indian regional airline based in Vijayawada, Andhra Pradesh. It started its flight operation from

October 2013 using two Embraer E-170 aircraft. It is part of LEPL Group. The airline initially planned to start operations with a fleet of Q400 aircraft. It received its Air Operators' Permit (AOP) from the Directorate General of Civil Aviation (DGCA) in September 2013. It has announced operations to Ahmedabad, Bangalore, Chennai, Hyderabad, and Jaipur and Vijayawada.

DECCAN SHUTTLES

Deccan Shuttles is a specialized wing under Deccan Charters Pvt. Ltd. Formerly known as Deccan

Aviation Limited, the flight service aims at providing improved connectivity to these important industrial hubs in Gujarat, for the value conscious business community. It endeavors to fuel their business by cutting their travel time, thereby enabling increased efficiency. Secure, committed and prompt, it is a one of its kind service that Deccan Charters brings to the state of Gujarat. Deccan Shuttles has 2 aircraft at present; one of them is fitted with amazingly comfortable 8 single-class seats and the other one with 7 single class seats and a bench seat for 2 in a single-aisle layout to provide you with even greater comfort.

Deccan Shuttles Sectors

Deccan Shuttles as of today connects key economic centers across Gujarat, the first phase of connectivity connects Ahmedabad to Surat, Jamnagar, Bhavnagar and Kandla.

INDIGO

IndiGo is for people who prefer the best, but at an affordable price. IndiGo is an enterprising venture set up by Interglobe Technologies who is already well versed with travel and hospitality industry in India. The airline believes in the power of 'one', which is

reflected in one type of fare—cheap; one type of aircraft—latest Airbus A320s; one type of customer support—proficient; and one type of dealing—honestly. The airline offers facilities and services that are beneficial for its passengers, in addition to affordable flights. These include options like online booking that saves time, low fare that saves money and web check in that saves energy, to name a few. There are numerous other offers provided by the airline, such as arranging wheelchair if required and special assistance for infants and unaccompanied minors. Additionally, it has introduced a new mobile application—goIndiGo that provides information about flight status, schedule, booking and much more at ease. Safe, comfortable and timely—best describe the services rendered by

IndiGo, which is counted amongst the preferred airlines to travel within or/and outside India.

IndiGo Domestic Sectors

With a commitment to offer low-fare flights to travelers, IndiGo has managed to make air travel an affordable experience. It boasts of a well-connected network that spans across different parts of the country. Delhi, Mumbai, Hyderabad, Kolkata, Chennai and Trivandrum are served by IndiGo flights. These destinations also connect to one or more international cities served by the airlines. There are various other Indian cities like Goa, Jaipur, Agartala and Coimbatore, which are connected by direct or/and indirect IndiGo flights. The airline's flights also serve many capital cities like Chandigarh, Lucknow, Patna and Bangalore, along with Jammu and Srinagar. Online booking can be done easily for flights operating on these routes.

IndiGo International Sectors

IndiGo is one of the low-fare airlines that provide travelers the opportunity to fly on international routes at a reasonable tariff. The airline serves Dubai, Muscat, Kathmandu, Singapore and Bangkok international airports through its comfortable flights. Passengers can avail the online booking facility for traveling to any of these international destinations.

JETLITE

JetLite Airlines is owned by Jet Airways and operates 110 flights on a daily basis connecting metro cities of India. JetLite flights connect 25 domestic destinations and 2 international cities. JetLite is one of the very few airlines that have an interesting

history of mergers and takeovers. The airline, former Air Sahara, was based in the commercial capital of India, Mumbai and offered flights to metropolitan centers. It was also engaged in providing helicopters for charter service and aerial photography. However, in 2007, the takeover of Air Sahara by Jet Airways was completed and was renamed as JetLite. In 2012, the airline merged with Jet Konnect brand (a subsidiary of Jet Airways) and presently, does business under this brand. Operating under the Jet Konnect brand, JetLite offers cheap flights to numerous cities across India, along with efficient ground and on-board services. Besides, the airline boasts of well-trained and courteous staff that takes care of varied needs of travelers, from the point of booking the ticket to deboarding the aircraft. All these make the overall travel experience simple, safe and pleasant for passengers.

JetLite Domestic Sectors

The JetLite airlines flights have ceased to operate independently and do so under the Jet Konnect brand of Jet Airways. Travelers can make online booking for flights operating on the domestic routes served by Jet Konnect. The well-connected network of Jet Konnect adds to the convenience of travelers as they can travel to numerous Indian cities at low cost. The destinations served by the airline include the former hubs of JetLite (Delhi Mumbai and Kolkata), besides many other cities across the country. Names of some of the other cities served by the airlines are Chennai, Ahmedabad, Bangalore, Hyderabad, Goa, Dehradun, Jaipur, Leh and Tirupati.

JetLite International Sectors

Jet Konnect is a domestic airline that offers flights to cities across the country and as JetLite operates flights under the Jet Konnect brand, it caters to Indian cities only.

Air International

AIR INDIA | JET AIRWAYS | AIR INDIA express | JetLite | SAUDI ARABIAN AIRLINES | Delta

malaysia | AIR MAURITIUS | GULF AIR | KUWAIT AIRWAYS | SINGAPORE AIRLINES | BRITISH AIRWAYS | Emirates

air arabia | SriLankan | AEROFLOT | Lufthansa | OMAN AIR | Thai | EVA AIR

QANTAS | Alitalia | AIR CANADA | KLM | QATAR | KOREAN AIR | SOUTH AFRICAN AIRWAYS

Asiana Airlines | Biman Air | AERO ASIA airlines | AIR FRANCE | nwa north west airlines | CHINA EASTERN AIRLINES | Continental

EGYPT AIR | Air Asia Now Everyone Can Fly | SOUTHWEST | AERO MEXICO | JAL JAPAN AIRLINES | SHAHEEN AIR INTERNATIONAL | Ethiopian

KENYA AIRWAYS | TURKISH AIRLINES | CSA CZECH AIRLINES | AIR NEW ZEALAND | SWISS | VARIG Brasil | AIR MACAU

Valuair Way Smarter | EASTERN | Elal Isreal | SILK AIR | BM British Midland | OLYMPIC AIRLINES | ROYAL BRUNEI

AA AmericanAirlines | INDUS AIR | Iraqi airways | Spanair | Royal Nepal Airlines | Turkmenistan Airlines | IBERIA

UZBEKISTAN Airlines | air astana | SAS Scandinavian Airlines | Austrian | MANDARIN AIRLINES | BWIA West Indies Airways | BULGARIA AIR

XIAMEN AIRLINES | Ukraine International Airlines | PIA PAKISTAN INTERNATIONAL | KYRGHYZSTAN Airlines | ANA ALL NIPPON | LIBYAN ARAB AIRLINES | TAP PORTUGAL

Eastern airways | ROYAL JORDANIAN | Jet | AERO ASIA | Kazakstan | Air Seychelles | Yemenia

virgin atlantic | Bangkok Airways | MIDWEST AIRLINES | CATHAY PACIFIC | SYRIANAIR | JAT | FINNAIR

AIR MANDALAY | Kartika Airlines | Asian Spirit THE PEOPLE AIRLINES | PBair | AeRoSviT | ETIHAD AIRWAYS | tiger airways.com

TAROM ROMANIAN AIR TRANSPORT | HAWAIIAN AIRLINES | AIR CHINA | One-Two-Go | Riau Airlines | Air Philippines | CEBUPACIFIC

UNITED | Philippine Airlines | SEAIR SOUTH EAST ASIAN AIRLINES | DRAGONAIR | Skywest Airlines | SHANGHAI AIRLINES | US AIRWAYS

SKYLINE NEPC | RYANAIR | LOT POLISH AIRWAYS | Aer Lingus | LANCHILE | Myanmar Airlines | Maldivian

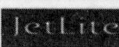

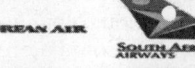

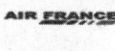

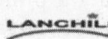

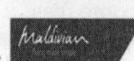

Tourist Attractions in India

INDIA AS A LAND OF CONTRASTS

There is a ring in the name of India. The sounds of bells, conch shells, religious chanting, put you on a transcendental high. The roar of the tiger, song of the nightingale, and dance of the peacock are part of a truly messmerizing ambience. The humming of industrial activity and futuristic development reveal a buoyant economy. Add to this the call of vibrant and color costumes, decorations, festivities, lilting music, folklore renditions, hand-crafted artifacts, puppetry, and you have an incredible wonderland that is India.

India covers an area of 32,87,263 sq.km, extending from the snow-covered Himalayan heights to the tropical rain forests of the south. Bounded by the Great Himalayas in the north, it stretches southwards and at the Tropic of Cancer, tapers off into the Indian Ocean between the Bay of Bengal on the east and the Arabia Sea on the west. The total length of coastline of the mainland, Lakshadweep Islands, and the Andaman and Nicobar Islands is 7,516.6 km.

Its geographical features give it a seasonal quality quite different from what one may have witnessed anywhere else. In fact the monsoons, or the rainy seasons, summer and winter are one of a kind.

WHY TRAVEL TO INDIA

India is a country popularly known for extending its lavish hospitality to all visitors, no matter where they come from. Due to its belief in the philosophy of 'Vasudeva Kutumbkam', its visitor friendly traditions, varied life styles, vast cultural heritage and colorful fairs and festivals, it holds multiple attractions for the tourist. The other attractions include beautiful sun drenched bathing beaches, forests and wildlife, majestic rivers, glorious architecture, fascinating fauna and flora and beautiful landscapes for eco-tourism, snow-clad mountain peaks, etc. for adventure tourism, technological parks and science museum for science tourism, centers of pilgrimage for spiritual tourism, heritage trains and hotels for heritage tourism, yoga, ayurveda and natural health resorts and hill stations also attract tourists. Indian handicrafts particularly jewelery, carpets, leather goods, ivory and brass work are the main shopping items of foreign tourists. India has also seen surfacing of some of the greatest religions on earth, like Buddhism, Hinduism, Jainism, and Sikhism.

The major tourist attractions in India can be explained by dividing India into five zones.

NORTH

Delhi

While it is a modern metropolis, it also showcases the rich and diverse history of the several cities that existed before Delhi acquired its modern day look. It unfurls a picture rich with culture, architecture, human diversity, monuments, museums, galleries and gardens. People from all parts of India and indeed the world live in Delhi, giving it a rainbow hue.

Places to Visit

Akshardham Temple, Birla Temple, Gurudwara Bangla Sahib, Humayun's Tomb, India Gate, Jama Masjid, Jantar Mantar, Lotus Temple, Mehrauli Archaeological Park, Purana Qila, Qutub Minar,

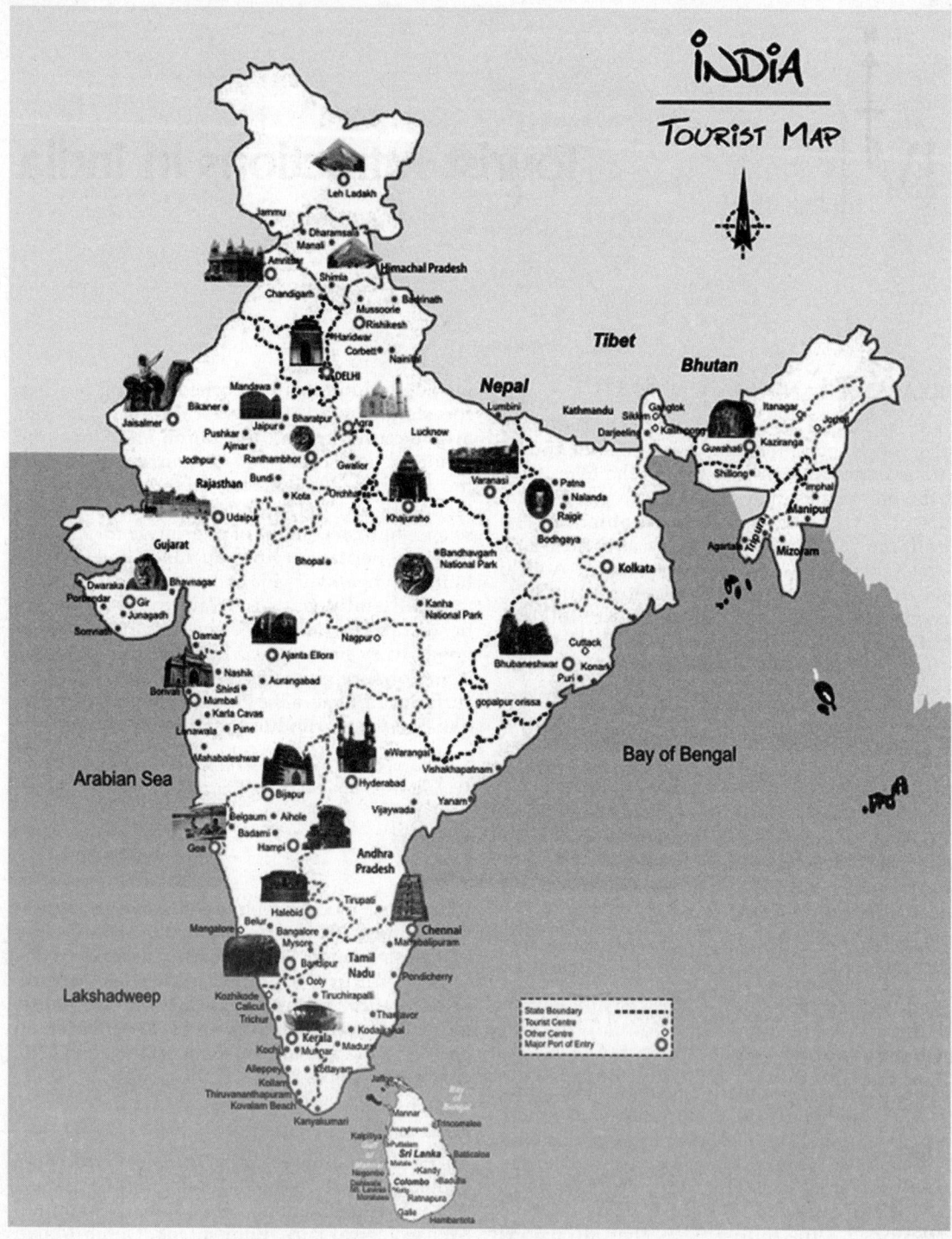

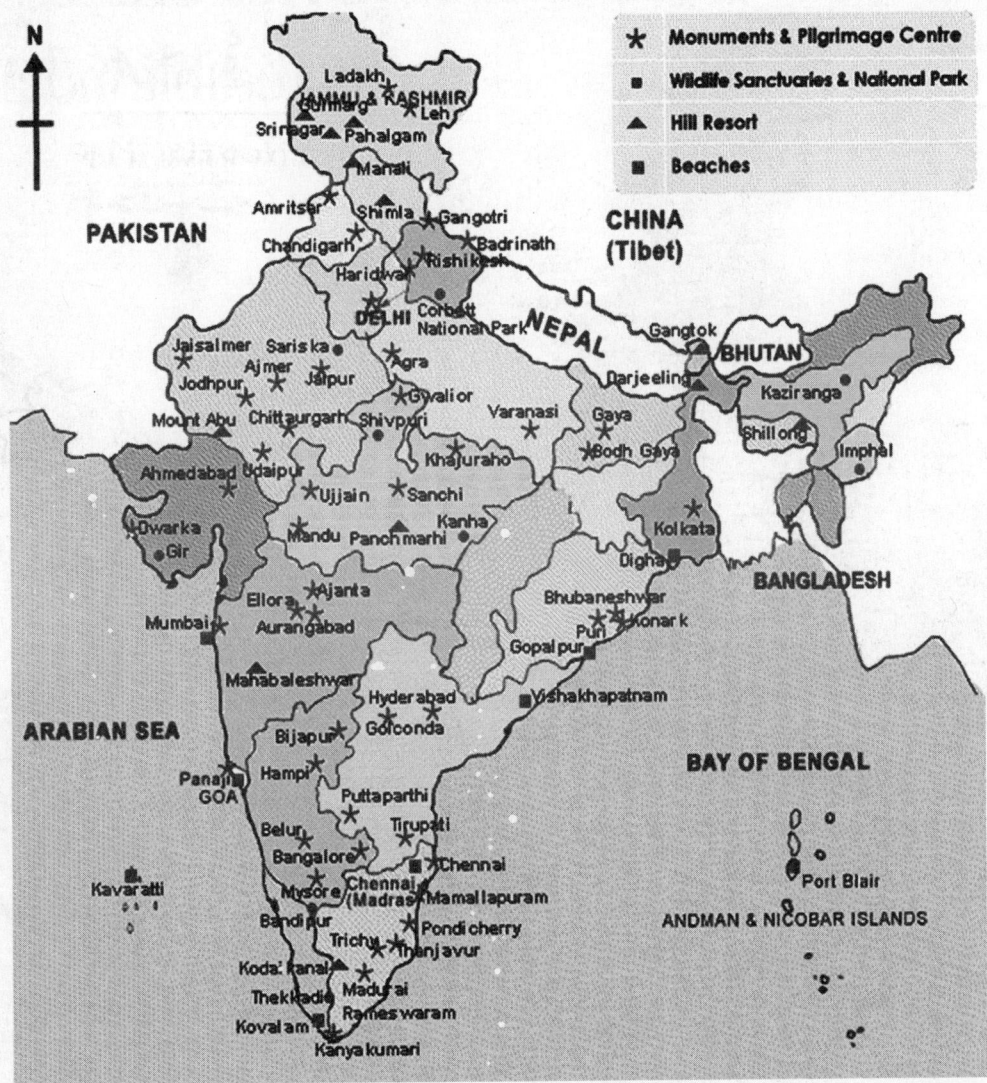

Red Fort, Rashtrapati Bhawan, Rajpath, Raj Ghat, Gurudwara Sisganj, etc. Connaught Place, Chandni Chowk, Dilli Haat, Janpath, State Emporia and Cottage Industries Emporium are good for shopping.

Haryana

Known for its robust people and rural agro-based lifestyles, Haryana is dotted with several places of historical and religious interest. Events and festivals include the Surajkund Crafts Fair, The Mango Mela, Lohri, Teej, Gangaur, Baisakhi, Heritage Festival, etc.

Places to Visit

Chandigarh, Ambala, Gurgaon, Faridabad, Hissar, Kurukshetra, Morni Hills, Panipat, Pinjore, Panchkula, Sultanpur Bird Sanctuary, Surajkund, etc.

Himachal Pradesh

Known for its apples and snow capped mountains, Himachal Pradesh boasts of some of the busiest

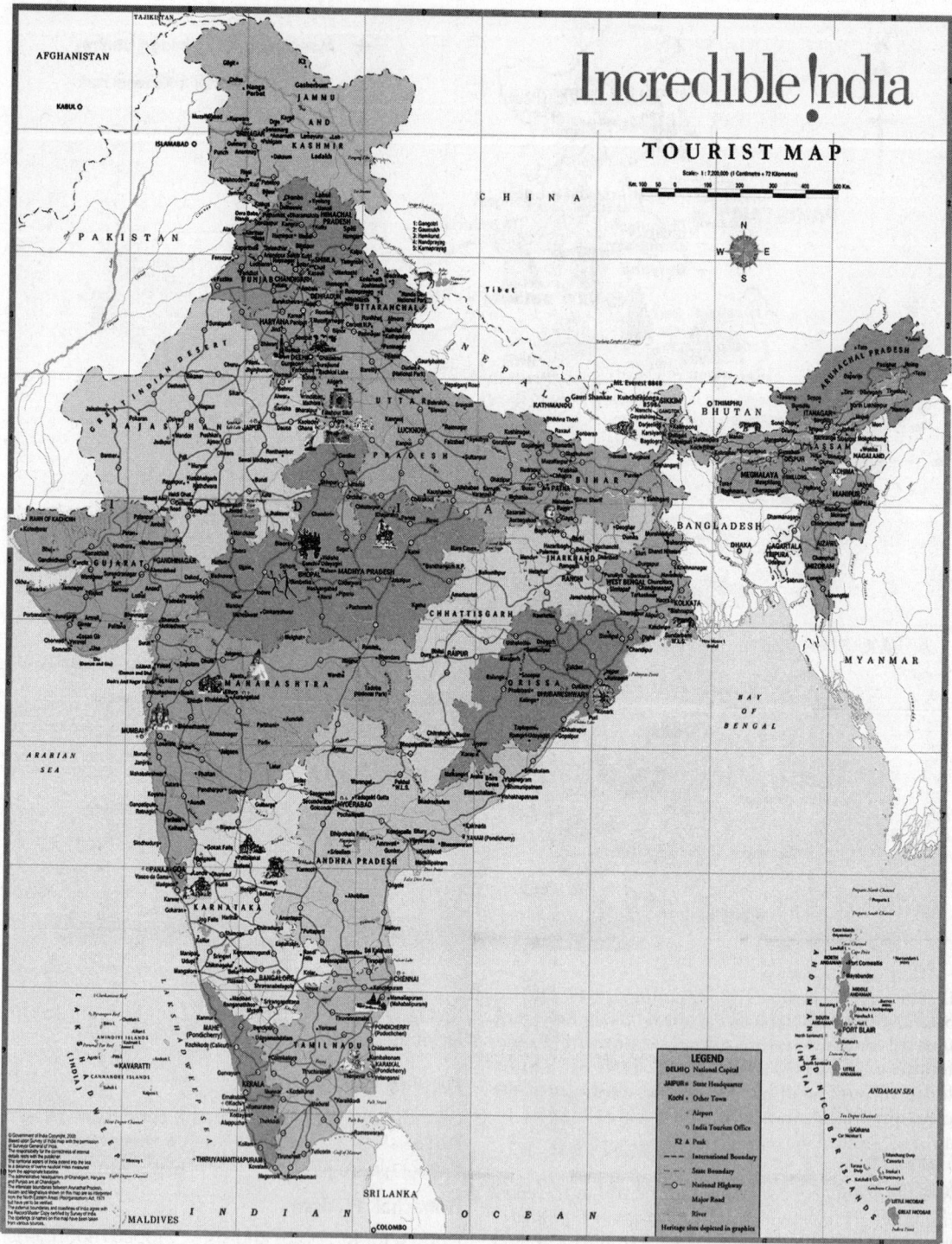

INDIA

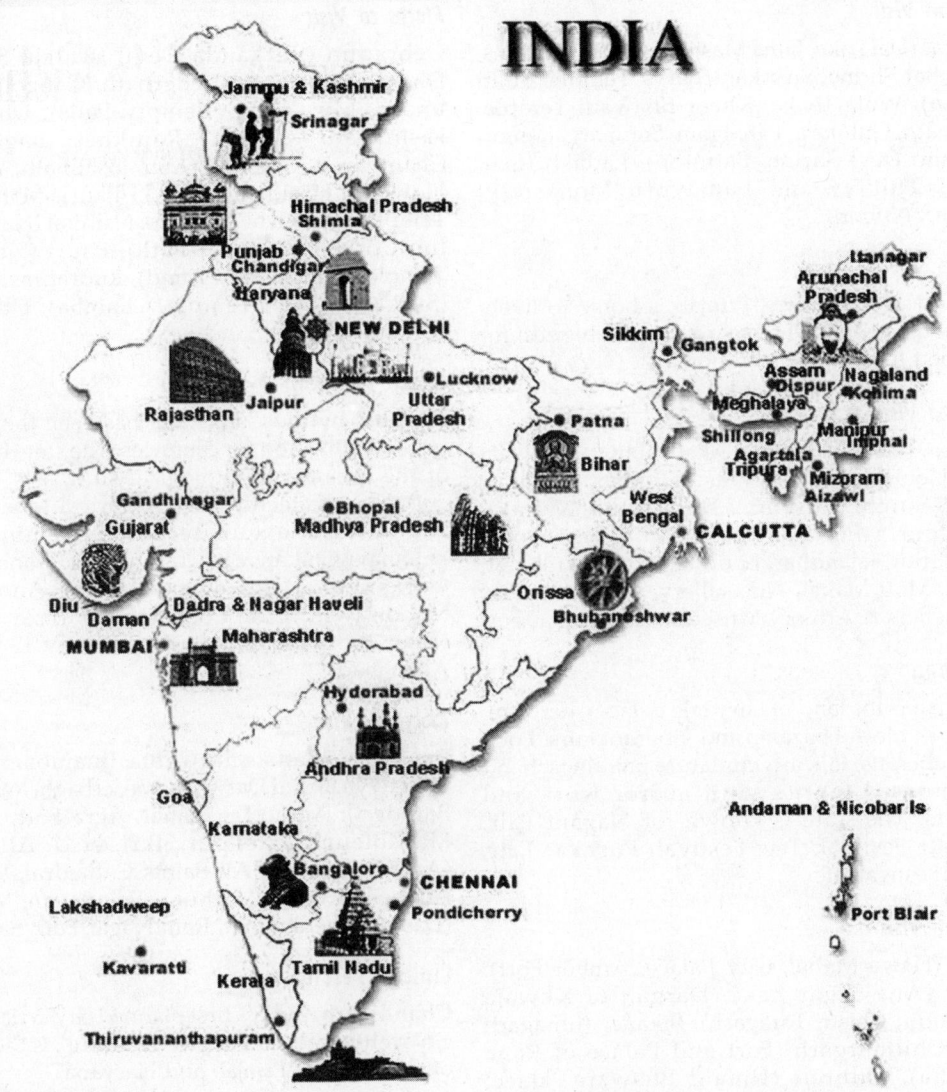

hill stations of India. Tibbetan Buddhist culture abounds in the ancient monasteries. The state also offers adventure sports. You can pick up wood and metal handicrafts, Himachali jackets, caps and plenty of agri-products. Major festivals include Losar, Pori festival, Minjar, Lavi Fair, Renuka Ji Fair and Kullu Dussehra.

Places to Visit

Shimla, Chail, Narkanda, Kufri, Chamba, Dalhousie, Dharamsala, Kullu Valley, Kangra, Kinnaur, Lahaul Spiti, Manali, Manikaran, GHNP, Nahan, etc.

Jammu & Kashmir

The valley is distinctively marked by lakes, streams, luscious fruits, magnificent forests, mighty mountains and lush meadows which make it a paradise on Earth. Jammu is the summer capital of the state and the Ladakh region is known for its barren and beautiful mountains and Buddhist lifestyle. Look for woolens, papier mache, pashmina shawls, embroidered suits, woodcarving, hand knotted carpets, dry fruits and nuts, etc. The major festivals of Kashmir are Losar, Eid, Baisakhi, Shivaratri, Hemis festival, etc.

Places to Visit

Srinagar (Dal Lake, Jama Masjid, Mughal Gardens, Hazratbal Shrine, Shankaracharya Temple, Tulip Garden), Wular Lake, Kheer Bhawani Temple, Amarnath, Gulmarg, Pahalgam, Sonmarg, Jammu (Vaishno Devi Shrine, Patnitop), Ladakh (Leh, Hemis Thiksey and Lamayuru Monastery, Zanskar, Nubra).

Punjab

The land of five rivers, Punjab is home to lively and robust people who have an incredible zest for life, good food and celebrations.

Places to Visit

Amritsar (Golden Temple, Akal Takht, Jallianwala Bagh, Gobindgarh Fort, Wagah Border, Maharaja Ranjit Singh Museum, Dera baba Nanak), Anandpur Sahib, Fatehgarh Sahib, Ferozepur, Hoshiarpur, Jalandhar, Ludhiana, Patiala (Sheesh Mahal, Moti Mahal, Art Gallery, Qila Mubarak, Chandeliers & Armory Museum), Rupnagar, etc.

Rajasthan

Rajasthan is the land of fairy tale palaces, heroism, chivalry, colorful bazaars and vibgrant fairs. Look for handicrafts, folk art, miniature paintings, hand embroidered fabrics with mirror work and puppets. The state is known for Nagaur Fair, Gangaur Festival, Teej Festival, Pushkar Fair, Desert Festival, etc.

Places to Visit

Jaipur (Hawa Mahal, City Palace, Amber Fort), Ajmer (Ana Sagar Lake, Dargah of Khwaja Moinuddin Chisti, Taragarh), Bikaner (Junagarh Fort), Chittaurgarh (Fort and Palace of Rana Kumbha), Jodhpur (Umaid Bhawan Palace, Mehrangarh Fort, Jaswant Thada), Jaisalmer (fort, sand dunes), Mount Abu (Dilwara Temples), Udaipur (Pichola Lake and City Palace) and for nature lovers there is the Bharatpur Bird Sanctuary, Ranthambore National Park, and Sariska Tiger Reserve.

Uttarakhand

Uttarakhand has numerous pilgrimage sites and adventure sports opportunities set in the midst of scenic mountains and natural beauty.

Places to Visit

Dehradun (Surkanda Devi, Jwalaji, Sahastra Dhara), Badrinath, Kedarnath, Gangotri, Yamunotri, Mussoorie (Kempty Falls), Dhanaulti, Rishikesh, Almora, Ranikhet, Bageshwar, Champawat, Chamoli (Auli, Joshimath, Valley of Flowers), Haridwar (Har ki Pauri, Mansa Devi Temple, Sapt Rishi Ashram), Nainital (Naini Lake, Jim Corbett National Park), Pauri (Siddhibali Temple, Khirsu, Tara Kund), Rudraprayag, Tehri (Surkanda Devi Temple, Chamba), Uttarkashi (Givind Wildlife Sanctuary).

Uttar Pradesh

Irrigated by the Ganga, UP has been the melting pot of civilization for centuries. The state has some of the greatest mythological sites, arts, culture, cuisine, architecture and historical monuments. Famous for exquisite Banarsi embroidery, Lakhnavi Chikan, carpet, rugs, wood and leather works, etc. Major events are the Annual Taj Mahotsav in Agra, Lucknow Mahotsav and the Magh and Kumbh Melas held every 12 years in Allahabad.

Places to Visit

Lucknow (Bada and Chota Imambara, Bhool Bhulaiyya, Rumi Darwaza, Kaiserbagh, Residency, Baradari), Agra (Taj Mahal, Agra Fort, Akbar's Mausoleum, Fatehpur Sikri, etc.), Allahabad (Anand Bhawan, All Saints Cathedral, Sangam, Allahabad Fort), Mathura, Vrindavan, Varanasi (Dasaswamedh Ghat, Ram Nagar Fort, Sarnath)

Union Territory

Chandigarh, India's first planned city, a rich, green city rightly called 'The city beautiful', is the capital of the states of Punjab and Haryana.

Places to Visit

Capitol Complex, Government Museum and Arts Gallery, Leisure Valley, Pinjore Gardens, Rock Garden, Rose Garden, Sukhna Lake, etc.

WEST

Chattisgarh

The state is covered with nature's canopy of sal, teak and bamboo trees, and carpeted with lush vegetation. Innumerable monuments, finely carved temples, forts and palaces speak of dynasties and kingdoms of great warriors. It is well known for exotic handicrafts using wood, metal and bamboo, bell metal figurines, jewelery, hand-woven and hand-printed fabrics. The major festivals are Madai, Narayanpur Mela, Bhoramdeo Festival, Goncha Festival, Champaran Mela, Bustar Dussera, etc.

Places to Visit

Raipur, Barsur (with several temples), Chirakote (Mendri Ghoomar), Dantewada (Danteshwari Temple), Kanger (Kanger Valley National Park, Kutumsar, Dandak and Kailashgufa Caves, Shivani Temple), Kawardha (Palace, Bhoramdeo Jungle Retreat), Malhar (famous for ancient temples), Raigarh (Singhanpur Rock Shelters), Sirpur (Laxman Temple and Gandeshwar Temple), Rajim. There are several waterfalls at Tirathgarh, Mandawa and Chitrakote in Jagdalpur.

Goa

Famous for its beaches, churched, lush green fields and a distinct lifestyle, Goa is a paradise for holiday makers from across the globe. Owing to its location on the Western Ghats range it also has rich flora and fauna. Goa is known for the Carnival, Ganesh Chaturthi, Shigmo, Feast of St. Francis Xavier, etc.

Places to Visit

- **Beaches:** North (Anjuna, Arambol, Baga, Calangute and Candolim) Central (Dona Paula, Marimar and Bambolim) South (Bogmalo, Majorda, Betalbatim, Colva, Cavelossim, Benaulim, Palolem, Agonda and Patnem)
- **Forts:** Aguada Fort, Terekol Fort, Cabo-de-Rama Fort, Chapora Fort, Fort of Reis Magos and Corjuem Fort
- **Museums:** Goa State Museum, Archaeology Museum and Portrait Gallery and Naval Aviation Museum
- **Churches:** Basilica of Bom Jesus, Se Cathedral, Basilica of St. Francis of Assisi, Reis Magos Church, The Holy Spirit Church and Church of Mary Immaculate Conception
- **Wildlife:** Dr. Salim Ali Bird Sanctuary, Bhagwan Mahaveer Wildlife Sanctuary and Mollen National Park, Cortigao Wildlife Sanctuary, Bondla Forest
- **Waterfalls:** Doodhsagar and Arvalem Waterfalls

- **Temples:** Mahadevi and Maha Laxmi Temples, Chandreshwar Bhutnath Temple, Shantadurga Temple
- **Mosques:** Safa Shouri and Jama Masjid

Gujarat

Gujarat is unique in its geological and topographical landscape and is famous for diamond work, textiles, Jamnagar tie & dye work, Kutch Embroidery with mirror work, the Tarnetar Festival and its vibrant folk dance Garba Ras. Gujarat is also the state where the Father of the Nation, Mahatma Gandhi was born and this is where his ashram is located at Sabarmati.

Places to Visit

Ahmedabad (Gandhi Ashram, Sidi Sayyed Mosque, Kankaria Lake, Jhulta Minarets, Akshardham Temple, Hutheesingh Jain Temple, Tomb of Ahmad Shah), Bhavnagar, Champaner (Jama Masjid, Tomb of Sikander Shah, Pavagadh Fort and early temples, Stepwell, Dhanpari Eco campsite, Jambughoda Wildlife Sanctuary), Dwarka, Junagadh, Jamnagar, Kachchh, Porbandar, Palitana, Rajkot, Surat (Dandi, Old Fort), Somnath, Vadodara (Palace, Dabhoi, Sayaji Baug, Museum, Sursagar Talav).

Madhya Pradesh

Madhya Pradesh occupies perhaps the oldest part of the subcontinent—called the Gondwana. Close to Bhopal at Bhimbetka are the prehistoric caves that preserve some fascinating paintings dating back to Neolithic times. Madhya Pradesh offers rich handicrafts made of bamboo and cane, dolls, folk paintings, iron and metal craft, paper mache, stone carvings, etc. and is known for the Bhagoria Haat (Jhabua), Khajuraho Festival of Dances and Tansen Music Festival (Gwalior).

Places to Visit

Bhopal (Bhojpur, Islam Nagar, Samasgarh, Manua Bhan ki Tekri), Bhimbetka, Gwalior (Fort, Man Mandir Palace, Tansen's Mausoleum, Jai Vilas Palace, Teli Ka Mandir), Indore (Laal Bagh Palace, Chattris, Bada Ganapati, Kaanch Mandir), Jabalpur (Marble Rocks), Khajuraho, Mandu (Jahaz Mahal, Hindola Mahal, Hathi Mahal, Rewa Kund, Hoshang Shah's Tomb), Orchha, Pachmarhi, Sanchi Stupa, Rewa, Ujjain, Jain Temples at Badoh. There are wildlife sanctuaries at Kanha, Bandhavgarh, Panna and Shivpuri.

Maharashtra

The scenic beauty of the sprawling Western Ghats, majestic forts, and beautiful temples sculpted out of rocks and endless silver beaches will leave you spellbound. It is famous for Paithani and Narayan Peth Sarees, Kolhapuri footwear, and Warli paintings. Mumbai, the capital of Maharashtra has been the country's commercial and film capital and is dotted with heritage architecture and monuments. The state is known for the famous Alphonso mangoes. Ganesh Chaturthi is a major festival of the state.

Places to Visit

Mumbai (Crawford Market, Chhatrapati Shivaji Terminus, Siddhi Vinayak Temple, Juhu and Chowpatty Beaches, Elephanta Caves, Kanheri caves, Marine Drive, Haji Ali & Gateway of India, Sanjay Gandhi National Park), Aurangabad (Ajanta & Ellora Caves, Buddhist Caves, Bibi ka Makbara, Panchakki), Kolhapur (Chhatrapati Sahu Museum, Old Palace, Mahalaxmi Temple), Nanded, Nagpur, Nashik (Trimbakeshwar, Pandaveini Caves, Coin Museum), Shirdi, etc. In addition, there are several beaches (Mandwa, Kihimj, Manori, Dahanu Bordi, Alibaug, Ganpati Phule, Sindhudurg, etc.), forts (Daulatabad, Sinhagad, Murud Janjira, Raigad, Pratapgad, Jaigarh, Shivneri), hill stations (Lonavala, Mahableshwar, Matheran, Panchgani, Khandala, etc.) and wildlife (Pench National Park, Bor Wildlife Sanctuary in Wardha, Tadoba Tiger Reserve in Chandrapur) to see.

Union Territories

Dadra and Nagar Haveli is endowed with lush green forests, winding rivers, unimaginable waterfronts and the gentle gurgle of streams. Calm and serene, it offers scenic beauty and monuments.

Places to Visit

Dadra Vanganga Lake and Garden, Hirwavan Garden, Piparia, Tribal Museum, Silvassa, Vanvihar Tourist Complex, Chauda

Daman and Diu is famous for historical monuments, museums and beaches. This union territory is scenic and extremely inviting.

Places to Visit

Daman (forts of Moti and Nani Daman, Jampore Beach, Devka Beach, Mirasol Lake Garden, Pargola Garden, Satya Sagar Udhyan, Daman Ganga Water Park, Kadaiya Lake Garden, Jetty Garden, Vaibhav Water World), Diu (Nagoa & Ghoghla Beaches, Jallandhar Shrine, Panikotha).

EAST

Bihar

Ancient India's centre of power, learning and culture, Bihar is a profusion of historic treasures. It is the birthplace of India's first greatest empire—the Mauryan Empire as well as one of the world's greatest pacifist religions, Buddhism. Famous for stone and bead jewelry, hand-painted wall hangings (Madhubani paintings), appliqué work and stone pottery. The main festivals and events of Bihar are Makar Sankranti, Rajgir Mahotsav, Chhath, Sonepur Fair, etc.

Places to Visit

Patna (Golghar, Martyr's Memorial Museum, Harmandir Sahib Gurudwara, Patthar ki Masjid, Agam Kuan, Mahatma Gandhi Setu), Bodhgaya (Mahabodhi Stupa, Vishnupad Temple, Bodhi Tree, Barabar Cave, Karan Chaupa Cave), Nalanda (Nalanda University Archaeological Complex, Hiuen Tsang Memorial Hall), Pawapuri (Jalmandir and Samosharan), Vaishali (Ashokan Pillar, Bawan Pokhar Temple, Buddha Stupa, Abhiskek Pushkarn), Rajgir (Jivakameavan Gardens, Ajatshatru's Fort, Swarna Bhandar), Bihar Sharif, Munger, etc.

Jharkhand

The steel city of Jamshedpur and the coal city of Dhanbad are the better known towns of nature's most brilliant displays of forests, hills and wildlife. Paitkar paintings, bamboo handicrafts, tribal jewelery and wood carvings are all reflections of Jharkhand's cultural and natural vibrancy. The main festivals and events are Sarhul, Karma, Tusu, Sohrai, Chhath, etc. The state is known for its energetic people and their colorful clothes and handicrafts.

Places to Visit

Ranchi (Tagore Hill, Rock Garden, Nakshatra Van, Dassam Fall, Sun Temple, Panch Gagh, Hundru Fall), Jamshedpur (Jubilee Park, Dimna Lake & Dam, Hudco Lake, Dalma Wild Life Sanctuary, Tata Steel Zoological Park), Deoghar (22 temples), Parasnath Hills, Hazaribagh, Maithon, Bokaro Steel City (Garga Dam, City Park, Nehru Park, City Centre, etc.)

Orissa

Orissa—a land of serpentine rivers, virgin beaches, mighty waterfalls and blue hills, basks in the Sun of the eastern coast of India. Sambalpuri silk and handicrafts like Ikkat, Patta Chitra, stone carvings, and silver filigree work are some signature attractions. It is also the birthplace of the classical dance form—Odissi. Major festivals include the Kalinga Mahotsava, Konark Dance Festival, Chariot Festival of Lord Jagannath.

Places to Visit

Bhubaneshwar (Rajarani Temple, Dhaulagiri, Lingaraj Temple, Nandankanan, Atri), Cuttack, Jeypore, Puri (Jagannath Temple), Ratnagiri, Sambalpur (Hirakud Reservoir), Pipli, Konark Sun Temple, Chandrabhaga, Chilka Lake (Asia'a largest brackish lake famous for Irrawaddy Dolphins), Berhampur (Taptapani), Kebdrapadra (Bhitar Kanika National Park), Udayagiri and Khandagiri Caves, and the beaches at Gopalpur-on-Sea, Chandipur, Ramchandi, Balighai, Astaranga, Paradeep, etc.

West Bengal

Well known as the state of Rabindranath Tagore, the Guru who has written India's National Anthem, the artistic streak of West Bengal does not at its lyrical music or literature. It finds its undertones extended to everything from the Baluchari or Cotton handloom sarees, conch shell products to precious silks. Kolkata, or 'the City of Joy', is synonymous with mouth-watering street food, love for soccer, 'Baul' style of folk music, and centuries of Indian heritage blending effortlessly into a cosmopolitan culture. The fragrant mangrove forests of Sundarban are the habitat of the famous West Bengal tiger. The state comes alive with colors and festivities during Durga Pooja.

Places to Visit

Kolkata (Birla Planetarium, Fort William, Howrah Bridge, Marble Place, Kalighat, National Library, Nicco Park, Shaheed Minar, Writers Building, Victoria Memorial, Eden Gardens, St. Paul's Cathedral, Dakshineshwar Temple, Diamond Harbor), Darjeeling (Himalayan Mountaineering Institute, Natural History Museum, Toy Train, Japanese Peace Pagoda, Batasia Loop, Buddhist Monasteries, Tiger Hill and Tea Gardens), Murshidabad (Nawab's Palace, Katra Mosque, Khushbagh Cemetery, Moti Jheel), Bankura (Bishnupur Temple, Mukutmanipur lake), Shantiniketan (Vishwa Bharti University), Sunderbans Delta, Digha, Shankarpur and Bakkhali.

Union Territory

The Andaman and Nicobar Island is an archipelago of more than 572 island that offer peace, nature and beauty at its very best. The hues spill over into local forms of handicrafts made from corals and seashells.

Places to Visit

Andaman (Cellular Jail National Memorial, Ross Island, Viper Island, Niel Island, Mayabunder, Long Island, Little Andaman Island) Port Blair (Chidiya Tapu, Mahatma Gandhi Marine National Park, Madhuban), Nicobar (Katchal, Car Nicobar, Great Nicobar).

Buddhist Circuit

Experience the spiritual magnetism exuded by various destinations that were landmarks in Gautama Buddha's journey to Truth. It starts with Bodhgaya (Bihar), where Prince Siddhartha sat under the 'Bodhi Tree', saw the first glimpse of light and achieved enlightenment. Primary points of homage are the Mahabodhi Temple, the Vajrasan throne donated by King Ashoka, the Bodhi Tree, Animeshlochana Chaitya, Ratnachankramana, Ratnagaraha, the Ajapala Nigrodha Tree, the Muchhalinda Lake and the Rajyatna Tree. Sarnath (near Varanasi) witnessed the first ever sermon given by the Buddha after achieving enlightenment. The magnificent Dhamekha Stupa at Sarnath was originally erected by King Ashoka, as was famous lion capital pillar, now a symbol of the Indian nation. Lord Buddha's journey ended at Kushinagar (near Gorakhpur), where he attained mahaparinirvan in 543 BC. Khushinagar is home to the famous Mukatanabandhana Stupa and the reclining Buddha statue that dates back to the Gupta period. Other places of significance include Vaishali (near Patna), Rajgir and Nalanda University (near Bodhgaya) and Shravasti (150 km from Lucknow).

SOUTH

Andhra Pradesh

Located in the Indian peninsula, Andhra Pradesh, the 'Rice Bowl of India', is also renowned for its colorful Banjara needle craft (silver inlary on metal), bronze idols, Budithi brass ware, Pembarthi metal crafts, Durgi stone craft, Kondapalli toys and Dharmavaram & Mangalagiri sarees. The region's Kalamkari Fabrics are famous for their vivid colors printed with vegetable dyes.

Places to Visit

Hyderabad (Charminar, Golconda Fort, Qutub Shahi Tombs, Taramati Baradari, Mecca Masjid, Salar Jung Museum, Legislative Assembly, Archaeological Museum, Nizam's Silver Jubilee Museum, Falaknuma Palace, Chow Mohalla, Shilparamam, Osmansagar and Hussainsagar Lakes, Mahavir Harina Vanasthali National Park), Nagarjunasagar (Nagarjunasagar Dam, Nagarjunakonda, Buddhist Museum, Anupa, Ethipothala Waterfalls), Puttaparthy (Prasanthi Nilayam), Penukonda Fort (Anantpur), Lepakshi, Tirupati (Sri Venkateswara Temple in Tirumala, Padmavathy Temple, Kapilatheertham, Govindaraja Swamy Temple, Chandragiri Fort, Kanipakam, Kalyani Dham, Kailasanathakonda, Nagalapuram and several other temple), Visakhapatnam (Rushikonda, Bhemunipatnam and Ramakrishna Beaches, Dolphin's Nose, Buddhist Sites, Borra Caves, Submarine Museum), Warangal (Thousand Pillars Temple, Warangal Fort, Ramappa Temple & Lake, Khush Mahal, Pakhal Wildlife Sanctuary and Lake, etc.), Sri Kalahasti Srisailam.

Karnataka

Karnataka or 'karu nadu', meaning elevated land, has an antiquity that dates back to the Paleolithic era. Renowned for its art, music and literature, Karnataka has contributed significantly to both forms of Indian classical music, the Carnatic and Hindustani, and it has a rich historical heritage. Look for exquisite Mysore silk sarees, Sandalwood handicrafts (in State emporiums) and Rosewood handicrafts. Bengaluru is one of the most tech-savvy cities in the country and is positioned as the software centre of the world. The Dussera festival at Mysore is a major event.

Places to Visit

Bengaluru (Vidhan Soudha, Tipu's Palace, Government Museum, Bangalore Palace, Lalbagh Botanical Garden, Indira Gandhi Musical Foundation, Cubbon Park, Bull Temple, Doda Ganesh Temple), Aihole (Rock Cut Temples), Badami (Cave Temples), Belur (Hoysala Temple), Bijapur (Gol Gumbaz, Ibrahim Rauza), Gulbarga (Chor Gumbaz, Fort, Juma Masjid, Dargah of Khwaja Banda Nawaz, etc.), Hampi (King's Palace, Lotus Mahal, Queen's Bath, etc.), Halebid & Somnathpur (Hoysala Temples), Mysore (Mysore Palace, Rail Museum, St. Philomena's Church, Chamundi Hills, Srirangapattanam, Shivansamudra Falls, Ranganathittu Bird Sanctuary, Brindavan Gardens, zoo), Pattadakal (monuments), Shravanabelagola (Gomateshwara). There are beautiful waterfalls at Jog, Gokak, Unchalli and Magod.

Kerala

Known as 'God's Own Country', Kerala is situated on the lush and tropical Malabar Coast, and is one of the most popular tourist destinations in India. Kerala is especially known for its eco-tourism initiatives, spices and native performing arts like Koodiyattam, Kathakali, Kaliyattam, Thullal, Theyyam, Chavittunadakam, Mohiniyattam, Padayani and the famous martial art. The Thrissur Pooram and Boat races are major events.

Places to Visit

Thiruvananthapuram (Akkulam & Veli Tourist Villages, East Fort, Kovalam Beach, Koyikkal & Kuthiramalika Palaces, Natural History Museum, Padmanabha Swamy Temple, Padmanabhapuram Palace, Neyyar & Peppara Wildlife Sanctuaries), Alapuzha (Backwater Cruises, Chakkulathukavu Bhagavathy Temple, Chettikulangara Bhagavathy Temple, Krishnapuram Palace, Kuttanad, Edathua Church), Ernakulam (Chennamangalam Synagogue, Cherai Beach, Fort, Hill Palace Museum, Kaladi, Kodanad, Jew Town, etc.), Idukki (Chinnar, Periyar & Idukki Wildlife Sanctuaries), Munnar, Eravikulam National Park, Kannur (St. Angelo Fort, Thalasseri Fort), Kasaragod (Bekal Fort, Chandragiri), Kollam (Shenduruny Wildlife Sanctuary, Police Museum, Thangasseri Fort), Kottayam, Kozikhode (Thusharagiri Falls, Peruvannamuzhi Dam), Palakkad (Parambikulam Wildlife Sanctuary, Malampuzha Dam), Sabrimala, Thrissur (Guruvayur, Peechi-Vazhani Wildlife Sanctuary), Wayanad (Wayanad Wildlife Sanctuary, Edakkal Caves).

Tamil Nadu

Enchanting Tamil Nadu—a land of innumerable and incomparable temple of Dravidian architecture, exotic hill stations, serene beaches, multi-religious pilgrimage sites. Look for its renowned arts and crafts such as Tanjore paintings, musical instruments, stone and ambellished jewelery, brass and copper metalware, pottery, woodcraft and stone sculpture carvings. Major events and festivals include Pongal (harvest festival), Chithirai Festival, Karthigai Deepam, Chennai Music and Dance Festival, Chennai Sangamam, etc.

Places to Visit

Chennai (Fort St. George, Marina Beach, Elliot Beach, etc.), Chidambaram (Nataraja Temple, Asyappan Temple), Coimbatore (Siruvani Falls, Velliangiri Hills, etc.), Cuddalore (Lignite Mines, Port Novo, Bhuvanagiri), Dharmapuri (Hogena-kkal Falls), Erode (Kodiveri Dam), Kanchipuram (Kailasanathar Temple, Ekambareswarar Temple and several others), Kanyakumari (Southern-most tip of India), Kodaikanal (Lake, Bryant's Park, scenic beauty), Krishnagiri (Krishnagiri Dam), Madurai (Meenakshi Temple, Thirumalai Nayak Mahal, etc.), Mamallapuram (Rock cut cave temples), Nagapattinam (Nagore, Velankanni Church, excavations at Poompuhar, etc.), Namakkal (Kolli Hills), Pudukkottai, Rames-waram (Sri Ramanathaswamy Temple), Salem (Mettur Dam), Sivagangai, Thanjavur (Sri Brahadeeswarar Temple, Palace, etc.), Theni, Thiruchirapalli (Rock Fort, etc.), Tirunelveli, Thiruvannamalai (Ramana Ashram , etc.), Thuthu Kudi, Udhagamandalam (Heritage Train, Ooty Lake), Velankanni (Shrine of Our Lady of Health), Vellore (Sripuram Golden Mahalaxmi Temple near Vellore, Fort, Museum etc,), Virudunagar, Yercaud (Killiyur Falls, Lake).

Union Territories

Lakshadweep is a group of islands 200 to 300 km off the coast of kerala in the Arabian Sea. It is known for adventure sports such as Scuba diving, Pedal boating, Kayaking, Canoeing and Snor-keling.

Places to Visit

Marine Museum, Light House and many others.

Puducherry: Known as French Riviera of the East, it has some beautiful beaches and a spectacular coastline drive from Chennai.

Places to Visit

Auroville, Aurobindo Ashram, Arikamedu Archaeological excavation, Manakula Vinayagar Temple, Sri Gokilambal Thirukameswara Temple and many others.

NORTHEAST

Arunachal Pradesh

Arunachal Pradesh offers picturesque peaks and scenic rivers. Beautiful Buddhist monasteries and almost impenetrable deciduous forests add to the mysticism possessed by this 'Land of Dawn Lit Mountains'. It also provides excellent oppor-tunities for adventure sports like camping and rafting. While here, you can enjoy a sip of the local drink 'Apong' or be a part of events like Losar and Saka Dawa.

Places to Visit

Itanagar (Itafort, Ganga Lake, Jawaharlal Nehru State Museum, Buddhist Temple), Tawang (Buddhist Monastery, Nuranang Waterfalls, Sela Pass, Sangetser Lake), Tezu (Parshuram Kund), Bhismak Nagar (archaeological ruins), Akash-iganga, Namdapha (National Park), Bomdila (crafts centre), Ziro (fish-cum-paddy culture).

Assam

A hot-spot for wild life tourism with over 25 national parks and sanctuaries. Assam is home to the Indian Rhinoceros, Tiger and lush green tropical rainforests. It spans the fertile plains of the mighty Brahmaputra. Fine silks, woven mats, cane and bamboo furniture, and world-class tea are some added attractions. Local events and festivals include Bihu, Baishagu, Ambubachi Mela, etc.

Places to Visit

Guwahati (Kamakhya Temple, Nabagraha Temple, Umananda Temple, Madan Kamdev Ruins, Museum, Zoo), Sivasagar, Majuli (Vaishnav Temple), Tezpur and wildlife sanctuaries in Kaziranga, Manas, Orang, Nameri and Pobitora.

Manipur

Manipur's rich culture excels in martial arts, dance (the Ras Leela) and theatre. Apart from breathtakingly beautiful orchids and Sangai, the state is famous for its hand-woven textiles, local pottery and Manipuri dolls. One can savor the Kabok cuisine and enjoy the festivities during Yaoshang, Cheiraoba (New Year), Heikru Hito-ngba (boat race festival), Ningol Chakouba, KUT (festival of Kuki-Chin-Mizo), etc.

Places to Visit

Imphal (Ima Market, Sri Govindaji Temple, Lagthabal's Palace, zoo, war cemeteries, museum), Thoubal, Bishnupur (Keibul Lamjao National Park, INA Memorial, Loktak Lake, Loukoipat Ecological Park, Leimaram Falls), Churachandpur, Tamenglong (Tharon Cave, Zeilad Lake, Bunung Meadow, known for Orange Festival), Chandel (Yangoupokpi-Lokchao Wildlife Sanctuary), Ukhrul.

Meghalaya

With its enchanting hill stations and picturesque landscaping, Meghalaya which means 'Abode of the Clouds', is an once-in-a-lifetime experience. The festivals of the state include Shad Suk Mynsiem, Behdeinkhlam, Nongkrem Dance Festival, Wangala Dance.

Places to Visit

Shillong (Ward's Lake, Shillong Peak, Elephant Falls, Sohpetbneng Peak, Butterfly Museum, Umiam lake), Sohra, Mawkdok, Nohkalikai Waterfalls, the Grand Canyon of Cherrapunjee, Mawsmai Cave, Nohsngithiang and Dainthlen Waterfalls, Mawphlang, Ranikor, Tura, Balpakram National Park, etc.

Mizoram

The poetic juxtaposition of mountains, valleys, rivers and lakes seems to be the inspiration behind Mizoram's beautiful music and dance. Even the traditional handloom shawls, hats and bags live up to its exotic reputation. The state is known for its dances like Cheraw or the Bamboo Dance while the events and festivals include Chapchar Kut, Thalfavang Kut and Anthurium festival.

Places to Visit

Aizawl (Bara Bazar, Luangmual Handicrafts Centre, State Museum, Durtlang Hill, Mini Zoo, Bung, Paikhai) lakes at Tamdil, Rungdil, Rengdil and Palak, Vantawng (Waterfall), Kolasib, Chaphai, Lunglei, Dampa Tiger Reserve, Phawngpui Blue Mountain National Park and wildlife sanctuaries in Ngenpui, Tawi, Thorangtlang.

Nagaland

Nagaland lies nestled amongst the three mountain ranges of Naga, Patkai and Barail that rise from the Brahmaputra Valley. Unique crafts like colorful woven shawls, mats, bamboo works, decorative spears and tribal song and dance distinguish the various tribes. It has a rich tribal culture with 16 major tribes. The major events of the state are Sekrenyi, Moatsu, Tuluni, Tsukheny, Hornbill Festival, etc.

Places to Visit

Kohima (World War II Cemetery, Kohima Cathedral, Musuem, Naga Heritage Village, Intaki Wildlife Sanctuary), Dimapur (medieval ruins of the Kachari Kingdom, Diezephe Craft Village), Kiphire (Fakim Wildlife Sanctuary, caves), Longleng, Mokokchung (lakes & springs), Mon, Peren, Phek (Shilloi Lake, waterfalls, caves) Tuensang, Wokha, Zunheboto (Ghosho Bird Sanctuary).

Sikkim

Infused with ancient history and Buddhist mysticism, Sikkim's mountains are a paradise for trekkers and adventurers. Trekking trails lead to hidden lakes and ancient monasteries. The state has some of the world's wildest rafting spots.

Places to Visit

East Sikkim (Gangtok: White Hall, Do Drul Chorten, Ridge Garden, Rumtek, Fambong La Wildlife Sanctuary, Kyongnosla Alpine Sanctaury, Tsongo Lake, Nathula Pass), North Sikkim (Singba Rhododendron Sanctuary, Kanchendzonga National Park, Chungthang, Yumthang, Gurudongmar Lake), South Sikkim (Tendong Hill, Namchi), West Sikkim (Rabdentse Ruins, Pelling, Yuksom).

Tripura

Tripura has a long historic past. It is delicately beautiful with a leagcy of fine palaces, Hindu and Buddhist monuments, protected forests and offers numerous cultural experiences. One can also enjoy festivals and events such as Kharchi, Orange Festival, etc.

Places to Visit

Agartala (Ujjayanta Palace, State Museum, Tribal Museum, Uma Maheswar Temple, Jagannath Temple, Benuban Bihar, Gedu Mian Mosque, Malanch Niwas, Rabindra Kanan, Fourteen Goddess Temple, Portuguese Church), Unakoti, Pilak, Udaipur, Tripurasundari Temple, Neermahal, Jampui Hill, Bhavaneswari Temple, Sepahijala, Kamalasagar, Deotamura, Dumboor Lake, etc.

CITY TRAVEL

Indian cities are bubbling with captivating and confusing major tourist attractions. One can see living breathing splendors of a bygone era as Lake Palace at Udaipur. If you have never seen a love story in marble, then you cannot really wait to experience Taj Mahal in the full moon night. Warm and sunny Southern Indian cities are distinctly different from their North Indian counterparts. India's glorious past residing amicably with the present in India's very own 'City of Joy' Kolkata. Visit India—as the name alone is worth the journey.

Delhi: Delhi is the proud capital of India spreading to the whole world the message of beauty and grandeur. The city is rich with architecture, culture and human diversity. It was seven times that the city went through the pain of being built and rebuilt. Steeped in history, Delhi is a complete amalgamation of new and old. It is in fact the best starting point to make a tour to north India. The monuments that speak of the sagas of the bygone era still stand majestically in Delhi.

Places to see: Red Fort, Jama Masjid, Chandni Chowk, Raj Ghat, India Gate, Lotus Temple, Bahai Temple, Rashtrapati Bhawan, Qutab Minar, Humayun's Tomb, Parliament House, newly built Akshardham temple, Jantar Mantar, Iskcon temple.

The city also provides a number of shopping options. Chandni Chowk, Palika Bazzar, Janpath, Lajpat Nagar, Sarojini Nagar, Connaught Place, Dilli Haat are some of the shopper's paradise of Delhi.

Jaipur: This 300-year-old city is much dear to the tourists for its regal forts, lavish havelis, gorgeous lakes, sand dunes and colorful bazaars. The charm of Jaipur has attracted Rajput emperors, Mughals and common men equally. Its proximity to the cities like Delhi and Agra has further added to its appeal. Also famed as *Pink City*, Jaipur has always been touted as a destination of tourist interest.

Places to see: City Palace, Hawa Mahal, Jantar Mantar, Sawai Mansingh Museum, Amber Fort, Jaigarh Fort, Gaitore and Nahargarh Fort.

Udaipur: This jewel of Mewar is considered to be one of the most romantic destinations in India. Three lakes form such an attraction of the city whose focal points are the towering elegant palaces, temples and havelis, which have flanked the shores of these lakes. In the heart of the Lake Pichola, Lake Palace Hotel is located, which creates a picture perfect vision in white.

Places to see: Fateh Sagar Lake, Lake Pichola, Shilpgram (Craft Bazaar), Bhartiya Lok Kala Museum, Lake Palace, City Place, etc.

Jaisalmer: In Jaisalmer, one can notice echoes of the past in its sandstone palaces, temples, forts and cenotaphs. A glance of the rich kaleidoscopic Rajasthani culture and heritage can be seen in the tough terrains of Jaisalmer. You can see how your fantasy meets reality while you will be enjoying camel safari in the wild beauty of the Thar Desert.

Places to see: Jaisalmer Fort, Nathmaliji-Ki-Haveli, Gadsisar Lake, Parswanath Temple, etc.

Agra: For ages, Agra has been synonymous with the Taj Mahal. Home to the three generations of one of the most vibrant empires in the medieval time, Agra reflects the finest examples of Mughal architectures in India. Visiting Agra is like taking a stroll through the glorious history of the Mughal era. It forms a major part of the *Golden Triangle* in addition to Jaipur and Delhi. The city of Agra livens up during the festive time of the Taj Mahotsava which is held in the month of February with Taj as the backdrop. The festival lasts for 10 days in Shilpgram, where several artists from various fields perform their art in front of large audience. Agra no doubt is included on every first time visitor's itinerary.

Places to see: Taj Mahal, Fatehpur Sikri, Agra Fort, Sikandra, Dayalbagh Temple, Jama Masjid, etc.

Srinagar: Kashmir's capital city Srinagar offers delightful holidays on the lakes with their shikaras or houseboats. It is the base for any holiday in the Kashmir Valley. This paradise on earth awe-inspiring beauty will surely sweep you off your feet. Beat the heat in the Srinagar under the shadow of the Chinar trees.

Places to see: Dal Lake, Shikaras, Hazratbal Shrine, House Boats, Hari Parbat Fort, Shankaracharya Temple, Jama Masjid, Chatti Padshahi, Mughal

Gardens, Nagin Lake, Chashmashahi Garden, Nishat Garden, Shalimar Garden, etc.

Kozhikode: Kozhikode is a kaleidoscope of swaying palm trees, alluring backwaters, emerald paddy fields, tranquil beaches, historical monuments, wildlife sanctuaries, lakes, waterways, mountains and a number of other popular tourist attractions. This cultural capital of Kerala, in the past was an important trade and commerce centre.

Places to see: Pazhassiraja Museum and Art Gallery, Kalipoika, Kozhikode Beach, Velliyamkallu, Kadalundi Bird Sanctuary, Pishakarikavu, Kurishupalli, Mishkal Masjid, Lokanarkavu Temple, Tali Temple, Mannur Temple, Ponmeri, Varakkal Devi Temple, St Mary's Church, etc.

Hyderabad: In Hyderabad, the Muslim culture of the Nawabs and Sultans has blended with the dominant Hindu culture, clearly visible in their traditions, cuisines and handicrafts. This city is also famous for its pearls, bidri work and bangles embellished with sparkling, semi precious stones set in lacquer.

Places to see: Charminar, Salar Jung Museum, Mecca Masjid, Golconda Fort, Shilparamam, Brahmanda Reddy National Park, Hyderabad Botanical Gardens, Durgam Cheruvu, Dhola-ridhan, Ramoji Film City, Kotaguda Reserve Forest, etc.

Mumbai: This commercial capital of India pulsates with energy. This incredible city is different with its varied cultures and amazing contradictions. Mumbai's Bollywood is the place where dreams are chased, broken and made. India's Little Paris, Mumbai's throngs with versatile designers, hip boutiques and stars of yesteryears.

Places to see: Gateway of India, Chowpatty Beach, Mahalaxmi Temple, Hanging Gardens, Haji Ali Shrine, Jehangir Art Gallery, Sanjay Gandhi National Park, etc.

Kolkata: Kolkata is home to the intelligent, sensitive and cultured Bengalis, who are equally passionate about music, literature, politics, Durga Puja, football and cricket. Also referred to as the *City of Joy*, this 300 years old historical city is recognized for its gorgeous colonial constructions and an exclusive and discrete cultural heritage.

Places to see: Eden Gardens, Vidyasagar Setu, Salt Lake Stadium, Raj Bhawan, Nalban Boating Complex, Dhakuria Lake, Kali Ghat, Victoria

Memorial, Birla Mandir, Belur Math, Science City, Academy of Fine Arts, Fort William, Dalhousie Square and Kumartuli, etc.

Khajuraho: Khajuraho is an ancient town located in the provincial state of Madhya Pradesh. It is famed all over the world for its temples, sex, sculpture and architecture built during the reign of the Chandela dynasty. The temples of Khajuraho have also been designated as a UNESCO World Heritage Site. The temples of Khajuraho are highly erotic and sensual depicting the graceful forms of love making. These temples display the whole range of human emotions and relationships.

Places to see: Ghantai temple, Vamana temple, Javari temple, Brahma temple, Chitra Gupta temple, Vishwanatha temple and Lakshmana temple, Duladeo temple and Chaturbhuja temple.

HILL STATION TRAVEL

To beat Indian summer what can be more soothing than a cool vacation in the Indian hill stations. The beauty and its very own local flavor of these hills stations attract tourists from far and near. Each of these hills stations of India has preserved their own cultural heritage, which has not yet being discolored even with the passage of time. Moreover one can appreciate nature at her best in a hill station only. With many glaciers and valleys, a huge range of wildlife, lush forests and cascades Indian hills stations are a breathtaking experience.

Mussoorie: This fairyland hill station is much adored by the tourist for its lush hills and diverse flora and fauna. Mussoorie is aptly known as the *Queen of Hill Stations*. This popular hill resort is also ideal retreat for the trekkers and adventure enthusiasts.

Places to see: Camel's Back Road, Gun Hill, Kempty Fall, Municipal Garden, Childer's Lodge, Cloud End, Nag Devta Temple, etc.

Auli: The amazingly beautiful mountain resort, Auli provides the adventure freak tourists a wide range of entertainment. Auli attracts flock of tourists from both India and abroad due to its unmatched scenic beauty, lively social life and sources of recreations.

Places to see: Bhavishya Badri, Vanishinarayan Kalpeshwar, Joshimath, Chenab Lake, Kwani Bugyal, Chattrakund, Gurso Bugyal, etc.

Dharamsala: A much sought after tourist resort, Dharamsala presents a pretty picture during summer. Worldwide tourists come here to enjoy attractive glimpses of the Himalayan ranges and to breathe in the fresh mountain air.

Places to see: Kangra Art Museum, St. John's Church, McLeodganj, etc.

Manali: This prime holiday destination, surrounded by high mountains, wild flowers, small picturesque hamlets and fruit laden orchards, is an ideal base for skiing, paragliding, rafting and trekking in India.

Places to see: Kothi, Solang Valley, Rahla Falls, Manali Sanctuary, Rohtang Pass, etc.

Shimla: Pine, deodar, oak and rhododendron forests, quaint cottages have created a fantastic

ambience in Shimla. Besides its unrivaled beauty Shimla also attracts tourist for shopping, sports and entertainments.

Places to see: The Ridge, Summer Hills, Daranghati Sanctuary, Jakhoo Hills, Mashobra, etc.

Ponmudi: The name Ponmudi means golden crown. It is so named because of the golden glow cast over these hills by the evening sun. This place is renowned for wild orchids, natural springs and beautiful picnic spots.

Places to see: Ponmudi Hills, Spice/Tea Plantations, etc.

Ooty: Ooty is one of the most popular and superbly gorgeous hill stations in South India. Leafy hills, spilling waterfalls and radiant brooks, thick forests, sprawling grasslands and extensive tea gardens have made its flawless beauty.

Places to see: Doddabetta Peak, Botanical Garden, Lake, Doddabetta Peak, Pykara Falls, Hindustan Photo Film, Cosmic Ray Laboratory, Radio Telescope, Tribal Research Center, etc.

Mount Abu: This picturesque hill resort serves as an emerald retreat in the sterile wasteland of Rajasthan. Here one can find interesting residues of the bygone Rajputana and the Raj period.

Places to see: Dilwara Jain Temples, Gaumukh Temple, Nakki Lake, Mansarovar Lake, Guru Shikhar, Mount Abu Wildlife Sanctuary, Sun Set Point, etc.

Darjeeling: Popularly known as the *Crowned Princess* and *Land of Celestial Thunderbolt*, Darjeeling is well known for its distinct good looks, fresh air, pleasant conditions and awesome landscapes. The hill station lies in the foothills of the Himalayas and is covered all the year round by snow. One can savor the spectacular views of the world's third highest mountain range, Mount Kanchenjunga. Also famous for tea gardens, Gothic Victorian ambience, Darjeeling indeed a true crowd puller and one of the favorite haunts of the honeymooners. For adventure enthusiasts the Darjeeling Toy Train and Tiger Hill are the most luring attractions.

Places to see: Batasia Loop, Observatory Hill, Senchal Lake, Tea Gardens, Dhirdham Temple, Buddhist Monasteries

Ladakh: Located in Jammu & Kashmir, Ladakh is one of the most beautiful regions of India. It is also famous as *Little Tibet*. Though bitterly cold Ladakh is one of the favorite jaunts of the tourists. Ladakh is popular for its old culture and monuments, monasteries, oral literature, art forms and various fairs & festivals. The mountain beauty and the Buddhist culture that exists here has made Ladakh an outstanding abode for the tourists. Ladakh is a perfect delight for the adventure lovers. An adventure buff can enjoy Trekking, Mountain Climbing, Cycling, Jeep Safari and Camel Safari.

Places to see: Hemis Gompa, Sankar Gompa, Shanti Gompa, Likir Gompa, Spituk Gompa, Cave Monastery, Leh Palace and Leh Mosque

Munnar: Munnar, located in the state of Kerala is one of the most luring hill stations of India. It is also a major centre of Kerala's tea industry. Munnar basically mean 'three rivers'. It is located in the confluence of three mountain ranges- Muthirapusha, Nallathani and Kundala. Picture book towns, winding lanes, vast stretch of tea plantations are the added attraction of this picturesque hill station. The forest of Munnar is extremely rich in wildlife.

Places to see: The Christ Church, Devikulam lake

Nainital: Nainital is a glittering jewel in the Himalyan necklace, blessed with scenic natural splendor and varied natural resources. Dotted with lakes, Nainital has earned the epithet of *Lake District* of India. The most prominent of the lake is Naini Lake ringed by hills.

Places to see: Naini Lake, Naina Devi Temple, Snow View, Naini Peak, Tiffin Top/Dorothy's Seat

RELIGIOUS TRAVEL

Mystical India opens doorways to a spiritual sphere. India is glorified by the myth and legend and sanctified by the religion. It has been attracted by a large number of pilgrims and worshippers from time immemorial. Here one can find every religion that is practiced on the earth. A journey to the pilgrimages in India will take you to the journey to Nirvana. You will find such celestial zones where the Gods bestow their blessings. An overwhelming experience is waiting for all you people, who are searching for the eternal bliss.

Varanasi: Varanasi also known as *Kashi* is one of the revered pilgrimage destinations for the Hindus. It lies on the western bank of the Ganga in eastern

Uttar Pradesh. This magnificent city with myriad attractions is visited by a large number of pilgrims from all across the world every year. It is believed by millions of pilgrims that a dip in the Ganges here is a quick solution for getting rid of one's sins. The city is very commonly referred to as 'City of Temples', 'Holy City of India', 'Religious Capital of India', 'City of Lights', and 'City of Learning'.

Places to see: Dashashwamedha Ghat, Asi Ghat, Barnasangam Ghat, Panchganga Ghat, Manikarnika Ghat, Saranath, Kashi Vishwanatha Temple, etc.

Rishikesh: It is a holy city for Hindus and it is located in the foothills of the Himalaya in northern India. A gateway to the four pilgrimage sites, i.e. Badrinath, Kedarnath and Haridwar, is formally known as Mayapuri.

Places to see: Lakshman Jhula, Ram Jhula, Bharat Temple, Rishi Kund and Raghunath Temple, Shkar Temple, Shatrughan Temple, Muni-Ki-Reti, Swarg Ashram, Lakshman Temple, Sadanand Jhula, Triveni Ghat, Shivanand Ashram, Kailash Ashram, etc.

Char Dham: Located in the Garhwal section of the state of Uttaral hand, it consists of four sites— Badrinath, Ked rnath, Gangotri and Yamunotri. These sacred pl ces, located at a height of 3,000 m above sea level, a e considered to be the places of Nirvana according to Hinduism. Today, the circuit receives hundreds of thousands of visitors in an average pilgrima e season, which lasts from approximately Apr l 15 until Diwali.

Vaishno Devi: Vaishno Devi Mandir is one of the holiest Hindu temples dedicated to Shakti, located in the hill of Vaishno Devi, Jammu. The shrine is at an altitude of 5200 feet and a distance of approximately 12 kilometres from Katra. The Shri Mata Vaishno Devi Shrine Board maintains the shrine. To see the holy shrine of Mata Vaishno Devi millions of devotees visit the temple every year.

Mathura-Vrindavan: Being the birth place of Lord Krishna, Mathura is celebrated as one of the most sacred places in Hinduism. Only 15 miles from Mathura, Vrindavan is famous for bhaktas and sagas related to Lord Krishna.

Places to see: Shri Krishna Janmbhoomi, Ranghabhumi, Iskon Temple, Radha Vallabha Temple, Mathura Krishna Balrama Mandir, Radha Damodara Temple, Shahji Temple, Jami Masjid, etc.

Tirupati: On Tirumala, the exquisitely carved gold gopurams of the Lord Venkateshwara Temple is placed. It is one of the richest temples in the country. This temple is shining in the sun, is a unique piece of Dravidian art.

Places to see: Kailasanatha Kona, Chandragiri Fort, etc.

Amarnath: The temple is reported to be around 5000 years old and is a popular pilgrimage destination for Hindus. Located in the 'Paradise', Kashmir, an ice Shivalinga (along with two other ice formations representing Ganesh and Parvati) is the presiding deity of this place. It is said to grow and shrink with the phases of the moon, reaching its height during the summer festival.

Ayodhya: Lying on the banks of the river Ghagra, this archaic city is believed to be the birth place of Lord Rama, the 7th incarnation of Lord Vishnu.

Places to see: Ram Janmabhumi, Treta Ka Mandir, Hanuman-Garhi Temple, Ramkot, etc.

Haridwar: Haridwar is one of the principal holy cities of India. During the Kumbha Mela, thousands of pilgrims troop into this place for ritual bath. Each evening, the sunset is celebrated with the traditional aarti at Har ki Pauri.

Places to see: Chandi Devi, Bharat Mata Temple, Mansa Devi Temple, Vaishno Devi Temple, Daksh Mahadev Temple, etc.

Yamunotri: Yamunotri is the first halt on the way to Chardham yatra. From this holy place, the sacred river Yamuna originates. It is also a much popular Hindu pilgrimage.

Madurai: Madurai is one of the pilgrimage centers of India located in the state of Tamil Nadu. The city of Madurai is said to be around 2500 years old and lies on the banks of Vaigai river. It is also famous as a temple land. One gets to see the best of temple architecture here which showcase a distinctive aesthetics of various rulers.

Places to see: Meenakshi temple, Koodal Azhagar temple, Tirumalai Nayal Palace, Gandhi Museum.

Shirdi: 122 km from Nashik is the abode of one of Maharashtra's most revered saint—Sai Baba of Shirdi. Popularly known as the *Child of God*, Sai Baba preached tolerance towards all religions and the message of universal brotherhood. Devotees start queuing up in the early hours of dawn to catch a glimpse and seek the blessings of the life-size statue of Sai Baba.

Places to see: Dwarkamani Mosque, Dhuni, Gurusthan, Kandoba Temple, Shani Mandir, Narsimha Mandir, Changdev Maharaj Samadhi, Sakori Ashram.

Golden Temple: The Golden Temple, located in the city of Amritsar, is a place of great beauty and sublime peacefulness. Originally a small lake in the midst of a quiet forest, the site has been a meditation retreat for wandering mendicants and sages since deep antiquity. The Golden Temple is not only a central religious place of the Sikhs, but also a symbol of human brotherhood and equality. Everybody, irrespective of cast, creed or race can seek spiritual solace and religious fulfillment without any hindrance. It also represents the distinct identity, glory and heritage of the Sikhs. The Golden Temple has a unique Sikh architecture. Built at a level lower than the surrounding land level, the Gurudwara teaches the lesson of egalitarianism and humility. The four entrances of this holy shrine from all four directions, signify that people belonging to every walk of life are equally welcome.

Kailash Manasarovar Yatra: Mount Kailash is claimed to be the apex of the Hindu religious axis, is also one of the highest mountain in Tibet at 22000 feet. A mythological story says that Lord Shiva stays here with his whole family including his wife Goddess Parvati and children Lord Ganesha and Lord Kartikiya and the other Shiv Ganas like Nandi and others.

Ajmer Sharif: Ajmer is the place where only faith matters and that faith is represented by the paradoxical delicacy of the threads tied to the shrine of a Sufi saint. Each thread tied is a wish and when it is granted, and it inevitably is, the person who tied it returns to untie a thread.

WILDLIFE TRAVEL

The forests of India are wild, magnificent and diverse. Strewn across the country, the wild trails of India can be found in varied landscapes, from the high Himalayas in the north to the rich Terai region in its foothills, from salt marshes of Kutch in the west to the mangroves in the east and the unexplored and impenetrable greens in northeast, to the fascinating landscape of the peninsular region in the south.

While the pride of Indian forests is the Royal Bengal tiger, these wild tracts are home to a thousand other species of fauna, avifauna, reptiles, bees and butterflies. Several endangered species such as the great Indian bustard, Bengal florican, paradise flycatcher, barasingha, Himalayan tahr, gharial, blue sheep, kiang, red panda call it home.

India boasts to be a haven over 390 mammals, 455 reptiles, 210 amphibians, 1230 bird species. The favorable climate and topography of India support several national parks and wildlife sanctuaries of India. See the king of the jungle in his own terrain, hiding behind the tall vegetation besides the lake and waiting for its prey.

Major National Parks and Wildlife Sanctuaries of India

North India

- Dachigam National Park – Jammu & Kashmir
- Hemis National Park–Jammu & Kashmir
- Pin Valley National Park–Himachal Pradesh
- Great Himalayan National Park–Himachal Pradesh
- Valley of Flowers National Park–Uttarakhand
- Corbett National Park–Uttarakhand
- Rajaji National Park–Uttarakhand
- Dudhwa National Park–Uttar Pradesh
- National Chambal Wildlife Sanctuary–Uttar Pradesh
- Desert National Park–Rajasthan
- Ranthambore National Park–Rajasthan
- Keoladeo Ghana National Park–Rajasthan
- Sultanpur National Park–Haryana

Central India

- Bandhavgarh National Park – Madhya Pradesh
- Kanha National Park – Madhya Pradesh
- Pench National Park – Madhya Pradesh
- Panna National Park – Madhya Pradesh
- Kanger Ghati National Park–Chhattisgarh
- Achanakmar Wildlife Sanctuary - Chhattisgarh

East India

- Betla National Park–Jharkhand
- Hazaribag National Park–Jharkhand
- Simlipal National Park–Odisha
- Chilka Wildife Sanctuary–Odisha
- Bhitarkanika National Park–Odisha
- Sunderbans National Park–West Bengal
- Jaldapara National Park–West Bengal
- Neora Valley National Park–West Bengal

Northeast India

- Khangchendzonga National Park–Sikkim
- Kaziranga National Park–Assam
- Manas National Park–Assam
- Hollongapar Gibbon Wildlife Sanctuary–Assam
- Nameri National Park–Assam
- Namdapha National Park–Arunachal Pradesh
- Eaglenest Wildlife Sanctuary–Arunachal Pradesh
- Pakke Tiger Reserve–Arunachal Pradesh
- Keibul Lamjao National Park–Manipur
- Murlen National Park–Mizoram

West India

- Gir National Park–Gujarat
- Velavadar National Park–Gujarat
- Marine National Park–Gujarat
- Kutch Desert Wildlife Sanctuary–Gujarat
- Wild Ass Sanctuary–Gujarat
- Tadoba-Andhari National Park–Maharashtra
- Navegaon National Park–Maharashtra

South India

- Bandipur National Park–Karnataka
- Nagarhole National Park–Karnataka
- Mudumalai National Park–Tamil Nadu
- Point Calimere Bird Sanctuary–Tamil Nadu
- Silent Valley National Park–Kerala
- Periyar National Park–Kerala
- Sri Venkateshwara National Park–Andhra Pradesh

Andaman and Nicobar Islands

- Mahatma Gandhi Marine National Park–Andaman and Nicobar Islands
- Campbell Bay National Park–Andaman & Nicobar Islands

Bharatpur Bird Sanctuary: If you are an avid bird watcher then you should not miss the Bharatpur Bird Sanctuary. 176 km from Delhi, this sanctuary is today a heaven for winged creatures, which have chosen the sheltered protection of the park to breed in.

Attractions: Cranes, Pelicans, Geese, Ducks, Eagles, Hawks, Shanks, Stints, Wagtails, Warblers, Wheatears, Flycatchers, Buntings, Larks and Pipits, etc.

Corbett National Park: Seven hour drive from the Delhi, Corbett National park is placed in the foothills of the Himalayas in Uttarakhand. This Jim Corbett's land of man-eaters provides an unforgettable experience to the nature lovers.

Attractions: Tigers, Gharial Crocodile and the 'Mugger' Crocodile, Himalayan Palm Civet, Indian Gray Mongoose, Common Otter, Blacknaped Hare and Porcupine, Elephants, etc.

Ranthambore National Park: This one-time hunting preserve of the Maharajas of Jaipur was also the venue for royal hunting parties. Today, it is the best place in the world to see a tiger.

Attractions: Tiger, Leopard, jungle Cat, Striped Hyena, Sloth Bear, Patel, Sambar, Spotted Deer, Nilgai, Chowsingha, Wild Boar, Indian Pangolin, Small Indian Mongoose, Lesser Spotted Eagle, Pallas's Fish Eagle, Eurasian Eagle Owl, Brown Fish Owl, Collared Scops Owl, etc.

Sunderbans National Park: Stretched over an area of 1330 sq. km., Sunderbans is the largest habitat of the Royal Bengal Tiger in the world. You can find around 300 of them roaming and swimming in that area.

Attractions: Stork, Kingfishers, Eagles, White Ibis, Swamp Francolin, Asian Dowitcher, White-bellied Sea Eagle, Purple Heron, Egrets, Brown Fish Owl, Osprey, Peregrine Falcon, Northern Pintail, Little Porpoise, Indian Fox, Fishing Cat, Common Grey Mongoose, Indian Flying Fox, Pangolin, Small Indian Civet, Rhinoceroses, Indian Python, etc.

Kaziranga National Park: Kaziranga National Park is home to the 1000 one-horned Rhinos who are considered among the endangered species. This population of this species which is found here is known to be the highest in number.

Attractions: Indian Bison, Hog Deer, Sloth Bears, Tigers, Leopard Cats, Jungle Cats, Otters, Hog Badgers, Capped Langurs, Hoolock Gibbons, Wild Boar, Jackal, Wild Buffalo, Pythons, etc.

Manas National Park: In the Manas National Park, one can enjoy a wild experience while relishing some of the best of natural beauty. This place is a dear home to 20 highly endangered species like Panda, Hispid Hare, Pigmy Hog and Golden Langur.

Attractions: Tigers, Elephants, Rhinoceros, Wild Buffalo, Wild Boar, Sambhar, Swamp Deer, Hog Deer, Riverchats, Forktails, Cormorants, Indian Hornbill, Pied Hornbill, etc.

Bandhavgarh National Park: White tigers of Rewa are the main attractions here. As this park is set amidst the Vindhyan ranges, you will love to traverse the park on elephant back.

Attractions: Nilgai, Chausingha, Chital, Chinkara, Wild Boar, Jackal, Peacock, Paradise fly catchers, Jungle Fowl, Golden and Black headed Orioles, Yellow Ioras, Red-vented Bulbuls, Blue Jays, Purple Sunbirds, Green Barbets, etc.

Gir National Park: Gir National Park is placed in the West Indian state of Gujarat. It is renowned as the sole habitat of the Asiatic Lion in India. One can even find Leopards along with the lions there.

Attractions: Sambar Deer, Chital Spotted Deer, Nilgai Antelope, Chowsingha Four-Horned Antelope, Chinkara Gazelle, Wild Boar, Langur Monkey, Jackal, Hyena, Paradise Flycatcher, Bonneli's Eagle, Crested Serpent Eagle, Woodpeckers Flamingo, etc.

Sariska Tiger Reserve: Barely one hours drive from Bharatpur is the Sariska Tiger Reserve sited. Sariska Tiger Reserve was once the royal reserve of the Alwar rulers. Sariska is a picturesque park, with plenty of nilgai and other deer species.

Attractions: Pea Fowl, Gray Partridge, Quail, Sand Grouse, Tree Pie, White Breasted Kingfisher, Golden-backed Woodpecker, Crested Serpent Eagle, Great Indian horned Owl, Sambhar, Chital, Wild Boar, Hare, Nilgai, Civet, Four-horned Antelope, Gaur (Indian Bison), Porcupine, etc.

Bandipur National Park: The 874 sq km Bandipur Park is also a tiger reserve. Bandipur is one of the finest habitats of the Asian elephants.

Attractions: Bonner Macaque, Nilgiri Langur, Tiger, Wild Boar, Chital Gaur, King Cobra, Common Cobra, Python, Adder, Viper, Rat Snake, Water Snake, Marsh Crocodile, Lizard, Chameleon, Monitor Lizard, Frog, Tree frog, Toad and Tortoise, Tiger, Four horned Antelope, Gaur, Elephant, Panther, Sloth bear, Crocodiles, Mouse deer, Python, Osprey, etc.

BEACH TRAVEL

Surrounded by glorious mountains and blessed with crystal clear waters, and a wonderful marine life and spread across more than 7500 km along coastline, Indian beaches offer wide choices for all kinds of travelers. Indian beaches attract tourists in galore, with soul-warming sun, crystal-clear waters, and fragrant sea air. Be it beach resorts, beach shacks and small restaurants serving yummy but inexpensive sea foods, or beach bazaars providing shopping delights, beach activities or isolated retreats perfect for un-winding—Indian beaches have it all for you.

Goa: *Pearl of the Orient* and *Tourist Paradise* are certain titles that have been bestowed on the land of captivating beauty, Goa. The place is located on the western coast of India. The magnificent scenic beauty, the architectural splendor is the real delights of Goa. The unique history, rich culture and picturesque scenery are the major crowd puller of this place. Apart from the beaches and seas there is so much to see and do in Goa. Goa is indeed a hot spot tourist destination in India. It is a perfect heaven for all those who look for peace and relaxation. The sun-baked beaches, sensuous silvery sands, fabulous assortment of flora and fauna, rich cultural heritage, Gothic churches, forts, ferry rides and not to forget the exciting Goa Carnival. A sip of Feni will also transfer you an entirely different world. Goa is said to be the hub of beaches.

Attractions: Anjuna Beach, Calangute Beach, Baga Beach, Colva Beach, Majorda Beach, Se Cathedral church, The holy shrine of Basilica of Bom Jesus in Old Goa.

Marina Beach: The second largest beach in the world, Marina Beach is a pictorial coast sited along 12 km long seashore in *Chennai*. Blonde sand, a shining blue sea and open avenue create that sight which is not to be missed.

Attractions: Aquarium, Anna and MGR Samadhis, University of Madras, Senate-House, Chepauk Palace, Presidency College, PWD office, Ice House, Beach Market, etc.

Digha Beach: This beach of *West Bengal* is breathtakingly beautiful and a true delight for all the nature lovers. Sun, sea, surf and the sand, in nutshell, is Digha beach.

Attractions: Shankarpur, Chandaneswar, etc.

Kovalam Beach: Owing to its natural beauty of amazing Arabian Sea, Kovalam offers a picture perfect holiday. It is located in *Kerala*.

Attractions: Sunbathing, swimming, herbal/Ayurvedic body toning massages, cultural programs, Catamaran cruising, etc.

Corbyn Cove: For all the adventurers and romantic couples this *Andaman and Nicobar Islands'* beach is an isolated paradise. The beach provides array of entertainments, activities, fun and relaxation, which include strolls on the beach, forest walks, scuba diving, wind surfing, etc.

Attractions: Elephant Safari, Trekking, Lagoon Cruises, Island Camping, Cellular Jail, Ross Island, Swimming, Scuba Diving, Viper Island, etc.

CULTURE TRAVEL

Pushkar Fair: This fair occurs every year for 12 days in the month of Oct/Nov at Pushkar, Rajasthan. This cultural, trade and religious fair is an attractive and lively spectacle with Rajasthani men and women in their colorful traditional attire, saffron-robed and ash smeared Sadhus (holy men) and thousands of bulls, cows, sheep, goats, horses and camels in richly decorated saddles. Perhaps the largest cattle fair in the world, it attracts more than 1,00,000 people, from all over India and abroad. Apart from the religious rituals and trading, people participate in a number of cultural and sporting events.

Taj Mahotsav: This 10-day extravaganza of art, craft, culture and cuisine, known as the Taj Mahotsav started in year 1992 and since then its grandeur has reached to greater heights. This carnival is organised by UP Tourism. It is a culturally vibrant platform that brings together the finest Indian crafts and cultural nuances. It is celebrated from 18th to 27th February every year at Shilpgram near the Eastern gate of 'Taj Mahal'. Renowned artists, musicians, dancers and chefs put their best foot forward to make you experience a spectacular show. About 400 legendary artisans from different parts of the country get an opportunity to display their exquisite works of art. To name a few among them are the wood/stone carvings from Tamil Nadu, Bamboo/cane work from North East India, Paper mash work from South India and Kashmir, the marble and zardozi work from Agra, wood carving from Saharanpur, brass wares from Moradabad, hand made carpets from Bhadohi, Pottery from Khurja, Chikan work from Lucknow, silk & zari work from Banaras, shawls & carpets from Kashmir/Gujarat and hand printing from Farrukhabad and Kantha stitch from West Bengal, etc.

Surajkund International Crafts Fair (Mela): It is one of the most famous fairs, organized every year in Surajkund, Faridabad, from 1st to 15th Feb, by Haryana Tourism Department to promote handicrafts and handloom items. Many talented artists come from India and across the World to showcase their handlooms, handicrafts, traditional handicraft items of their respective states. Not only this, but also to entertain and amuse large audience at Surajkund Mela many shows, cultural events, folk dance shows, folk music shows, are also organized at the venue. With authentic fragrances and rich rustic flavors the fair is a wonderful opportunity to accompany the connoisseurs of art and crafts and also taste the Indian cuisines to finally pamper your taste buds. Besides, in the sheer rural ambience it makes an appeal to come and get exposed to some great talents involved in paintings, wood stock, textiles, pottery, ivory work, stonework, terracotta and grass work—all highlighting the vibrant cultural range of their native state.

Khajuraho Dance Festival: Khajuraho Dance Festival is celebrated at a time when the hardness of winter begins to fade and the king of all seasons, spring, takes over. Started regularly since 2002, this weeklong festival has already become legendary with its outlandish classical dance performances presented in a dreamlike setting of splendidly illuminated temples. The best classical dancers from all across India give their performances in the open-air auditorium in front of the Chitragupta Temple dedicated to the Sun God and the Vishwanatha Temple dedicated to Lord Shiva. Some of the best artists and performers that have marked themselves in their fields come from the various states of India to participate in the festival and the performances including some of the best known dance styles such as the intricate footwork of Kathak, highly stylized and sophisticated Bharatnatyam, soft lyrical temple dance of Odissi, the dance dramas of Kuchipudi, Manipuri, the dance of rare and ancient civilization and Kathakali stage fights with elaborate masks. Recently, modern Indian dances have also found their place in the Khajuraho Dance festival. Along with the dance performances one can also see a number of craftsmen trading off their indigenous arts and crafts to the visitors.

Goa Carnival: The best time to travel to Goa is during February because this is the time when one witnesses Goa getting decked up for the carnival. Goa carnival is an eagerly awaited occasion where one forgets his worries and tensions and simply plunges into merriment and joy-making for three whole days. Goa carnival is all about fun and frolic. Colorful processions and lavish floats parade the streets. Singing, dancing and masked people mark the uproarious and flamboyant Goan celebration. Most of the countries have carnivals but Goa carnival stands apart as the people of Goa have inculcated different items in the carnival that makes us experience the different shades of Goa. Street plays, songs, dances, and impromptu travesties jeer and display their dances and songs before a crazy and amendable audience. The popular cradle songs and nursery rhymes with colorful attires worn make the atmosphere as well streets more jovial. Innumerable competitions and cultural functions are held right through the three days. The contestants wear colorful costumes and designer masks. In such fun-filled surroundings people smear color on each other, instead of the flour, eggs, fruit and water that used to be used in past. The carnival in Goa had its birth during the era of King Momo. He ushered in the Goa carnival just before the Lent season (Lent is the period of fasting and penance in the Christian calendar and corresponding somewhat to the Mohammedan fast before Ramzan Id). This festival of Goa usually starts off on Sabado Gordo (Fat Saturday) and concludes on Shrove Tuesday (Fat Tuesday)—the eve of Ash Wednesday, which is the first day of the season of Lent.

Kala Ghoda Arts Festival: Kala Ghoda Arts Festival is one of the most significant cultural fairs that enrich the commercial capital of India—Mumbai. This vibrant arts festival is held annually at Kala Ghoda area in Southern Mumbai, in the month of February. The festival is being conducted from the first Saturday of the month till the next Sunday since 1999 by the Kala Ghoda Association, a non-profit seeking organization.

For the past 15 years, Kala Ghoda Arts Festival has fostered the spirit of arts and culture in the hustle and bustle of India's financial hub. Poets, artists, painters, dancers, musicians and craftsmen form a city of art inside the real city. The mime-artists, open-air performances, street food, pavement art and street musicians bring a carnival-like atmosphere to the busy metro.

Shekhawati Fair: Shekhawati is a cultural festival celebrated for two days in the month of February every year, at the Shekhawati region, well-known for its painted havelis, lavishly constructed mansions and patio, in Rajasthan. The festival aims to provide an extensive view of Shekhawati and its culture for the tourists. This festival is jointly organized by the State Department of Tourism and the Morarka Foundation. The festival is celebrated in several regions that share the cultural likeness and include Nawalgarh, Sikar, Jhunjhunu and Churu. It is at this occasion that the people celebrate their annual productive output. The festival as such marks an equal response throughout, from rural to urban segments of population. The programs include a one day tour of the region, camel and jeep safaris, farm visits, rural games, cultural programs, haveli competitions and fireworks.

ADVENTURE TRAVEL

If you are one of those, whom the spirit of adventure has always lured and want to do something off beat, then India is the place you need to head to. A terrain stuffed with golden sands, twisting rivers, thriving hills, spilling waterfalls and slopping deserts India has enormous prospects of adventurous activities including safaris, water sports, aero sports and many more. Not only India has something for every level of competence—the beginner and the expert, but also prices here are extremely low by global standards.

Mountaineering in India: India, with its tough topography tests the physical strength of a person to its limits. Replete with several towering peaks it challenges even the best of mountaineers. The best time for mountaineering in India is roughly from July to mid-October.

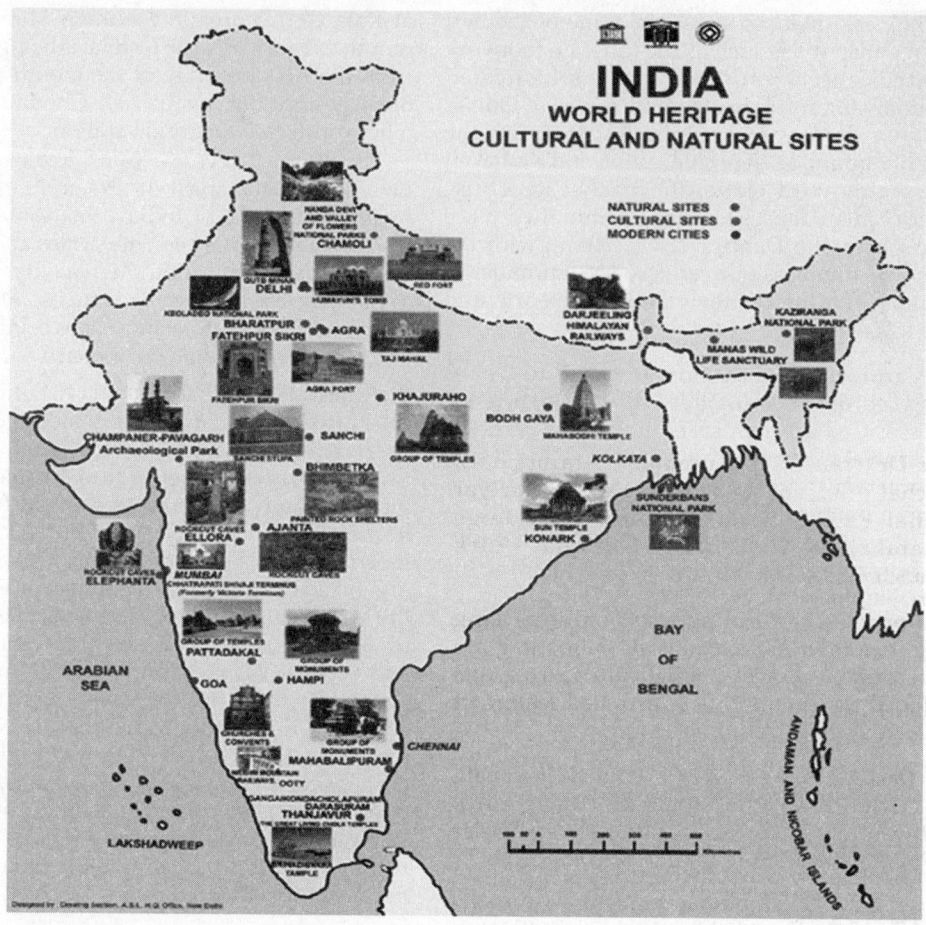

Major Destinations: Himachal Pradesh, Garhwal and Kumaon regions of Uttar Pradesh, Ladakh region of Jammu and Kashmir, Northeast India, Sikkim, Lahaul and Spiti as well as the Kullu valley, etc.

Trekking in India: For you India has untold trekking trails where you can always challenge your own self. During your trekking tour you can explore historical structures, ancient places, monasteries and many other unexplored attractions.

Major Destinations: Ladakh-Zanskar Via Lahaul, Manali to Beas Kund, Garhwal Trekking, Himachal-Manikaran To Spiti, Dodi Tal, Khatling Saharatal Trek, Gangotri Nandvan Trek, Kinner-Kailash Parikrama, Manimahesh Chui Yatra, Jagatsukh To Base of Deo Tibba, Adi Kailash Trek, Hemkund Trek, etc.

Water Sports in India: When it comes to water sports, Indian Beaches are completely exclusive in their own ways. For all the water babies they have plenty of activities like sailing, boating, fishing, angling, yachting, wind surfing, snorkeling, diving, etc.

Major Destinations: Ladakh and the Zanskar and Chenab in Kashmir. Sutlej, Beas (in Himachal Pradesh), Teesta (in Sikkim), Ganga, Yamuna, Brahmaputra, Kaveri, Narmada, Yamuna, Nagoa beach, beaches of Goa, Andaman and Nicobar Islands and Lakshadweep, etc.

Camping: Camping is known as a cluster of activities combined into one. In camping one can enjoy wildlife safari, trekking fishing, jungle walk, photography, picnicking, while enjoying complete solitude in the lap of the nature.

Major Destinations: Kumaon (Uttaranchal), Himachal Pradesh, Rajasthan, Kerala, Periyar National Park (Kerala), Kabini River Loge (Karnataka) and Corbett National Park (Uttaranchal), etc.

Aero Sports: Para gliding, para sailing, hang gliding, hand gliding and hot air ballooning are the most famous and most sought after aero sports in India. These sports give unmatched feeling of flying like a free bird.

Major Destinations: Bangalore, Jaipur, Guwahati, Agra (Uttar Pradesh), Pushkar (Rajasthan), Baneshwar (Rajasthan), Nagaur (Rajasthan), Shelar Hill, Pavna, Towerl Hill, Billing, Kanifnath, etc.

HEALTH TRAVEL

Are you looking for an opportunity for your prolonged ailment that goes beyond the expensive harmful chemical treatments? Then you can find your solution in Indian Ayurveda, yoga and meditations, which stress on the prevention of a disease rather than curing it. These 5000-year-old traditional sciences, offer unending list of wholesome treatments for all your problems whether you are suffering from chronic disease or you need just a simple rejuvenation after an exhausting week, without burning a hole in your pocket.

Ayurveda: Ayurvedic treatments like Ayurvedic Facial, Ayurvedic Scalp Treatment, Mother Earth Science, Yemana Ocean Therapy and Neurotherapy will give you ultimate solutions for all your maladies. The rejuvenation therapy of Ayurveda helps to revitalize both the body and the soul. It provides a wholesome healthy living exclusive of any side effects.

Major Destinations: Uttaranchal, Kerala, Goa, Rajasthan, etc.

Yoga and Meditation: Yoga and meditation synchronize mind, body, heart and soul. It helps to attain elevated spiritual insight and tranquility. It ensures a life blessed with an elevated physical, mental, emotional spirit. The quick effect of Yoga will make you feel completely at peace with your inner self.

Major Destinations: Kerala, Uttaranchal, Bihar, Parmarth Niketan, Sivanand Ashram, Yoga Niketan, Omkaranand Ashram, Vanprastha Ashram, Ved Niketan Dayanand, Vedanta Ashram and Vanmali Gita yogashram, Aurovila, etc.

GOURMET TRAVEL

The unforgettable aroma of India is not just the heavy scent of jasmine and roses on the warm air; it is also fragrance of the spices which are an indispensable part of Indian cooking. The variety of Indian cooking is immense, it is colorful and aromatic. Being so diverse in nature, each region has its own cuisine with its very own preparing style. The culinary delights of India are inexpensive even in the top class hotels. No wonder it is the third most popular cuisine in the world.

Northern cuisine: Dairy products like ghee, yogurt are used fairly extra in the North Indian cuisines. Mughlai, tandoori foods are two dishes which are appreciated by the world foodies. Shammi kebabs, Reshmi kabab, Biriyani are rich, deliciously spiced and liberally sprinkled with nuts and saffron.

Southern cuisine: In the south, curries are mainly vegetable and inclined to be hotter. Specialties to look out for are Bhujiya, Dosa, Idli and Sambar, dumpling with pickles.

Bengali cuisine: Fish is the main feature of Bengali cuisine. Traditional Bengali curries and sweet dishes are difficult to find outside Bengal.

Western cuisine: The Western India cuisines offer you a wide range of fish dishes. The influences of Parsi and Saraswat cuisine also can be seen in the important food items.

Sweet dishes: Be it marriage, any religious festival, family functions, simply anything one cannot visualize Indian life without sweets. Kaju barfi, halva, kheer, rasgulla, laddu, sandesh, rashmalai, firni, malpoa—every region has its own sweets.

LUXURY TRAIN TRAVEL

A luxury train tour lets visitors explore Indian people, culture, traditions in a very royal way. Luxury trains like Palace-on-Wheels, Deccan Odyssey, etc. have now become most admired attractions of Indian tourism. These luxury trains fairytale journey will take you to that sojourn, where you yourself will feel like a royal of the yesteryear. During the travel period of yours, while traveling to the royal destinations in India, you will have a complete imperial value for money experience.

Palace-on-Wheels: First luxury train in India, Palace-on-Wheels started as a joint venture of Rajasthan Tourism Development Corporation and Indian Railways in 26 January 1982. This royal train is inspired by the private carriages of erstwhile kings and rulers of the yore.

A train with historic charm and modern conveniences. A royal train beckoning you to step aboard for a week of splendor. It takes you on a royal odyssey. Bringing alive the luxury of princely travel of the age of the Raj, the Palace-on-Wheels has also been designed to suit the modern needs with central air-conditioning, soothing four channel music, intercom telephone system, a separate pantry and lounge in each coach, comfortable with privacy of attached bathrooms, two specialty restaurants—'Maharaja' and 'Maharani' serving mouth watering Continental, Chinese, Indian and Rajasthani cuisines that would definitely make way to your heart through your stomach. A well-stocked bar serves wine, liquor and spirits of the Indian and international make, to keep you in 'High Spirits'. A pampering spa saloon is the recent addition in the train with a variety of therapeutic massages on offer to relax and rejuvenate the guests. Expert therapists offer specialized services to the guests; both therapeutic and beauty treatments.

The 14 coaches of the Palace-on-Wheels with rich decor that evoke the age of Rajput chivalry, are named after former Rajput states—Kota, Jaipur, Udaipur, Jaisalmer, Jodhpur, Bikaner, Alwar, Sirohi, Kishangarh, Bundi, Dungarpur, Bharatpur, Jhalawar and Dholpur.

As most of the traveling is done at night, your days are left free—for discovering the delights of

Rajasthan and Uttar Pradesh covering the itinerary of Delhi-Jaipur-Sawai Madhopur-Chittaurgarh-Udaipur-Jaisalmer-Jodhpur-Bharatpur-Agra-Delhi.

Royal Rajasthan on Wheels: After successfully running Palace-on-Wheels for more than two and a half decades, the Indian Railways in association with Rajasthan Tourism Development Corporation Ltd. launched a new train 'Royal Rajasthan on Wheels' with added facilities compared to Palace-on-Wheels which itself is the benchmark in its segment, on the itinerary of Delhi, Jodhpur, Udaipur, Chittaurgarh, Sawai Madhopur, Jaipur, Khajuraho, Varanasi and Agra.

Deccan Odyssey: The Deccan Odyssey is more than a mere luxury train cruise. A lavishly decorated train that transports its guests on an unforgettable sojourn of a land shining in legions' grandeur, serene beaches, magnificent forts and palaces, and experiencing heavenly tales etched in colossal rocks in Mumbai- Kudal-Karmali-Verna-Madgaon-Vasco-Kolhapur-Pune-Daulatabad-Aurangabad-Bhusawal-Deolali-Mumbai. Facilities onboard include channel music, intercom, CD/mp3 player, wall-to-wall carpeting, money exchange/credit card facility, safety & security arrangements, health spa, bar, conference equipments. The journey is also available on the itinerary of Mumbai, Chalisgaon, Bhusawal, Udaipur, Sawai Madhopur, Jaipur, Bharatpur, Agra, Delhi and back.

The Golden Chariot: The lands once ridden by formidable armies is now traversed by the world's unique luxury train. The Golden Chariot beckons you to discover worlds that are enriched with history and culture. Recline and relax as the magnificent scenery unfolds outside your window with luxury on the inside, thus far reserved for royalty. The Golden Chariot presents a fine balance between a glorious yesterday and an omnipresent tomorrow in a magical land ... Karnataka, Goa, Tamil Nadu, Kerala and Puducherry on the following itineraries.

Itinerary I-Bangalore, Mysore, Hassan, Hospet, Badami, Goa, Bangalore.

Itinerary II-Bangalore, Chennai, Puducherry, Tanjavur, Madurai, Nagercoil, Trivandrum, Ernakulam, Bangalore.

Maharajas' Express: A lifetime experience, the journey on Maharajas' Express brings one closer to bygone era of imperial India. It offers several unique features from having most spacious cabins to having personal valets attached to each cabin to take care of the guests. State of the art facilities like live TV, Wi-Fi internet, individual temperature control in each cabin, environment friendly toilet systems, CCTV cameras in public areas to ensure security, direct dialing telephones to make international calls, make the experience all the more enjoyable. With hair dryers, electronic safe deposit boxes in each room, bath tubs and mini bars (in suites and presidential suite), the train offers almost everything that one would imagine to travel in luxury. Savoring the culinary delights, sitting in either of the two restaurants—'Mayur Mahal' and 'Rang Mahal', one can have a glimpse of the picturesque hinterland from the large windows specially created for the Maharajas' Express. The guests can relax in the Safari Bar or the Rajah Club and choose from the selection of choicest wines and liquors. With 4 categories of accommodation to choose from, guests can select the cabin they would wish to spend their sojourn—all with en suite facilities. The train can accommodate 88 passengers. There are 14 guest carriages featuring 43 cabins in total, which include 20 Deluxe Cabins, 18 Junior Suites, 4 Suites and one truly classical Presidential Suite.

It operates on a number of itineraries, mainly

- Delhi- Agra-Sawai Madhopur-Jaipur- Bikaner-Jodhpur-Udaipur- Mumbai
- Mumbai- Bhusawal-Udaipur-Jodhpur-Bikaner-Jaipur-Sawai Madhopur-Fatehpur Sikri-Agra-Delhi
- Delhi-Jaipur-Sawai Madhopur-Fatehpur Sikri-Agra-Gwalior-Khajuraho-Varanasi-Lucknow-Delhi
- Delhi-Agra-Sawai Madhopur-Jaipur-Delhi (operates on Wednesday)
- Delhi-Agra-Sawai Madhopur-Delhi (operates on Sunday)

RESTRICTED TOURIST AREAS

Some areas in India are designated as Restricted Areas where tourists require authorization to visit. Tourists need to ensure that they get the appropriate documentation before visiting such places.

Listed below are the provinces in India for which foreign tourists require a permit to visit. Also listed are the places that have the authority to issue these permits.

Manipur

Loktak Lake, Imphal, Moirang INA Memorial, Keibul Deer Sanctuary and Waithe Lake

Authority that can grant permit:

- All Indian Missions abroad
- All FRROs and MHA
- State Government of Manipur (Home Commissioner, Manipur)

Mizoram

Vairangte, Thingdawl and Aizawl

Authority that can grant permit:

- Home Commissioner, Government of Mizoram, Aizawl
- All FRROs at Delhi, Mumbai and Kolkata
- Chief Immigration Officer, Chennai
- All Indian Missions abroad

Arunachal Pradesh

1. Itanagar, Ziro, Along, Pasighat, Miao, Namdapha and Sujesa (Puki) Bhalukpong

Authority that can grant permit:

- Home Commissioner, Government of Arunachal Pradesh, Itanagar
- All FRROs at Delhi, Mumbai, Kolkata
- Chief Immigration Officer, Chennai
- All Indian Missions abroad

2. Gangtok, Rumtek, Phodang, Pemayangtse Khecheperi and Tashiging

Authority that can grant permit:

- MHA, All FRROs
- All Indian Missions abroad
- Immigration Officers at airports at Mumbai, Kolkata, Chennai and New Delhi
- Chief Secretary/Home Secretary/Secretary (Tourism)
- Government of Sikkim, Gangtok
- Inspector General of Police, Government of Sikkim, Siliguri
- Deputy Directors (Tourism), Sikkim Government, New Delhi
- Assistant Resident Commissioner; Government of Sikkim, Kolkata
- Tourism Officer, Rangpo,
- Deputy Commissioner, Darjeeling
- Deputy Secretary/Under Secretary Home Department, Government of West Bengal, Kolkata

Zongri (West Sikkim)

Authority that can grant permit:

MHA and all issuing authorities of Government of Sikkim and its representatives at Kolkata, Siliguri and Rangpo

Tsangu (Chhangu Lake in East Sikkim)

Authority that can grant permit:

Home Secretary, Government of Sikkim.

Mangan, Tong, Singhik, Chungthang, Lachung and Yumthang

Authority that can grant permit:

• Home Secretary, Government of Sikkim

• Secretary (Tourism), Government of Sikkim

Kashmir

1. Khaltse Sub-division (Drokahpa Area) Khaltse-Dunkhar-Sroduchan Hanudo-Biana-Dha

2. Nubra Sub-division: (a) Leh-Khardung La-Khalsar-Tirit up to Panasik (b) Leh-Khardung La-Khalsar up to Hunder (c) Leh-Sabo-Digar La-Digar-Labab-Khungru Gampa-Tangar

3. Nyona Sub-division: (a) Leh-Upshi-Chu-sathang-Mahe-Puga-Tso-Moari Lake/Kozok (b) Leh-Upshi-Debring-Puga-Tso-Moari Lake/ Korzok (c) Leh-Karu-Chang La-Durbuk-Tangtse-Lukung-Spanksik. (d) Pangong Lake up to Spanksik

Authority that can grant permit:

MHA / District Magistrates concerned

Andaman Nicobar Islands

Municipal Area, Port Blair, Havelock Island, Long, Neil Island, Jolly Buoy, South and North Cinque, Red Skin, Mayabunder, Diglipur, Rangat, Mount Harriet, Madhuban

Authority that can grant permit:

• MHA and All FRROs

• All Indian Missions abroad

• Immigration Officer, Port Blair

Himachal Pradesh

Poo-Khab-Sumdho-Dhankar-Tabo-Gompa-Kaza, Morang-Dabling

Authority that can grant permit:

• MHA /Government of HP/DM/SDM concerned/ITBP

• Special Commissioner (Tourism)

• Resident Commissioner, Government of HP, New Delhi

• Director General of Police, HP Shimla

Uttar Pradesh

1. Nanda Devi Sanctuary, Niti Ghati and Kalindi Khal in Chamoli, Uttar Kashi Districts

2. Adjoining areas of Milam Glacier

Authority that can grant permit:

• MHA/Government of UP DM/SDM concerned ATBP

• MHA/Government of UP/DM/SDM concerned/ITBP

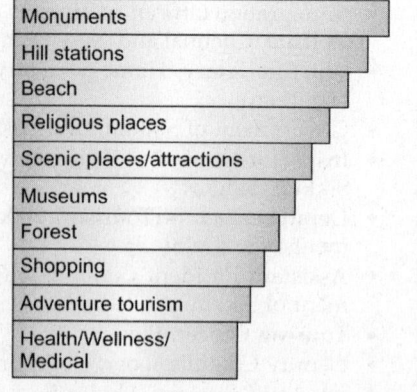

Domestic Tourists

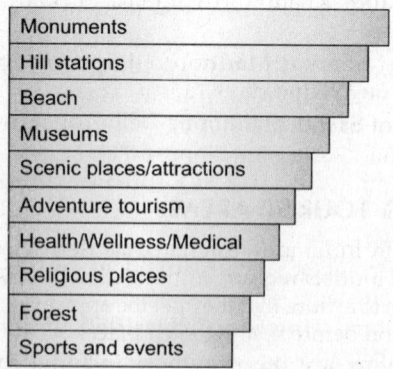

Foreign Tourists

Top 10 preferred Indian Tourism Attractions

Guidelines for Approval of Hotels at Project Stage and Classification and Re-classification of Hotels

1. Hotels are an important component of the tourism product. They contribute in the overall tourism experience through the standards of facilities and services offered by them. With the aim of providing contemporary standards of facilities and services available in the hotels, the Ministry of Tourism, Government of India has formulated a voluntary scheme for classification of operational hotels which will be applicable to the following categories:

   i. **Star Category Hotels**: 5 Star Deluxe, 5 Star, 4 Star, 3 Star, 2 Star and 1 Star

   ii. **Heritage Category Hotels**: Heritage Grand, Heritage Classic and Heritage Basic

2. The guidelines have been aimed to provide a higher level of services in hotels and also making the information about customers' rights available to them on and even before their arrival.

3. It shall be mandatory for all classified hotels to show the classification status on their website under a separate icon on the opening page, which on click will display the order of classification issued by the Ministry of Tourism, Government of India.

4. The Hotel and Restaurant Approval and Classification Committee (HRACC) inspects and assesses the hotels based on the facilities and services offered.
   - Hotel Projects are approved at implementation stage
   - Operational Hotels are classified under various categories

5. Details of the criteria for Project Approval/Classification along with the documents required for this purpose are given in this document.

Applications for project approvals under the category of Heritage, 4 star and 5 star as well as applications for classification of operational hotels in the category of 4 Star, 5 Star and 5 Star Deluxe as well as Heritage (Basic, Classic and Grand) categories along with the requisite fee (paid vide Demand Draft) may be sent to:

Member Secretary (HRACC)/Hotel and Restaurants Division, Ministry of Tourism C-1, Hutments, Dalhousie Road, New Delhi 110011

6. For project approval/classification in 3, 2 and 1 Star categories, application along with the requisite fee (paid vide Demand Draft) may be forwarded to the Regional Director, India Tourism Office in whose region the hotel/project is located. The offices of the Regional Directors are as under:

   i. Regional Director, India Tourism, West and Central Region, 123, Maharshi Karve Road, Mumbai–400 020

   ii. Regional Director, India Tourism, Northern Region, 88 Janpath, New Delhi–110 001

   iii. Regional Director, India Tourism, Southern Region, 154, Anna Salai, Chennai – 600 002

   iv. Regional Director, India Tourism, 4, Shakespeare Sarani, Kolkata–700 071

   v. Regional Director, India Tourism, North Eastern Region, Amarawati Path, Christian Basti, GS Road, Guwahati–781 005

7. The detailed Guidelines for Project Approval are at **Annexure I** and that for Classification/Re-Classification at **Annexure II**.

8. The Ministry of Tourism reserves the right to modify the Guidelines/Terms and Conditions from time to time.

ANNEXURE I

GENERAL TERMS, CONDITIONS AND APPLICATION FORMAT FOR APPROVAL OF HOTELS AT THE PROJECT LEVEL

Approval of Hotel at the Project Stage

1. The Ministry of Tourism will approve hotels at project stage based on documentation. Project approval is given to 1, 2, 3, 4, 5 Star and Heritage (Basic) categories. Hotel projects approved under 5 Star and Heritage category after becoming operational may seek classification under 5 Star Deluxe/Heritage Classic/Heritage Grand category if they fulfill the prescribed norms.

2. Project Approvals will be valid for 5 years. The Project Approval would cease 3 months before the date of expiry of project approval or from the date the hotel becomes operational, even if all its rooms are not ready. The hotel must apply for Classification within 3 months of commencing operations. The application for Project Approval will be submitted complete in all respect as per details given below. **Incomplete applications will not be accepted.**

3. Application Form should have the following details:
   i. Proposed name of the Hotel
   ii. Name of the promoters with a note on the business antecedents in not more than 60 words
   iii. Complete postal address of the promoter with Telephone, Fax and Email address
   iv. Status of the owner/promoter
      a. If Public/Private limited company with copies of Memorandum and Articles of Association
      b. If Partnership, a copy of Partnership Deed and Certificate of Registration
      c. If Proprietary concern, name and address of proprietor/Certificate of Registration
   v. Location of hotel site with postal address
   vi. Details of the site
      a. Area (in sq. meters)
      b. Title – owned/leased with copies of sale/lease deed
      c. Copy of Land Use Permit to construct Hotel from local authorities
      d. Distance (in km) from (a) Railway station (b) Airport (c) Main Shopping Centre

   vii. Details of the project:
      a. Copy of Feasibility Report
      b. Star category planned
      c. Number of rooms (with attached bathrooms) and size for each type of room (in sq.ft.)
      d. Size of bathrooms (in sq.ft.)
      e. Details of public areas with size in sq. ft. – lobby/lounge; restaurants; bar; shopping; banquet/conference halls; business centre; health club; swimming pool; parking facilities (no. of vehicles)
      f. Facilities for the differently abled guests (room with attached bathroom earmarked for this purpose, designated parking, ramps for free accessility in public areas and to at least one restaurant, designated toilet (unisex) at the lobby level, etc.).
      g. Eco-friendly practices (a) sewage treatment plant (b) rain water harvesting (c) waste management (d) pollution control method for air, water and light (e) introduction of non-CFC equipment for refrigeration and air conditioning.
      h. Energy/water conservation (use of CFL lamps, solar energy, water saving devices/taps)
      i. Details of fire fighting measures/hydrants, etc.
      j. Date by which project is expected to be completed and become operational
      k. Any other additional facilities
      l. Security related features
      m. The architecture of the hotel building in hilly and ecologically fragile areas should incorporate creative architecture keeping in mind sustainability and energy efficiency and as far as possible in conformity with local art and architecture with use of local materials.

4. Blue prints/Building plans signed by the owner, the architect and approved by the competent authority showing:
   i. Site plan
   ii. Front and side elevation
   iii. Floor plans for all floors
   iv. Detail of guest rooms and bath rooms with dimensions in sq.ft.
   v. Details of fire fighting measures/hydrants, etc.

vi. Air-conditioning details for guest rooms, public areas
5. Local approvals by
   i. Municipal Authority
   ii. Concerned Police Authority
   iii. Any other local authority as may be applicable/required (viz. Pollution Control Board/Ministry of Environment and Forests, etc.)
   iv. Approval/NOC from Airport Authority of India for projects located near the Airport
6. **Note:** The above mentioned approvals/NOCs are the responsibility of the promoter/concerned company as the case may be. The Ministry's approval is no substitute for any statutory approval and the approval given is liable to be withdrawn in case of any violation without notice.
7. Proposed capital structure
   a. Total project cost
   b. Equity component with details of paid up capital
   c. Debt—with current and proposed sources of funding
8. Submission of 'Undertaking' for observance of regulatory conditions/terms and conditions to be furnished by the applicant. (Format enclosed at **Annexure III**).
9. The application should indicate whether a few rooms or all rooms are to be let out on a Timeshare basis. Hotels which propose to let out part of or all its rooms on Timeshare basis will not be eligible for Classification under this scheme.
10. Application fee in the form of a Demand Draft payable to **'Pay and Accounts Officer, Department of Tourism, New Delhi'.**
11. In the event of any change in the project plan, the applicant should apply afresh for approval under the desired category.
12. Authorized officers of the Ministry of Tourism should be allowed free access to inspect the premises from time to time without prior notice
13. The hotel must immediately inform the Ministry of the date from which the hotel becomes operational and apply for Classification within 3 months from the date of operation.
14. The fee payable for the project approval and subsequent extension, if required is as under. The Demand Draft may be payable to **'Pay and**

Accounts Officer, Department of Tourism, New Delhi'.

Star Category	Amount in ₹
5- star	15,000
4-star	12,000
3-star	8,000
2-star	6,000
1-star	5,000
Heritage category	12,000

15. The promoter must forward quarterly progress reports failing which the project approval is liable to be withdrawn.
16. All documents must be valid at the time of application. All copies of documents submitted must be duly attested by a Gazetted officer/Notary. Documents in local language should be accompanied by a translated version in English which should also be duly certified.
17. Projects, where it is proposed to let out part or whole of the hotel on Timeshare basis, will not be covered under these guidelines.
18. Any change in the project plan or management should be informed to the Ministry of Tourism (for 5-D, 5, 4 Star and Heritage categories) or Regional Director's Office (for 3, 2 and 1 Star categories) within 30 days, failing which the approval will stand withdrawn/terminated.
19. The project approval is only applicable for new hotels coming up and not for additional rooms coming up in existing hotels.
20. The minimum size of rooms and bathrooms for all categories have been specified in the Guidelines. Hotels of 1, 2, 3 and 4 star categories availing subsidy/tax benefits/other benefits from the Central/State Government would be subject to a lock-in period of 8 years so that these hotels continue to serve as budget category hotels. Hotels would be permitted to apply for up-gradation to a higher star category after the completion of the lock-in period.
21. Applicants are requested to go through the **'CHECKLIST' OF FACILITIES & SERVICES** contained in this document before applying for project approval of new hotel projects/classification of operational hotels.
22. Application for Hotel Project Approvals forwarded through post will not be accepted if incomplete and applicant will be asked to complete the application and furnish required documents/information.

MINISTRY OF TOURISM

Proposal Submission Form for Hotel Project Approval
<u>General Information</u>

सत्यमेव जयते

For Official Use Only:

Proposing Agency _____

Scheme* _____

Application Date* (in dd/mm/yyyy format)   ____ / ____ / _____

Plan Year _____   **File No.** _____

<u>Project Details</u>

Project / Hotel Name* _____

Project /Hotel Address* (Please give complete Address of the Site):

_____

_____

_____

City* _____

State* _____   **District** _____

PIN Code ⬜⬜⬜⬜⬜⬜

Nodal Officer Name _____   **Mobile No:** _____

Phone _____ **Fax** _____

E-mail* _____

Website _____

Owner Status (Public/Private/Partnership/Proprietary)*: _____

Site Area (in Sq. Mts.)   : _____

Tourism Sector*   : _____

Hotel Category Planned*   : _____

Hotel Sub Category Planned*   : _____

MINISTRY OF TOURISM
Proposal Submission Form for Hotel Project Approval

Project Location Details:

Nearest	Name	Distance (in Km)
Airport		
Railway Station		
Sea Port		
Bus Stand		
Shopping Centre		

Fees Details*

Application Fee Amount		
Mode	☐ DD	☐ Cheque
Drawn in Favor of		
DD Number		
DD Date		
Bank		

Proposed Capital Structure:

Equity*	Amount (₹ in Crores)
Loan*	
Other	
Total Project Cost*	

* Attach additional sheets for more components

MINISTRY OF TOURISM
Proposal Submission Form for Hotel Project Approval

सत्यमेव जयते

Project Completion Related Dates

Expected Date of Completion*   :   ___ / ___ /_____

Expected Date of Commissioning :   ___ / ___ /_____

(in dd/mm/yyyy format)

Company /Promoter Details:

Promoter Name	Note on Business Antecedents	Address	Phone	Fax	E-mail

Room Details:

Room Type	No. of Rooms*	Min. Room Area	Min. Bathroom Area
Single			
Double			
Suite			
Serviced Apartment			
Others			
Total No. of Rooms*			

*Please see Guidelines for required documents

ANNEXURE II

GENERAL TERMS, CONDITIONS AND APPLICATION FORMAT FOR CLASSIFICATION/RE-CLASSIFICATION OF OPERATIONAL HOTELS

1. Classification for newly operational hotels if approved by Ministry of Tourism at project stage, must be sought within 3 months of completion of the project. Operating hotels may opt for Classification at any stage. However, hotels seeking Re-classification should apply for re-classification at least six months prior to the expiry of the current period of classification.

2. If a hotel fails to reapply six months before the expiry of the classification period, the application will be treated as a fresh case of classification.

3. Once a hotel applies for Classification/Re-classification, it should be ready at all times for inspection by the inspection committee of the HRACC. **No request for deferment of inspection will be entertained.**

4. Classification will be valid for a period of **5** (Five) years from the date of approval of Chairman HRACC or in case of Re-classification, from the date of expiry of the last classification, provided that the application has been received within six months prior to the expiry of the current period of classification, along with all valid documents. Incomplete applications will not be accepted.

5. The application should indicate whether a few rooms or all rooms are to be let out on a Timeshare basis. Hotels which propose to let out part of or all its rooms on Timeshare basis will not be eligible for Classification under this scheme.

6. Hotels applying for Classification must provide the following documentation:
   i. Name of the Hotel
   ii. Name and address of the promoter/owner with a note on their business antecedent in not more than 60 words
   iii. Complete postal address of the hotel with Telephone, Fax and Email address
   iv. Status of the owner/promoter
      a. If Public/Private limited company with copies of Memorandum and Articles of Association
      b. If Partnership, a copy of Partnership Deed and Certificate of Registration
      c. If Proprietary concern, name and address of proprietor/Certificate of Registration
   v. Date on which the hotel became operational
   vi. Details of hotel site with postal address and distance (in km) from (a) Airport (b) Railway Station (c) City-centre/downtown shopping area

7. Details of the hotel:
   a. Area of Hotel site (in sq. metres) with title – owned/leased with copies of sale/lease deed
   b. Copy of Land Use Permit from local authorities
   c. Star category being applied for
   d. Number of rooms and size for each type of room in sq.ft. (single/double/suites-all rooms to have attached bathrooms)
   e. Size of bathrooms in sq.ft.
   f. Air-conditioning details for guest rooms, public areas
   g. Details of public areas: (a) lobby/lounge (b) restaurants with no. of covers (c) bar (d) shopping area (e) banquet/conference halls (f) health club (g) business centre (h) swimming pool (i) parking facilities (no. of vehicles which can be parked)
   h. Facilities for the differently abled guests: dedicated room with attached bathroom, designated parking, ramps, free accessibility in public areas and at least to one restaurant, designated toilet (unisex) at the lobby level, etc.
   i. Eco-friendly practices (a) sewage treatment plant (b) rain water harvesting (c) waste management (d) pollution control method for air, water and light (e) introduction of non-CFC equipment for refrigeration and air conditioning and other eco-friendly measures and initiatives.

      A sewage treatment plant will not be a mandatory condition for hotels which have obtained completion certificate for construction before 01.04.2012
   j. Measures for energy and water conservation, water harvesting (use of CFL lamps, solar energy, water saving devices/taps, etc.)

k. Details of fire fighting measures/hydrants, etc.

l. Security features viz. CCTV, X-ray check, verification of staff, etc.

m. The architecture of the hotel building in hilly and ecologically fragile areas should incorporate creative architecture keeping in mind sustainability and energy efficiency and as far as possible in conformity with local art and architecture with use of local materials.

n. Any other additional facilities

8. **Copies of Certificates/No Objection Certificates to be furnished (copies should be current/valid and duly attested by a notary/gazetted officer):**

a. Certificate/licence from Municipality/Corporation to show that the establishment is registered as a Hotel mandatory for applying for Classification/Re-classification. It should be current and valid.

b. No objection certificate from concerned Police Department authorizing the running of the Hotel

Mandatory for applying for Classification/Re-classification. It should be current and valid.

c. No objection certificate from Municipal Health Officer/Sanitary Inspector giving clearance to the establishment from sanitary/hygiene point of view

d. No objection certificate from the Fire Service Department (Local Fire Brigade Authority)

e. Public liability insurance (optional)

f. Bar Licence (necessary for 4, 5, 5 Star Deluxe, Heritage Classic and Heritage Grand categories)

h. Building plans sanctioned by the competent authority and occupancy/completion certificate by the competent authority.

i. If classified earlier, a copy of the Classification Order issued by Ministry of Tourism

j. For Heritage property, certificate from the local authority stating the age of the property and showing the new and old built up areas separately

k. Clearance/NOC/approval required from any other local authority (viz. Pollution Control Board/Ministry of Environment and Forests, etc.) whichever is applicable

l. Approval/NOC from Airport Authority of India for projects located near the Airport

m. Application fees

The above-mentioned approvals/No Objection Certificates are the responsibility of the owner/promoter/concerned Company as the case may be. The approval of the Ministry of Tourism is no substitute for any statutory approval and the approval given is liable to be withdrawn without notice in case of any violations or misrepresentation of facts.

9. All applications for Classification and Re-Classification must be complete in all respects—application form, application fee, prescribed clearances, NOCs, certificates, etc. **Incomplete applications will not be accepted.**

10. Hotels will qualify for classification as Heritage Hotels provided a minimum of 50% of the floor area was built before 1950 and no substantial change has been made in the façade. Hotels, which have been classified/re-classified under Heritage categories prior to issue of these Guidelines will continue under Heritage categories even if they were built between 1935–1950.

11. The application fees payable for classification/re-classification are as follows. The Demand Draft may be payable to **'Pay and Accounts Officer, Department of Tourism, New Delhi'**

Star category	Classification/Re-classification fees in ₹
1-Star	6,000
2-Star	8,000
3-Star	10,000
4-Star	15,000
5-Star	20,000
5-Star Deluxe	25,000
Heritage(Grand, Classic, Heritage categories)	15,000

12. Upon receipt of application complete in all respects, the hotel will be inspected by a classification committee which will be constituted as follows:

a. For 4, 5, 5 Star Deluxe and Heritage (Basic, Classic and Grand) categories –

• Chaired by Additional Director General (Tourism), Govt. of India/Chairperson (HRACC) or a representative nominated by him

• Representative from FHRAI

• Representative from HAI

• Representative from IATO

- Representative from TAAI
- Principal-Institute of Hotel Management
- Regional Director, India Tourism Office/ local India Tourism office
- Member Secretary HRACC
- In case of Heritage category, a representative of Indian Heritage Hotels Association (IHHA)

The HRACC representatives/nominees of FHRAI, HAI, IATO and TAAI should have requisite expertise and experience of the hospitality and tourism industry (hands on experience)

b. For 1, 2 and 3 Star hotels
- Chairperson, Secretary (Tourism) of the concerned State Govt. or his nominee who should not be below the rank of a Deputy Secretary to the Government of India. In his absence the Regional Director, India Tourism who is also Member Secretary, Regional HRACC will chair the committee
- Regional Director, India Tourism Office/ local India Tourism office
- Representative from FHRAI
- Representative from HAI
- Representative from IATO
- Representative from TAAI
- Principal-Institute of Hotel Management

The HRACC representatives/nominees of FHRAI, HAI, IATO and TAAI should have requisite expertise and experience of the hospitality and tourism industry (hands on experience)

c. The Chairperson and any 3 members will constitute a quorum.

d. The recommendations duly signed by the committee will be sent to HRACC Division (Ministry of Tourism, Government of India) by next day through speed post and the recommendation of the HRACC inspection committee will be approved by the Chairperson (HRACC)/Addl. Director General (Tourism) expeditiously.

e. Appellate Authority: In case of any dissatisfaction with the decision of HRACC, the hotel may appeal to Secretary (Tourism), Government of India for review and reconsideration within 30 days of receiving the communication regarding Classification/ Re-classification. No request will be entertained beyond this period.

13. Hotels will be classified following a two-stage procedure:
    a. The presence of facilities and services will be evaluated against the enclosed checklist available at **Annexure IV.**
    b. The quality of facilities and services will be evaluated by the HRACC inspection committee against the mark sheet available at **Annexure V.**

14. The hotel is expected to maintain required standards at all times. The Classification Committee may inspect a hotel at any time without prior notice. The Committee may request that its members be accommodated overnight to inspect the level of services.

15. Any deficiencies/rectifications pointed out by the HRACC must be complied with within the stipulated time, which has been allotted in consultation with the hotel representatives during inspection. Failure to comply within the stipulated time will result in rejection of the application.

16. The committee may assign a star category lower but not higher than that applied for.

17. The hotel must be able to convince the committee that they are taking sufficient steps to conserve energy and harvest water, garbage segregation, and disposal/recycling as per Pollution Control Board (PCB) norms and following other eco-friendly measures.

18. For any change in the Star/Heritage category, the promoter must apply afresh along with requisite fee.

19. Any changes in the Building Plans or management of the hotel should be informed to the HRACC, Ministry of Tourism, Govt. of India within 30 days, otherwise the classification will stand withdrawn/terminated. In case of change of company name/hotel name, a copy of the fresh 'Certificate of Incorporation' or a copy of the 'Resolution of the Board of Directors' regarding the name change alongwith any other relevant documents may be submitted.

20. The minimum size of rooms and bathrooms for all categories have been specified in the Guidelines. Hotels of 1, 2, 3 and 4 star categories availing subsidy/tax benefits/other benefits from the Central/State Government would be subject to a lock-in period of 8 years so that these hotels continue to serve as budget category hotels. Hotels would be permitted to

apply for upgradation to a higher star category after the completion of the lock in period.

21. Applicants are requested to go through the **'CHECKLIST' OF FACILITIES & SERVICES** contained in this document while applying for Classification/Re-classification. The checklist may be duly filled up and signed and stamped on each page which should be submitted along with the application.

22. The Hotel should adhere to the tenets of the **Code of Conduct for Safe and Honorable Tourism** for which the following action would have to be taken.

   i. A signed copy of the Pledge and Under-taking of commitment towards 'Safe and Honorable Tourism' should be attached with the application. The format of the Pledge and Undertaking—Code of Conduct for Safe and Honorable Tourism are attached at **Annexure VI** and **Annexure VII** respectively.

   ii. On the day a new staff member joins the hotel, he/she would be required to take/sign the pledge. The pledge would be incorporated in the appointment letter/joining report of the staff.

   iii. Two focal points/Nodal Officers would be nominated (i.e. from HRD, security side, etc.) at the time of applying for approval by the hotel in the case of hotels which have more than 25 personnel. In the case of hotels with less than 25 personnel, one focal point would have to be nominated.

   iv. The training would be provided to the staff of the classified/approved hotels by Ministry of Tourism under its Capacity Building of Service Providers (CBSP) scheme in connection with 'Safe and Honorable Tourism'. The focal points of the hotel would be trained first within first six months of MOT approval. Subsequently the trained focal points in turn would impart in-house training to the staff which would be arranged within next six months.

   v. The Pledge of Commitment towards Safe and Honorable Tourism would have to be displayed prominently in the staff areas/back areas of the hotels/restaurants, etc. and in the office premises of all the Head of the Departments (HODs).

   vi. The signatures of the Code of Conduct would be required to maintain a record of action taken by them in compliance of the provisions of this para, which shall be kept in their office and shown to the Committee, at the time of Classification/Re-classification.

23. It will be mandatory for the hotel to participate in the skill development initiative of the Ministry of Tourism to meet the manpower needs for the tourism and hospitality industry. For this, the following action would have to be taken.

   i. Classified hotel would be required to train a minimum number of persons, in every calender year in the short duration Skill Development Courses under 'Hunar Se Rozgar' scheme as per following norms:

Rooms per Hotel	Ist Year No. of persons to be trained	2nd Year No. of persons to be trained	3rd Year No. of persons to be trained	4th Year No. of persons to be trained	5th Year No. of persons to be trained
100+	20	20	25	25	30
50 to 100	10	10	15	15	20
20 to 50	5	5	5	5	5

   ii. A minimum of ten persons will constitute a training class. Since a hotel with rooms between 20 and 50 will not be expected to have facilities/infrastructure necessary for the conduct of trainings an arrangement can be worked out between 2 and 5 hotels to conduct this obligatory training (only the theory part) in one cluster and the practical part being carried out in the respective hotels.

   iii. Operational guidelines for the training program will be circulated separately.

   iv. Each hotel would achieve the above mentioned yearly target and submit it to Ministry of Tourism in the re-classification application so as to be considered for re-classification.

Incomplete applications will not be considered. Efforts will be made to ensure that all cases of classification are inspected within three months from the date of application if complete in all respects and Classification Order will be issued within 30 days subsequently.

MINISTRY OF TOURISM
Proposal Submission Form for Hotel Classification/Re-classification
General Information

For Official Use Only:

Proposing Agency _____

Scheme* _____

Application Date* (in dd/mm/yyyy format)   ____ / ____ / _____

Plan Year _____   File No. _____

Is Proposing Hotel Classified/Re-classified by MOT   Yes   No

Is the Hotel Project Approved by MOT*   Yes   No

Current Category   _____ Sub Category _____

Current Classification Validity Period: From _____   To _____

Hotel Details

Hotel Name* _____

Hotel Address* (Please give complete Address of the Site):

_____

_____

_____

City* _____

State* _____ District _____

PIN Code | | | | | | |

Nodal Officer Name _____   Mobile No: _____

Phone   _____ Fax _____

E-mail* _____

Website _____

MINISTRY OF TOURISM

Proposal Submission Form for Hotel Classification / Re-classification

Owner Status (Public/Private/Partnership/Proprietary) : _____

Site Area (in Sq. Mts.)   : _____

Tourism Sector*   : _____

Hotel Category*   : _____

Hotel Sub Category*   : _____

Date of Commencement*   : ___ / ___ / _____

                                                        (in dd/mm/yyyy format)

Hotel Location Details:

Nearest	Name	Distance (in km)
Airport		
Railway Station		
Sea Port		
Bus Stand		
Shopping Centre		

Fees Details*

Application Fee Amount	_____
Mode	☐ DD ☐ Cheque
Drawn in Favor of	_____
DD Number	_____
DD Date	_____
Bank	_____

MINISTRY OF TOURISM
Proposal Submission Form for Hotel Classification / Re-classification

Company / Promoter Details:

Promoter Name	Note on Business Antecedents	Address	Phone	Fax	E-mail

Room Details:

Room Type	No. of Rooms*	Min. Room Area	Min. Bathroom Area
Single			
Double			
Suite			
Serviced Apartment			
Others			
Total No. of Rooms*			

*Please see Guidelines for required documents

ANNEXURE III

Format for Undertaking
(To be on official company letterhead)

To

The Secretary (Tourism)
Govt. of India
Ministry of Tourism
New Delhi

UNDERTAKING

I have read and understood all the terms and conditions mentioned above with respect to Project Approval/Classification-Re-classification under the Star/Heritage categories and hereby agree to abide by them. The information and documents provided are correct and authentic to the best of my knowledge.

I understand that the Ministry's approval is no substitute for any statutory approval and the approval given is liable to be withdrawn in case of any violation or misrepresentation of facts or non-compliance of directions that may be issued by the Ministry of Tourism, Govt. of India, without notice.

It is to certify that the hotel would not seek upgradation to a higher category for a period of eight (8) years in the event the hotel avails of subsidy/tax benefits/other benefits from the Government.

In case of any dispute/legal measure, the same may be eligible in the jurisdiction falling under the NCT of Delhi.

Signature

Name in BLOCK LETTERS

Seal of the applicant

Place:
Date:

ANNEXURE IV

CHECKLIST OF FACILITIES FOR CLASSIFICATION/RE-CLASSIFICATION OF HOTELS

FACILITIES AND SERVICES	1*	2*	3*	4*	5*/ 5* D	Yes/ No	COMMENTS
GENERAL							
Full time operation 7 days a week in season	N	N	N	N	N		
Establishment to have all necessary trading licenses	N	N	N	N	N		Documents as detailed in General Terms and Conditions.
Establishment to have public liability insurance	D	D	D	D	D		
24 hr. lifts for buildings higher than ground plus two floors	N	N	N	N	N		Mandatory for all hotels. Local laws may require a relaxation of this condition. Easy access for the differently abled guests.
Bedrooms, Bathroom, Public areas and kitchen fully serviced daily	N	N	N	N	N		
All floor surfaces clean and in good repair	N	N	N	N	N		Floor may be of any type.
GUEST ROOM							
Minimum 10 lettable rooms, all rooms with outside windows/ventilation.	N	N	N	N	N		
Minimum size of bedroom excluding bathroom in sq. ft	120	120	140	140	200		Single occupancy rooms may be 20 sq ft less. Rooms should not be less than the specified size.
Air-conditioning	25%	25%	50%	100%	100%		Air-conditioning/heating depends on climatic conditions and architecture. Room temperature should be between 20 and 28°C. For 4, 5 and 5 Star Deluxe (the percentage is of the total no. of rooms).
A clean change of bed and bath linen daily and between check-ins	N	N	N	N	N		Definitely required between each check-in. On alternate days for 1and 2 Star hotels.
Minimum bed width for single 90 cm and double 180 cm	D	N	N	N	N		
Mattress thickness minimum 10 cm	D	D	N	N	N		Coir, foam or spring foam
Minimum bedding 2 sheets, pillow and case, blanket, mattress protector/bed cover	N	N	N	N	N		
Suites	D	D	D	N	N		2% of room block with a minimum of 1 suite room

Contd.

Contd.

FACILITIES AND SERVICES	1*	2*	3*	4*	5*/ 5* D	Yes/ No	COMMENTS
Hairdryers	D	D	N	N	N		
Safe keeping/in-room safe	D	D	D	N	N		
Mini bar/Fridge	D	D	N	N	N		
Drinking water with minimum one glass tumbler per guest	N	N	N	N	N		
Guest linen							Good quality linen to be provided
Shelves/drawer space	N	N	N	N	N		
Wardrobe with minimum four clothes hanger per bedding	N	N	N	N	N		
Sufficient lighting, 1 per bed	N	N	N	N	N		
A 5 Amp earthed power socket	N	N	N	N	N		
A bedside table and drawer	N	N	N	N	N		
TV – cable if available			N	N	N		3*, 4*, 5* and 5* Deluxe must have remote
A writing surface with sufficient lighting			N	N	N		
Chairs	N	N	N	N	N		
A waste paper basket	N	N	N	N	N		
Opaque curtains or screening at all windows	N	N	N	N	N		
A mirror at least half length	N	N	N	N	N		
A stationery folder containing stationery	D	D	N	N	N		
A 'Do Not Disturb' notice	N	N	N	N	N		
Night spread/Bed cover	D	D	N	N	N		
Energy saving lighting	N	N	N	N	N		
Linen room	N	N	N	N	N		
BATHROOM							
Number of rooms with attached bathrooms	All	All	All	All	All		It is mandatory w.e.f. 01.09.2010 for all 1 and 2 Star category hotels to have attached bathrooms. All bathrooms to have sanitary bin with lid.
Minimum size of bathroom in sq. ft	30	30	36	36	45		25% of bathroom in 1 and 2 Star hotels to have western style WC. No higher ceiling/cap on the maximum size.
1 bath towel and 1 hand towel to be provided per guest	N	N	N	N	N		
Bath mat	D	D	N	N	N		

Contd.

Contd.

FACILITIES AND SERVICES	1*	2*	3*	4*	5*/ 5* D	Yes/ No	COMMENTS
Guest toiletries to be provided-minimum 1 new soap per guest	N	N	N	N	N		Quality products depending on the star category.
A clothes–hook in each bath/shower room	N	N	N	N	N		
A sanitary bin	N	N	N	N	N		These must be covered.
Each western WC toilet to have a seat with lid and toilet paper	N	N	N	N	N		
Floors and walls to have non–porous surfaces	N	N	N	N	N		
Hot and cold running water available 24 hours	N	N	N	N	N		It is mandatory w.e.f. 01.09. 2010 for all 1 and 2 Star category hotels to provide hot and cold running water.
Shower cabin	N	N	N	N	N		Where shower cabin is not available, a shower with shower curtain will suffice.
Bath tubs				D	D		In 4 Star and above hotels, some rooms should offer this option to guests.
Water saving taps/shower							
N N	N	N	N				
Energy saving lighting	N	N	N	N	N		
PUBLIC AREA							
Lounge or seating area in the lobby	N	N	N	N	N		Doorman on duty. Lobby shall have furniture and fittings which shall include chairs/arm chairs, sofa, tables and fresh floral display.
Reception	N	N	N	N	N		Manned minimum 16 hours a day. Call service 24 hrs. Local directions to hotel including city/street maps to be available.
Valet (parking) services	D	D	N	N	N		
Availability of Room, F&B and other tariff	N	N	N	N	N		
Heating and cooling to be provided in public areas				N	N		Temperatures to be between 20 and 28°C.
Public restrooms for ladies and gents with soap and clean towels, a washbasin with running hot and cold water, a mirror, a sanitary bin with lid in unisex and ladies toilet	N	N	N	N	N	N	

Contd.

Contd.

FACILITIES AND SERVICES	1*	2*	3*	4*	5*/ 5* D	Yes/ No	COMMENTS
ROOM AND FACILITIES FOR THE DIFFERENTLY ABLED GUESTS							
At least one room for the differently abled guest	N	N	N	N	N		Minimum door width should be one metre to allow wheelchair access with suitable low height furniture, low peep hole, cupboard to have sliding doors with low cloth hangers, etc. Room to have audible and visible (blinking light) alarm system.
Ramps with anti-slip floors at the entrance. Minimum door width should be one metre to allow wheelchair access.	N	N	N	N	N		To be provided in all public areas. Free accessibility in all public areas and at least one restaurant in 5 Star and 5 Star Deluxe.
Bathroom	N	N	N	N	N		Minimum door width should be one metre. Bathroom to be wheelchair accessible with sliding door, suitable fixtures like low wash basin, low height toilet, grab bars, etc. No bath tub required.
Public restrooms	N	N	N	N	N		Unisex. To be wheelchair accessible with low height urinal (24" maximum) with grab bars. Minimum door width should be one metre.
FOOD AND BEVERAGE							
1 and 2 Star category							1 and 2 Star categories should have minimum one dining room serving all meals. Room service not necessary.
3 Star category							One multi-cuisine restaurant cum coffee shop open from 7 am to 11 pm and 24 hr. room service.
4 Star category							**Grade A cities:** One multi-cuisine restaurant cum coffee shop open from 7 am to 11 pm, one speciality restaurant and 24 hr. room service. **Grade B cities:** One multi-cuisine restaurant open from 7 am to 11 pm and 24 hr. room service.
5 Star category							**Grade A cities:** One multi-cuisine restaurant cum 24 hr. coffee shop/all day diner, one speciality restaurant and 24 hr. room service.

Contd.

Contd.

FACILITIES AND SERVICES	1*	2*	3*	4*	5*/ 5* D	Yes/ No	COMMENTS
							Grade B cities: One multi-cuisine restaurant cum coffee shop open from 7 am to 11 pm, one speciality restaurant and 24 hr. room service.
5 Star Deluxe category							**Grade A cities:** One multi-cuisine restaurant cum 24 hr. coffee shop/all day diner, one speciality restaurant and 24 hr. room service. **Grade B cities:** One multi-cuisine restaurant cum coffee shop open from 7 am to 11 pm, one speciality restaurant and 24 hr. room service.
Grade–A: Delhi,* Mumbai, Kolkata, Chennai, Bangalore, Pune, Hyderabad/Secunderabad. **Grade–B:** Cities in the rest of the country excluding Grade 'A' cities.					**Note:** The Ministry of Tourism may review and revise the cities falling under the Grade 'A' and Grade 'B' from time to time *Delhi would include the hotels falling in Gurgaon, Faridabad, Ghaziabad, Noida and Greater Noida.		
Crockery and Glassware	N	N	N	N	N		Plasticware accepted in pool area.
Cutlery to be at least stainless steel	N	N	N	N	N		All categories should use good quality metal cutlery. Aluminium cutlery prohibited.
Bar	D	D	N	N	N		
KITCHEN AND FOOD PRODUCTION AREA							
Refrigerator with deep freeze	N	N	N	N	N		Capacity based on size of F&B service.
Segregated storage of meat, fish and vegetables	N	N	N	N	N		Meat, fish and vegetables in separate freezers.
Color coded synthetic chopping boards	N	N	N	N	N		Wooden chopping boards prohibited.
Tiled walls non-slip floors	N	N	N	N	N		
Head covering for production staff	N	N	N	N	N		
Daily germicidal cleaning of floors	N	N	N	N	N		
Good quality cooking vessels/utensils	N	N	N	N	N		Use of aluminium vessels prohibited except for bakery.
All food grade equipment containers	N	N	N	N	N		
Ventilation system	N	N	N	N	N		
Garbage to be segregated—wet and dry	N	N	N	N	N		To encourage recycling.
Wet garbage area to be air-conditioned			N	N	N		
Receiving areas and stores to be clean and distinct from garbage area	N	N	N	N	N		

Contd.

Contd.

FACILITIES AND SERVICES	1*	2*	3*	4*	5*/ 5* D	Yes/ No	COMMENTS
Six monthly medical checks for production staff	N	N	N	N	N		
First-aid training for all kitchen staff	N	N	N	N	N		
Pest control	N	N	N	N	N		
STAFF							
Staff uniforms for front of the house	N	N	N	N	N		Uniforms to be clean and in good condition.
English speaking front office staff	D	D	N	N	N		This may be relaxed outside the metros/sub-metros for 1 and 2 Star category hotels.
Percentage of supervisory staff	20%	20%	40%	40%	80%		Hotels of 4 Star category and above should have formally qualified Heads of Departments. The supervisory or the skilled staff may have training or skill certification as follows: Degree/Diploma from Central or State IHM's/FCI's or from NCHMCT affiliated IHM's or from other reputed Hospitality schools.
Percentage of skilled staff	20%	20%	30%	30%	60%		The supervisory or the skilled staff may have training or skill certification as follows: i. Degree/Diploma from Central or State IHM's/FCI's or from NCHMCT affiliated IHM's or from other reputed Hospitality schools. ii. Skill training certificate issued under the guidelines and scheme of the Ministry of Tourism.
STAFF WELFARE FACILITIES							
Staff Restroom	D	D	N	N	N		Separate for male and female employees with bunk beds, well lighted and ventilated.
Staff locker room	D	D	N	N	N		
Toilet facilities	N	N	N	N	N		Full length mirror, hand dryer with liquid soap dispenser.
Separate dining area	D	D	N	N	N		
GUEST SERVICES							
Provision of wheelchair for the differently abled guest	N	N	N	N	N		Wheel chair to be available on complimentary basis in hotels of all categories

Contd.

Contd.

FACILITIES AND SERVICES	1*	2*	3*	4*	5*/ 5* D	Yes/ No	COMMENTS
Valet (parking) services	D	D	N	N	N		
Dry- cleaning/laundry	D	D	D	D	N		In house for 5 star Deluxe hotels. For 5 Star category and below, maybe outsourced.
Tea/Coffee making facility in the room	D	D	D	N	N		
Iron with Iron Board Facility	D	D	D	N	N		Iron and iron board to made available on request in 1 to 4 Star category hotels on complimentary basis. For 5 and 5 Star Deluxe categories, to be available in the room.
Linen room	N	N	N	N	N		Well ventilated
Paid transportation on call	D	D	N	N	N		Guest should be able to travel from hotel.
Shoe cleaning service, shoe horn and slippers	D	D	D	N	N		Free facility to be provided for in house guests.
Ice (from drinking water) on demand	D	D	N	N	N		Complimentary on request.
Acceptance of common credit cards	D	D	N	N	N		
Assistance with luggage on request	N	N	N	N	N		
A public telephone on premises. Unit charges made known	D	D	N	N	N		There should be at least one telephone no higher than 24" from floor level in 5 and 5 Star Deluxe (to also cater to differently abled guests).
Wake-up call service on request	N	N	N	N	N		
Messages for guests to be recorded and delivered	N	N	N	N	N		A prominently displayed message board will suffice for 1 and 2 Star categories.
Name, address and telephone numbers of doctors with front desk	N	N	N	N	N		Doctor on call in 3, 4, 5 and 5 Star Deluxe.
Stamps and mailing facilities	D	D	N	N	N		
Newspapers available	D	D	D	N	N		This may be placed in the lounge for 1, 2 and 3 Star hotels.
Access to travel desk facilities	N	N	N	N	N		This need not be on the premises for 1, 2 and 3 Star categories.
Left luggage facilities	D	D	N	N	N		This must be in a well secured room/ 24 hour manned area.
Provision for emergency supplies toiletries/first-aid kit	D	D	N	N	N		May be chargeable.
Health/Fitness facilities	D	D	D	D	N		Indian system of treatments should preferably be offered.
Beauty salon and barber shop	D	D	D	D	D		

Contd.

Contd.

FACILITIES AND SERVICES	1*	2*	3*	4*	5*/ 5* D	Yes/ No	COMMENTS
Florist	D	D	D	D	D		
Utility Shop/Kiosk	D	D	D	D	N		5 and 5 Star Deluxe category hotels to have one utility and one souvenir shop. 4 Star to have minimum one utility shop.
Money changing facilities	D	D	D	D	D		Money changing facility to be made available.
Bookshop	D	D	D	D	N		
SAFETY & SECURITY							
Metal detectors (door frame or hand held)	D	D	N	N	N		
CCTV at strategic locations	N	N	N	N	N		
X-ray Machine	D	D	D	D	N		For 5 Star Deluxe category, it would be 'Necessary' to have an X-ray Machine at the guest entrance for screening of baggage. Manual checks may be conducted for staff and suppliers at designated entry points.
Under belly scanners to screen vehicles	D	D	D	N	N		
Verification	N	N	N	N	N		All hotels should conduct a verification of their staff and suppliers by the police/private security agencies.
Staff trained in fire fighting drill	N	N	N	N	N		All hotels to conduct periodic fire drills and maintain 'Manuals' for Disaster Management, First-Aid and Fire Safety. Quarterly drill or as per law.
Security arrangements for all hotel entrances	N	N	N	N	N		
Each bedroom door fitted with lock and key, viewport/peephole and internal securing device	D	D	N	N	N		A safety chain/wishbone latch is acceptable in place of viewport /peephole.
Smoke detectors	N	N	N	N	N		These can be battery operated.
Fire and emergency procedure notices displayed in room behind door	N	N	N	N	N		
Fire and emergency alarms should have visual and audible signals	N	N	N	N	N		
First-aid kit with over the counter medicines with front desk	N	N	N	N	N		

Contd.

Contd.

FACILITIES AND SERVICES	1*	2*	3*	4*	5*/ 5* D	Yes/ No	COMMENTS
Fire exit signs on guest floors with emergency/back up power	N	N	N	N	N		
CODE OF CONDUCT FOR SAFE & HONORABLE TOURISM							
Display of Pledge	N	N	N	N	N		
Training for 'Code of Conduct for Safe & Honorable Tourism'	N	N	N	N	N		
Maintenance of action taken report with regards to compliance of the provisions of the Code	N	N	N	N	N		
Focal Points/Nodal Officers	N	N	N	N	N		
COMMUNICATION FACILITIES							
Telephone facility within arm's reach of the toilet seat	D	D	D	N	N		
Provide at least two multi-purpose sockets	N	N	N	N	N		
A telephone for incoming and outgoing calls in the room	D	N	N	N	N		4 star and above should have direct dialing and STD/ISD facilities. 1, 2 and 3 Star category hotels may go through a telephone exchange.
PC available for guest use with internet access	D	D	N	N	N		This can be a paid service. Up to 3 Star, PC can be in the executive offices. Internet subject to local access being available.
E-mail service	D	D	N	N	N		Subject to local internet access being available.
Fax, photocopy and printing service	N	N	N	N	N		
In room internet connection/Data Port	D	D	D	N	N		Subject to local internet access being available. Wi- Fi wherever possible.
Business Centre	D	D	D	N	N		This should be a dedicated area. (This provision may be relaxed for resort destinations, tourist and pilgrimage centres).
Swimming Pool	D	D	D	D	N		This can be relaxed for hill destinations. Mandatory that trained Life Guard to be available. Board containing do's and don'ts, no diving sign, pool depth, etc. should be displayed at a strategic location in the pool area.

Contd.

Contd.

FACILITIES AND SERVICES	1*	2*	3*	4*	5*/ 5* D	Yes/ No	COMMENTS
Parking Facilities	D	D	N	N	N		Should be adequate in relation to the number of room and banquet/convention hall capacities. Exclusively earmarked accessible parking nearest to the entrance for differently abled guests.
Conference Facilities	D	D	D	D	N		
No. of people to be trained under 'Hunar Se Rozgar'	N	N	N	N	N		
Eco-friendly Practices:							
a. Sewage Treatment Plant	N	N	N	N	N		
b. Rain Water Harvesting	N	N	N	N	N		
c. Pollution control, methods for air, water and light	N	N	N	N	N		
d. Waste Management	N	N	N	N	N		
e. Introduction of non-CFC equipment for refrigeration and air-conditioning and other eco-friendly measures and initiatives	N	N	N	N	N		

Display of Classification status by the hotel:

Note 1

All hotels should clearly indicate on their websites the facilities and amenities provided to guests 'free of cost' like complimentary breakfast, iron and iron board facility, shoe cleaning facility, shoe horn and slippers, dental kit, shaving kit, etc. If any facility is provided only 'on request' but is included in the room rent, this should be mentioned on the hotel's website under the head 'Facilities and Amenities provided on complimentary basis' and also mentioned to the guest when the hotel staff introduces the room to the guest on arrival.

Note 2

It is mandatory for all the hotels classified under the categories 1 to 4 Star to display their classification status prominently outside the hotel and at the reception from 01.04.2014 as per a scheme to be evolved in consulation with FHRAI and HAI

Note 3

D-Desirable

N-Necessary

There is no relaxation in the necessary criteria except as specified in the comment column.

ANNEXURE V

Mark sheet for quality

Government of India
Department of Tourism (H&R Cell)
Hotel Classification

Criteria	Max Marks	Score	Comments
Exterior and Grounds	8		Exteriors, Approach 2/Landscaping 2/Exterior lighting 2/Parking 2
Guest Rooms	10		Furniture 2/Furnishings 2/Décor 2/Room facilities and amenities 2/Linen 2
Bathrooms	8		Facilities 2/Fittings 2/Linen 2/Toiletries 2
Public Areas	8		Furniture 2/Furnishings 2/Décor 2/Restrooms 2
Food & Beverage	8		Choice of cuisine, menu 3/décor 2/food quality 3
Kitchens	8		Equipment 3/State of repair 2/food storage 3
Cleanliness	8		Overall impression
Hygiene	8		Pot and Dish Washing 2/drinking water 2/staff facilities 1/pest control 2/garbage disposal 1
Safety & Security	8		Fire fighting equipment 2/signage 2/awareness of procedures 2/public area and room security 2
Communications	6		Phone service 2/e-mail access 2/internet access 1/PC and other equipment 1
Guest Services	5		Overall impression
Eco-friendly practices	5		Waste management, recycling, no plastics 1/Water conservation, Harvesting 1/pollution control-air, water, sound, light 2/Alternative energy usage 1/
Facilities for physically challenged persons	5		At least a room for physically challenged persons 1/public toilet in lobby 1/telephone in public places 1/ramps, etc 1/facilities for aurally or visually handicapped 1
Staff quality	5		Overall impression
TOTAL	100		

Comments

_____

_____

_____

HRACC Members

1.   2.   3.   4.   5.   6.

_____

Qualifying Score		
5*D	90 %	
5*	80 %	
4*	75 %	
3*	65 %	
2*	55 %	
1*	50 %	

ANNEXURE VI

PLEDGE FOR COMMITMENT TOWARDS SAFE & HONORABLE TOURISM AND SUSTAINABLE TOURISM (For internal circulation and use of the hotel)

I/We solemnly pledge and reiterate our commitment to conduct our business in a manner that befits the culture and ethos of our rich and ancient civilization, and the tolerant and accommodating nature of our multicultural society and protects all individuals, especially women and children from all derogatory acts which are contrary to the spirit of our country. I/We hereby commit to abide by the Code of Conduct for Safe & Honorable Tourism.

Recognizing that every earth resource is finite and fragile, I/We further pledge to fully implement sustainable tourism practices, consistent with the best environment and heritage protection standards, such that my/our present tourism resource requirements optimize both local community benefit and future sustainable uses.

Signature

Name in BLOCK LETTERS

Place:.....................

Date:......................

ANNEXURE VII

Format of **UNDERTAKING** in respect of the **'Pledge for Commitment towards Safe & Honorable Tourism'**
(To be on official company letterhead)

To
The Secretary (Tourism) Govt. of India
Ministry of Tourism
New Delhi

UNDERTAKING

It is to hereby confirm that I/We have read and understood the 'Code of Conduct for Safe & Honorable Tourism' adopted on 1st October 2010 as per copy attached with application with respect to Project Approval/Classification – Re-classification of hotels under the Star/Heritage categories and hereby agree to abide by them.

That I/We have read and solemnly pledge and reiterate our commitment to conduct our business in a manner that befits the culture and ethos of our rich and ancient civilization, and the tolerant and accommodating nature of our multicultural society and protects all individuals, especially women and children from all derogatory acts which are contrary to the spirit of our country. I/We hereby commit to abide by the Code of Conduct for Safe & Honorable Tourism.

Recognizing that every earth resource is finite and fragile, I/We further pledge to fully implement sustainable tourism practices, consistent with the best environment and heritage protection standards, such that my/our present tourism resource requirements optimize both local community benefit and future sustainable uses.

Signature

Name in BLOCK LETTERS

Place: _____

Date: _____

APPENDIX

VI | Front Office Standard Operating Procedures

A Standard Operating Procedure (SOP) is a set of written instructions that document a routine or repetitive activity followed by a hotel. SOP helps in maintaining quality and consistency of service and standards in a hotel.

The development and use of SOPs are an integral part of a successful quality system as it provides individuals with the information to perform a job properly, and facilitates consistency in the quality and integrity of a product. It is a must that all newly recruited hotel staff should be given training on hotels SOP.

Standard Operating Procedure No: 1	
Process :	How to report to work/shift
Objective :	To be a thorough professional as you enter the department
Responsibility :	All staff

S.No.	Procedure	Standard (Measurement, Number, Time, Quantity, Precautions, Phraseology, etc.)
1.	Arrive at the staff gate 20 minutes prior to the beginning of the shift	• Arrive at the hotel well in advance to the duty timing • Punch your arrival at the staff gate • Go to the Lockers and collect the soiled uniforms
2.	Get the uniform issued from Linen Room	• Exchange the soiled uniform for fresh ones • Ensure that the uniform is in good repair before leaving the uniform exchange counter
3.	Change in Locker Room	• Come to locker room with the fresh uniform • Change into uniform • Ensure that the body odor is addressed using deodorants • Ensure all accessories which are part of uniform are worn • Ensure shoes are shining • Ensure grooming is up to the mark and as per standards • After changing, move to briefing area

Standard Operating Procedure No: 2

Process:	How to report to work briefings
Objective:	To be a thorough professional as you enter the department
Responsibility :	Front office staff

S.No.	Procedure	Standard (Measurement, Number, Time, Quantity, Precautions, Phraseology etc.)
1.	Punctuality	• Staff should report at least 10 minutes prior to the stipulated time of duty. Morning Shift at 06:45 hrs for 07:00 hrs Afternoon Shift at 13:15 hrs for 13:30 hrs Night Shift at 22:15 hrs for 22:30 hrs
2.	Grooming Standards for Ladies	• Uniform should be clean and well ironed. • Hair above shoulder length for Ladies has to be neatly tied • No flashy hairpins, rubber bands or hair accessories allowed • Hair color/hair bleach is not permitted. • Only black well polished closed shoes with skin color socks or stockings to be worn. • Apply light make up to suit your skin color • Nail needs to be clean and properly trimmed • Only transparent gloss nail paint allowed. • Long nails or flashy nail paint is not permitted • Use only mild perfumes/deodorants • Long nails not permitted • One ring in each hand (inclusive of wedding band) should be sober, conservative and not too large. • A simple gold chain or a mangal sutra or a pearl chain is allowed around the neck. • Matching ear studs (one in each ear) to the neckwear is allowed. Dangling earrings are a big no. • Gold bangles not more than two numbers to be worn in one wrist only. No dangling bracelets to be worn. • The other wrist should have a business style watch, should be sober, conservative and not too large.
	For Men	• Uniform should be clean and properly ironed • Hair should be trimmed short and neatly set using hair gel • Long hair, hair color is not permitted • Shoes should be black, well polished with black shocks to be worn • Use only mild perfumes/deodorants • Long nails not permitted • Only one wedding ring is allowed. • A business watch. No flashy watches. • No bracelets or bands (except for religious reasons).
3.	Staff must update themselves with the following information about the hotel	• Hotel occupancy, revenue, average room rate for last night • Expected occupancy, revenue and average room rate for the day • VIP in house and Long staying guest • VIP arrivals for the days

Contd.

Contd.

S.No.	Procedure	Standard (Measurement, Number, Time, Quantity, Precautions, Phraseology, etc.)
		• Groups in-house • Groups expected for the day • Special traces, comments if any • City comparison • Staff must know the events and the promotions going on in the hotel
4.	Staff must read the log book	• Banquet functions of the day • Special promotion in any of the restaurants

Standard Operating Procedure No: 3

Process:	What are the openings and closing duties of a front desk associate
Objective:	To ensure smooth and efficient operations of the shift
Responsibility:	Front office staff

S.No.	Procedure	Standard (Measurement, Number, Time, Quantity, Precautions, Phraseology, etc.)
1.	Opening duties of a front desk associate	• To report on duty on time and be well groomed • To read log book and acknowledge same by signing the log book • To take key handover from the previous shift • To take message or any specific handover • To go through the arrivals for the day • To check the room blocking for VIP arrivals • To ensure that the desk is equipped with required stationery
2.	Closing duties of a front desk associate	• To prepare log book handover for next shift • To prepare key handover for next shift • To prepare message or any specific handover for next shift • To update A & D (arrival and departure) register
	To file following reports	• Arrivals checked in report • Managers report • Shift checklist

Standard Operating Procedure No: 4

Process:	How to welcome the guest on arrival
Objective:	Every guest entering the hotel must be given a warm welcome and departure.
Responsibility:	Front Office Staff/Guest Relations

S.No.	Procedure	Standard (Measurement, Number, Time, Quantity, Precautions, Phraseology, etc.)
1.	To welcome the guest on arrival	• Body Language : When you see guest entering the hotel from the main porch stand straight with head up, smile on your face and confidence in yourself, take pride in your job

Contd.

Contd.

S.No.	Procedure	Standard (Measurement, Number, Time, Quantity, Precautions, Phraseology etc.)
2.	Acknowledge the guest within 30 seconds of the arrival	• Eye Contact: Make eye contact and wish guest with smile and refer to the time of the day. " Good Morning/Afternoon/Evening Mr.....Welcome to The___ "

3.	Meet guests requirement	• Check with the guest what he requires,
		  I. If the guest is checking in the hotel; then proceed with registration process, ensure you take the guests ID or Passport Copy for registration.
		  II. If guest is a visitor then escort him to his destination

4.	Information about the hotel	• Inform guest about the hotel, its facilities and the various promotions going on in the hotel
5.	Guest Recognition	• Always address guest by his last name and show genuine interest in his talks
6.	To give warm departure to the guest	• Always give warm farewell to the guest and wish him a wonderful/pleasant day • Always thank the guest • Always invite the guest to return to the hotel

Standard Operating Procedure No: 5

Process:	How to welcome a guest on his first time visit
Objective:	To create everlasting impression about the hotel.
Responsibility :	Front Office Staff

S.No.	Procedure	Standard (Measurement, Number, Time, Quantity, Precautions, Phraseology, etc.)
	Case study of a first time visit guest on arrival when he approaches the reception on his own.	• Guest reaches the hotel porch • The doorman opens the door with a smile and wishes guest, "Good Morning/Afternoon/Evening, Welcome to the ___ "
1.	Doorman greets the guest	
2.	Bell boy greets the guest and offers baggage assistance	• Bell boy wishes guest the time of the day and assist him with his baggage. "Good Morning/Afternoon/Evening Sir/Madam, Allow me to take care of your baggage"
3.	Bell boy confirms the count of baggage	• Bell boy confirms the number of baggage with the guest. • Guest approaches the reception
4.	Reception greets the guest and offers assistance	• Reception associate wishes the guest, "Good Morning/Afternoon/ Evening, Welcome to The___ Sir/Madam. How may I assist you?" • Guest: I am checking in
5.	Reception offers seat assistance, makes guest comfortable and request for his last name	• Reception Associate: "Please make yourself comfortable, May I request you for your Last name under which the reservation is being made" • Guest: I am Mr. Singh.

Contd.

Contd.

S.No.	Procedure	Standard (Measurement, Number, Time, Quantity, Precautions, Phraseology, etc.)
6.	Receptionist takes out the registration card of the guest	• Receptionist will then take out the registration card of Mr. Singh
7.	Guest registration card will have the following details taken at the time of reservation	• Guest Name • Guest Co. Name • Guest Arrival and Departure date • Guest room category (booked for) • Guest room rate • Guest credit card details • Guest contact details
8.	Registration card is printed on arrival if the reservation is made on same day	• Guest registration card is to be printed on arrival if registration card is not printed earlier, which may happen if the reservation is made on same day
9.	Guest registration formalities	• Receptionist completes the guest registration formalities. {For registration refer SOP No: 8 and 9}
10.	Programing of guest room key	• Receptionist will then make the key for Mr. Singh. {For Programing of guest key refer SOP No: 11}
11.	Guest relations to escort the guest to his room	• Receptionist will introduce Mr. Singh to the guest relations and request her to escort Mr.Singh to his room "Mr. Singh, my guest relations Ms. Bhumika will escort you to your room." {For escorting refer SOP No. 12}
12.	Wish guest a pleasant stay	• Receptionist wishes Mr Singh a very pleasant stay "Mr. Singh, have a pleasant stay with us"
13.	Delivery of guest baggage to the room	• Receptionist will inform the bell boy Mr.Singh's room number for baggage delivery
	Case study of a first time visit guest on arrival when guest relations meets the guest in the lobby	• Guest reaches the hotel porch
1.	Doorman greets the guest	• A doorman opens the door for the guest and wishes him with a smile "Good Morning/Afternoon/Evening, Welcome to The _____."
2.	Bell boy greets the guest and offers baggage assistance	• Bell boy wishes guest the time of the day and assist guest with his baggage
3.	Guest relations greets the guest in the lobby and offers assistance	• Guest relations meets the guest in the lobby and wish him "Good Morning/Afternoon/Evening, Welcome to The _____. How may I assist you?" • Guest: I have a reservation
4.	Guest relations request guest for his last name and escorts the guest to the reception	• Guest Relations will ask the guest under what name the reservation is being made. "May I have your last name please?" Guest: "I am Mr. Singh." • Guest Relations escorts Mr. Singh to the reception
5.	Guest relations informs reception of the guest checking in	• Guest relations informs receptionist of Mr. Singh checking in
6.	Reception offers seat assistance, makes guest	• Receptionist will wish Mr. Singh the time of the day, make him comfortable and will take out his registration card "Good Morning/

Contd.

Contd.

S.No.	Procedure	Standard (Measurement, Number, Time, Quantity, Precautions, Phraseology, etc.)
	comfortable, takes out the registration card	Afternoon/Evening Mr. Singh, Welcome to The _____ please make yourself comfortable"
7.	Guest registration formalities	• Receptionist will complete Mr. Singh's registration. {For registration refer SOP No. 8 and 9}
8.	Programing of guest room key	• Receptionist will then make the key for Mr. Singh. {For Programing of guest key refer SOP No: 11}
9.	Guest relations to escort the guest to his room	• Receptionist will introduce Mr. Singh to the guest relations and request her to escort Mr. Singh to his room "Mr. Singh, my guest relations Ms. Bhumika will escort you to your room."{For escorting refer SOP NO. 12}
10.	Wish guest a pleasant stay	• Receptionist wishes Mr. Singh "Have a pleasant stay with us"
11.	Delivery of guest baggage to the room	• Receptionist will inform bell boy of Mr. Singh's room number for baggage delivery
12.	On multiple check-ins	• In case of multiple check-in guest relations will request the guest to identify his baggage and informs the same to bell boy before proceeding to his room

Standard Operating Procedure No: 6

Process:	How to welcome a repeat guest on arrival
Objective:	To give warm and recognized welcome to the guest
Responsibility:	Front Office Staff/Guest Relations

S.No.	Procedure	Standard (Measurement, Number, Time, Quantity, Precautions, Phraseology, etc.)
1.	Case study of a repeated guest on arrival when he approaches the reception on his own	• Guest reaches the hotel porch
2.	Doorman greets the guest by his name wishing him welcome back	• The doorman opens the door and wishes guest "Good Morning/Afternoon/Evening, Mr. Chaddha Welcome back to The _____"
3.	Bell boy greets the guest by his name wishing him welcome back and offers baggage assistance	• Bell boy wishes the guest "Good Morning/Afternoon/Evening Mr. Chaddha, Welcome back to The_____" and assist him with his baggage
4.	Bell boy confirms the count of baggage	• Bell boy confirms the number of baggage with Mr. Chaddha. • Mr. Chaddha approaches the reception
5.	Receptionist greets the guest by his name and wishes him welcome back	• Reception associate stands up and wishes, "Good Morning/Afternoon/Evening, Mr. Chaddha, Welcome back to The _____. Please make yourself comfortable"
6.	Receptionist will take out the registration card of the guest	• The Receptionist will then take out the registration card of Mr. Chaddha

Contd.

Contd.

S.No.	Procedure	Standard (Measurement, Number, Time, Quantity, Precautions, Phraseology, etc.)
7.	Receptionist will confirm the departure date and time	• Receptionist will confirm the departure date and time with Mr. Chaddha and will request him only for his signature as she already has the details of Mr. Chaddha (being a repeated guest)
8.	Receptionist will request the guest for his signature on the registration card	• "Mr. Chaddha, May I request you for your signature on the registration card"
9.	Programing of guest room key	• Receptionist will then make the key for Mr. Chadda.{For Programing of guest key refer SOP No: 11}
10.	Guest introduction to the guest relations	• Receptionist will introduce Mr. Chadda to the guest relations and request her to escort Mr. Chadda to his room. "Mr. Chadda, my guest relations Ms. Bhumika will escort you to your room."{For escorting refer SOP No. 12}
11.	Wish guest a pleasant stay	• Receptionist wishes Mr. Chaddha, "Have a pleasant stay with us Mr. Chaddha"
12.	Delivery of guest baggage to the room	• Receptionist will inform the bell boy Mr. Chaddha's room number for baggage delivery
	Case study of a repeated guest on arrival when guest relations meets the guest in the lobby	• Guest reaches the hotel porch
1.	Doorman greets the guest by his namewishing him welcome back	• A doorman opens the door for the guest and wishes him "Good Morning/Afternoon/Evening, Welcome back to The_____ Mr. Chaddha."
2.	Bell boy greets the guest by his name wishing him welcome back and offers baggage assistance	• Bell boy wishes the guest "Good Morning/Afternoon/Evening Mr. Chaddha, Welcome back to the _____." and assist him with his baggage
3.	Guest relations greets the guest by his last name and wishes him welcome back	• Guest relations meets the guest in the lobby and wish him "Good Morning/Afternoon/Evening, Welcome back to The_____, Mr. Chaddha"
4.	Guest relations escorts the guest to the reception	• Guest Relations escorts Mr. Chaddha to the reception
5.	Guest relations inform the receptionist of guest checking-in	• Guest relations inform receptionist of Mr. Chaddha checking in
6.	Receptionist greets the guest by his name and wishes him welcome back	• Receptionist will wish Mr. Chaddha the time of the day "Good Morning/Afternoon/Evening Mr. Chaddha, Welcome back to The _____"
7.	Receptionist will take out the registration card of the guest	• The Receptionist will then take out the registration card of Mr. Chadda
8.	Receptionist will confirm the departure date and time	• Receptionist will confirm the departure date and time with Mr. Chaddha and will request him only for his signature as she already has the details of Mr. Chaddha being a repeated guest
9.	Receptionist will request the guest for his signature on the registration card	• "Mr. Chaddha, May I request you for your signature on the registration card"

Contd.

Contd.

S.No.	Procedure	Standard (Measurement, Number, Time, Quantity, Precautions, Phraseology, etc.)
10.	Programing of guest room key	• Receptionist will then make the key for Mr. Chadda. {For Programing of guest key refer SOP No: 11}
11.	Guest relations to escort the guest to his room	• Receptionist will introduce Mr. Chadda to the guest relations and request her to escort Mr. Chadda to his room "Mr. Chadda, my guest relations Ms. Bhumika will escort you to your room."{For escorting refer SOP No. 12}
12.	Wish guest a pleasant stay	• Receptionist wishes: "Mr. Chaddha, Have a pleasant stay with us"
13.	Delivery of guest baggage to the room	• Receptionist will inform bell boy of Mr. Chaddha's room number for baggage delivery
14.	On multiple check-ins	• In case of multiple check-in guest relations will request the guest to identify his baggage and informs the same to bell boy before proceeding to his room

Standard Operating Procedure No: 7

Process:	How to prepare for group arrival
Objective:	To ensure smooth check-in of the group
Responsibility:	Front Office Staff/Guest Relations

S.No.	Procedure	Standard (Measurement, Number, Time, Quantity, Precautions, Phraseology, etc.)
1.	Group Information Sheet	• Group Information Sheet is circulated to all concerned departments and HOD's a day or two prior to the group arrival date by the reservations
2.	Preparation of rooming sheet of the group	• As per the arrival time of the group previous night shift at the reception prepares the rooming sheet of the group
3.	Rooming list gives following information of the group	• Group Name • Name of the Group/Tour leader • Guest name with room number. • Arrival and Departure details of the group • Total number of paid and complimentary room, if any • Total number of single
4.	Room blocking for a group	• Rooms are blocked a day prior or by previous shift duty manager depending on the group arrival time and the availability of the rooms
5.	Point to be kept in mind while blocking the rooms for the group	• While blocking the rooms for group, try and accommodate the group in same room category and on the same floor
6.	Check for room rate adjustment	• Room rate is checked for any adjustment if required
7.	Routing for the group	• Group routing is done as per the billing instructions of the group
8.	Fruit and Flower (F&F) Voucher for the group	• F&F voucher for the group welcome drink or any other amenities (as per GIS) is sent on the morning of the day of group arrival

Contd.

Contd.

S.No.	Procedure	Standard (Measurement, Number, Time, Quantity, Precautions, Phraseology, etc.)
9.	Information of room blocked for the group to the house keeping department	• House keeping is informed of the rooms blocked for group and the expected time of arrival of the group.
10.	Preparation of room key and key jackets for the group	• Once the rooms are blocked, keys are prepared and kept ready in key jackets with guest name and room number on it by the previous shift at the reception
11.	Registration card is prepared for group leader	• One registration card is prepared in the name of the Group/Group leader, who signs on behalf of the group
13.	Restaurant booking for the group	• Restaurant booking is done for the group as per the group information sheet
14.	Note – Check for staffing	• Ensure that we have enough staffing at bell desk for smooth handling of group baggage

Standard Operating Procedure No: 8

Process:	What is the procedure of registration for a domestic guest
Objective :	To record and maintain proper and accurate information of the guest
Responsibility:	Front Office Staff

S.No.	Procedure	Standard (Measurement, Number, Time, Quantity, Precautions, Phraseology, etc.)
1.	Case study of registration for a domestic guest Receptionist greets the guest, offers the seat assistance	• Wish guest "Good Morning/Afternoon/Evening Sir/Madam. Welcome to The_____, Please make yourself comfortable"
2.	Request guest for his last name	• Politely ask guest for his last name under which the room is reserved. "May I have your last name please?" • Guest: I am Mr. Singh
3.	Trace guest reservation with following details	• If could not trace the guest reservation with his last name, then ask for other details like the company name, confirmation number. "I am sorry Mr. Singh, I am not able to trace your reservation, May I have your company name or confirmation number"
4.	Check for comments and traces	• Check for guest comments, traces, message, if any
5.	Registration card to be taken out	• Take out the registration card from Piano file
6.	Registration card is printed on arrival if the reservation is made on the same day	• If the guest reservation is made on the day of arrival, print the registration card. {For printing of registration card refer SOP No. 35}

Contd.

Contd.

S.No.	Procedure	Standard (Measurement, Number, Time, Quantity, Precautions, Phraseology, etc.)
7.	Registration card to be presented in folder	• Neatly keep the registration card on the leather folder.
8.	Request guest for his business card and credit card	• Politely ask guest for his business card/credit card/ID. (all required documents together) "Mr. Singh, May I request you for your business card/ID and credit card?"
9.	Note down the details	• Note down the details on the registration card using black pen only
10.	In case guest is not carrying his business card, request him to write down the same on registration card	• If guest is not having his business card, politely request him to write down his address, contact numbers and email address on the registration card "Mr. Singh, May I request you to write your Address, Contact number and E-mail address"
11.	Confirm the departure date, time and hotel transfer with the guest	• Confirm the departure date and time with the guest and the hotel transfer if required for airport drop on departure "Mr. Singh, your departure date is 1st June, 2016, that means you will be staying with us for 3 nights, and would you be requiring the hotel car to drop you to the airport on departure"
12.	Confirm the room category, room rate and the package inclusive of with the guest	• Confirm the room category and the rate at which the room is booked by pointing towards the rate printed on the registration card and package inclusive of "Mr. Singh, the room category you have been booked for is deluxe room and the rate for same is ₹. ____ + Taxes inclusive of morning breakfast in our All day dinning restaurant, i.e. Orient at lobby level"
13.	Inform guest if the room is blocked as per his liking	• Inform guest if the room is blocked as per his special request "Mr. Singh, As per your liking your room has a beautiful view of the pool"
14.	Confirm the arriving and departing destination with the guest	• Confirm the arriving and departing destination with the guest "Mr. Singh, May I know the place you are coming from and place you are heading towards"
15.	Confirm the mode of payment with the guest, If the mode of payment is bill to co/travel agent voucher	• Confirm the billing instructions with the guest. If the mode of payment is bill to co./travel agent voucher check if you have correct correspondence for the same. If not please follow up the same with the concerned sales account manager "Mr. Singh, Your bills for room and taxes will be taken care by the company/travel agent and extras will be charged to you directly"
16.	If the mode of payment is cash	• If the mode of payment is cash, request the guest to make advance deposit for at least a night plus ₹ 5000 additional "Mr. Singh, I request you to make the advance payment for one night which is ₹ ____ plus ₹ 5000 additional for extras"
17.	If the mode of payment is credit card	• If the mode of payment is credit card than politely ask him for his credit card "Mr. Singh, May I request you for your credit card?"
18.	Guest and signature to be tallied	• Please tally the guest name and signature on the given credit card.

Contd.

Contd.

S.No.	Procedure	Standard (Measurement, Number, Time, Quantity, Precautions, Phraseology, etc.)
		• Note down the credit card number on registration card
19.	Request guest for his signature on the registration card	• Once you finished with all guest details on the registration card. Present registration card in leather folder along with a black pen to the guest for his signature "Mr. Singh, May I request you for your signature on registration card"
20.	Programing of the guest room key	• Prepare key for the guest and present it in key jacket along with guest name and room number written on it. {For programing of key refer SOP No. 11}
21.	Guest relations to escort the guest to his room	• Introduce guest to Guest Relations and request her to escort the guest to his room "Mr.Singh, My Guest Relations Ms. Bhumika will escort you to your room"{For escorting refer SOP No.12}
22.	Wish guest a pleasant stay	• Wish guest a pleasant/wonderful stay with us "Mr. Singh, Have a pleasant stay with us"

Standard Operating Procedure No: 9

Process:	What is the procedure of registration for an International guest
Objective:	To ensure proper and accurate information of the guest for legal documentation
Responsibility:	Front Office Staff

S.No.	Procedure	Standard (Measurement, Number, Time, Quantity, Precautions, Phraseology, etc.)
	Case study of registration for an International guest	
1.	Receptionist greets the guest	• Wish guest "Good Morning/Afternoon/Evening Sir/Madam. Welcome to the ____,"
2.	Receptionist offers seat assistance to the guest	• Make guest comfortable on seat. "Please make yourself comfortable Sir/Madam"
3.	Request guest for his last name	• Politely ask guest for his last name under which the room is reserved. "May I have your last name please?" • Guest: I am Mr. David
4.	Trace guest reservation with following details	• If could not trace the guest reservation with his last name, then ask for other details like the company name, confirmation number "I am sorry Mr. David, I am not able to trace your reservation, May I have your company name or confirmation number"
5.	Check for comments and traces	• Check for guest comments, traces, message, if any
6.	Registration card to be taken out	• Take out the registration card from Piano file
7.	Registration card is printed on arrival if the reservation is made on the same day	• If the guest reservation is made on the day of arrival, please print the registration card. {For printing of registration card refer SOP No. 35}

Contd.

Contd.

S.No.	Procedure	Standard (Measurement, Number, Time, Quantity, Precautions, Phraseology, etc.)
8.	Registration card to be presented in folder	• Neatly keep the registration card on the leather folder
9.	Request guest for his passport, business card and credit card	• Politely ask guest for his Passport/Business card and Credit card. "May I request you for your Passport/Business Card and Credit card" (Ask for all required documents together)
10.	Note down the details	• Note down following details on the registration card using a black pen only.
11.	Points to be kept in mind while taking down passport details	• Tally Mr. David's name on registration card with name mentioned on the passport • Date of Issue of passport • Date of Expiry of passport • Place of Issue of passport • Date of Birth of guest • Date of Arrival in the Country • Purpose of visit in the Country • Arriving from and Next destination. • Guest credit card number and the date of expiry of the credit card
12.	In case guest is not carrying his business card, request him to write down the same on registration card	• If guest is not having his business card, politely request him to write down his address, contact numbers and email address on the registration card "Mr. David, May I request you to write your Address, Contact number and E-mail address"
13.	Confirm the departure date, time and hotel transfer with the guest	• Confirm the departure date and time with the guest and the hotel transfer if required for airport drop on departure "Mr. David, your departure date is 1st June, 16, that means you will be staying with us for 3 nights, and would you be requiring the hotel car to drop you to the airport on departure"
14.	Confirm the room category, room rate and the package inclusive of to the guest	• Confirm the room category and the rate at which the room is booked by pointing towards the rate printed on the registration card and the package inclusive of "Mr. David, the room category you have been booked for is deluxe and the rate for same is ₹/US $ + Taxes inclusive of morning breakfast only in our All day dinning restaurant, i.e. Orient at lobby level"
15.	Inform guest if the room is blocked as per his liking	• Inform guest if the room is blocked as per his/her preference "Mr. David, As per your liking your room has a beautiful view of the pool"
16.	Confirm the arriving and departing destination with the guest	• Confirm the arriving and departing destination with the guest "Mr. David, May I know the place you are coming from and place you are heading towards"
17	Confirm the mode of payment with the guest, if the mode of payment is bill to co/travel agent voucher	• Confirm the billing instructions with the guest. If the mode of payment is bill to co./travel agent voucher check if you have correct correspondence for same. If not, please follow up the same with the concerned sales account manager. "Mr. David, Your bills for room andTaxes will be taken care by the company/travel agent and extras will be charged to you directly"

Contd.

Contd.

S.No.	Procedure	Standard (Measurement, Number, Time, Quantity, Precautions, Phraseology, etc.)
18.	If the mode of payment is cash	• If the mode of payment is cash, request the guest to make the advance deposit for at least one night "Mr. David, I request you to make the advance payment for one night which is ₹ plus ₹ 5000 additional for extras."
19.	If the mode of cash payment is foreign currency	• If guest pays by foreign currency, same needs to be converted into local currency and posted to the room account
20.	If the mode of payment is credit card	• If the mode of payment is credit card than politely ask him for his credit card "Mr. David, May I request you for your credit card?"
21.	Guest and signature to be tallied	• Please tally the guest name and signature on the given credit card • Note down the credit card number on registration card
22.	Request guest for his signature on the registration card	• Once you finished with all guest details on the registration card. Present registration card in leather folder along with a black pen to the guest for his signature "Mr. David, May I request you for your signature on registration card."
23.	Programing of the guest room key	• Prepare key for the guest and present it in key jacket along with guest name and room number written on it to the Guest Relations. {For programing of key refer SOP No. 11}
24.	Guest relations to escort the guest to his room	• Introduce guest to Guest Relations and request her to escort the guest to his room "Mr. David, My Guest Relations Ms. Bhumika will escort you to your room."{For escorting refer SOP No. 12}
25.	Wish guest a pleasant stay	• "Mr. David, Have a pleasant stay with us."

Standard Operating Procedure No: 10

Process:	What is the procedure of registration for a group
Objective:	To ensure proper and accurate information of the group
Responsibility:	Front Office Staff

S.No.	Procedure	Standard (Measurement, Number, Time, Quantity, Precautions, Phraseology, etc.)
1.	Check for staffing	• Ensure you have enough staffing at the front office to handle group efficiently and smoothly
2.	Traditional welcome for the group	• When group checks-in Guest Relations to do Arti Tika (traditional welcome) for the Group, if required as per Group Information Sheet at the porch
3.	Wish group the time of the day	• Wish group members the time of the day and a very warm welcome to the _____ .
4.	Make group comfortable	• Make group sit and comfortable in the lobby
5.	Welcome drink to be served	• Inform in room dinning to get welcome drink in the lobby
6.	Registration of the group	• For group there is only one registration card

Contd.

Contd.

S.No.	Procedure	Standard (Measurement, Number, Time, Quantity, Precautions, Phraseology, etc.)
		• Present registration card neatly in black leather folder along with a black pen to the group leader/tour leader
7.	Signature of group leader/ tour leader on the registration card	• Group leader/tour leader is the authorized person to sign on the group registration card
8.	Group details sheet to be taken from the group leader	• Please collect group detail sheet from the group leader on check In (if not received prior to their arrival) containing following details:
9.	Details to be taken for domestic group	• Name of the group members • Date of arrival and departure of group members • Room sharing information
10.	Details to be taken for international group	• Name of the group members • Passport details of the group members • Date of arrival and departure of the group • Group arriving from • Next destination of the group • Date of arrival in India of the group
11.	Travel agent voucher	• Collect the travel agent voucher from group leader, if not received earlier
12.	Details to be confirmed from Group leader on check in	• Time for breakfast/lunch/dinner booking of the group in the hotel restaurant, if any, as per the group information sheet • Group departure date • Wake up call for the group • Reminder wake up call for the group • Time for baggage down of the group • Checkout time of the group • Request to deposit the keys at the bell desk on checkout
13.	Request group leader for group contact address and details	• Politely request group leader to give the correspondence address of the group, his contact numbers and email address for any future requirement
14.	Allocation of group room numbers	• Allocate the room numbers to the group and give them the keys as per the group rooming sheet
15.	Information on hotel facilities and services	• Brief the group leader on the various facilities and services available in the hotel
16.	Wish group a wonderful stay	• Wish group members a very memorable/wonderful stay with us
17.	Group baggage to be tagged	• Bell boy has arranged the baggage of the group with baggage tag on one side of the lobby
18.	Request group members to identify their baggage	• After the room allocation of the group is completed duty manager/ guest relations request the group to identify their baggage to ensure the delivery of the bags to the correct room
19.	One point contact for group	• Group leader is one point contact for any group movement activity
20.	The group information sheet is prepared by the previous	• Name of Group Leader and Room Number • Name of the Group Members and Room Number

Contd.

Contd.

S.No.	Procedure	Standard (Measurement, Number, Time, Quantity, Precautions, Phraseology, etc.)
	night shift at the reception, giving the following details:	• Arrival and Departure Date • Wake Up Call for the Group • Time for Baggage Down • Breakfast time • Checkout time
21.	Group Information Sheet is circulated to the concerned departments:	• Telephones – For group wake up call • Bell Desk – For group baggage down • Housekeeping – For Checking the mini bar on departure • IRD – For group breakfast • Cashier – For group departure

Standard Operating Procedure No: 11

Process:	Explain the procedure for programing and issuing the guest key
Objective:	To program and issue the keys to the guest only for the number of days that the guest would be staying with us in order to prevent for the key being misused
Responsibility:	Front Office Staff

S.No.	Procedure	Standard (Measurement, Number, Time, Quantity, Precautions, Phraseology, etc.)
1.	Steps to be followed for programing the guest room key in CISA machine	• Guest room number • Number of days, guest would be staying with us, i.e. duration of stay of the guest • Number of keys to be made, depending on the number of Pax in the room • Guest name and then press enter • The screen will play write card, then enter the key in the CISA slot • In case the screen displays Pass Enter to Continue, then press enter and then put key in CISA slot • The key will be made
2.	For programing the multiple keys	• In case of making more than one key, after the first key is made, the screen will show again Write Card, and then put another fresh key in the CISA slot • As soon as the guest has checked in, a guest key is programed for the number of days the guest would be staying with us
3.	Issuing of guest room key	• Only one key is issued to the guest. In case of sharer or a joiner another key may be issued • The key should be collected from the guest at the time of checkout
4.	In case guest extends the stay key needs to be reprogramed. Steps to be followed for existing key being extended on account of guest departure date is extended	• The screen will be on a display of Read Card after programing of any key, at that particular note, the key to be put in CISA slot, it will show the following: • The Guest room number • Guest Name • Number of keys made for that particular room • Number of days the key is been made for

Contd.

Contd.

S.No.	Procedure	Standard (Measurement, Number, Time, Quantity, Precautions, Phraseology, etc.)
		• Scroll the arrow key in the screen by keys on the keyboard to the number of days column • Edit the number of days as per the guest extension as required • Press Enter and screen will display Write Card • Put the key card in the CISA slot and the key will be programed for the extended number of days
5.	Steps to be followed for programing the duplicate key in CISA machine Method I	• Put the room number in the CISA machine displaying read card • Press down button and copy • The CISA will search for the program of the main key and display the option of copy card • Put the blank key card in the CISA slot • Duplicate key for the given number will be programed
	Method II	• We can also make duplicate key, while programing the main key by entering the number of keys required to be made in the option of number of keys • Maximum two keys are issued to the guest

Standard Operating Procedure No: 12

Process:	How to escort a first time visit guest to his room
Objective:	To orient the guest to the hotel and his room
Responsibility:	Front Office Staff

S.No.	Procedure	Standard (Measurement, Number, Time, Quantity, Precautions, Phraseology, etc.)
1.	Case study of escorting of a first time visit guest to the room Open hand gestures to be used On entering the guest room- For a domestic guest For an International guest	• Guest is escorted to the room by Guest Relation Executive • Open hand gestures be used while directing guest. "This way please Mr./Ms. ..." • Here is The _____ Welcome letter for you Mr./Ms. .... • Here is The _____ Welcome letter and a small souvenir (showing The Tea Box) for you Mr./Ms. ....
2.	Explain the facilities of the hotel to the guest Orient	• Amongst the dining options we have Orient, the multi cuisine all day dining which is open 24 Hrs. Your breakfast will be served here from 0700 Hrs to 1030 Hrs
3.	Dhaba and Mehak	• Mehak serves authentic Chinese from Mainland China. Dhaba is our signature Indian restaurant serving Highway cuisine from North India. Both are open for lunch from 1230 Hrs to 1445 Hrs and dinner from 1930 Hrs to 2330 Hrs
4.	Gardenia	• Gardenia which serves authentic and inspired Mediterranean cuisine has an indoor-outdoor ambience and is open only for dinner from 1930 Hrs–0030 Hrs and closed on Tuesdays

Contd.

Contd.

S.No.	Procedure	Standard (Measurement, Number, Time, Quantity, Precautions, Phraseology, etc.)
5.	Theka the Bar	• We also have a lounge bar—Theka The Bar stocking over 70 brands of vodka from across the world, besides other premium spirits. It is open daily from 1600 Hrs to 0100 Hrs with Happy Hours from 1700 to 2000 Hrs daily
6.	Ye Old Bakery	• The patisserie Ye Old Bakery has a selection of cakes, pastries, chocolates and savories to choose from
7.	Offer assistance for restaurant reservation	• "Would you like me to make a lunch/dinner reservation for you" If Yes go ahead and note down the reservation details and if guest says No then inform guest "If you wish to dine at any of the restaurants we request you to make a prior reservation at the In–Room Dining number (5082)"
8.	Explain other facilities of the hotel Business Centre	• The Business Centre is accessible to you 24 Hrs a day, we also have wireless internet connectivity through out the hotel as well as in your room, to access it your last name is the User-Id and room number is the password
9.	Beauty Salon and Health Club	• We have a beauty salon with expert make-over professionals providing a host of beauty treatments. The Health Club is equipped with gymnasium, sauna, steam and Jacuzzi. You can also go for rejuvenating massage therapies at the Health Club. There is also an outdoor swimming pool. The pool is accessible from 0700 Hrs – 1900 Hrs and the Health Club is accessible from 0630 Hrs- 2130 Hrs (Timing for ladies is from 1000 Hrs to 1500 Hrs and for gentlemen, from 0630 Hrs to 0930 Hrs and 1530 Hrs to 2130 Hrs)
10.	Explain if the hotel renovation is going on	• We are in the process of upgrading our products so there is some renovation work taking place, however, we will ensure minimum disturbance during the day all renovation is stopped after 1900 Hrs in the evening
11.	Guest Relation Executive opens the door and keycard is inserted for power.	• Welcome to your room no. _____ • Welcome to your room which is part of our newly renovated wing. (In case of new wing) • Guest Relation "May I assist you with your room." If yes, Guest Relations explains various features of the room
12.	View of the guest room Note	• While showing around the room care be taken with the • View (in case of pool facing or landscape) Only if the room has a view else please avoid. "You have a beautiful view of the pool side from your room".
13.	Power saving unit	• The keycard needs to be inserted in order to control power to the room, (in case the guest enquires mention—on removal all lights go off and the air conditioner goes on to blower mode)
14.	Make my room	• Would you want your room to be cleaned at any time please indicate here
15.	Reliance Interactive TV	• Show usage of the TV. We have newly introduced the interactive television. The main menu will offer you options for you to choose with the number keys button • The shortcut buttons will also help you to choose options as displayed on the screen.

Contd.

Contd.

S.No.	Procedure	Standard (Measurement, Number, Time, Quantity, Precautions, Phraseology, etc.)
		• We offer movies on demand, internet connection and world radio which are chargeable
16.	DVD Player	• We have also placed a DVD player. For DVD's, Information on shopping, City guide please contact concierge Extn. _____ • This blue button will also allow you to go directly to television mode. At any time you can also get back to the main menu. You may also use the keypad for internet and games
17.	Minibar	• The Mini Bar has been placed for you (indicate where) and this is the rate list. Should you require additional amenities or replenishment, kindly call housekeeping on Extn. _____
18.	Tea/Coffee maker	• This is with compliments from us. For replenishment, kindly contact housekeeping on Extn. _____
19.	Indicate folder for Hotel Compendium	• The Hotel Compendium contains all information on the hotel as well as the In-Room Dining Menu
20.	Mixon panel	• This is a Mixon Panel. It controls all the electrical facilities in the room, the lights are controlled from here (entrance, rooms, chandelier, and nightlight). It controls the opening and closing of the drapes. The air conditioning temperatures can be increased or reduced from here.
21.	Do not disturb	• Should you require not to disturbed kindly indicate here it will display outside the room
22.	Bathroom amenities	• For regular rooms "We have placed for you special Kama products, which are ayurvedic" • For Suites "We have Molten Brown products for you." • "The hair dryer is kept here" • "This is the music knob. By turning it you can listen to music or watch television in here"
23.	Laundry bag and bathrobes	• The laundry bag, list and bathrobes have been placed for you here
24.	Wardrobe and CISA safe	• The safe is also placed for you; the instructions are mentioned here would you like me to explain how to use it?
25.	Fire Plan and Emergency Exit	• This is the fire plan. You are here right now. In case of emergency this is the nearest exit door
26.	Offer assistance to the guest	• Guest Relations will ask the guest before leaving the room. "Is there anything else I may help you with?"
27.	Wish guest a wonderful stay and appreciate guest for giving his valuable time	• Wish the guest a wonderful stay before leaving the room. "Thank you for your time, I wish you a wonderful stay with us. Should you require any assistance, please feel free to call reception at Extn_____"
28.	While escorting a repeated guest offer assistance to explain the room features	• In case of a repeated guest, Guest relations escorts the guest to the room and ask the guest "Mr./Ms. _____you wish me to explain you the features of the room".
29.	If guest says yes	• If the answer is yes, guest relations explain the room features as mentioned above

Contd.

Contd.

S.No.	Procedure	Standard (Measurement, Number, Time, Quantity, Precautions, Phraseology, etc.)
30.	Offer assistance for restaurant reservation	• If No, Guest Relations will offer lunch/dinner reservation for guest "Would you like me to make a lunch/dinner reservation for you" If Yes go ahead and note down the reservation details and if guest says No then inform guest "If you wish to dine at any of the restaurants we request you to make a prior reservation at the In-Room Dining number_____"
31.	Wish guest a wonderful stay	• Guest Relations will come out of the room wishing guest a very wonderful stay. "Have a wonderful stay with us Mr/Ms. _____" "For any assistance please call at Extn. _____".

Standard Operating Procedure No: 13

Process:	How to escort a VIP guest to the room
Objective:	To make the guest feel special/important and recognized
Responsibility:	Guest Relations/Front Office Staff

S.No.	Procedure	Standard (Measurement, Number, Time, Quantity, Precautions, Phraseology, etc.)
	Case study of a VIP arrival having airport pick up from the hotel	
1.	Airport representative pages duty manager from the hotel giving details about the guest	• Name of the guest • Number of bags • Baggage tag needs to be ready before the guest checks in
2.	Guest Relations welcomes guest at the hotel porch	• Car arrives at the porch • Guest Relations will be ready at the Porch with the folder which contains: 1. Guest Registration Card 2. A black pen 3. The welcome letter 4. Guest room key in key jacket with guest name and room number written on it
3.	Guest relations greets guest	• Guest relations will welcome the guest at the porch "Good Morning/Afternoon/Evening Mr./Ms _____ Welcome to The _____ ."
5.	Guest relations escorts guest from the porch to his room	• Guest Relations will escort the guest from the porch to his/her room
6.	Tea box to be given on check in for an International VIP arrival	• In case of a International VIP arrival Tea box is given to the guest on arrival by guest relations • Guest Relations to escort the guest to the room
7.	In-room check-in to be done by guest relations	• Guest relations will do in-room checking for VIP guest

Contd.

Contd.

S.No.	Procedure	Standard (Measurement, Number, Time, Quantity, Precautions, Phraseology, etc.)
8.	Registration formality	• Guest Relations proceeds for registration formality of Mr./Ms. _____ {For guest registration in the room refer SOP No: 8 and 9} (To explain room feature refer SOP No:12)
9.	Note	• For VIP arrivals not scheduled for airport transfers, trace is been left in the reservation to inform the guest relations on check-in for guest in-room check-in.

Standard Operating Procedure No: 14

Process:	How to welcome a guest having airport pick up
Objective:	The correct way to do a checking for a guest who has an airport pickup
Responsibility:	Guest Relations/Front Office Staff

S.No.	Procedure	Standard (Measurement, Number, Time, Quantity, Precautions, Phraseology, etc.)
1.	Case study of a guest arrival having airport pick from the hotel Airport representative informs the duty manager from the hotel	• Name of the guest • Number of bags • Baggage tag needs to be kept ready before the guest checks in
2.	Guest Relations welcomes guest at the hotel porch	• Car arrives at the porch • Guest Relation Executive welcomes the guest at the porch
3.	Guest relations greets guest	• All guests having airport pick up, must be met and greeted at the main porch. "Good Morning/Afternoon/Evening Mr. Thomas. Welcome to The _____."
4.	Guest relations greets the guest by his last name and wishes him welcome back	• All repeat guests must be met and greeted by name at the main porch. "Good Morning/Afternoon/Evening Mr. Thomas. Welcome to The _____. It's good to see you again."
5.	Unload the guest baggage	• Baggage is unloaded and brought up to the lobby by the bell boy. • Guest baggage should be offloaded and placed neatly at the right entrance to the lobby on left of the bell desk. "Please come this way Mr./Ms. ..." (If room not reserved.)
6.	Tagging of guest baggage	• Baggage is tagged and room number allotted. "May I tag your baggage Mr./Ms...?"
7.	Guest relations escorts the guest to the room	• Guest is escorted to the room by Guest Relation Executive for in-room check in. Guest Relations proceeds for registration formalities of the guest{For registration in room refer SOP No. 8 and 9} {To explain room feature refer SOP No.12}

Standard Operating Procedure No: 15

Process:	How to handle a check-in when guest room is not ready
Objective:	To handle the situation in a professional and cordial manner by providing alternatives to the guest
Responsibility:	Front Office Staff/Guest Relations/Duty Manager

S.No.	Procedure	Standard (Measurement, Number, Time, Quantity, Precautions, Phraseology, etc.)
1.	Case study to handle a guest when the room is not ready on arrival	• When a guest arrives at the hotel before the assigned room is ready, offer a suitable alternative in a cordial manner
2.	Points to be kept in mind when the guest room is not ready on arrival. Apologize to the guest	• Apologize guest for the inconvenience caused
3.	Make guest comfortable	• Make guest comfortable in the lobby lounge and offer a refreshing beverage (Tea/Coffee/Fruit Punch)
4.	Offer guest an alternative room	• Find and offer guest alternative room in different category
5.	Time guarantee to be given to the guest	• Time to be given to the guest for room and keep guest informed of room status
6.	Make guest feel important and his need on priority	• Show the guest that every effort is being made to make room ready for him on priority basis
7.	Offer food and beverage service to the guest	• Guest might have been on a long flight: Depending on the time of the day guest may be offered breakfast or tea/coffee/soft beverages in coffee shop
8.	Guest relations to entertain the guest	• Guest Relations to entertain the guest in coffee shop
9.	Room to be made on priority basis	• Inform Housekeeping Supervisor to make room ready for the guest on urgent basis
10.	Follow up with housekeeping on room status	• Constantly follow up with housekeeping on room status
11.	Update duty manager with room status	• Keep duty manager informed about the same
12.	Guest relations/duty manager to speak to the guest	• Guest Relations or Duty Manager should speak to the guest for the inconvenience caused for the room was not ready on the arrival of the guest
13.	Apology note from the hotel	• Flowers and a personalized card from the General Manager of the hotel with an apology to be placed in the room when it was next serviced by housekeeping
14.	Update guest history	• Guest history to be updated for future reference
15.	Email to be marked	• Email to be sent to all concerned persons whenever guest is made to wait for a room

Standard Operating Procedure No: 16

Process:	How to show check-in and profile updating on system
Objective :	To know the check-in process and to maintain the records for audit purpose
Responsibility :	Front Office Staff

S.No.	Procedure	Standard (Measurement, Number, Time, Quantity, Precautions, Phraseology, etc.)
1.	Steps to show check-in on the system	• Go to the front Desk • Press enter on arrivals • Enter the last name of the guest • Click on check-in option
2.	Search for arrivals by either of the followings	I. Guest First Name II. Guest Last Name III. Guest Company Name IV. Travel Agent Name V. Group Name VI. Confirmation Number, etc. • Select correct option. • Press Enter on Check-In • Check-in should be done on system so that other departments are aware of the guest's occupancy • Show check-in of guest in the computer in order to ensure that all other departments are aware of the guests who have checked in and to release the telephone lines
3.	Updation of guest profile in the system	• All personal information collected at check-in will be updated on the guest's profile
4.	Guest information to be entered in the system accurately	• All information will be entered accurately in the system and completely without error • Update folio and guest information to have information that is correct and to record it accurately
5.	Following information is must for profile updation	• Guest Name • Room Number • Arrival and Departure Date • No. of Adults and Children • Room Type (Upgraded reason if any) • Rate Code • Room Rate • Source of Business • Market Segment • Discount and reason • Source of booking • Credit card number and expiry date • Comments/Traces • Designation, Company Name • Address Telephone Number • Fax and E-Mail Id • Passport number, Date and Place of Issue • Nationality and date of Birth. Frequent flyer program membership details (If reservation is made through world hotels)

Contd.

Contd.

S.No.	Procedure	Standard (Measurement, Number, Time, Quantity, Precautions, Phraseology, etc.)
6.	Steps to update the profile in the system	• Go to the Front Desk • Go to the Arrival option/Guest In house • Enter the last name of the guest or room number • Click on profile and update profile in the system • Click on save button • Click on OK to confirm to save the data.
7.	Check for guest mode of payment	• To have clear knowledge of guest mode of payment is very important to ensure that the checkout is carried out in quick and efficient manner
8.	If guest mode of payment is bill to co./travel agent voucher	• If payment instruction has been given specifying the whole or part of the bill, will be settled by either a company, a travel agent or a third party
9.	Check the correspondence	• Check the correspondence to see whether the hotel management has approved the arrangement • Make sure that the correspondence, reservation order and letter has been attached • Double check the voucher or correspondence presented by guests and check for any discrepancy in guest name, period of stay • For any discrepancy please make a note of same and inform to the duty manager and the concerned sales account manager
10.	Routing Instructions	• Set appropriate routing for agent/company billing on the system
11.	Updation of arrival and departure register	• All registration cards entry has to be made in Arrival and Departure register shiftwise on daily basis
12.	Check for the guest information on the registration card tallies with the information fed in the system	• Double check whether information on the registration card matches in the system, i.e. Room Rate, Room Type, Checkout Date, etc. • Set up appropriate routing according to correspondence
13.	Note	• In case there is any information missing such as address, payment, departure time, etc. • Call guest to confirm or leave a trace for the next shift to follow up

Standard Operating Procedure No: 17

Process:	How to do a room change
Objective:	To ensure room change in a prompt and efficient manner
Responsibility:	Front Office Staff/Guest Relations/Duty Manager

S.No.	Procedure	Standard (Measurement, Number, Time, Quantity, Precautions, Phraseology, etc.)
1.	Point to be kept in mind for an occupied guest room change	• In the event, an occupied guest room needs to be changed, it will be done in an efficient manner in the presence of duty manager and security supervisor ensuring all the guests belongings are transferred, all the concerned departments are intimated and the billing is accordingly adjusted

Contd.

Contd.

S.No.	Procedure	Standard (Measurement, Number, Time, Quantity, Precautions, Phraseology, etc.)
2.	To meet guest request	• In-house guest expect to have immediate access to the rooms and the request to be met immediately
3.	Guest request reception for a room change because of various reasons	• Smell in the room • Room has defect • Different category of room given on arrival than what the guest was booked for • Guest did not like the room, etc.
4.	Assign new room	• Check with the duty manager to assign new room for the guest
5.	Apologize to the guest	• Duty manager to apologize the guest in case of hotels fault leading to room change
6.	Offer assistance to pack guest baggage	• Offer assistance to guest to pack his baggage, which is to be shifted to the newly assigned room
7.	Meet guest expectation	• Ensure that the new room meets the expectation of the guest
8.	Guest relations to escort the guest	• Guest Relations to escort the guest to the new room and ensure that the guest is satisfied with the new room
9.	New room key to be given in place of changed room key	• Give the new key to the guest and old key to be taken back from the guest
10.	Offer assistance to move guest baggage	• Bell boy to be present to move the guest baggage immediately
11.	Room move to be done in the system	• As soon as the guest is moved physically to the new room, room change has to be shown in the system • Go to the reservation folio and show a change of room in the system
12.	Room status to be changed in the system	• Please change the status of old room dirty in the system for house keeping to clean that room.
13.	Concerned departments to be informed about the guest room change	• House keeping • Laundry • Telephone Operator • Cashier
14.	Staff to be informed	• All staff should be informed about the change of guest room
15.	Update guest history	• Update the guest history as well as the reason for room change. • Ensure all guest records are updated.
16.	Room change voucher	• Room change voucher to be made and acknowledged by all departments
17.	Guest registration card to be moved in the back office	• Guest registration card to be moved to new room number slot in the pigeon hole in back office
18.	Room number to be updated on the guest registration card	• Mention new room number on the registration card
19.	Steps to do room change in the system	• Go to the front desk • Go to the Guest In-house option • Enter the guest room number • Go to the options

Contd.

Contd.

S.No.	Procedure	Standard (Measurement, Number, Time, Quantity, Precautions, Phraseology, etc.)
		• Select the room move option • Enter the new room number in which the guest is moved physically • Click OK button to confirm the room move of the guest
20.	Screen will display following options	• Change room status to dirty • Change room status to clean • Do not change room status
21.	Select required option	• Select required option and press enter on that, room change is done on system

Standard Operating Procedure No: 18

Process:	What is the procedure for rate change
Objective :	To maintain records and documentation to avoid discrepancies and for audit purpose
Responsibility :	Front Office Staff

S.No.	Procedure	Standard (Measurement, Number, Time, Quantity, Precautions, Phraseology, etc.)
1.	Explain room rate	• Rate is the amount charged to the guest for the room he is staying in and is printed on the guest registration card
2.	Confirm room rate with the guest	• It is important to confirm the Room rate with the guest on his Check-in
3.	Documentation of room rate change	• Any changes in guest's room rate should be documented and communicated to all concerned
4.	Updation of room rate change	• The rate change should be updated in the guest records and the cashiers should be notified about the same • In case of any changes that has to be done in the rates that is quoted to the guest, reason for rate change has to be mentioned on Registration Card
5.	Room rate may be changed for various reasons	• Discount given to the guest • Up-selling a room to a guest • Change in rate from a rack rate to corporate rate
6.	Steps to do room rate change in the system	• Obtain proper documentation • Change rate in the Registration Card • Mention the reason for rate change in Registration Card • Change rate in the Opera system • Mention the reason for rate change in the Opera system • Inform to the concerned department • Select guest profile for which room rate is to be changed • Click on edit • Enter the new room rate in rate column • Click save • Click OK button to confirm the rate change • Mention the reason for rate change in comments
7.	Note	• Any rate change has to be updated on the registration card and must also mention the reason for rate change

Standard Operating Procedure No: 19

Process:	What is the procedure for receiving the guest wake up call at the reception
Objective :	To attend wake up call requests immediately
Responsibility :	Front Office Staff

S.No.	Procedure	Standard (Measurement, Number, Time, Quantity, Precautions, Phraseology, etc.)
1.	Request for guest wake up call at the reception	• Few requests for wake up call will be taken at the front desk
2.	Guest wake up call request to be met immediately	• All wake up call requests should be attended immediately
3.	Offer reminder wake up call to the guest	• Whenever guest requests for wake up call at the reception, front office associate must check with guest if a reminder wake up call is required
4.	Points to be kept in mind while taking guest wake up call request	• Name of the guest • Room number • Wake up call time

5.	Offer Tea/Coffee assistance to the guest	• Front office associates must also ask if any Tea/Coffee required along with the wake up call
		• If required, then front office associate must ask further questions like:
		   I. Regular coffee or Decaffeinated coffee
		   II. Coffee with milk or black coffee
		   III. Readymade tea or everything separate
		   IV. Regular tea or a Masala tea, etc.

6.	Wish guest a good night sleep	• Front office associate must wish guest "Good Night Mr./Ms. _____"
7.	Inform In room dinning for guest order, any	• Front office associate will give guest order to in-room dinning to be served along with the wake up call
8.	Wake up call sheet is filled giving following details	• Guest Name • Guest Room No. • Wake call time • Reminder wake call time, if any • Any in room dining request • Any other request • Wake call taken by • Wake call given by
9.	Wake call sheet to be acknowledged by the operators	• This sheet is then given to the operator who in turn, calls up the guest at the requested time for the wake up call • A duplicate copy of wake up call sheet with guest IRD request is send to the in-room dining by 01:00 hrs. by the operators acknowledged by IRD associate
10.	Operators to be updated with daily weather report	• Bell desk must give daily weather report to the operators
11.	Guest to be informed about city weather temperature	• Operators must give weather details to the guest along with wake up call
12.	Guest to be acknowledged using their name	• Operators must use guest name

Standard Operating Procedure No: 20

Process:	How to create a sharer in the system
Objective:	To give accurate occupancy status of the room and details of the sharer guest
Responsibility:	Front Office Staff

S.No.	Procedure	Standard (Measurement, Number, Time, Quantity, Precautions, Phraseology, etc.)
1.	Steps to create sharer in the system before Primary guest checks-in	• Go to Front Desk • Go to arrivals
2.	To select Primary guest use following options	• Guest first name • Guest last name • Guest company name, etc. • Go to the options and select shares • Click on Combine on right hand side • Feed last name or first name of the Sharer guest • Press Enter • Fill the required details • Press Enter and the sharer are created • Rate code for sharer is SHARER and the rate must be zero
3.	Steps to create sharer in the system after Primary guest checks-in	• Go to front desk • Go to guest In- house • Enter guest name or guest room number • Go to the options and select share • Select combine and enter the name of the sharer • Fill the details and select the profile, if any • Rate code for sharer is SHARER and the rate must be zero • Click OK • Click close button • Click Yes to save the changes
5.	Note	• Please remember to update the profile of the sharer guest

Standard Operating Procedure No: 21

Process:	How to handle a sharer with advance notice
Objective:	To welcome the notified sharer in a warm and cordial manner
Responsibility:	Front Office Staff

S.No.	Procedure	Standard (Measurement, Number, Time, Quantity, Precautions, Phraseology, etc.)
1.	Explain Sharer	• Sharer: All guest who joins after the first guest has checked in smoothly are called Sharer.
2.	To handle sharer with an advance notice Welcome sharer in a warm manner	• Notified sharer will be welcomed in a warm and cordial manner
3.	Meet guests expectation	• With advance notice the guest expects to have immediate access to the room

Contd.

Contd.

S.No.	Procedure	Standard (Measurement, Number, Time, Quantity, Precautions, Phraseology, etc.)
4.	Name verification of the sharer guest	• Verify the name against arrival information in the system
5.	Information for profile updation	• Collect all information required for profile updation
6.	Show check-in in the system	• Check in the guest into the system
7.	Inform primary guest in the room	• Offer to phone the primary guest in the room to inform about the sharer is arrived
8.	Points to be kept in mind for a Sharer guest	• The rate code of sharer must be "SHARER" • The rate amount of sharer must be "ZERO" • The rate amount must feature in PRIMARY guest room account
9.	Note	• Above things must not be changed unless specified by the guest in specific manner

Standard Operating Procedure No: 22

Process:	How to handle a sharer without advance notice
Objective :	To ensure the privacy of the guest and avoid inconveniences
Responsibility:	Front Office Staff

S.No.	Procedur	Standard (Measurement, Number, Time, Quantity, Precautions, Phraseology, etc.)
1.	To handle sharer without advance notice (Unnotified sharer guest)	• Unnotified sharers/join in would be accommodated in a helpful and tactful manner while preserving the security of the registered guest
2.	Inform duty manager	• Inform duty manager about the Unnoticed Sharer
3.	Call the in-house guest to confirm about the sharer	• If the guest arrives and says that he/she is going to share room with in-house guest but there is no instructions received concerning his/her arrival, call the in-house guest to clarify
4.	Points to be kept in mind if the in-house guest is not available in the room	• Request guest to have seat in lobby • Leave message to the in-house guest, asking the guest to contact front office • Advice sharer guest that he/she will be contacted as soon as the guest returns. Ask sharer to be comfortable in the lobby or in any of the restaurants • Do not allow another person to enter the guest room unless an instruction is received from the guest • On confirmation with the primary guest, check in the sharer guest after completing his registration formalities • Update no. of pax in system • Inform housekeeping to place double amenities in the room

Standard Operating Procedure No: 23

Process:	How to handle guest telephone messages
Objective:	To ensure efficient and prompt delivery of messages to guest rooms
Responsibility:	Front Office Staff

S.No.	Procedure	Standard (Measurement, Number, Time, Quantity, Precautions, Phraseology, etc.)
1.	Messages for the guest to delivered correctly and promptly	• Messages left for a guest over the telephone will be accurately recorded and delivered to the guest • Staff is required to leave text messages for in-house guest and all messages to be delivered to guest promptly
2.	Guidelines to be followed for guest messages, in case guest is not available in the room	• The guest expects to get the telephone message while they are not in the room • In case guest is not in the room then a message can be left on the voice mail, which the guest will receive when he comes back to his room. There is an option in the guest telephone called message by which the guest can retrieve all his voicemail messages. e.g. "Hi I am Peter, please call me back at xxxxxxxxx Thank you." However, if the caller wishes to leave a text message, the operator will transfer the call to the front desk • The front desk personnel should realize that the caller has been online for sometime and must ensure that the message is taken promptly • The received message should then be typed on the system in the in-house screen
3.	Contents of messages	• Name of the caller (correctly spelled) and the company name • The contact details of the calling party • Date and time of call
4.	Steps for printing guest messages in system	• Once the message is typed for an in-house guest, then it should be printed on message slip and the same should be updated on system • Go to main menu • Go to front office option • Press enter on the in-house option • Press enter on either of the following for which the message is to be typed: I. Guest name. II. Room number. • The screen will display the room profile of the guest with options at the bottom • Press enter at the message option • Screen will display previous messages, if any • If there are no previous messages Click on New to type new message • Type new message on the blank screen • After typing the message, click Ok on right hand side. • The screen will have options of print if message is required to be printed

Contd.

Contd.

S.No.	Procedure	Standard (Measurement, Number, Time, Quantity, Precautions, Phraseology, etc.)
		• Click on print button and the message is printed • Once the message is delivered to the room, click on receive button on right hand side to confirm the delivery of message to the guest room
5.	Bell boy to deliver message to the guest room	• Bell boy to deliver the message to the guest in silver tray
6.	Update bell desk control sheet	• Entry of the guest message is made in bell desk control sheet
7.	For urgent delivery of message	• In case the message is an urgent one, please check if the guest is in any of the public areas before the sending the message to the room

Standard Operating Procedure No: 24

Process:	How to do departure control
Objective:	To know actual hotel position
Responsibility:	Front Office Staff

S.No.	Procedure	Standard (Measurement, Number, Time, Quantity, Precautions, Phraseology, etc.)
1.	Objective of departure control	• Departure control must be done for all guest as it helps to know actual hotel position • Departure control helps us to know the exact availability of room
2.	Confirmation of guest departure details on check-in	• At the time of check-in along with other details of the guest receptionist must confirm: – Date of departure of the guest – Time of departure of the guest – Try to sell the hotel transportation for airport drop
3.	Departure control is done a day prior to the guest departure date	• One day prior to the date of departure of the guest, guest relations will call the guest and reconfirms the departure date and time with the guest. "Good Evening Mr./Ms. _____ This is Bhumika calling from the guest relations, How are you today?" Guest: I am good. Bhumika: "Mr./Ms._____, May I reconfirm your departure date and time for tomorrow."
	Note: If a guest wish to have his breakfast before 07:00 am, guest can order for it through in-room dining.	• If guest is departing before 07:00 am and wish to have breakfast before that, guest can avail the facility of having complimentary breakfast through In room dining • Complimentary breakfast facility is offered only to the **FIT's** departing before 07:00 am. Guest: I am leaving at 06:30 hrs. Guest Relations: "Mr./Ms. _____ our breakfast at coffee shop starts at 07:00 hrs and you are checking out at 06:30 hrs, if you wish to have breakfast at that hour you can order for it through in-room dining"
4.	Up-sell hotel transportation	• Guest relations also checks with the guest, if any transportation required on departure "Mr./Ms. _____ would you be requiring the hotel car to drop you to the airport tomorrow on departure"

Contd.

Contd.

S.No.	Procedure	Standard (Measurement, Number, Time, Quantity, Precautions, Phraseology, etc.)
5.	Offer wake up call assistance to the guest	• Guest relations also checks with the guests, if any wake call required "Mr./Ms. Would you wish to place a wake up call for tomorrow." Guest: Yes at 7.00 am
6.	Offer assistance for reminder wake up call and for tea/coffee	Guest Relations: Certainly Mr./Ms., do you wish to have a reminder wake up call and Would you like to have tea/coffee along with your wake call? Guest: No, I don't need the reminder wake up call and I need coffee at 07:45 am. Guest Relations: Certainly Mr./Ms., would you like to have Decaffeinated coffee or a Regular coffee? Guest: No I want regular coffee Guest Relations: Certainly Mr./Ms._____, you wish to have black coffee or coffee with milk. Guest: I want black coffee
7.	Reconfirm wake up call details with the guest	• Guest Relations: Certainly Mr./Ms., I repeat your room no. is 301 and your wake up call is at 07:30 am with no reminder wake up call and with black regular coffee at 07:45 am. Have a nice sleep Mr./Ms
8.	Importance on sold out dates	• On sold out dates departure control plays a very important role to control room reservations
9.	Point to be kept in mind on sold out date	• No extensions or late checkouts to be given to the guest on sold out dates, unless authorized by Rooms Division Manager (RDM)
10.	Steps to enter the departure time of the guest in the system	• Go to the front desk • Go to guest in-house • Enter the guest room number • Click on edit on right hand side • Go to the more fields • Departure time to be entered in C/O time column

Standard Operating Procedure No: 25

Process:	How to handle In-house guest stay extension
Objective :	To handle stay extension in an efficient manner keeping guest informed of the situation
Responsibility :	Front Office Staff

S.No.	Procedure	Standard (Measurement, Number, Time, Quantity, Precautions, Phraseology, etc.)
1.	In house guest stay can be extended depending on the availability of the rooms. Guest stay extension also depends on the type of the reservation.	• All guest extension depends on the availability of rooms and will be communicated to guests
2.	For a Travel Agent reservation	• If rooms are available, request guest to ask the agent to send the amended voucher for the extended number of days/Stay can be extended on direct payment at the best available rate
3.	For a Bill to company reservation	• If rooms available speak to the concerned sales account manager/ reservations manager for the amended bill to company letter for

Contd.

Contd.

S.No.	Procedure	Standard (Measurement, Number, Time, Quantity, Precautions, Phraseology, etc.)
		the extended number of days/Stay can be extended on direct payment at the availed rate
4.	For a Complimentary/House use room	• To check with the Rooms Division Manager
5.	For a World Hotel reservation	• Stay can be extended on direct payment at the best available rate
6.	Points to be kept in mind while handling in-house guest request for stay extension	• In house guest stay extension must be handled in the efficient manner and the guest must be kept informed of the situation • Check room availability and house position to see whether the extension of the stay is possible
7.	Apologize to the guest, if could not meet his request for stay extension	• If available, to be accommodated and if not possible the guest has to be informed immediately about the same by the duty manager "I apologize Mr./Ms. ____ the hotel is completely sold out today and we are not in the position to extend your stay." • If extension is not possible then inform the guest that he would be kept on priority waiting list "We will surely keep you in our priority waiting list, in case of any cancellations or amendments will be informed to you immediately"
8.	Check with duty manager before making commitment to the guest	• Before extending the stay or making any commitments to the guest please check with the duty manager
9.	Guest registration card to be up-dated and room key to be re-programed	• Whenever guest stay is extended, please make sure to change the departure date on the guest registration card and also program guest key for the extended number of days
10.	Offer guest assistance for alternative arrangement on sold out dates	• On sold out dates assistance is offered for making alternative arrangements in another "I sincerely apologize for not being able to extend you stay in our hotel, would you like me to check the availability and reserve you in another five star hotel?"

Standard Operating Procedure No: 26

Process:	How to handle late checkouts
Objective:	Late checkouts should be accommodated according to the set guidelines.
Responsibility:	Front Office Staff

S.No.	Procedure	Standard (Measurement, Number, Time, Quantity, Precautions, Phraseology, etc.)
1.	For late checkout check on the availability and inform guest accordingly	• Late checkouts will be extended subject to availability of rooms and same needs to be communicated to guests in advance. "I apologize Mr./Ms. ____ as the hotel is full we will not be able to accommodate your request for late checkout. The hotel checkout time is 12:00 hrs, at the most we can extend it till 14:00 hrs. Beyond that there will be half day charge till 18:00 hrs and after that there will be full day charge. We also have left luggage facility if you wish to keep your baggage with us"

Contd.

Contd.

S.No.	Procedure	Standard (Measurement, Number, Time, Quantity, Precautions, Phraseology, etc.)
2.	Points to be kept in mind for late checkouts	• Check availability of rooms • Guest request to be fulfilled. Avoid inconvenience to the guest • Check room blocking status and see if there are any rooms, which have been blocked for another guest • As much as possible, try to accommodate guest request • If room is available, confirm extension to the guest • If room is blocked, check if you can change the block and do the needful
3.	Check comments on system for late departure charges	• Explain the late departures charges, only when necessary and applicable • Check if applicable, according to reservation system
4.	Guidelines to be followed for late checkout	• Normal Checkout Time: 12:00 Noon • Late Checkout till 14:00 Hrs: Without any charges • Late Checkout till 18:00 Hrs : Half Day of Room Rate • Late Checkout after 18:00 Hrs : Full Day charge
5.	Authorization for late checkout	• Late checkout has to be authorised by the duty manager or any authorised signatory

Standard Operating Procedure No: 27

Process:	How to do a front office courtesy calling
Objective:	To check if the guest is comfortable and to get feedback
Responsibility:	Front Office Staff

S.No.	Procedure	Standard (Measurement, Number, Time, Quantity, Precautions, Phraseology, etc.)
1.	Courtesy call is made for the following guests	• All VIP In-house guest • All long staying In-house guest depending on the length of their stay • All guest a day prior to their departure
2.	For a VIP guest	• For VIP guest, Guest Relations will make first courtesy call after the ten minutes of the check-in of the guest in the room
3.	Points to be kept in mind while making courtesy calling to a VIP staying guest	Guest Relations will ask guest following questions related to the experience of their stay in the hotel • Is he comfortable in the room? • How did he find the room? • Does he wish to reserve himself for lunch/dinner in the hotel restaurants? • Does he wish to use the health club and beauty parlor services?
4.	Explain a long staying guest	• Any assistance he requires, please call guest relations at extn: _____ • A long staying guest is one who stays in the hotel for a minimum period of 10 nights
5.	For a long staying guest	• For a long staying guest, guest relations will give minimum three courtesy calls to the guest and will have a courtesy meet with the

Contd.

Contd.

S.No.	Procedure	Standard (Measurement, Number, Time, Quantity, Precautions, Phraseology, etc.)
		guest. The process will be repeated every ten days for guest staying for more than ten days • Guest Relations will take the print out of long staying in-house guest report. • Guest Relations will call the guest in their room post 17:00 hrs.
6.	Points to be kept in mind while making courtesy calling to a long staying guest	Guest Relations will ask guest following questions related to the experience of their stay in the hotel. • How is their stay in the hotel? • How is the food and the service of the hotel? • Has he visited all the restaurants of the hotel? • Has he experienced the health club and the beauty parlor services of the hotel? • Any incident occurred in the hotel, which they would like to share with us • Any special feedback the guest wish to give to us • Any thing guest wishes to add to improve/enhance the services of the hotel
7.	First courtesy call for all guests	• Guest Relations will make first courtesy call in ten minutes of the check-in of the guest
8.	Points to be kept in mind while making first courtesy call to the guest	Guest Relations will ask following questions to the guest: • Is he comfortable in the room? • How did he find the room? • Does he wish to place a wake up call? • Any assistance he requires, please call guest relations at Extn: ____.

Standard Operating Procedure No: 28

Process:	How courtesy meet is done
Objective:	To give personalized feel to the guest and to win guest's loyalty.
Responsibility :	Front Office Staff

S.No.	Procedure	Standard (Measurement, Number, Time, Quantity, Precautions, Phraseology, etc.)
1.	Explain Courtesy Meet	• Guest Relations will meet guest personally on appointment as per the guest convenience to gain more information about the guest, his preferences and the feedback on his stay with the hotel.
2.	Guest Relations to call the guest	• Guest Relations will call guest in his room.
3.	Fix up an appointment for courtesy meet as per guest convenience	• Guest Relations will ask guest if its possible to have courtesy meet with him at any time as per his convenience. Preferably for an evening tea in coffee shop • Guest Relations will fix up an appointment with the guest
4.	Offer assistance to remind for courtesy meet	• Guest Relations will check with guest if a reminder in the evening is required

Contd.

Contd.

S.No.	Procedure	Standard (Measurement, Number, Time, Quantity, Precautions, Phraseology, etc.)
5.	Table reservation in coffee shop	• Guest Relations will reserve a table in the coffee shop for a courtesy meet
6.	Guest relations must update herself with the following guest details before going for courtesy meet	• Guest company name • Company details, if possible • Guest designation • Duration of stay in the hotel • Restaurants experienced by the guest so far. (By checking the guest bill folio.) • Special comments or traces, if any
7.	Welcome guest in the lobby	• At the said time Guest Relations has to be ready for the guest in the lobby for courtesy meet • Guest Relations must welcome the guest in the lobby
8.	Introduction with the guest	• Guest Relations must introduce herself to the guest and to proceed towards the coffee shop
9.	Point to be kept in mind for courtesy meet	• During courtesy meet, Guest Relations has to gain maximum information about the guest and provide maximum information about the hotel
10.	Purpose of courtesy meet	• The purpose of courtesy meet is to build a relationship with guest and to win guests loyalty and also to sell the hotel services in a smarter way
11.	Guest feedback to be informed to the concerned departments	• Any good or bad experience of the guest in the hotel is circulated to the concerned departments through Email and updated in guest history
12.	Service recovery to be provided	• If required, service recovery is provided
13.	Update guest relations courtesy meet log book	• Guest relations have to update the Guest Relations courtesy meet log book on daily basis

Standard Operating Procedure No: 29

Process:	What are the various welcome letters
Objective:	To provide information about hotel services and facilities
Responsibility:	Front Office Staff

S.No.	Procedure	Standard (Measurement, Number, Time, Quantity, Precautions, Phraseology, etc.)
1.	The welcome letter for all first time arrival guest	• Welcome letter is placed in the guest rooms prior to guest arrival. Welcome letter gives complete information about the hotel services and facilities and is signed by the Rooms Division Manager • Welcome letter given to all the guest on check in by the front desk associate before proceeding the registration formalities

Contd.

Contd.

S.No.	Procedure	Standard (Measurement, Number, Time, Quantity, Precautions, Phraseology, etc.)
2.	The welcome back letter for all repeated guest	• Welcome letter given to all repeated guest on check-in before proceeding the registration formalities
3.	The welcome letter for all VIP guests	• Welcome letter given to all VIP arrivals by Guest Relations in their room before proceeding In-room registration/check in formalities
4.	The welcome back letter for all repeated VIP guests	• Welcome letter given to all repeated VIP arrivals by Guest Relations in their room before proceeding in-room registration/check-in formalities
5.	The VIP checkout letter	• VIP checkout letter is a letter from the General Manager of the hotel to all VIP guests personally thanking them to have stayed with the hotel and to welcome them back in the near future. This letter is sent to all VIP guests on the day of their departure.

Standard Operating Procedure No: 30

Process:	What is co-ordination meeting
Objective :	To share the information of In house guest and to prepare for next arrivals
Responsibility :	Front Office Staff

S.No.	Procedure	Standard (Measurement, Number, Time, Quantity, Precautions, Phraseology, etc.)
1.	Explain Co-ordination meeting	Co-ordination meeting is a meeting conducted by the Guest Relation with the representatives of the following departments: • Housekeeping • Engineering • In Room Dining
2.	Points to be discussed in co-ordination meeting	• Arrivals for next day • Departures for next day • Group movement for next day • Projected occupancy for next day • Special comments or traces, if any • GSTS feedback • Guest feedback during courtesy call and courtesy meet • General sharing of information.

Standard Operating Procedure No: 31

Process:	What are the different amenities vouchers (F&F) used at Front Office
Objective :	To ensure amenities placed in the room as per requirement
Responsibility:	Front Office Staff

S.No.	Procedure	Standard (Measurement, Number, Time, Quantity, Precautions, Phraseology, etc.)
1.	Explain F&F (Fruits and Flower) Voucher	• F&F voucher is prepared by front office to inform the concerned departments for the amenities to be placed in the guest rooms
2.	Morning F&F reports to in-room dining	• Fruit and Flower (F&F) reports are printed twice a day. • Morning 0800 Hrs: F&F reports: • VIP arrivals • VIP in-house • All arrivals • All in-house
3.	Afternoon F&F reports to in-room dining and Housekeeping	• Afternoon 15:00 Hrs : F&F reports: • VIP In-house for Chocolate drops • Single lady in-house for Chocolate drops and Moisturizer
4.	In case of birthday/ anniversary/guest not well	• F&F is prepared for ½ kg bitter chocolate cake • F&F is prepared for flowers. (Housekeeping) • F&F is also prepared for wine or other amenities to concern departments: I. As per guest request mentioned in the reservation II. Comments or traces in the reservation.
5.	In case of group arrivals welcome drink is prepared for in-room dining with following details	• Name of the group • Arrival time of the group • Number of person • Any preference for welcome drink • Any request mentioned in traces or comments

Standard Operating Procedure No: 32

Process:	What is room discrepancy
Objective:	To ensure there is no discrepancy in the room by the end of the day
Responsibility :	Duty Manager/Front Office Staff

S.No.	Procedure	Standard (Measurement, Number, Time, Quantity, Precautions, Phraseology, etc.)
1.	Explain Occupancy report	• Occupancy report is a report prepared by the housekeeping department giving the physical status of the rooms as per their record and is given to the front office department
2.	Occupancy report is prepared twice a day	• House keeping also updates occupancy report on the system Morning at 10:30 hrs Evening at 21:30 Hrs
3.	Tallying of the occupancy report	• The housekeeping occupancy report will be checked against the front desk status to ensure that there are no discrepancies in the room status
4.	Explain Room discrepancy	• If there is difference in the status of room in housekeeping report and as per front office it is called Room discrepancy • Ensure there is no discrepancy in the room by the end of the day
5.	Points to be kept in mind while	• Scanty Baggage

Contd.

Contd.

S.No.	Procedure	Standard (Measurement, Number, Time, Quantity, Precautions, Phraseology, etc.)
	checking the occupancy report for the room status	• Sleep out/bed not used • Privacy card • Out of order/out of service
6.	Steps to be followed for checking the occupancy report	• Duty Manager will check the status of room in housekeeping report • Duty Manager will check the status of room in the Opera system • If the status of room in housekeeping report is different to the room status in the Opera system then there is discrepancy
7.	Discrepancy can be of following types	• Number of person in the room • Status of the room (Occupied/Vacant/Out of order/Out of service) • Sleep Out • Scanty Baggage • No luggage in the room, etc.
8.	Steps to be followed in case of any discrepancy in the occupancy report	• Any discrepancy in the report, Duty Manager will check with the Asst. Manager/HouseKeeper • Duty Manager will call the guest in the room • If no reply from the guest room, Duty Manager will check the physical status of the room along with the security supervisor • Duty Manager to update the report on the system, if any variance. Same to be resolved within 15 minutes of receiving the report from the housekeeping • Once the status is updated and tallied, the Duty Manager staples both the reports together and files them for future references
9.	Filing of discrepancy report	• Discrepancy report of morning and evening shift is filed in the "Discrepancy Report File" by the Duty Manager on daily basis
10.	Authorize the report	• Two discrepancy reports for the day is signed by the front office manager and the executive housekeeper the next day

Standard Operating Procedure No: 33

Process:	Explain the procedure of room allocation for a VIP arrival
Objective :	To ensure room is ready before VIP arrival and to provide in-room check in to the guest
Responsibility :	Duty Manager/Front Office Staff

S.No.	Procedure	Standard (Measurement, Number, Time, Quantity, Precautions, Phraseology, etc.)
1.	Room allocation for VIP arrivals	• Guest Relations to print the list of VIP arrivals for the next day • Room allocation for VIP guest is to be done by guest relations in co-ordination with the duty manager
2.	Points to be kept in mind for room allocation of the guest Check for guest preferences/comments/traces	• Guest Relations to check for the traces, comments, preferences, if any

Contd.

Contd.

S.No.	Procedure	Standard (Measurement, Number, Time, Quantity, Precautions, Phraseology, etc.)
3.	Allocate best available room	• Guest Relations to allocate the best available room for a VIP guest
4.	Room status to be made do not move	• Allocated room status to be made DNM (Do not move)
5.	Time of arrival of the guest	• Guest Relations to check the expected time of arrival of a VIP guest to ensure that the room is ready before his arrival
6.	Amenities voucher for a VIP guest	• Guest Relations to send the amenities voucher for a VIP guest indicating expected time of arrival to ensure amenities ..re placed before guest arrival
7.	Trace for In room check	• Leave trace for in room check in the Fidelio system
8.	Information to the concerned departments	• Any special instruction related to VIP arrival is informed to concerned department
9.	Note	• Room allocation for a VIP arrival is to be done a day before by 18:00 hrs

Standard Operating Procedure No: 34

Process:	Explain the procedure of room allocation for guests
Objective:	To assign room according to guest request and changes should be kept to a minimum. To ensure the room is ready before guest arrival
Responsibility:	Duty Manager/Front Office Staff

S.No.	Procedure	Standard (Measurement, Number, Time, Quantity, Precautions, Phraseology, etc.)
1.	Room allocation for a regular guest	• Room allocation for a regular guest is done by night manager
2.	Points to be kept in mind for room allocation of the guest Check for guest preferences/comments/traces	• Room allocation is done keeping in consideration guest preferences, comments, traces, if any
3.	Room status to be made do not move	• Allocated room status to be made DNM (Do not move) if done as per guest preference
4.	Time of arrival of the guest	• Room allocation is done considering the time of arrival of the guest to ensure room is ready before guest arrival
5.	Allocation of rooms on sold out dates	• On sold out dates room blocking is done as per the guest's expected time of arrival keeping in mind back to back room

Standard Operating Procedure No: 35

Process:	Explain the procedure for printing guest registration card
Objective:	To provide smooth and efficient check-in to the guest
Responsibility :	Front Office Staff

S.No.	Procedure	Standard (Measurement, Number, Time, Quantity, Precautions, Phraseology, etc.)
1.	Explain registration card	• Registration card gives the details about the guest staying in the hotel • Its a legal requirement to fill the registration card for all the guest staying in the hotel
2.	Registration card for next days	• Registration card for next day arrivals is arrivals printed by the previous day night shift front desk associate

3.	Steps for printing registration card for future date	• Go to the reservations
		• Go to update reservations on left hand side
		• Enter the guest name and select it
		• Click on edit
		• Click on option on right hand side
		• Select the option of register card
		• Click on print and the registration card is printed
		• Put either of the following details for which registration card is to be printed :
		   I. Reservation/Arrival Date
		   II. Guest Name
		   III. Company Name
		   IV. Travel Agency Name
		   V. Confirmation Number

4.	Registration cards are printed and attached with	• Correspondence, if any • Hotel Welcome Letter • Fax, message, if any
5.	Arrangement of registration card	• Registration cards are then arranged in Piano file in alphabetical order.
6.	Registration card for same day arrivals may be printed for the following reasons	• Registration card for the same day is printed when the guest reservation is made on the same day of arrival • Registration card for same day is printed in case of walk-in guest
7.	Steps to be followed for printing registration card for same days arrivals	• Go to the arrivals • Select the profile for which registration card is to be printed • Click on registration card option on right hand side and the registration card is printed • For In-house guest select the guest room number • Click edit • Go to the options • Select the option of register card • Click on print and the registration card is printed
8.	Put either of the following details for which registration card is to be printed	• Guest Name • Company Name • Travel Agency Name • Confirmation Number

Standard Operating Procedure No: 36

Process:	What is the procedure for Master key
Objective:	To avoid misuse of Master key
Responsibility:	Front Office Staff

S.No.	Procedure	Standard (Measurement, Number, Time, Quantity, Precautions, Phraseology, etc.)
1.	Master key in front office department	• The front office department has two Master keys. • One with Duty Manager and second with the Rooms Division Manager/Housekeeping
2.	Function of master key	• It opens all guest rooms • It opens all corporate offices • It opens all double lock rooms
3.	Validity of master key	• Master key is activated for the period of 90 days. On 90th day at 12:00 Noon card gets de-activated automatically
4.	Programing of master key	• Only IT department has the rights to program the Master key
5.	Usage of master key by authorized personnel only	• Master key to be used only by the authorized personnel, i.e Duty Manager • Employees authorized by the duty manager can only use Master key
6.	Master key control register	• Each time Master key is used to open a guest room for any reason, entry has to be made in Master key control register, e.g.: Guest Relations for inspecting the guest rooms, in-room dining attendant for placing amenities in the room, etc.
7.	Master key control register has the following details	• Date • Name of the employee who has taken the Master key • Department of the employee • Time of issue of Master key • Purpose of issuing the Master key • Signature of the employee • Time of depositing back the Master key • Signature of the employee at the time of returning back the Master key • Signature of the Duty Manager
8.	History of CISA lock	• To check the history of CISA lock, we have CISA machine and data cord with data card attached to it
9.	Steps for taking the reading of CISA lock	• Insert the data card to the guest room (for which reading needs to be taken) • Press menu button • Go to data reading • Enter number from 1 to 100 to record the data. • The record of that many numbers of last readings of the lock will be recorded in the card • It will give the reading of various keys used to open the room lock • Take the print out of the card reading

Standard Operating Procedure No: 37

Process:	What is C-form
Objective :	To maintain hotel records and for legal requirement
Responsibility :	Front Office Staff

S.No.	Procedure	Standard (Measurement, Number, Time, Quantity, Precautions, Phraseology, etc.)
1.	Explain C- form	• C- forms are maintained for all foreign nationalities staying in the hotel as a legal requirement and the same is to be submitted to the nearest FRRO office (Foreigner's Regional Registration Office), within 24 hours of next working day.
2.	Contents of C-form	• Name of the guest • Nationality • Passport No • Date of issue of passport • Place of issue of passport • Address • Date of arrival in India • Arrived from • Employed in India • Purpose and duration of stay in India • Certificate of registration with date of issue and place of issue • Date of arrival in hotel • Proceeding to which place
3.	C-form is printed in 3 copies by night shift reception.	• For Front Office record • For FRRO Office • For acknowledged copy for Front Office
4.	Steps to be followed for printing C-form in the system	• Go to the Miscellaneous • Go to the Reports • Enter C-form in Report column. There is a option of two C form reports • C-form report will give you the option of date for which the report is to be printed • C-form yesterday will print the report for previous date
5.	For Pakistan nationals	• C-form of Pakistan nationality must reach FRRO within 12 hours of time of arrival of guest • A guest with Pakistan nationality has to report to the nearest police station.

Standard Operating Procedure No: 38

Process:	What are the back-up reports
Objective :	To ensure smooth functioning of front desk activities when system is shut down
Responsibility :	Front Office Staff

S.No.	Procedure	Standard (Measurement, Number, Time, Quantity, Precautions, Phraseology, etc.)
1.	Following back-up reports will be printed at the front desk for the emergencies by all shifts	• Go to the Miscellaneous • Go to the Reports on left hand side. Print following reports: • Guest Open balance report • Arrival report • VIP arrival report • Departure report for Due outs • Vacant report • Guest In-house report • Housekeeping status report • Trace report
2.	Note	• Obtain information for system shut down in night shift before running night audit. • Emergency may also arise to shut down the system for system upgradation by IT department.

Standard Operating Procedure No: 39

Process:	What is First-aid box
Objective :	To provide first-aid in emergencies
Responsibility:	Front Office Staff

S.No.	Procedure	Standard (Measurement, Number, Time, Quantity, Precautions, Phraseology, etc.)
1.	Explain first-aid box	• First-aid box consist of basic medicines and first-aid kit is required to provide first-aid to the guest or to the hotel staff • First-aid Box is kept at the concierge with bell desk at all times

2.	First-aid box comprises the following	• Medicines
		  I. Disprin – For headache.
		  II. Saridon – For headache.
		  III. Pudinhara – For stomach upset.
		  IV. Imodium – For stomach upset.
		  V. Avil – For allergy.
		  VI. Crocin – For cold and fever.
		  VII. Eno – For gastric problems.
		  VIII. Digene – For proper digestion.
		• Band aid
		• Savlon
		• Burnol
		• Cotton
		• Bandage
		• Thermometer
		• Scissor
		• Relief Spray

Glossary of Terms Used in Hotel Front Office

0-Call (Zero-call)

A telephone call placed with an operator's assistance. Examples may include calling and credit card calls, collect calls, and third-party calls.

AAA

American Automobile Association

AARP

American Association of Retired Persons

Accessible

A guest room that is designed to accommodate persons with disabilities by removing barriers that otherwise limit or prevent them from obtaining the services that are offered.

Account posting machine

A device used to post, monitor and balance charges and credits to guest account. There is a key pad in the account posting machine which is used by the cashier to enter the room number of the guest, department key (i.e. room, tax, food) and also the type of transaction (i.e. debit, credit and transfer).

Accounts payable

Financial obligations the hotel owes to private and government-related agencies and vendors.

Accounts receivable

Amount of money owed to the hotel by guests

Adjacent rooms

Rooms close to each other, perhaps across the hall.

Adjoining rooms

Guest rooms located side by side (one common wall) without a connecting door between them. Not all adjoining rooms are connecting, however, every connecting rooms are adjoining.

Advance deposit guarantee

A type of reservation guarantee which requires the guest to furnish a specified amount of money in advance of arrival.

Adventure tour

A tour designed around an adventurous activity such as rafting or hiking.

Affiliate reservation network

A hotel chain's reservation system in which all participating properties are contractually related. Each property is represented in the computer system database and is required to provide room availability data to the reservation center on a timely basis.

Affiliated hotel

A hotel that is a member of a chain, franchise, or referral system. Membership provides special advantages, particularly, a national reservation system.

After-departure charge

Expenses such as telephone charges that do not appear on a guest's folio at checkout.

Aging of accounts

A method for tracking past due accounts according to the date of charges originated. It is an indication of the stage of the payment cycle—such as 10 days old, 30 days overdue, 60 days overdue.

AIDA

A management term referring to marketing management and the abbreviation stands for Attention, Interest, Desire and Action.

Air travel

The use of air transport to get customers to their destination.

Airline fare

The price that is charged for an airline ticket. Some of the categories are as follows:

- Advance Purchase Excursion (APEX): Heavily discounted excursion fare available on many international routes. Reservations and payment will be required well in advance of departure, with varying penalties for cancellation.
- Excursion: Individual fares that require a round-trip within time limits, discounted from coach fare, limited availability.
- Group: Discounts from regular fares for groups.
- Regular or Normal: Any unrestricted fare.

Airline Reporting Conference (ARC)

A consortium of airline companies who, by agreement, provide a method of approving authorized agency locations for the sale of transportation and cost-effective procedures for processing records and funds to carriers. Not all airlines are ARC companies.

Airport hotel

A hotel located near a public airport. Airport hotels vary widely in size and service level.

Ala carte menu

A food and beverage menu in which each item is listed and priced separately.

All suite

Hotel type that offers suites as the primary room type.

All-inclusive

A form of package holiday where the majority of services offered at the destination are included in the price paid prior to departure (e.g. refreshments, excursions, amenities, gratuities).

Allocentric

A term used to describe a person who is more adventurous and willing to travel to exotic destinations, and who travels more frequently and by more modern or unusual forms of transportation. Allocentric travelers are apt to spend more money than psychocentric travelers.

Allowance voucher

A voucher used to support an account allowance.

Allowance

A monetary reduction given on a guest bill as compensation for unsatisfactory service or a correction of a posting error detected after the close of business. Also known as *Rebate*.

Amenities

Something, such as a swimming pool or shopping centre that is intended to make life more pleasant or comfortable for the people in a hotel or city.

American Bus Association (ABA)

A trade organization consisting of bus companies, operators and owners.

American Hotel and Lodging Association (AHLA)

A professional association of hotel owners, managers, and related occupations.

American Hotel and Motel Association (AH &MA)

A federation of state and regional hotel associations that offers benefits and services to hospitality properties and suppliers. AH&MA reviews proposed legislation affecting hotels, sponsors seminars and group study programs, conducts research, and publishes Lodging magazine. The Educational Institute of AH&MA is the world's largest developer of hospitality industry training materials, including textbooks, videotapes, seminars, courses, and software.

American Plan (AP)

A type of room rate that includes the price of the room, breakfast, lunch and dinner. Also known as *Full Pension, En-pension* or *Full Board Basis*.

American Society of Travel Agents (ASTA)

The oldest and largest travel agent organization in the world with travel agents being the primary members.

Americans with Disabilities Act (ADA)

A US law enacted in 1990 that protects people with disabilities from being discriminated against when seeking accommodations, public transportation and employment.

Apartment Hotel

Accommodation in apartment-style units rather than rooms: With minimum or expanded in-suite cooking facilities.

Arrival/Departure Report

A night audit report that summarizes all the check-ins and checkouts that occurred in the course of a day.

Assets

Items that have monetary value.

Association of British Travel Agents (ABTA)

It represents the interests of the larger UK tour operators and travel agents and operates a bonding scheme whereby customers booking with ABTA members have their holidays protected should the operator/agent in question collapse.

Atrium concept

A design in which guest rooms overlook the lobby from the first floor to the roof.

Atrium

A guest room floor configuration in which rooms are laid out off a single-loaded corridor encircling a multi-storey lobby space; also the multi-storey lobby space, usually with a skylight.

Attractions

A natural or man-made facility, location or activity that have visitor appeal, like museums, historic sites, performing arts institutions, preservation districts, theme parks, entertainment and national sites.

Attrition

The difference between the original request and the actual purchases of a group. For example, a group might reserve one hundred rooms, but actually use only fifty. The hotel's standard group contract may, in such a case, stipulate that the group pay a penalty for 'over-reserving'.

Audiovisual equipment

Those items including DVD players, laptops, LCD projectors, microphones, sound systems, flip charts, overhead projectors, slide projectors, TVs, and VCRs that are used to communicate information to meeting attendees during the meetings.

Audit trial

An organized flow of source documents detailing each step in the processing of a transaction.

Audit work time

The period from the end of day until the completion of the audit.

Auditing

The process of verifying front office accounting records as to accuracy and completeness.

Authorization code

A code generated by an online credit card verification service, indicating that the requested transaction has been approved.

Automated Call Distributor (ACD)

Routes reservations calls to available agents.

Availability report

A report that lists, by room type, the number of available rooms each day. This report contains expected arrival and departure information for the next several days, typically prepared as part of the night audit.

Average occupancy

A ratio that shows rooms sold over a fixed period of time as a percentage of total available rooms in a property over the same period of time.

Average rate per guest/average person rate

An occupancy ratio derived by dividing net room revenue by the number of guests.

Average rate per room occupied

A very useful statistic that is calculated by dividing total sales of rooms during a set period by the total number of rooms sold during that period. This gives the average selling price of all guest rooms for a given time period. Also known as *Average Room Rate (ARR)* or *Average Daily Rate (ADR)*

Baby sitter

A person who takes care of babies while their parents are away from hotel/home and is usually paid for this service.

Back of the house

The functional areas of a hotel in which personnel have a little or no direct guest contact, such as engineering, accounting, and human resources divisions.

Back to back

A full house situation wherein there are equal numbers of arrivals and departures.

Bag pull

At a predetermined time in the day (usually when all group attendees are in session), the bell staff go into each room and 'pull' each attendee's luggage. This luggage is then stored until the group is ready to depart.

Banquet sheet

A listing of the details of an event at which food and beverages are served.

Bed and breakfast (B&B)

A type of room rate which includes the price of the room and breakfast. Also known as *Continental Plan.*

Bed and breakfast hotel

Overnight accommodation usually in a private home or boarding house, with full American-style or Continental breakfast included in the rate, often without private bath facilities.

Bell captain

The leader of the bell staff.

Bell cart

A large metal cart on rollers that bell staff use to carry luggage to and from a guest room.

Bell desk

A desk located in the lobby from where bell boys operate. Also called *Porters' Lodge*.

Bell staff

People who lift and tote baggage, familiarize guests with their new surroundings, run errands, deliver supplies, provide guests with information on in-house marketing efforts and local attractions, and act as the hospitality link between the lodging establishment and the guest.

Blacklisted guest

A list of unwanted guests. These may be those guests who make a lot of fuss and create problems for the hotel (non-payment of bills, damage to hotel property, misbehave with staff or other guests)

Blackout

Total loss of electricity.

Block booking

A number of hotel rooms held without deposit by wholesalers, tour operators, or receptive operators who intend to sell them as components of tour packages.

Block

An agreed-upon number of rooms set aside for members of a group planning to stay at a hotel.

Book

To reserve rooms ahead of time.

Booking lead time/reservation lead time

A measurement of how far in advance bookings are made. Time period between the request for reservation and the actual date of arrival.

Bottom up

A selling approach that involves presenting the least expensive rate first.

Boutique hotels

Sometimes called 'design hotels' or 'lifestyle hotels', they differentiate themselves from larger chain or branded hotels by their intense focus on the physical space through the design of their facilities. These properties have typically between 10 to 100 rooms and often contain luxury facilities in unique or intimate settings with full service accommodations.

Brand loyalty

The institutionalized preferences of a consumer for a product or service based on a brand name or logo.

Brownouts

Partial loss of electricity.

Bucket check

The night auditor's check of room rate postings on guest folio against the housekeeping department's occupancy report and the front desk room rack. Helps ensure that rates have been posted for all occupied rooms and helps reduce the occupancy errors caused when front desk agents do not properly complete check-in and checkout procedures.

Bundling

The process of combining one or more hotel products or services together to make the new entity more attractive; most commonly used with package rates.

Business travel

Traveling for work purposes or travel for a purpose to a destination determined by a business with leisure as a secondary motivation, and where all costs are met by that business.

Call accounting system

It is used by the telephone exchange section to automatically trace and bill the outgoing calls made by the guests during their stay in the hotel.

Cancellation

A reservation voided/withdrawn at the request of the guest.

Cancellation date

Indicates the date when the reservation was manually cancelled.

Cancellation hour

A specific time after which a property may release for sale all unclaimed non-guaranteed reservations, according to property policy. Also called *Cut-off hour*.

Cancellation number

A number issued to a guest who properly cancels a reservation, proving that a cancellation was received and acted upon.

Carrier

Any provider of mass transportation such as an airline, motor coach, bus, cruise line or railroad which carries passengers and/or cargo.

Cash advance voucher

A voucher used to support cash flow out of the hotel, either directly to or on behalf of a guest.

Cash bank

This is the amount of money which is assigned to the cashier by the accounts department to carry out various transactions smoothly during the shift.

Cash flow

Money available to meet the company's daily operating expenses, as opposed to equity, accounts receivable, or other credits not immediately accessible

Cash voucher

A voucher used to support a cash payment transaction at the front desk.

Cashier's report

A daily cash control report that lists cashier activity of cash and credit cards and machine totals by cashier shift

Casino hotel

A hotel that features legal gambling.

Catering manager

A hotel manager who promotes and sells hotel's banquet facilities and uses his expertise to plan, organize, and execute hotel banquets.

Central reservation office

A central reservation office typically deals directly with the public, advertises a central telephone number (usually toll-free), provides participating properties with necessary communications equipment, and bills properties for handling their reservations. Each participating property is required to provide room availability data to the CRO on a timely basis.

Centralized electronic locking system

An electronic locking system which operates through a master console at the front desk which is wired to every guest room door.

Certified Hospitality Housekeeping Executive (CHHE)

Certification for Executive Housekeepers offered through the Education Institute of AH&LA

Certified Tour Professional (CTP)

A designation conferred upon tour professionals who have completed a prescribed course of academic study, professional service, tour employment and evaluation requirements. It is administered by the National Tour Association.

Certified Travel Counselor (CTC)

A designation conferred upon travel professionals who have completed a travel management program offered by the 'Institute of Certified Travel Agents'.

Chain

A group of hotels that follow standard operating procedures such as marketing, reservations, quality of service, food and beverage operations, housekeeping, and accounting

Chain hotel

A hotel owned by or affiliated with other properties.

Chamber of Commerce (C of C)

Serves as an advocate for the community and business, as well as a resource for consumers and businesses. Chambers of commerce comprises local business people, council representatives, etc. They aim to improve local economic development through a variety of means, for example, professional development of members and promotion of the town as a business centre or tourism centre.

Charge back

Refusal of payment of a hotel submitted voucher by a credit card charges company.

Charge voucher

A voucher used to support a charge purchase transaction that takes place somewhere other than the front desk. Also referred to as an *Account Receivable Voucher*.

Charter

To hire the exclusive use of any aircraft, motor coach, or other vehicle

Charter group

Group travel, in which a pre-formed group travels together usually on a customized itinerary.

Chauffer

He is a person employed to drive a passenger motor vehicle by the car rental operator.

Check-in

The procedures for a guest's arrival and registration.

Checkout

The procedure for a guest's departure from hotel. It involves settlement of the guest's account and returning the room keys.

Checkout time

The hour by which departing guests must checkout of a property.

City Centre hotel

Full-service hotel located in a downtown area.

City ledger

The collection of all non-guest accounts, including house accounts and unsettled departed guest accounts. Tracks revenues due to the hotel.

Class A fires

The burning of ordinary combustibles such as wood, paper, and cloth; can be extinguished by the cooling action of water-based or general purpose chemicals.

Class B fires

Fires involving flammable liquids such as grease, gasoline, paints, and other oils; can be extinguished by eliminating the air supply and smothering the fire, not by using water.

Class C fires

Electrical fires, usually involving motors, switches, and wiring; can be extinguished with chemicals that do not conduct electricity, never with water.

Closed

The status of a date for which a hotel will not accept additional reservations.

Closed to arrival

A revenue management strategy that allows reservations to be taken for a particular date as long as the guest arrives before that date.

Cold calls

Sales calls on customers with no previously identified interest in a product.

Commercial hotel

A property, usually located in downtown, or business district, that caters primarily to business clients.

Commercial rate

A special room rate agreed upon by a company and a hotel for frequent guests based on volume of business. Usually the hotel agrees to supply rooms of a specified quality or better at a flat rate to corporate clients. The company signs a contract with the hotel guaranteeing a certain volume of business and accepts financial responsibility for no-shows. Also called *Corporate Rate or Company Volume Guaranteed Rate (CVGR)*.

Commission

A percentage of the total product cost paid to travel agents and other travel product distributors for selling the product to the consumer. Travel agents usually receive an amount averaging no less than 10% of the retail price. Wholesalers or inbound tour operators usually receive 20–30% of the advertised price. Commission levels for online travel agencies vary.

Commissioned tours

A tour available for sale through retails and wholesale travel agencies which provides for a payment of an agreed upon sales commission either to the retail or wholesale seller.

Common areas/public areas

Hotel spaces where most, if not all, guests may walk through.

Complimentary room

A complimentary or 'comp' room is an occupied room for which the guest is not charged.

Complimentary

No charge is made for the item or service offered.

Complimentary occupancy percentage

A ratio that shows the percentage of occupied rooms that are complimentary and generate no revenue; calculated by dividing complimentary rooms for a period by total available rooms for the same period.

Concierge

The individual within a full-service hotel responsible for providing guests with detailed information regarding local dining and attractions, as well as assisting with related guest needs such as restaurant reservation and travel arrangements. A person who provides an endless array of information on entertainment, sports, amusements, church services, and babysitting in a particular city.

Condominium hotel

A hotel in which an investor takes title to a specific hotel room, which remains in the pool to be rented to transient guests whenever the investor is not using the room. The investor expects to receive a gain from the increase in value of the hotel over time, as well as receive ongoing income from the rental of his room.

Conducted tour

A pre-arranged travel program, usually for a group, that includes escort service or a sight-seeing program, such as a city tour, conducted by a guide. Also called an *Escorted Tour*.

Conference call

A conversation in which three or more persons are linked by telephone.

Conference centre

A specialized hotel, usually accessible to major market areas but in less busy locations, that almost exclusively books conferences, executive meetings, and training seminars. A conference centre may provide extensive leisure facilities.

Confirmed reservations

Prospective guests who have a reservation for accommodations that is honored until a specified time (usually 6 PM)

Continental breakfast

A small meal served in the morning that usually includes juice, tea or coffee, fruit, sweet roll, butter, jam and/or cereal.

Continental Plan (CP)

A hotel rate which includes a continental breakfast with the overnight room stay.

Control folio

An accounting document used internally by a front office computer to support all account postings by department during a system update routine.

Convention and Visitors Bureau (CVB)

These organizations are local tourism marketing organizations specializing in developing con-ventions, meetings, conferences and visitations to a city, country or region. A CVB promotes tourism, encourages groups to hold meetings and trade shows in its city and assists groups before and during meetings. A non-profit organization, generally funded by taxes levied on overnight hotel guests that seek to increase the number of visitors to the area it represents.

Copyright

Literally 'the right to copy' an original creation and constitutes a set of exclusive rights to the use of an idea or information.

Corkage

A charge levied by a hotel when a guest brings a bottle (e.g. of a special wine) to the hotel for consumption at a banquet function or in the hotel's dining room.

Corporate guaranteed reservation

A type of reservation guarantee in which a corporation signs a contractual agreement with the hotel to accept financial responsibility for any no-show business travelers it sponsors.

Correction voucher

A voucher used to support the correction of a posting error which is rectified before the close of business on the day the error was made.

Cost center

A hotel department that incurs costs in support of a revenue center. Example, housekeeping and maintenance department.

Cost per key

The average purchase price of a hotel's guest room.

Cost Per Occupied Room (CPOR)

Those room-related costs incurred directly as a result of selling a guest room. Examples include labor costs, room supplies, and room amenities. Also referred to as *room-related occupancy cost or occupied room cost or revenue per occupied room.*

Cost rate formula

Method of determining initial hotel rates based on construction costs. This formula determines the average room rate by allocating $1 towards the rate for each $1,000 spent in construction.

Cover

Each diner at a restaurant.

Credit card guaranteed reservation

A type of guarantee supported by credit card companies. These companies guarantee participating properties payment for reserved rooms that remain unoccupied.

Credit card imprinter

The credit card imprinter makes an imprint of the credit card the guest will use as the method of payment. An imprinter presses a credit card voucher against a guest credit card. The impact causes the raised credit card details like the number, expiry date, name of the card holder, etc. onto the credit card voucher for use in credit card billing and collection procedure.

Credit card validator

The equipment is a computer terminal linked to a credit card data bank, which holds information concerning the validity of the credit card of the guest. The equipment assures the management that the guest has credit balance high enough to cover the projected charges and it also verifies that the card presented by the guest is not a stolen property.

Credit card voucher

The form designated by a credit card company to be used for imprinting a credit card and recording the amount charged. Also called a *credit card invoice*.

Credit facility

Facility given to guests to charge expenses to their account during their stay.

Crib/baby cot

A baby's bed

Cross-training

Training employees for performing multiple tasks and jobs.

Customer relationship management

A system that allows hotel managers to integrate technology to support customer service techniques that provides top-notch customer service. It is a widely implemented model for managing a company's interactions with customers, clients, and sales prospects.

Customized tour

A tour designed to fit the specific needs of a particular target market.

Customs

The common term for a government agency charged with collecting duty on specified items imported into that country. The agency also restricts the entry of persons and forbidden items without legal travel documents.

Cut-off date

The date agreed upon between a group and a hotel after which all unreserved rooms in the group's block will be released back to the general rooms inventory for sale.

Cut-off hour

That time at which the day's unclaimed reservations are released for sale to the general public. Also called *Cancellation hour*.

D card

This is the night auditor's report used in the semi-automated front office accounting system and is a formon which the totals of the front office posting machine are summarized and posted.

Daily flash report

Daily information provided to the GM that reports key financial information from the previous day (often accumulated for the month and/or year-to-date and compared to actual data from previous years).

Daily function board

The place in the lobby and other locations within the hotel used to post-information such as name of function, time, and banquet hall for events scheduled on a specific date.

Daily operations report

A report, typically prepared by the night auditor, that summarizes the hotel's financial activities during a 24-hour period and provides insight into revenues, receivables, operating statistics, and cash transactions related to the front office. Also known as the *Manager's Report*.

Daily sales report

A financial activity report produced by a department in a hotel that reflects daily sales activities with accompanying cash register tapes or point-of-sale audit tapes

Daily transcript

A detailed report of all guest accounts that indicates each charge transaction affecting a guest account for the day, used as a worksheet to detect posting errors.

Database interfaces

The term used to describe the process in which one data generating system shares its data electronically with another system.

Date roll

A certain point in the night to establish a change in date.

Day rate

A reduced rate granted for the use of a guest room during the daytime up to 5 pm, not overnight occupancy. Usually one-half the regular rate.

Day shift

A hotel workshift, generally 7 am to 3 pm.

Day use

A room status term indicating that the room will be used for less than an overnight stay.

Day-trippers

Tourists who visit a destination and return home on the same day. Also known as *Excursions*.

Debit cards

Embossed plastic cards with a magnetic strip on the reverse side that authorize direct transfer of funds from a customer's bank account to the commercial organization's bank account for purchase of goods and services

Deep cleaning

The intensive cleaning of a guest room, typically including the thorough cleaning of items such as drapes, lamp shades, carpets, furniture, walls, and the like.

Delegation

The process of assigning authority (power) to others to enable subordinates to do work that a manager at a higher organizational level would otherwise do.

Delinquent account

A city ledger account that has not been settled within the reasonable collection period, usually 90 days.

Demographic profile

Personal information about customers used to understand their buying or selling preferences, for example, age, income and gender.

Denial code

A code generated by an online credit card verification service, indicating that the requested transaction has not been approved.

Deposit

Money paid to secure a reservation.

Destination Management Company (DMC)

A company working in a specific destination to handle all bookings and arrangements for tours or conferences, including hotel accommodation, transfers, sightseeing, meetings and special events. Tour operators or conference planners are likely to use the services of a DMC because of their specialist local knowledge. Also referred to as a *ground operator*.

Destination Marketing Organization (DMO)

A non-profit marketing organization for a city, state, province, region or area whose primary purpose is the promotion of the destination.

Destination marketing

Advertising and promotions, aimed at consumer and trade, designed to build awareness and desire to travel to a particular location.

Destination

In the travel industry, any hotel, resort, attraction, area, city, state or country which can be marketed as a single entity for tourists.

Detained property

Personal property of a guest that is held by a hotel until payment is made for the purchase of lawful products/services.

Did Not Checkout (DNCO)

A room status term indicating that the guest made arrangements to settle his or her account (and thus is not a skipper), but has left without informing the front office.

Differential pricing

The practice of a seller charging different prices to different buyers for the same product or slightly different versions of the same product.

Direct billing

A credit arrangement, normally established through correspondence between a guest or a company and the hotel, in which the hotel agrees to bill the guest or the company for charges incurred.

Discount grid

A grid indicating the occupancy percentage needed to achieve equivalent net revenue, given different discount levels.

Discounted fare

Negotiated air fare for convention, trade show, meeting, group or corporate travel.

Displacement

The turning away of transient guests for lack of rooms due to the acceptance of group business. Also called *non-group displacement*.

DND

Do Not Disturb

Domestic tourism

Travel within the traveler's country of residence.

Doorknob menu

A type of room service menu that a housekeeper can leave in the guest room. A doorknob menu lists a limited number of breakfast items and times of the day that the meal can be served. Guests select what they want to eat and the time they want the food delivered, then hang the menu outside the door on the doorknob. The menus are collected and the orders are prepared and sent to the rooms at the indicated times.

Double bed

A bed approximately 54 inches by 75 inches or 57 inches by 81 inches

Double occupancy percentage

Double Occupancy Percentage

$$= \frac{\text{House Count} - \text{Room Occupied}}{\text{Number of Room Available}} \times 100$$

Double

A room assigned to two people; may have one or more beds.

Double-double

A room with two double (or perhaps queen) beds; may be occupied by one or more people. Also called a *twin-double*.

Double-locked room

An occupied room for which the guest has refused housekeeping service by locking the room from the inside with a dead bolt. Double-locked rooms cannot be accessed by a room attendant using a standard passkey.

Downgrade

To move to a lesser level of accommodations or a lower class of service.

Downsizing

Reducing the number of employees and/or labor hours for cost-containment purposes.

Due back

A situation that occurs when a cashier pays out more than he receives, the difference is due back to the cashier's cash bank. In the front office, due backs usually occur when a cashier accepts many cheques and large bills during a shift, such that he or she cannot restore the initial bank at the end of the shift without using cheques or large bills.

Due out

A room status term indicating, that the room is expected to become vacant after the following day's checkout time.

Duration control

Duration rules and limitations related to guests' arrival dates, departure dates and minimum stay lengths. Also known as *Stay controls*.

Dwell time

The length of time a visitor spends at an attraction or destination. Dwell time is often taken into consideration when setting entry fees as a way of ensuring perceived value for money.

Early arrival

A guest who arrives at a property before the time or date of his reservation. Also called *Waits or Early Check-in*.

Early departure

A guest who checks out of the hotel before his or her originally scheduled checkout date.

Early make-up

A room status term indicating that the guest has reserved an early check-in time or has requested his or her room to be cleaned as soon as possible.

Early out

A clause in a franchisee agreement that grants both the franchisor and the franchisee the right, with proper notification, to terminate the agreement after it has been in effect for a relatively short period of time. When this clause exists, a window may be granted after only one, two, or three years.

E-commerce

A system of conducting business activities using the Internet and other information technologies. Refers to using computer networks to conduct business, including buying and selling online, electronic funds transfer, business communications, and other activities associated with the buying and selling of goods and services online.

Economy/limited service

A level of service emphasizing clean, comfortable, inexpensive rooms that meet the most basic needs of guests. Economy or limited service hotels appeal primarily to budget-minded travelers.

Ecotel

Hotel with strict standards pertaining to preservation of the ecological system.

Eco-tourism

Low-impact tourism that avoids harming the natural or normal environment. In this relatively new approach to promoting enjoyment, as well as protection, of the environment, tourists seek out environmentally-sensitive travel and/or tours or vacations which, in some way, improve or add to their knowledge of an environment. It is a responsible travel to natural areas that conserves the environment and sustains the well-being of local people.

Educational tour

Tour designed around an educational activity.

E-fridge

A cabinet usually including both refrigerated and non-refrigerated sections designed with an electronic processing unit that allows direct PMS interface.

Electronic key system

A system composed of battery-powered or, less frequently, hardwired locks; a host computer and terminals; a keypuncher; and special entry cards that are used as keys.

Electronic key/card key

A plastic card, resembling a credit card, with electronic codes embedded on a magnetic strip, used in place of a metal key to open a guest room door.

Electronic locking system

A locking system that replaces traditional mechanical locks with sophisticated computer-based guest room access devices.

Emergency key (E-key)

A key which opens all guest room doors, even when they are double locked and is generally handled by the general manager. This key is also called *Grand master key*.

Emergency plan

A document describing a hotel's pre-determined, intended response to a safety/security threat encountered by the hotel.

Employee handbook

Publication that provides general guidelines concerning employee conduct. Written policies and procedures related to employment at a hotel. Also sometimes called an *Employee Manual*

Empowerment

Management's act of delegating certain authority and responsibility to employees to make key decisions.

End-of-day

An arbitrary stopping point for the business day, established so that the audit can be considered complete through that time.

Entrepreneur

A person who assumes the risk of owning and operating a business in exchange for the financial rewards the business may produce.

Equivalent occupancy

Given a contemplated or actual change in the average room rate, the occupancy percentage needed to produce the same net revenue as was produced by the old price and occupancy percentage.

Escort

A person, usually employed or subcontracted by the tour operator, who accompanies a tour from departure to return, acting as a troubleshooter. Also known as *Tour manager, Tour conductor and Courier* (in Europe).

Escorted tour

A group of travelers traveling with a guide who has travel experience and has set up an itinerary for the group.

ETA

Estimated time of arrival.

ETD

Estimated time of departure.

Ethnic menu

Menu featuring the cuisine of a particular nation or ethnic group, such as Chinese, Mexican, or Italian.

European plan

A rate that quotes room charges only. A billing arrangement under which meals are priced separately from rooms.

Exchange rate

The exchange rate between two currencies specifies how much one currency is worth in terms of the other. Also known as the *Foreign-exchange rate, Forex rate or FX rate.*

Excursion

It is trip by a group of people, usually made for leisure or educational purpose.

Executive floor

A floor of a hotel that offers exceptional world-class service to business and other travelers. Also called the *Tower concept or Concierge floor or Business floor.*

Executive housekeeper

The individual responsible for the management and operation of the housekeeping department.

Expected arrival list

A daily report showing the number and names of guests expected to arrive.

Expected departure list

A daily report showing the number and names of guests expected to depart as well as the number of stayovers.

Express checkout

A pre-departure activity performed by the front desk for all the guests expected to depart on a particular morning and involves the production and distribution of guest folios in the individual guest rooms early in the morning.

Extended-stay hotels

Hotels that caters mostly to persons who must be in an area for a week or longer. The guest rooms of extended-stay hotels have more living space than regular hotel guest rooms, and may also have cooking facilities. Guest rooms in these hotels tend to be less expensive than guest rooms in full-service or all-suite hotels.

External audit

An independent verification of financial records performed by accountants who are not employed by the organization operating the hotel.

Facilities

Core physical features such as accommodation, restaurants, bars, and meeting rooms.

Familiarization (FAM) tour

A reduced-rate, often complimentary, trip or tour offered to travel agents, tour operators, whole-salers, incentive travel planners, travel writers, broadcasters, or photographers to familiarize them with a specific destination or attraction, thereby helping to stimulate sales.

Family rate

A special room rate for parents and children occupying one guest room.

Fax (Facsimile) machine

This machine operates through telephone lines and is used to receive and send official documents. While sending a fax message, the operator dials the destination fax machine number and then sends the fax message by inserting the message page in the machine.

Feasibility study

A technique used to assess the financial potential of a proposed development. All aspects of the project are examined—financial, human resources, marketing, etc.

FF&E

The term used to refer to the furniture, fixtures, and equipment used by a hotel to service its guests. FF&E reserve funds set aside by management today for the future furniture, fixture, and equipment replacement needs of a hotel.

Fixed costs

Costs that remains constant in the short run even though sales volume varies.

Flag

A term used to refer to the specific brand with which a hotel may affiliate. Examples of currently popular flags include brands such as Comfort Inns, Holiday Inn Express, Ramada Inns, Hampton Inns, Residence Inns, Best Western, and Hawthorn Suites. The hotels affiliated with a specific flag are sometimes referred to as a chain.

Flat organization chart

The collapse/combination of positions within an organization to reduce the number of management layers in efforts to improve communication, increase operating efficiencies, and reduce costs.

Float

This is a term describing a situation when there is a delay in payment from an account after using a credit card or cheque.

Floor limit

A limit assigned to hotels by credit card companies indicating the maximum amount in credit card charges the hotel is permitted to accept from a card member without special authorization. In online surroundings it's been withdrawn from use.

FOC
Free of charge.

Folio tray/folio well/folio bucket
This equipment contains a large number of slots where the folios are arranged sequentially according to the room numbers. It is used by the front office cashier to store and track the guest folios of various registered guests of the hotel and is used to maintain the folios safely for future use and reference.

Folio
An itemized record of a guest's charges and credits, maintained in the front office till departure, and can be referred to as *Guest bill or Guest statement.*

Food and beverage division
The division in a hospitality organization that is responsible for preparing and serving food and beverages within the organization or property. Also includes catering and room service.

Food and beverage manager
A person who plans, directs, organizes, and controls all phases of the food and beverage departments of an establishment.

Foot patrol
Walking the halls, corridors, and outside property of a hotel to detect breaches of guest and employee safety.

Forecasting
The process of predicting future events and trends in business. Typical forecasts developed for the rooms division include room availability and occupancy forecasts.

Franchising
A method of doing business wherein one organization that has developed a particular pattern or method of doing business—the *Franchisor*—grants the right to another organization—the *Franchisee*—to conduct the same business provided it follows the rules and regulations in exchange for a fee.

Franking machine
This is a machine used for printing the postage stamp value on the envelope.

Free Independent Traveler (FIT)
A guest coming to the hotel as an individual and not as part of a group or company.

Free sale
A travel component (room or seat) allotted to a wholesaler, which they can sell and confirm directly with the client without checking availability.

Fringe benefits
Fringe benefits are various non-wage compensations provided to employees in addition to their normal wages or salaries. Examples of these benefits include: Housing (employer-provided or employer-paid), group insurance (health, dental, life, etc.), disability income protection, retirement benefits, daycare, tuition fee reimbursement, sick leave, vacation (paid and non-paid), social security, profit sharing, funding of education, and other specialized benefits.

Front desk
The focal point of activity within the hotel, usually prominently located in the hotel lobby. Guests are registered, assigned rooms, and checked out at the front desk.

Front of the house
The functional areas of the hotel in which employees have extensive guest contact, such as food and beverage facilities and the front office.

Front office accounting formula
The formula used in posting transactions to front office accounts;
Previous Balance + Debits – Credits = Net Outstanding Balance

Front office cash sheet
A form completed by front office cashiers which lists each receipt or disbursement of cash during a work shift. It is used to reconcile actual cash on hand with the transactions which occurred during the shift.

Front office
A department of rooms division which is the most visible department in a hotel, with the greatest amount of guest contact. It is usually situated in the lobby of a hotel, the main functions of which are to control the sale of guest rooms, providing keys, mail, and information, keeping guest accounts, rendering bills/payments, and providing information to the guests.

Full house/sold-out/booked to capacity
100 percent hotel occupancy; a hotel that has all its guest rooms occupied. A situation in which every room in the hotel has been booked.

Full-service hotel

A hotel with a full range of services and amenities which may include some or all; on-site restaurant and lounge, meeting facility, pool, fitness centre, business centre, etc.

Fully automated

A computer-based system of front office record keeping which eliminates the need for many handwritten and machine-produced forms common in non- and semi-automated systems.

Function room

Public space such as meeting rooms, conference areas, and ballrooms (which can frequently be subdivided into smaller spaces) that are available in the hotel for banquet, meeting, or other group rental purposes.

Galileo

Airline reservation system.

Garni hotel

Hotels without restaurant service, except for continental breakfast.

Gateway/gateway city

A major airport, rail or bus center through which tourists and travelers enter from outside the region.

Geotourism

Tourism that sustains or enhances the geographical character of a place—its environment, culture, aesthetics, heritage, and the well-being of its residents.

Global Distribution System (GDS)

This is a reservation network which provides worldwide reservation information in the various hotels and is established by connecting the hotel reservation system with the airline reservation system. Example, Sabre, Galileo, Apollo, Amadeus, Pegasus and Worldspan.

Globalization

The condition in which countries and communities within them throughout the world are becoming increasingly inter-related.

Goal

A definition of the purpose of a department or division, which directs the actions of employees and the functions of the department or division toward a hotel's mission.

Government rate

A special room rate available at some hotels for government employees.

Graveyard

A work shift beginning about midnight.

Green tourism

A concept that encourages development, usually in rural areas, that has regard to and respects the landscape of the area including its wildlife.

Green washing

Green washing is a term used to describe businesses, services, or products that promote themselves as environmentally friendly when they are not.

Gross Operating Profit (GOP)

It refers to hotel revenue less those expenses typically controlled at the property level. It is generally expressed on the income statement and in the industry as both a dollar figure and percent of total revenue.

Gross Operating Profit Per Available Room (GOPPAR)

$$= \frac{\text{Total revenue} - \text{Management controllable operating expenses}}{\text{Number of rooms available}}$$

Group history

The number of rooms blocked for and ultimately used by a group during similar events held in the past.

Group Inclusive Tour (GIT)

Group travel in which individuals purchase a group package in which they will travel with others along a pre-set itinerary.

Group rate

Room rates offered to large groups of people visiting the hotel for a common reason like convention, trade show, meeting, tour or incentive group.

Group résumé

A summary of all group's activities, billing instructions, attendees, recreational arrangements, arrival and departure patterns, and other impor-tant information. Usually stored in a binder at the front desk.

Group tour

A prearranged, prepaid travel program for a group usually including transportation, accommod-ations, attraction admissions and meals. Also referred to a *Package tour*.

Group travel

A prepaid tour usually with a set itinerary and number of travelers.

Guaranteed reservation

A reservation which assures the guest that a room will be held until checkout time of the day following the day of arrival. The guest guarantees payment for the room, even if he fails to arrive. Types of guaranteed reservations include prepayment, credit card, advance deposit, travel agent, and corporate.

Guest account

An itemized record of a guest's charges and credits, which is maintained in the front office until departure. Also referred to as a *Guest bill, Guest folio or Guest statement*. It is a form imprinted with the hotel's logo and a control number and allowing space for room number, guest name, date in and date out, and room rate in the upper left-hand corner; it allows for guest charges to be imprinted with a PMS.

Guest amenities

Items placed in guest rooms for the comfort and convenience of guests, and at no extra cost. The term given to the range of disposable items provided in guest room and includes items such as shampoo, lotion, hair conditioner, soap, toothpaste, toothbrush, mouthwash, shower caps, shoeshine, sewing-kit, etc. Amenities are designed to increase a hotel's appeal, enhance a guest's stay, and encourage guests to return.

Guest check average

The average amount spent by a guest in a room service or dining room order. The guest check average typically includes the food and alcoholic beverage sales.

Guest check average

$$= \frac{\text{Total room revenue}}{\text{Total number of available rooms}}$$

Guest check

The invoice presented to restaurant and bar patrons for food and beverage consumed during a visit. Also referred to as a *Waiter's check or Restaurant check*.

Guest comment card

Short questionnaires that lodging properties and food service establishments ask their guests to fill out. Guest comments are used by the property to define current markets and to improve the operation.

Guest cycle

The sequence of phases that begins with pre-sale events, continues through point-of-sale activities, and concludes with post-sale transactions. The phases identify the physical contacts and financial exchanges that occur between guests and various revenue centers within a lodging operation.

Guest essentials

Items serving for the guest's needs and convenience, normally not used up or expected to be taken away by guests.

Guest expendables

Items serving for the guest's needs and convenience, expected to be used up or to be taken away upon departure.

Guest history card

A record maintained for each guest who has stayed at the hotel with a separate entry for each visit with details of pertinent preferences and special needs. This is a valuable reference tool for reservations, marketing, and credit departments. This record is relevant to marketing and sales, and can help the hotel serve the guest should he or she return. State law may require retention of certain guest data for some period of time. Guest histories are now more readily available through the increased utilization of computers and technology.

Guest history file

A collection of guest history records constructed from expired registrations cards or created through sophisticated computer-based systems, that automatically direct information about departing guests into a guest history database. It is maintained for marketing purposes and is referred to for return visits.

Guest ledger

The set of accounts for all guests currently registered in the hotel. Also called the *Front office ledger, Transient ledger, or Room ledger*.

Guest loan items

Equipment loaned to guests upon request and at no charge. For example, hair-dryers and ironing boards.

Guest mix

The variety and percentage distribution of hotel guests—individual, group, business, leisure, and so on—who stay at a hotel or patronize a restaurant.

Guest profile

A list of the characteristics that a property's guests have in common. The guest profile helps management to identify which market segments the property appeals to and which segments the property wants to attract.

Guest Relations Executive (GRE)

The primary goal of a guest relations manager is to make guests feel welcome and ensure their satisfaction. A guest relations executive needs to work well with people, be able to handle stressful situations and maintain a positive attitude.

Guest relations

The establishment of personal rapport and goodwill with guests through service and attention to individual guest needs. In a narrower sense, the promotion of in-house products and services, the entertainment of VIPs, and the handling of social functions—especially in a resort hotel.

Guest service directory

A documented listing of all of the features of a hotel together with general and pertinent information about the community within which the property is located. Directories are usually provided within each guest room.

Guest

A hotel visitor. Most guests rent rooms and/or purchase food or beverages in a hotel outlet or a banquet function.

Guest room key

A key which opens a single guest room door if it is not double-locked.

Handicap room

A room with special features designed for handicapped guests.

Hard-key system/mechanical lock system

A security device consisting of the traditional hard key that fits into a keyhole in a lock; preset tumblers inside the lock are turned by the designated key.

Head in beds

Industry slang referring to the primary marketing objective of accommodations and most destinations—increasing the number of overnight stays.

High balance report/high risk account/high bills

A report that identifies guests who are approaching an account credit limit. Typically prepared by the night auditor.

High season

The period of the year when occupancy of a hotel or attraction is normally the highest. High usage invariably means higher prices for rooms. Also referred to as *On-season or Peak season.*

Horizon

The future time frame for which a property accepts reservations.

Hospitality industry

Lodging and food service businesses that provide short-term or transitional lodging and/or food.

Hospitality Sales and Marketing Association International (HSMAI)

A trade association for hotel sales and marketing professionals.

Hospitality

The cordial and generous reception and entertainment of guests, visitors, or strangers with liberality and goodwill.

Hotel chain

A group of hotels with the same brand name.

Hotel

Place where a bonafide guest receives food and shelter, provided he is in a position to pay for it and is in a fit condition to be received.

Hoteliers

Those who work in the hotel business.

House count

The number of guests registered in a hotel on a specific night.
House Count = House count of previous day brought forward + Today's arrival – Today's departure

House limit

A specified limit in guest bills established by the hotel in lieu of advance deposit beyond which the guest will be asked to settle his bills before raising more credit. Also called *Credit limit.*

House phone

A publicly located telephone within the hotel used to call the front desk, or in some cases, the front desk and guest rooms.

House position

It refers to the number of rooms available for sale at the beginning of the day. A 'plus position' implies that rooms available for sale exceed expected arrivals. A 'minus position' indicates that expected arrivals exceed the number of rooms available for sale.

House position = Vacant rooms + Expected departures – Expected arrivals

House use

A room status term indicating that the room is being used by someone in the hotel staff at no charge.

Housekeeper's room report/housekeeping status report

A report prepared by the housekeeping department which indicates the current housekeeping status of each room, based on physical check.

Housekeeping room status

Terminology that indicates availability of a guest room such as *available, clean*, or *ready* (room is ready to be occupied), *occupied* (guest or guests are already occupying a room), *dirty* or *stayover* (guest will not be checking out of a room on the current day), *on change* (guest has checked out of the room, but the housekeeping staff has not released the room for occupancy), and *out of order* (the room is not available for occupancy because of a mechanical malfunction)

Housekeeping

The department within the rooms division which inspects rooms for sale, cleans occupied and vacated rooms, and coordinates room status with the front office. Also responsible for public area cleaning.

HRACC

Hotel and Restaurant Approval and Classification Committee

Hubbart formula

A bottom-up approach to pricing rooms. This method is used to compute and determine room rates that consider factors such as the operating expenses, average price per room, desired profit, desired return on investment, expected rooms sold and income from various departments of the hotel.

Human resources manager

The person responsible for administering federal, state, and local employment laws as well as advertising, screening, interviewing, selecting, orienting, training, and evaluating employees.

Hurdle rate

In the context of yield management, the lowest acceptable room rate for a given date.

HVAC

A shorthand term for 'Heating, Ventilating, and Air-conditioning'.

IATA

International Air Transport Association.

Icon

A facility or landmark which is visually synonymous with a destination.

Immigration

The process by which a government official verifies a person's passport, visa or origin of citizenship

In-balance

A term used to describe an accounting situation when the total debit amounts and the total credit amounts of an account are equal.

Incentive rate

A special room rate for guests in affiliated travel and tourism organizations because of the potential referral business they can generate for the hotel.

Incentive travel/tour/trip

The term is used for travel offered by corporations as a reward for top performance by staff or by distributors/clients. Also the term used for the business that develops, markets and operates these programs.

Incidental charges

Guest charges on a folio or bill for items other than room and tax such as food, beverage, phone, movies, etc.

Inclusive tour

A tour program in which all specific elements—transportation, airfare, accommodations, transfers, sightseeing, and other costs—are offered for a flat rate. Also referred to as an *All expense tour*.

Independent hotel

A hotel with no chain or franchise affiliation. It may be owned by an individual proprietor or a group of investors.

Induction

The process of informing new employees about matters related to the department in which they will work.

Information directory

A collection of information kept at the front desk for front desk agents to use in responding to guest requests, including simplified maps of the area; taxi and airline company telephone numbers; bank, theatre, church, and store locations; and special event schedules.

Information rack

An alphabetical index of registered guests used in routing telephone calls, mail, messages, and visitor inquiries. The information rack normally consists of aluminium slots designed to hold information rack slips.

In-room beverage service system

A computer-based system which dispenses beverages within a guest room, monitors sales transactions, and determines inventory replenishments quantities. Two popular in-room beverage service systems are non-automated honor bars and microprocessor-based vending machines.

In-room checkout

This is an exclusive feature provided by the property management system in fully automated hotel systems which allows the in-house guest to checkout of the hotel by reviewing their folio and settling their bills through an in-house guest room television. Also known as *In-room folio review and checkout*.

In-room guest console

A multi-feature phone that may include such functions as two-way speakerphone capability; a jack for portable computer use; an alarm clock; radio; remote control of heating, ventilating, and air conditioning, television, and room lights; energy management; and a theft alarm.

In-room movie system

Guest room entertainment provided through a dedicated television pay channel. Charges for the use of this in-room entertainment are posted to the appropriate guest folio.

Intelligent hotels

Hotels that are identified because they have state-of-the-art technology systems for their operations. These hotels have replaced the traditional systems to reduce their energy cost and usually have integrated systems which join analog and digital systems to achieve an effective communication in their hotels. The return on investment is reflected in the energy-cost savings and the comfort they provide to their guests.

Inter-connecting rooms

Rooms with individual entrance doors from the outside and connecting door between. Guests can move between rooms without going through the hallway.

Interface applications

Stand-alone computer software packages that may be linked to a front office management system, including point-of-sale systems, call accounting systems, and electronic locking systems.

Interhotel property referrals

This is a system in which one hotel which is a member of the hotel referral group recommends another hotel property which is also the member of the group to the guests in case of an overbooking situation.

Internal control

This is the process of verification of the various financial transactions by reconciliation with the source documents such as vouchers and sales summary sheet.

Internal recruiting

Tactics to identify and attract currently employed staff members for job vacancies that represent promotions or lateral transfers to similar positions.

International tourism

Travel people make outside their country of residence.

Internet

A network of computer systems that share information over high-speed electronic connections

Intersell agency

A central reservation system that handles reservations for more than one product line, such as airline companies, car rental companies, and hotel properties.

Inventory

The process of keeping track of available tourism product.

Itinerary

The travel schedule provided by a travel agent or tour operator for the client. A proposed or preliminary itinerary may be rather vague or very specific. A final itinerary spells out all details, including flight numbers, departure times, and similar data, as well as describing planned activities.

Job analysis

This is a process of collecting, analyzing and synthesizing information about the job and is thus a job to identifying its component parts and circumstances in which it is to be performed.

Job breakdown

A specification of how each task on a job list should be performed.

Job description

A list of tasks that an employee working in a specific position must be able to effectively perform. It is a listing of the job title, reporting relationships, working conditions and tasks.

Job evaluation

The process of establishing the value of jobs in a job hierarchy.

Job list

A list of tasks that must be performed for a front office position.

Job redesign

These are the attempts and strategies developed by the hotel to improve the quality of work performance of the employees of the hotel.

Job sharing

An arrangement by which two or more part-time employees share the responsibilities of one full-time position.

Job specification

A list of the personal qualities judged necessary for successful performance of the tasks required by the job description.

Key fob

A decorative or descriptive plastic or metal tag attached to a hard key.

King bed

A king-size bed is the largest size of bed available with a dimension of 78 inches by 80 inches.

Lanai

A room with a patio or balcony usually over-looking garden or water.

Late arrival

A guest holding a reservation who plans to arrive after the property's designated cancellation hour and so notifies the property.

Late charge

This is an outstanding charge which has not been posted into the guest folio and reaches the front desk after the guest has already checked out of the hotel.

Late checkout

A guest who is being allowed to checkout later than the property's standard checkout time.

Late checkout fee

A charge imposed by some hotels on guests who do not checkout by the established checkout time.

Leisure travel

Travel for recreational, sightseeing, relaxation and other experiential purposes.

Lettable room

Room ready for sale.

Liabilities

Financial or other contractual obligations or debts

Limited service hotel

A lodging facility that offers no or very few amenities, services or extra facilities such as restaurants, pools, meeting rooms, etc. Generally an inn or motel is limited service.

Local Area Network (LAN)

A communication network that connects computers and other terminals within a geographically limited area (typically within adjacent buildings or complexes). A LAN requires a LAN server and allows resources to be shared.

Lock-out

A room status term indicating that the room has been locked so that the guest cannot re-enter until he is cleared by a hotel official. This situation generally occurs when the guest does not settle his bill in spite of repeated requests made by the front desk.

Log book

A journal in which important front office events and decisions are recorded for reference during subsequent shifts.

Low season

The time of year at any destination when tourist traffic and rates are at their lowest. Also referred to as *Off-peak or Off-season*.

Loyalty programs

These programs are structured marketing efforts that reward, and therefore encourage, loyal buying behavior—behavior which is potentially beneficial to the firm. Participants will be rewarded for their behavior in terms of points. These points can then be converted, for example into online gift vouchers.

Loyalty

A deeply held commitment to re-buy or rep-atronize a preferred product or service consistently in the future.

Luxury hotel

A hotel with high room rates that features exceptional service and amenities.

Mail, message and key rack

It is a wooden framework containing an array of pigeon holes with each pigeon hole used to store the various mails and messages received for in-house guest. This rack also contains the keys of guest room when he is not in the room. The rack is located underneath the counter of the front desk.

Management contract

An agreement between a hotel's owners and a professional hotel management company. The owner usually retains the financial and legal responsibility for the property, and the management company receives an agreed-upon fee for operating the hotel.

Manager On Duty (MOD)

The individual on the hotel property responsible for making any management decisions required during the period he is MOD.

Manager's report

A report typically prepared by the night auditor, which summarizes the hotel's financial activities during a 24-hour period and provides insight into revenues, receivables, operating statistics, cash transactions related to front office, occupancy percentage, yield percentage, average daily rate, revpar, and number of guests. Also known as *Daily Operations Report*.

Marginal cost

The variable or added cost of selling a product that is incurred only if the room is sold. (Fixed costs are incurred whether the product is sold or not). Also called *Cost per occupied room*.

Market condition appraoch

An approach to pricing that bases prices on what comparable hotels in the geographical market are charging for a similar product.

Market segmentation

The division of the total market into identifiable groups of customers with similar needs for products and services.

Market share

The percentage of business within a market category.

Marketing environment

Outside factors that affect a business and which an operator has no control over, such as government policy, technological changes, societal changes, competition and industry trends.

Marketing mix

The mix of media (radio, print, television, online, direct marketing, etc.) used to bring your product to the attention of customers.

Mark-up

It is the difference between the cost price and the selling price of a given product.

Mass marketing

Advertising products and services through mass communications such as television, radio, and the Internet

Mass tourism

Wide-scale travel by a large number of people—not just the elite—brought about by the increase in leisure time, discretionary income, and reliable and inexpensive modes of transportation such as the automobile and airplane.

Master folio

A folio used to chart transactions on an account assigned to more than one person or guest room, usually reserved for group accounts. A master folio collects charges not appropriately posted elsewhere.

Master key

A key that can open all guest room doors that is not double-locked.

Material Safety Data Sheets (MSDS)

A written statement describing the potential hazards of; and best ways to handle, a chemical or toxic substance. An MSDS is provided by the manufacturer of the chemical or toxic substance to the buyer of the product and must be posted and made available in a place where it is easily accessible to those who will actually use the product.

Message slip

A loose-leaf binder in which the front desk staff can record important messages for guests.

MICE

Umbrella term to refer to several aspects of business tourism: meetings, incentives, conventions and exhibitions.

Mid-range service

A modest but sufficient level of service that appeals to the largest segment of the traveling public. A mid-range property may offer uniformed service, airport limousine service, and food and beverage room service; a specialty restaurant, coffee shop, and lounge; and special rates for certain guests.

Mini-bar

A small, under-the-table unit that can be stocked with liquor, beer, and wine, usually located within a hotel room for the convenience of guests.

Minimum Length of Stay (MLOS)

A revenue management strategy that instructs reservationists to decline any room reservation request that does not equal or exceed the pre-determined minimum number of nights allowed.

Miscellaneous Charge Order (MCO)

This is a voucher which is issued by an airline company and thus authorizes the mentioned guests to avail the facilities of accommodation and food and other facilities according to the contract with the payment due to the airline company.

Mission

The unique purpose that sets a hotel apart from other hotels. A mission expresses the underlying philosophy that gives meaning and direction to the hotel's actions, and addresses the interests of guests, management and employees.

Modified American Plan (MAP)

Rate includes breakfast and lunch or dinner with the room. Also known as *Demi-Pension or Half Board Basis*.

Motel

Overnight accommodation originally targeted to automobile travelers and therefore, situated at roadside locations. Often referred to as *Motor hotel*.

Movement list

It is a document which contains the names of expected arrivals and expected departures for the next day.

Multiple occupancy percentage

The number of rooms occupied by more than one guest divided by the number of rooms occupied by guests.

Multiple occupancy ratio

Measurement used to forecast food and beverage revenue, to indicate clean linen requirements, and to analyze daily revenue rate. Derived from multiple occupancy percentage or by determining the average number of guests per rooms sold. Also called *Double occupancy ratio*.

Multiple occupancy statistics

Occupancy ratios, indicating either a multiple occupancy percentage or the average number of guests per room sold. Used to forecast food and beverage ratios, indicate clean linen requirements, and analyze average daily room rates.

Murphy bed

This refers to a bed that folds up into the walls and looks like a bookshelf or cupboard when folded away. Also known as *Sico bed* or *Foldaway bed* or *Wall bed*.

NB

No Baggage

Net cash receipts

This is the total amount of cash, cheques and other negotiable items in the cashier's drawer excluding the cash bank.

Net Operating Income (NOI)

The income before interest and taxes found on a restaurant or hotel income statement.

Net Operating Profit (NOP)

Net operating profit represents the profitability of a company after accounting for cost of goods sold and operating expenses.

Net rate

The rate provided to wholesalers and tour operators that can be marked up to sell to the customer.

Night audit

The process of verification of the various guest accounts for their accuracy and completeness conducted by the night auditor during the night shift at the front desk.

Night auditor

An employee who checks the accuracy of front office accounting records and compiles a daily summary of hotel financial data as part of the night audit. In many hotels, the night auditor is actually an employee of the accounting division.

No frills

No additional details.

Non-affiliate reservation network

A central reservation system which connects independent (non-chain) properties.

Non-guaranteed reservation

A reservation arrangement where the hotel agrees to hold a room for the guest until a stated reservation cancellation hour on the day of arrival. The property is not guaranteed payment, in the case of no-shows.

Non-guest account

An account created to track the financial transactions of a local business or agency with charge privileges at the hotel, a group sponsoring a meeting at the hotel, or a former guest whose account was not satisfactorily settled at the time of departure.

Non-guest folio

A folio used to chart transactions on an account assigned to a local business or agency with hotel charge purchase privileges, a group sponsoring a meeting at the hotel, or a former guest with an outstanding account balance.

No-post status

A term used to indicate a guest who is not allowed to charge purchases to his room account.

No-show

A guest who makes a confirmed room reservation but fails to cancel the reservation or arrive at the hotel on the date of the reservation. Also called *Did Not Arrive (DNA)*

No-show factor

Percentage of guests with confirmed or guaranteed reservations who do not show up

Occupancy management formula

Calculation that considers confirmed reservations, guaranteed reservations, no-show factors of these two types of reservations, predicted stayovers, predicted understays, and predicted walk-ins to determine the number of additional room reservations needed to achieve 100 percent occupancy.

Occupancy percentage

A commonly used measure of hotel performance; occupancy is calculated by dividing the number of rooms sold for a period by the number of rooms available for the same period and multiplying by 100.

Occupancy ratio

A measurement of success of the hotel in selling rooms. Typical occupancy ratios include average daily rate, average rate per guest, multiple occupancy statistics, and occupancy statistics.

Occupancy report

A report prepared each night by a front desk agent which lists rooms occupied that night and indicates those guests expected to checkout the following day.

Occupational Safety and Health Administration (OSHA)

A federal agency established in 1970 that is responsible for developing and enforcing regulations related to assuring safe and healthy working conditions.

Occupied

A room status term indicating that a guest is currently registered to the room.

Off peak

Period when business is slowest.

Off-the-street

Walk-in guest

On-change

A room status term indicating that the guests has departed, but the room has not yet been cleaned and readied for resale.

Online agency

Travel agencies who operate using the worldwide web to provide information to potential customers as well as allowing the customer to book travel and related products without the necessity of speaking to a sales person.

Online reservation system

An internet based system used by hotels that allows prospective hotel guests to check availability and make reservations at the hotel.

On-the-job training

A training process designed to enhance the skills in which the employee observes and practices a task while performing his job. OJT programs are typically offered by management with the intent of improving guest service and employee performance at the hotel.

Open

The status of a date for which a reservation system can still accept reservations.

Operational reports

These are the various reports generated by the various departments of a hotel in conjunction with the accounts department as these reports include important financial information which is critical to the operations of the hotel.

Optimal occupancy

Achieving 100 percent occupancy with room sales that yield the highest room rate

Option date

The date agreed upon when a tentative agreement is to become a definite commitment by the buyer

Organization chart

A schematic representation of the relationships between varoius positions within an organization, showing where each position fits into the overall organization and illustrating the divisions of responsibility and lines of authority.

Orientation

The introduction of new hires to the organization and work environment, in order to provide background information about the property. Also known as *Induction*.

Outbound tour

Any tour that takes groups outside a given country, opposite of inbound.

Out-of-balance

A term used to describe an accounting situation when the total debit amounts and the total credit amounts of an account are not equal.

Out-of-order

A room status term indicating that a room cannot be assigned to a guest. A room may be out-of-order for maintenance, refurbishing, deep cleaning, or other reasons.

Outsourcing

Provision of service to the hotel—for example, a central reservation system—by an agency outside of the hotel

Outstanding balance

The amount the guest owes the hotel or the amount the hotel owes the guest, in the event of a credit balance at settlement.

Over and short

A discrepancy between the cash on hand and the amount that should be on hand.

Overage

This is a type of discrepancy which occurs at the front office cash section when after removing the initial cash bank, the total amount of cash, cheques and other negotiable items in the cashier's drawer is more than the net cash receipts.

Overbooking

Accepting reservations for more rooms than are available by forecasting the number of no show reservations, stayovers, understays, and walk-ins, with the goal of achieving 100 percent occupancy. This process is also called *Overselling*.

Overdue account

A city ledger account that is unpaid beyond the current billing period, usually between 30 and 90 days.

Overflow facility

This is a term given to a hotel participating in the central reservation network system and receives reservation requests only when all the hotels of the particular chain in the same geographical region are booked for a particular requested date.

Overstay

This is a term given to a guest who has extended his stay at the hotel and thus continues to stay even after his scheduled date of departure from the hotel.

Package

A fixed price saleable travel product that offers a mix of elements such as transportation, accommodation, restaurants, entertainment, cultural activities, sightseeing and car rental. Packages make it easy for a traveler to buy and enjoy a destination or several destinations. The name is given to an assembly of components under a one price system.

Package rate

Room rates that include goods and services in addition to rental of a room.

Package tour

A travel product covering several travel components, e.g. transport, accommodation, catering and perhaps sightseeing activity which is sold to the consumer as a single product at a single price.

Paid In Advance (PIA)

A guest who pays his room charges in cash during registration. PIA guests are often denied in-house credit or charge privileges.

Paid-out

Cash disbursed by the hotel on behalf of a guest and charged to the guest account as a cash advance.

Paid-out slips

Numbered forms that authorize cash disbursement from the front desk clerk's bank for products on behalf of a guest or an employee of the hotel

Par level

Inventory levels of recycled items are measured in par numbers. Par means the standard number of inventoried item that must be on hand to support daily, routine housekeeping operations. One par is also referred to as a *House setup*.

Parlor

This is the term given to a living room or a sitting room. Also known as a *Salon*.

Passport

Government document permitting a citizen to leave and re-enter the country

PATA

Pacific Asia Travel Association.

Patent

Legal protection that prevents other companies from using a firm's innovation.

PAX

Industry abbreviation for passengers or number of persons.

Peak season

Primary travel season.

Peaks and valleys

The high and low end of the travel season. Travel industry marketers plan programs to build consistent year-round business and event out the 'peaks and valleys'.

Percent yield

The number of rooms sold at average daily rate versus number of rooms available at rack rate multiplied by 100.

Petty cash

A small amount of cash available on-site that is not co-mingled with cash banks for revenue centers and is used to make small, miscellaneous purchases.

Pick-up error

An error in manual and semi-automated systems that occurs when the user enters an incorrect previous balance or transaction value in the process of posting.

Pilferage

Petty theft of small, less than full package (case) amounts from inventory.

Point-of-sale

An outlet in the hotel where goods or services are purchased and that generates income, such as a restaurant, gift shop, spa, or garage.

Point-of-sale system

A computer network which allows electronic cash registers at the hotel's point-of-sale to communicate directly with a front office guest accounting module.

Port of entry

Point at which persons enter a country where customs and immigration services exist

Portion

A standard quantity of food or beverage served for one person.

Posting

The process of recording the various transactions between the guest and the hotel by debiting and crediting charges and payments to a guest account.

Potential average rate

A collective statistics that effectively combines the potential average single and double rates, multiple occupancy percentage, and rate spread to produce the average rate that would be achieved if all rooms were sold at their full rack rates.

Pre-key

Making an electronic key for a guest room prior to the actual arrival of the guest who will be assigned to that room.

Prepayment guaranteed reservation

A type of reservation guarantee which requires a payment in full made before the day of arrival.

Pre-registration

A process by which sections of a registration card are completed for guests arriving with reservations to facilitate arrival of guest by reducing time at reception counter. Room and rate assignment, creation of a guest folio, and other functions may also be part of pre-registration activity.

Private Branch Exchange (PBX)

The system within the hotel used to process incoming, internal, and outgoing telephone calls. Also known as *Telephone switchboard*.

Product

Term used to describe any place or service used by tourists, including hotel, motel, inn, lodge or other accommodation facility, as well as tour, attraction or activity.

Promotional rate

A special room rate offered to promote future business.

Property Management System (PMS)

A generic term for applications of computer hardware and software used to manage the day to day operations of a hotel by networking reservation and registration databases, point-of-sale systems, accounting systems, and other office software.

Proprietary booking engine

A internet reservation system that is owned and operated by an individual hotel or group of hotels to allow them to take reservation on their own website without paying a fee to the GDS, third party booking engines or franchise reservation systems.

Quad

A room assigned to four people, may have two or more beds.

Quality circlesIt is an important part of Total Quality Management program running in a hotel property and consists of executives of the various departments meeting together to solve the various organization problems regarding quality management.

Quality control

This is a process of assigning the staffs, the responsibility of maintaining the quality of the products and services at the time of their delivery to the guests.

Queen bed

A bed approximately 60 inches by 80 inches

Quick-service restaurants

Eat-in or take-out operations with limited menus, low prices, and fast service; these restaurants are commonly called *Fast-food* or *Fast-service restaurants*; examples include McDonald's, Burger King, KFC, and Taco Bell.

Rack rate

This is the maximum possible rate which is charged by the hotel from the guest for overnight accommodation. It is the highest room rate category offered by a hotel.

Rate potential percentage

The percentage of the rack rate that a hotel actually receives, found by dividing the actual average room rate by the potential average rate. Also called the *Achievement factor.*

Rate spread

The mathematical difference between the hotel's potential average single rate and potential average double rate.

Ratio analysis

The analysis of financial statements and operating results using ratios.

Receptive operator

A tour operator who provides local services, transfers, sightseeing, guides, etc. Many large receptive operators develop packages and sell them through wholesale tour operators in foreign countries.

Re-engineering

Reorganizing hotel departments or work sections within departments.

Referral group

A group of independent hotels that have banded together for business profitability. Hotels within the group refer their departing guests or those guests they cannot accommodate to other properties in the referral group.

Reflag

To change a hotel from one franchise brand to another

Refurbishment

A process that involves the major cleaning and redecoration of hotel areas.

Registration

The procedure by which an incoming guest signifies his or her intent to stay at a property by completing and signing a registration card.

Registration card

A form on which arriving guests record their names, addresses, and other details including mode of transportation used, nationality, purpose of visit, method of payment, and length of stay. A space is also provided for signature, room rate and room number. Additional questions may be included as a part of the hotel's market research platform.

Registration record

A collection of important guest information created by the front desk agent following the guest's arrival. The registration record includes the guest's name, address, telephone number, and company affiliation; method of payment; and date of departure.

Renovation

The process of making repairs that brings a building to a good condition.

Repeat business

Revenue generated from guests returning to a commercial operation such as a hotel as a result of positive experiences on previous visits.

Reservation

An agreement between the hotel and a guest or a company that the hotel will hold a specific type of room for a particular date and length of stay.

Reservation confirmation number

A code which provides a unique reference to a reservation record and assures the guest that the reservation record exists.

Reservation rack

A part of Whitney System, where all the information of the prospective guest is recorded. The slips are arranged in alphabetical order. The racks are filed chronologically by the guests' scheduled dates of arrival.

Reservation record

A collection of data that identifies a guest and his anticipated occupancy needs before arrival at the property, and enables the hotel to personalize guest services and accurately schedule staff.

Reservation status

Terminology used to indicate the availability of a guest room to be rented on a particular night, i.e., *open* (room is available for renting), *confirmed* (room has been reserved until 4:00 P.M. or 6:00 P.M.), *guaranteed* (room has been reserved until guest arrives), and *repair* (room is not available for guest rental)

Reservation transactions report

A summary of daily reservations activity in terms of record creation, modification, and cancellation.

Reservations agent

An employee, either in the front office or in a separate department, who is responsible for all aspects of reservations processing.

Reservations department

A department within a hotel's rooms division staffed by skilled telemarketing personnel who take reservations over the phone, answer questions about facilities, quote prices and available dates, and sell to callers who are shopping around.

Reservations manager

The person who takes and confirms incoming requests for rooms, noting special requests for service; provides guest with requested information; maintains an accurate room inventory; and communicates with marketing and sales

Resident manager

A manager in a large hotel who is directly responsible to the GM for the property's operating departments that include F&B, purchasing, engineering and maintenance, front office and security.

Residential hotels

Hotels that provide long-term accommodation for guests

Resorts

A hotel that caters primarily to vacationers and tourist and typically offers more recreational amenities and services, in a more aesthetically pleasing setting, than other hotels. These hotels are located in attractive and natural tourist destinations and their clientele are groups and couples that like adventure with sophistication and comfort. The attractions may vary depending on the region and some might offer golf, tennis, scuba diving and, depending on the natural surrounding, may also arrange other recreational activities.

Responsible tourism

A type of tourism which is practiced by tourists who make responsible choices when choosing their holidays. These choices reflect responsible attitudes to the limiting of the extent of the sociological and environmental impacts their holidays may cause.

Restoration

Returning a hotel to its original (or better than original) condition.

Retailer

Another term for travel agents who sell travel products directly to consumers.

Retrenchment

A turnaround strategy that involves tactics such as reducing the workforce, closing unprofitable plants, outsourcing unprofitable activities, implementing tighter cost or quality controls, or implementing new policies that emphasize quality or efficiency.

Revenue

Money the hotel collects from guests for the use of rooms or from the purchase of hotel goods and services.

Revenue center

A hotel division or department that sells goods or services to guests and thereby directly generates revenue for the hotel. The front office, food and beverage outlets, and retail stores are typical hotel revenue centers.

Revenue forecast report

A projection of future revenue calculated by multiplying predicted occupancies by current room rates.

Revenue management

A process of planning to achieve maximum room rates and most profitable guests (guests who will spend money at the hotel's food and beverage outlets, gift shops, etc.), that encourages front office managers, general managers, and marketing and sales directors to target sales periods and develop sales programs that will maximize profit for the hotel

Revenue manager

A management position that provides oversight to room inventory and room rates through various marketing channels

Revenue Per Available Room (RevPAR)

Revpar is the key indicator of performance for hotels. It is the average sales revenue generated by each guest room during a given time period.

$$Revpar = \frac{Total\ room\ revenue}{Total\ number\ of\ available\ rooms}$$

Revpar = Occupancy % × Average room rate per night.

Revenue Per Available Seat Hours (RevPASH)

It is the revenue generated during a specified time period divided by the number of seat hours available during that period (F&B).

Revenue potential

The room revenue that could be received if all the rooms were sold at the rack rate

Revenue realized

The actual amount of room revenue earned

Reward

A positive stimulus that can be presented in the process of reinforcing a repeat-buying behavior.

RNA

Room Not Assigned

Room assignment

The identification and allocation to a guest of an available room in a specific room category, finalized as part of the registration process.

Room attendant cart

A wheeled cart that contains all of the items needed to properly and safely clean and re-stock a guest room.

Room block

An agreed-upon number of rooms set aside for members of a group planning to stay at a hotel.

Room blocking

Process of holding or reserving a specific room for guest based on his room preferences and other factors.

Room forecasting

Projecting room sales for a specific period

Room inspection

A final review of the room to assure that all housekeeping tasks have been completed and room furnishings are in order.

Room key control system

An administrative procedure that authorizes certain personnel and registered guests to have access to keys

Room mix

The ratio of room types available in a hotel. For example, the number of double-bedded rooms compared with king-bedded rooms, the number of smoking permitted rooms to non-smoking rooms, and the number of suites compared with standard rooms.

Room night/guest night/bed night

It means stay of one guest for one night. One guest staying for 30 nights or 30 guests staying for one night means the same.

Room occupancy sensor

A device that uses infrared light or ultrasonic sound waves to sense the physical occupancy of a room. Sensors have the ability to turn on devices and appliances such as lights, air conditioning, and heating whenever a guest enters a space, and to turn these devices and appliances off when the guest leaves.

Room rack slip

This is a small form which is prepared from the registration card of the guests and is inserted in the guest rooms and shows some important information about the guest such as name of the guest, date of arrival and departure and rate offered to the guest.

Room rack

The room rack is a wooden framework with an array of metallic pockets designed to hold room rack slips arranged by room number. The room rack summarizes the reservation and house-keeping status of each guest room in the hotel.

Room rate variance report

A report listing rooms that have not been sold at rack rates.

Room rate

The price a hotel charges for overnight accommodation.

Room revenue and count report

A report which shows the rack rate for each room and the actual rate at which the room was sold, providing an opportunity to analyze room revenues.

Room revenue

The amount of room sales received

Room sales projections

A weekly report prepared and distributed by the front office manager that indicates the number of departures, arrivals, walk-ins, stayovers, and no-shows

Room service

The department within food and beverage division that is responsible for delivering food and beverages to guests in their guest rooms. Also known as *In Room Dining (IRD)*.

Room status discrepancy

A situation in which the housekeeping department's description of a room's status differs from the room status information being used by the front office to assign guest rooms.

Room status

The up-to-date (actual) condition (occupied, vacant, dirty, out of order, etc.) of the hotel's individual guest rooms.

Room variance report

A report listing any discrepancies between front desk and housekeeping room status.

Rooming

The procedures involved in greeting a guest, assigning a room, and escorting or directing the guest to the room

Rooming list

The list of names of passengers on a tour submitted to a hotel at the time of arrival The names are not alphabetized as on a flight manifest, but rather room by room indicating who is rooming with whom. Twin-bedded rooms, singles and triples are usually listed in separate categories.

Rooms activity forecast

Information on anticipated arrivals, departures, stay-overs, and vacancies. Managers use this forecast to determine staffing needs at the front desk and in housekeeping areas.

Rooms availability report

A report that lists, by room type, the number of available rooms each day. This report contains expected arrival and departure information for the next several days, typically prepared as part of the night audit.

Rooms discrepancy report

A report that notes any variances between front desk and housekeeping room status updates. It often alerts management to investigate the possibility of sleepers.

Rooms division

The largest, and usually most profitable, division of a hotel. It typically consists of front office department, housekeeping department, and the uniformed service department.

Rooms management module

A front office application of a computer-based property management system. The module maintains up-to-date information on the status of rooms, assists in the assignment of rooms during registration, and helps coordinate various guest services.

Rooms status report

A report that indicates the current status of rooms according to housekeeping designations, such as: on-makeup, on-change, out-of-order, clean, and ready for inspection.

Rule-of-thumb method

A method of fixing room tariff. According to this method; a hotel will charge $1 for every $1,000 of construction costs.

Run of the house

The term is used to indicate that an agreement has been made with a company to offer its corporate members the best available rooms in the hotel at a specifically quoted price.

Safety deposit boxes

Individual boxes provided for the safekeeping of guest valuables. Located either in a central, secure, and supervised location or in individual guest rooms.

Sales call

A meeting arranged for the purpose of selling the hotel's products and services.

Scanty Baggage (SB)

The term used for any guest with light baggage.

Science tourism

A subgroup of eco-tourism in which laypersons travel with scientists and students to help with scientific work at various sites throughout the world. Science tourists often work very hard (even though they are paying for the vacation) and make a contribution to a body of scientific knowledge.

Security monitor

A closed-circuit television monitor that allows front office employees to monitor security and safety throughout the hotel from a central location.

Self check-in/self registration

A computerized system which automatically registers a guest and dispenses a guest room key to complete the registration process, based on the guest reservation and credit card information.

Self checkout

A computerized system usually located in the lobby of a fully automated hotel, which allows the guest to review his folio and settle the account to the credit card used at check-in.

Sell up/upselling

Convince a prospective guest to take a higher priced room than what he had asked for during the time of reservation or registration. A sales technique whereby a guest is offered a more expensive room than what he reserved or originally requested, at an extra cost, by highlighting the added features, benefits and services of the room.

Sell-through

A yield management availability strategy that works like a minimum length of stay requirement except that the length of the required stay can begin before the date the strategy is applied.

Semi-automated

A system of front office record keeping characterized by the use of both handwritten and machine produced forms and electro-mechanical equipment such as posting machines.

Service charge

A percentage of the bill (usually 10% to 20%) added to the guest's bill for distribution to service employees in lieu of direct tipping.

Settlement

The collection of payment for an outstanding account balance, bringing the balance to zero. A guest can settle an account by paying cash, charging the balance to a credit card, deferring payment to a direct billing, or using some combination of these payment methods.

Shortage

This is a type of discrepancy which occurs at the front office cash section when after removing the initial cash bank, the total amount of cash, cheques and other negotiable items in the cashier's drawer is less than the net cash receipts.

Shoulder period or season

The time period between peak and low season. In a restaurant, the period just before or just after lunch is a shoulder period. In a resort, the weeks just before and just after the resort's peak season are shoulder periods.

Single bed

A bed approximately 36 inches by 75 inches or 39 inches by 81 inches

Single

A room assigned to one person; may have one or more beds.

Skipper

A hotel guest who has left the hotel without settling the bills.

Sleeper

A vacant room that is believed to be occupied because the room rack slip or registration card was not removed from the rack or the front office staff has failed to properly update the room's status when the previous guest departed.

Sleep-out

A room status term indicating that the guest is registered to the room, but the bed has not been used.

Smart card

An electronic device with a computer chip that allows a guest or an employee access to a designated area, tracking, and debitcard capabilities for the hotel guest

Smerf

Meetings acronym for a category of meeting market segments including social, military, educational, religious, or fraternal type groups. These organizations often are looking for 'value' when selecting a meeting destination.

Software

A set of programs that controls the operation of the hardware components of a computer system. Software tells the computer what to do, how to do it, and when to do it.

Spa

A mineral spring, or a locality or resort hotel near such a spring, to which people resorted for cures (from Spa, a watering place in eastern Belgium). Today, the word spa is used more loosely to refer to any fashionable resort locality or hotel.

Specialty restaurant

A theme restaurant that features certain types of food.

Split folio

An arrangement whereby a guest's charges are separated into two or more folios.

Standard Operating Procedure (SOP)

Established procedure to be followed in carrying out a given operation.

Star ratings

Five Star Hotel: Luxury hotels; most expensive hotels/resorts in the world; numerous extras to enhance the quality of the client's stay (for example: some have private golf courses and even a small private airport).

Four Star Hotel: First class hotels; expensive (by middle-class standards); has all of the previously mentioned services; has many "luxury" services (for example: massages or a health spa).

Three Star Hotel: Middle class hotels; moderately priced; has daily maid service, room service, and may have dry-cleaning, Internet access, and a swimming pool.

Two Star Hotel: Budget hotels; slightly more expensive; usually has maid service daily.

One Star Hotel: Low budget hotels; inexpensive; may not have maid service **or room service.**

No Category Hotels: These hotels include motels, cottages, bungalows and others with limited services. Nevertheless, these hotels represent 41% of the total hotel market share.

Statement of cash flows

A projection of income from income-generating areas of the hotel

Statement of income

A financial statement which provides important information about the results of hotel operations for a given period of time.

Stay restrictions

Duration rules and limitations related to guests' arrival dates, departure dates and minimum stay lengths. Also known as *Stay controls.*

Stay-over

A room status term indicating that the guest is not checking out today and will remain at least one more night; a guest who occupied the room previous night and intend to continue his stay current night.

Stay-sensitive hurdle rate

In the context of yield management, a hurdle (or minimum acceptable room) rate that varies with the length of the guest reservation.

Stop over guest/lay over guest

This refers to the guests who en route from one destination to another stops in between on a third destination and breaks his journey.

Strategy

A method or a plan developed to achieve a long-range goal.

Studio

A room with a studio bed—a couch that can be converted into a bed; may also have an additional bed.

Suburban hotel

A hotel that is somewhat smaller than a downtown hotel (typically 250 to 500 rooms), is usually part of a chain, and has restaurants, bars, and other amenities found at downtown hotels.

Suite

A parlor or living room connected to one or more bedrooms, and perhaps a kitchenette.

Supplemental transcript

A detailed report of all non-guest accounts that indicates each charge transaction that affected a non-guest account that day, used as a worksheet to detect posting errors.

Supplier

Those businesses that provide industry products like accommodations, transportation, car rentals, restaurants and attractions.

Support center

A hotel division or department that does not generate revenue directly, but supports the hotel's revenue centers. Support centers include the housekeeping, accounting, engineering and maintenance, and human resources divisions.

Sustainable tourism

It is a form of tourism in which all resources are managed in such a way that economic, social and aesthetic needs are fulfilled while maintaining cultural integrity, essential ecological processes, and biological diversity and life support systems. Sustainable development implies meeting the needs of the present without compromising the ability of future generations to meet their own needs.

System update

This is the term given to the night audit routine followed in fully-automated system and thus includes preparing back-up copies of the reports generated, file re-organization, system maintenance, generation of the reports and providing an end-of-day time frame.

Table D'Hôte Menu (TDH)

It refers to a menu of limited choice. It usually includes three or five courses available at a fixed price. It is also referred to as a *Fixed menu*. A table

D'hôte menu is a complete meal at a predetermined price.

T-account

A two-column account recording format (resembling the letter T) in which charges are posted to the left side and payments to the right side.

Tactic

An action or method used to attain a short-term objective.

Target markets

Distinctly defined groups of customers (market segments) at which hotels aim or target their marketing efforts.

Tariff

Rate of fare quoted and published by a travel industry supplier (i.e. hotels, tour operators, etc.) Usually an annual tariff is produced in booklet form for use in sales calls at trade shows.

Tea Coffee Maker (TCM)

An automatic machine that makes tea or coffee and dispenses it into individual cups.

Telephone operator

The person who handles incoming and outgoing calls, locates registered guests and management staff, deals with emergency communication, and assists the desk clerk and cashier when necessary

Theme restaurant

A restaurant distinguished by its combination of decor, atmosphere, and menu, all of which relate to a particular theme.

Third party booking engine

An internet site that provides a booking engine where a traveler can search a large number of lodging facilities for availability and reserve a room. The lodging facilities are not affiliated with the site and pays a fee for the business that the third party site generates. Examples of third party sites include hotels.com, priceline.com.

Time punching machine

It is used to record the check-in and checkout time of the guests and delivery time of any mail or message for the in-house guest. Folios, mails and other front office paper work are inserted into a time stand device to record the correct time and date.

Top-down selling method

A selling approach that seeks to sell an entity's highest priced items prior to the sale of its lower priced items.

Total Quality Management (TQM)

A management technique that encourages managers to look critically at processes used to produce products and services

Total revpar (TREVPAR)

The average rooms and non-rooms revenue generated by each available guest room during a specific period of time.

Tour

Any pre-arranged (but not necessarily prepaid) journey to one or more destinations and back to the point of origin.

Tour guide

A person qualified to conduct tours of specific localities or attractions. He provides assistance, information and cultural, historical and contemporary heritage interpretation to people on organized tours.

Tour leader

Usually a group leader.

Tour manual

A compendium of facts about a destination, including its attractions, accommodations, geography, and special events, used by destination marketing organizations to attract tour operators and visitors and their area

Tour operator

Develops, markets and operates group travel programs that provide a complete travel experience for one price and includes transportation (airline, rail, motorcoach, and/or ship), accommodations, sightseeing, selected meals and an escort. Tour operators market directly to the consumer, through travel agents and are beginning to be listed on computerized reservation systems.

Tour voucher

Documents issued by tour operators to be exchanged for accommodations, meals, sightseeing, admission tickets and other services. Also referred to as *Coupons* or *Tour orders*.

Tourism

The business of providing and marketing services and facilities for leisure travelers. Thus, the concept of tourism is of direct concern to governments, carriers, and the lodging, restaurant, and entertain-ment industries, and of indirect concern to virtually every industry and business in the world.

Tourist destination

A city that is dependent to a significant extent on the revenues accruing from tourism. It may have one or more tourist attraction.

Tourist/visitor/traveler

People who travel to and stay in places outside their usual environment for more than 24 hours but not more than one consecutive year for leisure, business, health, sports, holiday, study, religion and other purposes not related to the exercise of an activity remunerated from within the place visited.

Tower

A guest room floor configuration in which rooms are grouped around a central vertical core

Transfer

This refers to the transportation of visitors between their point of arrival and selected hotel, and back again on departure day.

Transfer voucher

A voucher used to support a reduction in balance on one folio and an equal increase in balance on another. Transfer vouchers are used for transfers between guest accounts and for transfers from guest accounts to non-guest accounts when they are settled by the use of credit cards.

Transient Occupancy Tax (TOT)

TOT or bed tax is a locally set tax on the cost of commercial accommodations and camp grounds.

Transient sales

Rooms and services sold primarily through the efforts of the front office and its staff.

Transit visa

Visa allowing the holder to stop over in a country to make a travel connection or brief visit

Transit

Process of changing planes without going through security and/or customs

Transposition error

An error which occurs when numerals in a figure are reversed. For example, 189 for 198.

Travel agency voucher

A type of travel agent guaranteed reservation in which the agent forwards a voucher to the hotel as proof of payment and guarantees that the prepaid amount will be sent to the hotel when the voucher is returned to the travel agency for payment.

Travel agent guaranteed reservation

A type of reservation guarantee under which the hotel generally bills the travel agency after a guaranteed reservation has been classified as a no-show.

Travel agent

An individual who arranges travel for individuals or groups. Travel agents may be generalists or specialists (cruises, adventure travel, conventions and meetings.) The agents receive a 10% to 15% commission from accommodations, transportation companies and attractions for coordinating the booking of travel. They typically coordinate travel for their customers at the same or lower cost than if the customer booked the travel on his/her own.

Travel and tourism industry

A variety of interrelated businesses that provide services to travelers; the tourism industry includes a broad range of businesses such as airlines, bars, cruise lines, car rental firms, casinos, entertainment firms, hotels, restaurants, travel agents, timeshares, tour operators, and recreational enterprises

Travel trade

The collective term for tour operators, wholesalers and travel agents.

Travel

Leisure and other types of travel, including business, medical care and educational travel. All tourism is travel, but not all travel is tourism.

Traveler's cheque

A prepaid cheque sold by banks and other financial institutions which is considered equivalent to cash.

Trial balance

In the night audit, the process of balancing front office accounts with transaction information by department.

Triple

A room assigned to three people, may have two or more beds.

Turnaround strategies

Sometimes called *retrenchment*; can involve workforce reductions, selling assets to reduce debt, outsourcing unprofitable activities, implementation of tighter cost or quality controls, or new policies that emphasize quality or efficiency; turnaround can occur at the corporate level of a company or on a property-by-property basis.

Turn-away report/refusal report

A report which tracks the number of guests who were refused accommodation because rooms were not available or the guest was not in a fit condition to be received. Turn-away guests are also called *Displacement*.

Turn-down service

An evening service rendered by the housekeeping department, which replaces soiled bathroom linen and prepares the bed for use.

Turn-in

This is the amount of money deposited by the front office cashier in the accounts division after the end of a particular shift.

Twin bed

A bed approximately 39 inches by 75 inches or 37 inches × 81 inches

Twin

A room with two twin beds; may be occupied by one or more people.

Typsy guest

A drunkard guest who may misbehave with staff.

Underbooking

An erroneous belief that all the rooms are sold while in fact they are not.

Understay

A guest who checks out before his stated departure date. Also called *Early checkout*.

Uniform System of Accounts for the Lodging Industry (USALI)

A standard set of accounting procedures used to record a hotel's financial transactions and condition.

Uniformed service

A department within the rooms division including parking attendants, door attendants, porters, limousine drivers, and bell persons.

Unique Sales Proposition (USP)

The sustainable, competitive edge a product has over other product. The special features or benefits and features that differentiate the product of one organization from the competitors and becomes the basis of promotion of sales. Hotel features that are unique to the property are also called *Signature attractions*.

Unity of command

Each employee should report to/be accountable to only one boss for a specific activity.

Unpaid account balance

Charges remaining in a guest account after the guest has left the hotel.

Upgrade

Move a reservation or registered guest to a better accommodation or class of service. A process whereby a guest is offered a higher category room with special features, benefits and services at no extra cost, for reasons of goodwill.

Vacant and ready

A room status term indicating that the room has been cleaned and inspected, and is ready for an arriving guest.

Valet

Those individuals responsible for parking guest vehicles.

Variable costs

Costs that vary in direct proportion to sales. For example, housekeeping costs with room sales and F&B costs with restaurant sales.

Visa

Stamp of approval recorded in a passport to enter a country for a specific purpose

Vision

Expresses what the organization wants to be in the future

Vision statement

A forward-looking statement of what a firm wants to be in the future; an ideal and unique picture of the future.

Visiting Friends and Relatives (VFR)

The same as tourists, however, they are usually staying in private homes and their principal reason for travel is to visit friends or relatives.

Visitor Information Centre (VIC)

An information centre located at a specific attraction or place of interest, such as a landmark, national park, national forest, providing information (such as trail maps, events, accommodation, staff contact, restrooms, etc.) and in-depth educational exhibits and artifact displays (e.g. about natural or cultural history) to the visitors. These are often operated by a convention and visitors bureau, chamber of commerce or tourism promotion organization.

Voice mail

A system that is part of the telephone equipment which provides for hotel guests and staff to retrieve a messages left by a caller. The device is capable of storing, recording, and playing back messages through the telephone system.

Voucher

A document detailing a transaction to be posted to a front office account, used to communicate information from a point of sale to the front office. Common vouchers include cash, charge, transfer, allowance, and paid-out vouchers. Also, a form provided by travel agent to their clients as a receipt for advance registration payments.

Waitlist

List of clients awaiting transportation or accommodations at times when they are not available, confirmed as a result of subsequent cancellations.

Wake-up call

A call made by front office, usually by telephone, to a guest room at the time requested by a room guest to be wakened.

Wake-up device/reminder-o-timer

A special alarm clock with pull-out pins used for wake-up calls. The guests' requests for wake-up calls are recorded in a wake-up sheet with the information of time, room number and name of the guest. In fully automated systems, the telephone exchange automatically places the various requests of the guests for wake-up calls by automatic voice recorded wake-up messages.

Walk through

Review of meeting details or inspection of function room or trade show floor prior to event.

Walked guests/walking a guest with a reservation

A situation in which a guest with a reservation is relocated from the reserved hotel to another nearby hotel because no room was available at the reserved hotel. This usually happens when the hotel is overbooked and a guest room is not available for a confirmed guest. This usually includes paying for transportation to the hotel and covering any difference in the room rate at the hotel the guest was 'walked' to. This situation is also called *Bounced reservation*.

Walk-in

A person who comes to the hotel without prior reservation. Also called *Chance guest* or *Off-the-street guest*.

Walk-out

When a guest moves out of the hotel because of improper services or standard. Also called *Did Not Stay (DNS)*

Wash factor

This is the process of removing all the unreserved rooms within a group block in to the general inventory pool after the cut-off-date for reselling them to other guests or groups. Also known as *Washing*.

Watch

Another term for work shift.

Watch down

Blocking fewer rooms than the number requested by a group, based on previous group history.

Weighted average contribution margin ratio

In a multiple product situation, an average contribution margin for all operated departments that is weighted to reflect the relative contribution of each department to the establishment's ability to pay fixed costs and generate profits. Abbreviated CMR_w.

Whistle-blower

An employee or manager who reveals wrong-doing; an attempt to force the organization to cease a behavior that society finds unacceptable or to incorporate a practice that is in keeping with a new social value, if value changes in society are not voluntarily incorporated into a firm.

Who

An unidentified guest occupying a guest room in a hotel that is vacant as per front desk record.

Wholesaler

This is a term given to an entrepreneur who purchases tours and packages from hospitality suppliers such as hotels and airlines and sells them either directly to the customers or through travel agents.

Window

A clause in a franchisee agreement that grants both the franchisor and the franchisee the right, with proper notification, to terminate the agreement.

Wireless Fidelity (WI-FI)

An Internet access technology that does not utilize a building's wiring system when providing users Internet access.

Word-of-mouth advertising

Informal conversations between persons as they 'discuss' their positive or negative experiences at a hotel.

Work order

A form used to initiate and document a request for maintenance.

World Tourism Organization (WTO)

An organization created to promote and develop tourism in the interest of the economic, social and cultural progress of all nations.

World-class service

A level of guest service that stresses personal attention. Hotels offering world-class service provide upscale restaurants and lounges, guest rooms with exquisite décor, concierge services, and abundant amenities above all a high staff-guest ratio.

X report

The term commonly used to indicate the total revenue generated by a revenue-producing department during one part of a specific time period.

Xenodogheionology

This is the study of the history, stories and myths about the hospitality establishments such as inns, hotels and motels.

Year to Date (YTD)

Used when comparing performance from the beginning of the year up through, and including, the present period.

Yield

This is the statistical information that gives information about the general performance of the hotel and is the product of the occupancy of the hotel on a particular date and the average daily rate. It is the ratio of actual revenue to potential revenue.

Yield management

A technique, based on the principle of supply and demand, used to maximize revenue generation of any hotel by lowering prices to increase sales during periods of low demand and raising prices during periods of high demand.

Yield percentage

This is the statistical figure that determines the effectiveness and efficiency of a hotel in selling all its rooms at the rack rate.

Yield statistic

The ratio of actual rooms revenue to potential rooms revenue.

Youth hostel

A budget accommodation with limited services generally catering to shoestring budget travelers.

Z report

The term commonly used to indicate the total revenues generated during an entire time period. Producing a Z report includes re-setting the continuous total feature of a POS or other electronic register system to 'zero' to begin recording the next period's revenues.

Zed bed

Fold-away beds with a slatted wood frame that folds into a double-hinged shape of three sections when used as a sofa, named after the letter 'Z' which it resembles because of its three jointed parts.

Zero out

This is the process of balancing the guest accounts by the front office cashier during the process of checkout of the guests and making arrangements for the settlement of the folio.

Abbreviations

AAA	American Automobile Association	DND	Do Not Disturb
ACD	Automatic Call Distributor	DNS	Did Not Stay
ADA	Americans With Disabilities Act	DOT	Department of Tourism
ADR	Average Daily Rate	ECO	Express Check Out
ADTOI	Association of Domestic Tour Operators of India	ECR	Exchange Currency Record/Encashment Certificate
AHMA	American Hotel and Motel Association	EDC	Electronic Data Capturing
AI	All Inclusive	EFTPOS	Electronic Fund Transfer Point of Sale
ALC	Advance Letting Chart	ELS	Electronic Locking System
ANS	Arrival Notification Slip	EMT	Early Morning Tea
AP	American Plan	EP	European Plan
APC	All Payment Cash	EPABX	Electronic Private Automatic Branch Exchange
ARR	Average Room Rate		
ASSOCHAM	Associated Chambers of Commerce and Industry of India	ESC	Entire Stay Complimentary
		ETA	Expected Time of Arrival
ASTA	American Society of Travel Agents	ETD	Expected Time of Departure
ATG	Aarti Tilak and Garlanding	FAITH	Federation of Associations of Indian Tourism and Hospitality
ATOI	Adventure Tour Operators of India		
BIT	Bulk Inclusive Tour	FAX	Facsimile Automated Xerox
BTC	Bill To Company	FFC	Flowers Fruits & Cookies
CAS	Call Accounting System	FFE	Furniture, Fixture & Equipment
CBC	Conventional Booking Chart	FFIT	Foreign Free Individual Traveler
CII	Confederation of Indian Industry	FHRAI	Federation of Hotel & Restaurant Association of India
CIP	Commercially Important Person		
CP	Continental Plan	FICCI	Federation of Indian Chambers of Commerce & Industry
CRO	Central Reservation Office		
CRS	Central Reservation System	FIT	Free Independent Traveler
CVGR	Corporate Volume Guaranteed Rate	FRRO	Foreigners' Regional Registration Office
DFIT	Domestic Free Individual Traveler	GDS	Global Distribution System
DG	Distinguished Guest	GHC	Guest History Card
DNA	Did Not Arrive	GIT	Group Inclusive Tour
DNCO	Did Not Check Out	GRC	Guest Registration Card

GWB	Guest Weekly Bill	MOD	Manager On Duty
HAI	Hotel Association of India	NB	No Baggage
HG	House Guest	NCR	National Cash Register
HRACC	Hotel and Restaurant Association of India	NTAC	National Tourism Advisory Council
HWC	Handle With Care	OOO	Out Of Order
IATA	International Air Transport Association	OTH	On The House
		PATA	Pacific Asia Travel Association
IATM	International Association of Tour Managers	PCO	Public Call Office
		PIA	Paid In Advance
IATO	Indian Association of Tour Operators	PIP	Politically Important Person
		PMS	Property Management System
ICAO	International Civil Aviation Organization	POS	Point Of Sale
		PSO	Passenger Service Order
ICPB	Indian Convention Promotion Bureau	RDM	Room Division Manager
		RNA	Room Not Assigned
IDD	International Direct Dialing	SB	Scanty Baggage
IFTO	International Federation of Tour Operators	SOP	Standard Operating Procedure
		SPATT	Special Attention
IHA	International Hotel Association	STD	Subscribers Trunk Dialing
IHCL	Indian Hotel Company Limited	TAAI	Travel Agents' Association of India
IHHA	Indian Heritage Hotel Association		
IHRA	International Hotel & Restaurant Association	TAFI	Travel Agents Federation of India
		TAC	Travel Agent Commission
IRCTC	Indian Railway Catering and Tourism Corporation Limited	TCM	Tea Coffe Maker
		TIPS	To Insure Prompt Service
IRD	In Room Dining	UR	Under Repair
IRO	Instant Reservation Office	USP	Unique Sales Proposition
IRS	Instant Reservation System	UFTAA	Universal Federation of Travel Agents Association
ISD	International Subscribers Dialing		
ITDC	India Tourism Development Corporation	VFR	Visiting Friends & Relatives
		VPO	Visitors' Paid Out
		VTL	Visitors' Tabular Ledger
ITTA	Indian Tourist Transport Association	WTO	World Tourism Organization
		WTTC	World Travel and Tourism Council
LIU	Local Intelligence Unit	WWF	World Wildlife Fund
LLR	Left Luggage Room	XO	Exchange Order
MAO	Meal and Accommodation Order	YMCA	Young Men's Christian Association
MAP	Modified American Plan		
MICE	Meeting Incentive Convention Exhibition	YWCA	Young Women Christian Association

Index